FORMAL METHODS FOR PROTOCOL ENGINEERING AND DISTRIBUTED SYSTEMS

IFIP - The International Federation for Information Processing

IFIP was founded in 1960 under the auspices of UNESCO, following the First World Computer Congress held in Paris the previous year. An umbrella organization for societies working in information processing, IFIP's aim is two-fold: to support information processing within its member countries and to encourage technology transfer to developing nations. As its mission statement clearly states,

IFIP's mission is to be the leading, truly international, apolitical organization which encourages and assists in the development, exploitation and application of information technology for the benefit of all people.

IFIP is a non-profitmaking organization, run almost solely by 2500 volunteers. It operates through a number of technical committees, which organize events and publications. IFIP's events range from an international congress to local seminars, but the most important are:

- The IFIP World Computer Congress, held every second year;
- open conferences;
- working conferences.

The flagship event is the IFIP World Computer Congress, at which both invited and contributed papers are presented. Contributed papers are rigorously refereed and the rejection rate is high.

As with the Congress, participation in the open conferences is open to all and papers may be invited or submitted. Again, submitted papers are stringently refereed.

The working conferences are structured differently. They are usually run by a working group and attendance is small and by invitation only. Their purpose is to create an atmosphere conducive to innovation and development. Refereeing is less rigorous and papers are subjected to extensive group discussion.

Publications arising from IFIP events vary. The papers presented at the IFIP World Computer Congress and at open conferences are published as conference proceedings, while the results of the working conferences are often published as collections of selected and edited papers.

Any national society whose primary activity is in information may apply to become a full member of IFIP, although full membership is restricted to one society per country. Full members are entitled to vote at the annual General Assembly, National societies preferring a less committed involvement may apply for associate or corresponding membership. Associate members enjoy the same benefits as full members, but without voting rights. Corresponding members are not represented in IFIP bodies. Affiliated membership is open to non-national societies, and individual and honorary membership schemes are also offered.

FORMAL METHODS FOR PROTOCOL ENGINEERING AND DISTRIBUTED SYSTEMS

FORTE XII / PSTV XIX'99
IFIP TC6 WG6.1 Joint International Conference on Formal Description Techniques for Distributed Systems and Communication Protocols (FORTE XII) and Protocol Specification, Testing and Verification (PSTV XIX)
October 5-8, 1999, Beijing, China

Edited by

Jianping Wu
Tsinghua University
China

Samuel T. Chanson
Hong Kong University of Science & Technology
China

Qiang Gao
IVT
China

SPRINGER-SCIENCE+BUSINESS MEDIA, B.V.

Library of Congress Cataloging-in-Publication Data

IFIP TC6 WG6.1 Joint International Conference on Formal Description Techniques for Distributed Systems and Communication Protocols and Protocol Specification, Testing, and Verification (1999: Beijing, China)

Formal methods for protocol engineering and distributed systems : FORTE XII/PSTV XIX'99: IFIP TC6 WG6.1 Joint International Conference on Formal Description Techniques for Distributed Systems and Communication Protocols (FORTE XII) and Protocol Specification, Testing, and Verification (PSTV XIX) : October 5-8, 1999, Beijing, China / edited by Jianping Wu, Samuel T. Chanson, Qiang Gao.

Includes bibliographical references.

ISBN 978-1-4757-5270-0 ISBN 978-0-387-35578-8 (eBook)

DOI 10.1007/978-0-387-35578-8

1. Formal methods (Computer science) Congresses. 2. Electronic data processing—Distributed processing Congresses. 3. Computer network protocols Congresses. I. Wu, Jianping, 1953- . II. Chanson, Samuel T. (Samuel Tin-hung). III. Gao, Qiang, 1963- . IV. IFIP TC6/WG 6.1 International Conference on Formal Description Techniques for Distributed Systems and Communication Protocols (12th : 1999 : Beijing, China). V. IFIP WG 6.1 International Symposium on Protocol Specification, Testing, and Verification (19th : 1999 : Beijing, China). VI. Title. VII.

QA76.9.F67I35 1999
005.13'1—dc21

99-40808
CIP

Originally published by Kluwer Academic Publishers in 1999
Softcover reprint of the hardcover 1st edition 1999

Printed on acid-free paper.

CONTENTS

Invited Talk II

PART FOUR FDTs II

PART FIVE Model Checking

PART SIX Verification

Invited Talk III

PART SEVEN Test III

* Short Papers

PREFACE

The IFIP TC6 WG 6.1 Joint International Conference on Formal Description Techniques for Distributed Systems and Communication Protocols (FORTE) and Protocol Specification, Testing and Verification (PSTV) will be held for the first time in China in the nineteen-year history of the conferences. It is hoped that this will stimulate work in this important field in China and Asia, and further foster international co-operation in protocol engineering.

As in the past, the conference has maintained a high standard in paper selection. Out of the 78 papers submitted, 22 full papers and 5 short papers were selected for inclusion in this Proceeding. Also for the first time, paper selection was conducted via the Internet followed by a Program Committee meeting in Beijing. We would like to thank the PC members for their hard work in reviewing the papers and in participating in the paper selection process. Their effort and expertise will guarantee the quality of the technical program. Thanks are also due to the other reviewers. A complete list of referees is given in the Proceedings. The conference will also include tutorials in several hot topics and three invited talks by leading experts from university/industry/government.

Many people have contributed to the organization of the joint conference. We would like to thank Guy Leduc, chairman of IFIP WG 6.1 for his continuous advice and support.

We would also like to thank the support from Tsinghua University, Hong Kong University of Science and Technology, The National Natural Science Foundation of China (NSFC) and China Computer Federation (CCF). We gratefully acknowledge the financial support to the tutorials from Networking Research Laboratories of Lucent Technologies.

Last, but not least, we would like to express our thanks to all people participating in the local organization.

We hope you will find the conference enjoyable and rewarding, both technically and socially.

Jianping Wu
Samuel Chanson
Qiang Gao

Committee Members

General Chairs

Samuel T. Chanson, Hong Kong University of Science & Technology, China
Chengwei Wang, China Computer Federation, China
Guanqun Gu, East South University, Nanji, China

Program Committee Co-Chairs

Jianping Wu, Tsinghua University, China
Samuel T. Chanson, Hong Kong University of Science & Technology, China
Qiang Gao, IVT, China

Program Committee

B. Baumgarten (GMD, Germany), G. v. Bochmann (Univ. of Ottawa, Canada), T. Bolognesi (IEI, Italy), H. Bowman (Univ. of Kent, UK), Ed Brinksma (Univ. of Twente, Netherlands), Stan Budkowski (INT,France), O.Catrina (Univ.of Bucharest,Romania), Ana Cavalli (INT,France), T.Y. Cheung (City Univ. Hong Kong), J.P. Courtiat (LAAS-CNRS,France), P. Dembinski (IPIPAN, Poland), M. Diaz (LAAS, France), R. Dssouli (Univ. of Montreal, Canada), C. J. Fidge (Univ. of Queensland,Australia), S. Fischer (International Univ, Germany), R. Gorrieri (Univ. of Bologna,Italy), R. Groz (CNET, France), R. Gotzhein (Univ. of Kaiserslautern, Germany), Teruo Higashino (Osaka Univ. Japan), Daoyuan Hu (Tsinghua Univ. China), D. Hogrefe (Univ. of Luebeck, Germany), H. Koenig (BTU, Cottbus, Germany), M. C. Kim (IC Univ., Korea), R. Lai (La Trobe Univ. Australia), G. Leduc (Univ. of Liege, Belgium), D. Lee (Bell Lab., USA), S. Leue (Univ. of Waterloo, Canada), Jiren Liu (East North Univ. China), Mike Liu (Ohio State Univ. USA), L. Logrippo (Univ. of Ottawa, Canada), T. Mizuno (Shizuoka Univ., Japan), Elie Najm (ENST, France), A. Petrenko (CRIM, Canada), J. Quemada (ETSI Telecom, Spain), Omar Rafiq (Univ. of Pau, France), N. Shiratori (Tohoku Univ, Japan), Meiling Shi (Tsinghua Univ. China), J.B. Stefani (CNET, France), K. Suzuki (KDD, Japan), K.C. Tai (N.Carolina State Univ. USA), R. Tenney (Univ. of Massachusetts, USA), K. Tarnay (Univ. of Budapest, Hungary), A. Togashi (Shizuoka Univ., Japan), Ken Turner (Univ. of Stirling, UK), S. Vuong (Univ. of British Columbia, Canada), N. Yevtushenko (Tomsk Univ., Russia), Shiyong Zhang (Fudan Univ., China)

Organization Committee Co-Chairs

Lizhu Zhou, Tsinghua University, China
Zhiyong Liu, National Natural Science Foundation of China, China
Zide Du, China Computer Federation, China

LIST OF REFEREES

Arnaud Bailly
Bernd Baumgarten
Jun Bi
Gregor V. Bochmann
Tommaso Bolognesi
H.Bowman
Ed Brinksma
Pascal Brisset
S. Budkowski
Ana Rosa Cavalli
A. Campo
Richard Castanet
Octavian Catrina
Antonio Cerone
Jason Chiang
Jean-Pierre Courtiat
Dan Craigen
Piotr Dembinski
L. Díaz
Michel Diaz
Rachida Dssouli
Fabrice Dubois
Funda Ergun
David de Frutos Escrig
Kazi Farooqui
Colin Fidge
Stefan Fischer
Arnaud Fontaine
Qiang Gao
Stefania Gnesi
Roberto Gorrieri
Reinhard Gotzhein
Jens Grabowski
Roland Groz
Alan G. Hamilton
Ruibing Hao
Olaf Henniger
Stefan Heymer
Teruo Higashino
Dieter Hogrefe
Gerard Holzmann
Daoyuan Hu
Guy Juanole
Moataz Kamel
Ferhat Kendek
Myungchul Kim
Beat Koch
H. Koenig
Richard Lai
P. Langendoerfer
Diego Latella
David Lee
Guy Leduc
Stefan Leue
Jiren Liu
Ming Liu
Ming T Liu
Liugi Logrippo
Mieke Massink
Juan Pavón Mestras
Tadanori Mizuno
Jean-Francois Monin
Akio Nakata
Elie Najm
R. Negulescu
Tomohiko Ogishi
Kozo Okano
Philippe Owezarski
Doron Peled
Alexandre Petrenko
Simon Pickin
Juan Quemada
Jean-Luc Raffy
Omar Rafiq
Charles M. I. Rattray
James Riely
C. Rinderknecht
Tomas Robles
H. Rust
Pierre de Saqui-Sannes
Fumiaki Sato
Michael Schmidt
Carron E. Shankland
Meilin Shi
Norio Shiratori
Rakesh Sinha
Jean-Bernard Stefani
Kenji Suzuki
Marine Tabourier
KC Tai
Katalin Tarnay
Richard L Tenney
Atsushi Togashi
Kenneth J. Turner
Piotr Tysowski
Son Vuong
Jianguo Wang
Ke Xu
Mingwei Xu
Keiichi Yasumoto
Nina Yevtushenko
Xia Yin
ShiyongZhang
Jun Zhou
Mohammad Zulkernine

Invited Talk I

PART ONE

Test I

Confirming configurations in EFSM

A. PETRENKO [1], S. BORODAY [1], and R. GROZ [2]
1 - *CRIM, Centre de Recherche Informatique de Montréal,*
550 Sherbrooke West, Suite 100, Montreal, H3A 1B9, Canada
phone: 1 (514) 840-1290, fax: +1 (514) 840-1244, Petrenko@crim.ca, Boroday@crim.ca;
2 - *France Télécom - R&D Centre (CNET)*
2 av. Pierre Marzin - F-22307 Lannion cedex, France
phone: +33 2.96.05.37.90, fax: +33 2.96.05.39.45, Roland.Groz@cnet.francetelecom.fr

Abstract: In this paper we investigate the problem of configuration distinguishability for the EFSM model, specifically, given a configuration and an arbitrary set of configurations, determine an input sequence such that the EFSM in the given configuration produces an output sequence different from that of the configurations in the given set or at least in a maximal proper subset. Such a sequence can be used in a test case to confirm the destination configuration. We demonstrate that the distinguishability problem could be reduced to the EFSM traversal problem, so that the existing methods and tools developed in the context of model checking become applicable. The theoretical framework for determining configuration-confirming sequences based on projections and products of EFSMs is presented. Our approach can be implemented in a number of heuristic test derivation strategies.

Key Words: Extended Finite State Machines, test derivation, state identification

1. INTRODUCTION

It is widely acknowledged that the model of extended FSM (EFSM) is a very powerful model for verification and test derivation. There exist a number of tools that support development activities around specifications based on the EFSM model. In particular, the commercial tools available on SDL [1, 2] now offer test generation facilities. Such tools may resort to reachability analysis to compute tests that will cover transitions of the EFSM and provide test preambles to reach specific configurations of the EFSM enabling the transitions to be tested. However, they currently do not check the tail state of transitions or the configurations reached after a test. This casts doubt on the confidence that the tests have really assessed the corresponding behavior of the tested machine (known as Implementation Under Test), let alone that they would reach a significant coverage of faults in tail states and configurations.

In contrast, the classical FSM model allows one to systematically derive tests with so-called complete fault coverage in the sense that such a test suite detects any fault modeled by a mutant FSM within an assumed bound on the number of states in implementations. The completeness stems from the fact that the implementation states can be pairwise distinguished, identified and compared with specification states using appropriate state identification or confirming sequences (such as a DS or W set). The corresponding methods are discussed in a number of surveys, see e.g., [3, 4].

An EFSM can often be viewed as a compressed notation of an FSM. It is possible to unfold it into a pure FSM by expanding the values of the parameters, assuming that all the domains are finite. The latter have to be reduced in order to alleviate the state and input explosion problem. The result of this unfolding is a pure FSM computed as the reachability graph of the EFSM. Henceforth, we shall use the term "configuration" to mean a state of the unfolded machine, i.e., a node of the reachability graph, which can be labeled with a couple consisting of a state and a vector value for the context variables of the EFSM (this will be detailed in Section 2).

It is not realistic to achieve perfect configuration identification as in testing of a flat FSM, mainly because of state/test explosion effect, widely understood today. Nevertheless, augmenting test cases by appending tail sequences, which could increase the confidence that the reached configuration is correct, would certainly pay off.

The crucial point for defining such an approach is the form in which the set of configurations, w.r.t. which the destination configuration has to be confirmed, is given. In a trivial situation, the set is just a singleton and to find a sequence separating the two configurations one may attempt "perebor", i.e., a brute force search, for example, by flattening the EFSM into a pure FSM. In the other extreme situation, the set includes all the possible configurations, when perebor most probably fails. In a more "realistic" or interesting situations, the test designer provides a list of states, for each of which a set of suspicious contexts is defined by assigning some values to certain context variables, while leaving the others free, i.e., unknown.

The rest of the paper is organized as follows. In Section 2, we define the EFSM model and make certain assumptions about the class of specification machines. Section 3 addresses the problem in a formal way; configuration distinguishability is defined, various settings of the configuration confirming sequence generation are discussed and unified in a general framework. Section 4 relates the obtained results to previous work. Section 5 concludes the paper.

2. EFSM MODEL

The model of a Mealy (finite state) machine extended with input and output parameters, context variables, operations and predicates defined over context variables and input parameters, is what is understood by an extended FSM in this paper. We first define all the objects of an extended machine, at the same time introduce necessary notations and then define the way it operates respecting the semantics of Mealy machines.

Definition 2.1. An *extended* finite state machine (EFSM) M is a pair (S,T) of a set of states S and a set of transitions T between states from S, such that each transition $t \in T$ is a tuple (s, x, P, op, y, up, s'), where

- $s, s' \in S$ are the initial and final states of the transition, respectively;
- $x \in X$ is input, X is a set of inputs, and D_{inp_x} is the set of input vectors, each component of an input vector corresponds to an input parameter associated with x;
- $y \in Y$ is output, Y is a set of outputs, and D_{out_y} is the set of output vectors, each component of an output vector corresponds to an output parameter associated with y;
- P, op, and up are functions, defined over input parameters and context variables V, namely,

  P: $D_{inp_x} \times D_V \rightarrow \{\text{True, False}\}$ is a predicate, where D_V is a set of context vectors $\vec{v}$;

  op: $D_{inp_x} \times D_V \rightarrow D_{out_y}$ is an output parameter function;

  up: $D_{inp_x} \times D_V \rightarrow D_V$ is a context update function.

❑

To define the operation of an EFSM, we first introduce some additional definitions.

Definition 2.2. Given input x and a (possibly empty) set of input vectors D_{inp_x}, a pair of input x and input vector from D_{inp_x} is called a *parameterized input*. A sequence of parameterized inputs is called a *parameterized input sequence*.

❑

Similarly, we define parameterized outputs and their sequences.

Definition 2.3. A context vector $\vec{v} \in D_V$ is called a *context* of M. A *configuration* of M is a pair of state s and context $\vec{v}$.

❑

Note that in case of an empty set of context variables, a configuration coincides with a state.

Definition 2.4. A transition is said to be *enabled* for a configuration and parameterized input if the transition predicate evaluates to true.

❑

The EFSM operates as follows. The machine receives input along with input parameters (if any) and computes the predicate that is satisfied for the current configuration. The predicate identifies enabled transitions. A single transition among those enabled fires. Executing the chosen transition, the machine produces output along with output parameters, which, if they exist, are computed from the current context and input parameters by the use of the output parameter function. The machine updates the current context according to the context update function, and moves from the initial to the final state of the transition. Transitions are atomic and cannot be interrupted. The machine usually starts from a designated configuration, called the *initial* configuration. A pair of an EFSM M and the initial configuration is called a *strongly initialized* EFSM.

To simplify the notations for transitions of EFSMs, we present a few conventions. Specifically, we normally use $(s\text{—}x,P/op,y,up\rightarrow s')$ to denote a transition $t \in T$. If, in t, P is a True constant, P can be dropped from the transition. Similarly, when the transition does not change the context, the update function up can be omitted. At the same time, the output parameter function can only be absent when output y has no output parameters at all. Notations $(s\text{—}x,P/y\rightarrow s')$, $(s\text{—}x/y,up\rightarrow s')$, $(s\text{—}x/y\rightarrow s')$ are examples of notations used for such situations. If present in a transition, the update and output parameter functions can take forms of operations on separate variables, such as assignments. Figure 1 gives an example of a machine specified using these notations.

To define the class of specification machines considered in this paper, we first list a number of properties of EFSMs.

Definition 2.5. An EFSM M is said to be :

- *consistent* if for each transition t, every element in $D_{inp_x}\times D_V$ evaluates exactly one predicate to true among all predicates guarding transitions with the start state and the input of t; in other words, the predicates are mutually exclusive and their disjunction evaluates to true.
- *completely specified* if for each pair $(s, x) \in S\times X$, there exists at least one transition leaving state s with input x.
- *deterministic* if any two transitions outgoing from the same state with the same input have different predicates.
- *observable* if for each state and each input, every outgoing transition with the same input has a distinct output.

❑

In this paper, we study the problem of deriving configuration-confirming sequences, or CCS, for short, for a class of specification EFSMs which are consistent, completely specified, deterministic, and observable. We use the abbreviation REFSM to refer to a machine from this Restricted class of EFSMs. At the same time, in our constructions, we will also use other types of EFSMs, which do not necessarily meet all these requirements.

3. CONFIGURATION CONFIRMING SEQUENCES

The problem we are dealing with can be informally stated as follows. We know the configuration reached in the REFSM M in response to some parameterized input sequence applied to the initial configuration (this is the tail configuration of the test up to that point). Our goal is to determine a *single* parameterized input sequence that can increase our confidence in the correctness of the configuration reached in any implementation under test derived from M. To that end, we try to ensure that we can ascertain the correct configuration has been reached in the implementation, or at least that no suspicious configuration has been reached. Typically, we might allow an implementation to have different values from those specified for non relevant context variables, but pay special attention to crucial variables or the control state.

The reason for having a single sequence rather than several as in the W-method for FSM is a practical one : in typical protocol testing, only a single check sequence would be associated to a given test case. Also, it is not easy to ensure that the same configuration would be reached if the test was rerun, as non relevant context variables are allowed to vary.

For a classical deterministic FSM, a simple i/o sequence (SIS) [5], known also as a UIO sequence [6] is a solution to the problem. Assuming that faults in any implementation under test neither increase the state number nor mask each other, such a sequence can ensure correctness of the tail state of any transition once it is executed immediately after the transition. The problem of UIO generation was studied in a number of works, as early as the 1960's, see, e.g., [7] and [6] for most recent results. In case of the EFSM model, we are dealing with a more general problem, which we call here a configuration confirming sequence (CCS) generation. The key issue here is configuration distinguishability.

Finally, for practical reasons again based on experience in protocol testing (and on acceptability by test experts), we should try to find confirming sequences that are not "too long". Basically, it would not make sense to claim that a 100-input long sequence would bring enough added confidence to justify appending it to test preambles of length 4 or 5, all the more so as a confirming sequence, to be fully trustable, has yet to be applied in all the other configurations (as in the UIOv-method), since faults may mask each other. Therefore, we will consider that we can set up an arbitrary limit l on the length of the sequence.

3.1 Configuration distinguishability

We define configuration distinguishability by generalizing a corresponding notion used in ordinary Mealy machines. Let M and N be two EFSMs defined over the same inputs and input parameters. M is REFSM, while N is also consistent, completely specified, and observable, but not necessarily deterministic. We assume that the output alphabets of the two machines intersect, but the sets of output parameters associated with each common output in M and N are not necessarily identical.

Definition 3.1. Two parameterized outputs of M and N are said to be *compatible* if the output symbols coincide and every common output parameter has the same value in both parameterized outputs. Two parameterized output sequences, $y_1 \ldots y_k$ of M and $y'_1 \ldots y'_k$ of N, are *compatible* if for all $i = 1, \ldots, k$, y_i and y'_i are compatible.

❑

Based on this notion, we now define distinguishability of configurations.

Definition 3.2. Given a parameterized input sequence α, configuration c of M and configuration c' of N are *distinguishable by* α if the parameterized output sequence, produced by (M,c) in response to α, is not compatible with any parameterized output sequence that can be produced by (N,c') in response to α. α is said to be a sequence *separating* c from c'.

❑

Indistinguishable configurations of REFSMs are also referred to as *equivalent* configurations. Two REFSMs are *equivalent* if their initial configurations are equivalent. Note that in this study, we assume that the given REFSM M may have equivalent configurations.

Definition 3.3. Let $(M,(s,\vec{v}))$ and $(N,(q,\vec{u}))$ be strongly initialized, consistent, completely specified, observable EFSMs defined over the same input alphabet and input parameters. M is deterministic, while N possibly not. A *distinguishing machine* of $(M,(s,\vec{v}))$ for $(N,(q,\vec{u}))$ is an EFSM with a state set $S\times(Q\cup\{fail\})$, where S and Q are state sets of M and N, respectively, $fail \notin Q$ is a designated symbol, and a set of transitions defined as follows.

- For every transition of M $(s \text{---} x,P_M/op_M,y,up_M \rightarrow s')$ and every $q\in(Q\cup\{fail\})$, if N has no transition from q with the input x and the output y then the transition

$$((s,q) \text{---} x,P_M/op_M,y,up_M \rightarrow (s',fail))$$

is defined; if N has a transition $(q\text{—}x,P_N/op_N,y,up_N\rightarrow q')$ then two transitions

$$((s,q)\text{—}x,P_M\wedge P_N\wedge(op_M^*=op_N^*)/op_M,y,(up_M,up_N)\rightarrow(s',q'))\text{ and}$$

$$((s,q)\text{—}x,P_M\wedge(\neg P_N\vee(op_M^*\neq op_N^*))/op_M,y,up_1\rightarrow(s',fail))$$

are defined, where op_M^* and op_N^* are projections of op_M and op_N into the set of common output parameters of the output y in M and N, respectively, and the predicate $(op_M^*=op_N^*)$ is satisfied if op_M^* and op_N^* have the same value for each common output parameter.

- The set of context variables is $V\cup U$, where V and U are the sets of context variables of M and N, respectively (assuming $V\cap U=\varnothing$).
- $((s,q), (\vec{v}, \vec{u}))$ is the initial configuration.

We use $[(M,(s,\vec{v}))\text{-}(N,(q,\vec{u}))]$ to denote the distinguishing machine of $(M,(s,\vec{v}))$ for $(N,(q,\vec{u}))$.

❑

If the outputs y in the M and N have no common output parameter then the predicate $(op_M^*=op_N^*)$ is a tautology and can be omitted. If, in addition, P_N is also a True constant then a predicate $P_M\wedge(\neg P_N\vee(op_M^*\neq op_N^*))$ is a contradiction and any transition guarded by such a predicate can be removed from $[(M,(s,\vec{v}))\text{- }(N,(q,\vec{u}))]$, as the transition can never be enabled. Note that we require the two machines to have disjoint sets of context variables. One can always rename variables to meet this requirement.

We formulate a few obvious properties of distinguishing machines.

Proposition 3.4. Given $(M,(s,\vec{v}))$ and $(N,(q,\vec{u}))$, where M and N are machines from Definition 3.3, the distinguishing machine $[(M,(s,\vec{v}))\text{-}(N,(q,\vec{u}))]$ is a consistent, completely specified, deterministic, and observable EFSM equivalent to $(M,(s,\vec{v}))$.

❑

A state is referred to as a *fail-state* of the machine $([M,(s,\vec{v}))\text{-}(N,(q,\vec{u}))]$ if its second substate is *fail*.

Proposition 3.5. A parameterized input sequence takes the distinguishing machine $[(M,(s,\vec{v}))\text{-}(N,(q,\vec{u}))]$ from the initial configuration to a fail-state if it is a separating sequence for $(s,\vec{v})$ and $(q,\vec{u})$.

❑

The last statement implies that the problem of determining a sequence separating two configurations of a given REFSM M can be reduced to a well-understood (though hard) reachability problem for EFSMs. In particular, any parameterized sequence that labels the shortest sequence of enabled transitions from the initial configuration to a fail-state of the distinguishing machine is a minimal solution to the original problem. For EFSMs with a finite reachability graph, the reachability analysis is theoretically tractable, though hard. The advantage of the approach to configuration distinguishability problem that we are developing in this paper, is the applicability of existing tools for model checking, i.e., EFSM traversal. In practical situations, using such a tool, one usually imposes some limit on the length of desired sequences to deal with complexity. In this case, only a fragment of the distinguishing machine is required.

Definition 3.6. Given an EFSM N, the *l-fragment* of N is defined as a submachine of N obtained from the graph of N by pruning nodes, whose distance from the initial node exceeds l, along with their adjacent transitions. ❑

Example. The working example of the paper is the ESFM M shown in Figure 1. It has four states and ten transitions that are labeled with two inputs a and b, three outputs x, y and z, the latter has a parameter, and four predicates. The machine has two context variables, u and w. It is consistent, completely specified, deterministic, and observable.

Assume we are required to find a sequence (if it exists) separating the configuration (1; 0,0) from (1; 0,5). Here 1 denotes state 1 and (0,5) means u=0 and w=5. The length of the sequence should not exceed two, i.e., l=2.

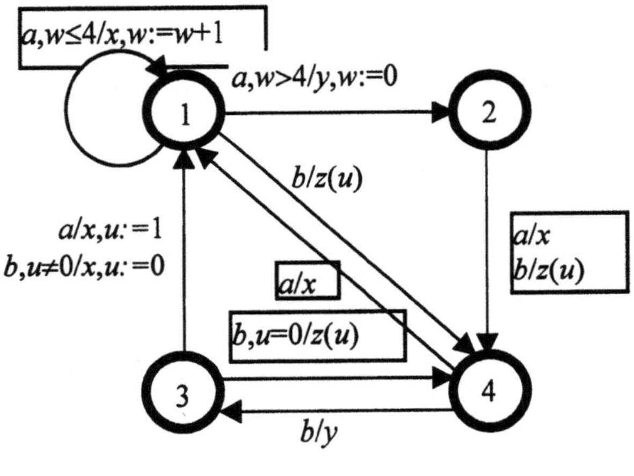

Figure 1. The EFSM *M*

Figure 2 shows the l-fragment of the distinguishing machine [(M,(1; 0,0))-(M,(1; 0,5))]. Note that in this fragment, we have not depicted any path of length two or more that does not lead to a fail-state. In this diagram, the fail-state represents all states of the l-fragment with a fail-substate, namely (1*fail*) and (4*fail*). The context variables of (M,(1; 0,0)) are u_0 and w_0. The context variables of (M,(1; 0,5)) are u_1 and w_1. The initial configuration of the distinguishing machine is (11; u_0=0, w_0=0, u_1=0, w_1=5). By direct inspection of the graph, one may see that input a separates the configurations

(1; 0,0) and (1; 0,5). The transition from state 11 guarded by the predicate $w_0 \leq 4 \wedge \neg(w_1 \leq 4)$ is enabled and takes the machine to the fail-state.

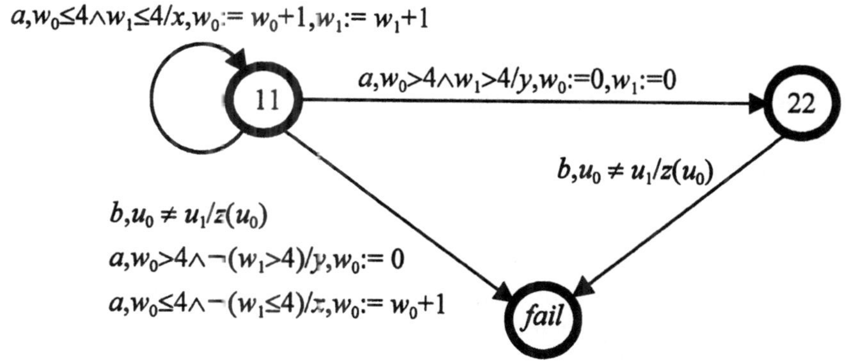

Figure 2. The fragment of the distinguishing machine [(M,(1; 0,0))-(M,(1; 0,5))]

Consider another situation when we are required to find a sequence of length at most two that separates the configuration (1; 0,0) from (1; 0,2). The l-fragment of the distinguishing machine [(M,(1; 0,0))-(M,(1; 0,2))] is exactly what we have just constructed (Figure 2). It has a different initial configuration (11; u_0=0, w_0=0, u_1=0, w_1=2), though. Reachability analysis shows that the fail-state can only be reached when a sequence of four inputs is applied. In other words, there is no sequence of length two that separates (1; 0,0) from (1; 0,2).

❑

Based on the above results, we next address the problem of configuration confirming sequence generation for an arbitrary set of configurations.

3.2 Configuration-confirming sequence for a set of configurations

We wish to distinguish a configuration from an arbitrary finite set of configurations of a known REFSM. Informally speaking, the set represents all possible "suspicious" configurations to which any implementation may go due to potential faults. It may be treated as a special fault model for a given transition or a fault function used in [8, 9].

Definition 3.7. Given configuration c and a configuration set C of an REFSM M, a parameterized input sequence α is said to *confirm c in* the set C if α separates c from every $c' \in C$ distinguishable from c.

❑

We use distinguishing machines to derive configuration confirming sequences. To this end, we first generalize the notion of a distinguishing machine of $(M,(s,\vec{v}))$ for $(N,(q,\vec{u}))$ (Definition 3.3) to a set of EFSMs. Specifically, let $\aleph$ be a finite set of strongly initialized, consistent, completely specified, and observable, but not necessarily deterministic EFSMs, i.e., $\aleph = \{(M_1,(s_1,\vec{v}_1)), \ldots, (M_k,(s_k,\vec{v}_k)\}$, $k \geq 1$.

Definition 3.8. A *distinguishing machine* of $(M,(s,\vec{v}))$ for $\aleph$ is defined as an EFSM constructed recursively, as follows.

- The distinguishing machine of $(M,(s,\vec{v}))$ for $\{(M_1,(s_1,\vec{v}_1))\}$ is $[(M,(s,\vec{v}))\text{-}(M_1,(s_1,\vec{v}_1))]$.
- Let $i<k$ and $[(M,(s,\vec{v}))\text{-}(M_1,(s_1,\vec{v}_1))\text{-}\ldots\text{-}(M_i,(s_i,\vec{v}_i))]$ be the distinguishing machine of $(M,(s,\vec{v}))$ for $\{(M_1,(s_1,\vec{v}_1)), \ldots, (M_i,(s_i,\vec{v}_i))\}$. Then the distinguishing machine of $(M,(s,\vec{v}))$ for $\{(M_1,(s_1,\vec{v}_1)), \ldots, (M_{i+1},(s_{i+1},\vec{v}_{i+1}))\}$ is the distinguishing machine of $[(M,(s,\vec{v})) - (M_1,(s_1,\vec{v}_1))\text{-}\ldots\text{-}(M_i,(s_i,\vec{v}_i))]$ for $(M_{i+1},(s_{i+1},\vec{v}_{i+1}))$.

We use $\Delta[(M,(s,\vec{v})), \aleph]$ to denote the distinguishing machine of $(M,(s,\vec{v}))$ for $\aleph$.

❑

The following property immediately follows from Definition 3.8 and Propositions 3.4 and 3.5.

Proposition 3.9. Given a finite set of machines $\aleph = \{(M_1,(s_1,\vec{v}_1)), \ldots, (M_k,(s_k,\vec{v}_k))\}$, $k \geq 1$, let $(s', s'_1, \ldots, s'_k)$ be the state, where a sequence α takes the distinguishing machine $\Delta[(M,(s,\vec{v})), \aleph]$ from the initial configuration. Then $s'_i = fail$ iff α is a separating sequence for $(s,\vec{v})$ and $(s_i,\vec{v}_i)$.

❑

If in the set $\aleph$, $M = M_1 = \ldots = M_k$ then we use a simplified notation for the distinguishing machine, namely, $\Delta[M, (s,\vec{v}), C]$, where $C = \{(s_1,\vec{v}_1), \ldots, (s_k,\vec{v}_k)\}$. Proposition 3.9 implies the following.

Corollary 3.10. Let $C = \{(s_1,\vec{v}_1), \ldots, (s_k,\vec{v}_k)\}$ contain only configurations distinguishable from $(s,\vec{v})$ and let $(s', s'_1, \ldots, s'_k)$ be the state, where a sequence α takes the distinguishing machine $\Delta[M, (s,\vec{v}), C]$ from the initial configuration. Then $s'_i = fail$ for all $1 \leq i \leq k$ iff α confirms $(s,\vec{v})$ in the set C.

❑

It is clear that if C contains configurations indistinguishable from $(s,\vec{v})$ then one can always omit the corresponding substates from the distinguishing machine $\Delta[M, (s,\vec{v}), C]$ without losing a CCS. To verify equivalence, a distinguishing machine for each configuration from C has first to be constructed. Once a configuration is found indistinguishable from

$(s,\vec{v})$, it is removed from the set C. This may simplify the machine $\Delta[M, (s,\vec{v}), C]$.

A CCS may not exist even if $(s,\vec{v})$ is distinguishable from every configuration in the set C. The situation is similar to SIS or UIO sequences of FSMs. A number of best-effort strategies can be applied when no CCS can be found. One may attempt the search for an input sequence such that it confirms $(s,\vec{v})$ in a maximal subset of C. In the graph of the distinguishing machine $\Delta[M, (s,\vec{v}), C]$, one has to find all the nodes with a maximal number of fail-substates and then determine a transfer sequence from the initial configuration to one of these node. If failed, one can resort to nodes with fewer fail-substates. It is also possible that one assigns certain weights to configurations in the set C that indicate how important is to distinguish each of them from the given configuration. Then the search has to be bounded for those states of a distinguishing machine whose weight exceeds a predefined limit. Alternatively, one may simply set a minimal percentage of configurations to be separated and solve a more modest task to reduce the complexity.

Another way of alleviating the state explosion effect is to impose a limit l on the length of CCS, as discussed in Section 3.1. To derive a CCS of at most l inputs, one can directly construct an l-fragment of the distinguishing machine rather than determining the whole machine.

The problem of CCS construction is a generalization of a well-known problem of deriving UIO for a minimal FSM to the class of EFSMs. The problem is PSPACE-complete even for FSMs [6]. Therefore, we have to look for approximation approaches to further alleviate the state explosion effect. Proper opportunities arise when all the configurations in the given set share a few common properties.

3.3 CCS for configurations of given state

Let $(s,\vec{v})$ be a configuration to be confirmed in a finite set $C = \{(s_1,\vec{v}_1), \ldots, (s_k,\vec{v}_k)\}$, such that $s_1 = \ldots = s_k = s'$ and all contexts $\vec{v}_i$ have an identical value of each context variable in $V^* \subseteq V$, i.e., for all $i=1, \ldots, k$, v_{ij} is constant for each $v_j \in V^*$, where v_{ij} denotes the j-component of the context $\vec{v}_i$. The approach of Section 3.2 can be tried to solve the problem. To avoid a potential state explosion effect, caused by the number k of substates in a distinguishing machine, we propose to approximate the behavior of k machines $(M,(s_1,\vec{v}_1)),\ldots, (M, (s_k,\vec{v}_k))$ by a single machine using an EFSM projection. The projection is based on common properties of configurations.

Definition 3.11. Given an REFSM M and a subset of context variables V^*, let $\overline{V^*}$ be the set of all context variables that depend on any variable from the set $V\backslash V^*$, defined by context update operations and let V^{**} denote the set $V\backslash\overline{V^*}$ (the set V^{**} is said to be a maximal *closed* subset of V^*). A V^{**}-*projection* of M is defined as an EFSM, obtained from the REFSM M by projecting out all context variables not in V^{**}, all the operations on them, output parameters defined by any operation on a variable not in V^{**}, and by replacing every predicate over such variables by a True constant. We use $M_{\downarrow V^{**}}$ to denote the V^{**}-projection of M.

❑

To define a projection of a strongly initialized machine, we also define a V^{**}-projection of context $\vec{v}_i$ as a vector, obtained from $\vec{v}_i$ by deleting variables that are not in the set V^{**}, denoted $\vec{v}_{i\downarrow V^{**}}$. Then the V^{**}-projection of $(M, (s_i, \vec{v}_i))$ is a pair $(M_{\downarrow V^{**}}, (s_i, \vec{v}_{i\downarrow V^{**}}))$. Obviously, given a set $C = \{(s', \vec{v}_1),\ldots, (s', \vec{v}_k)\}$, such that v_{ij} is constant for all $v_j \in V^*$ and all $i=1, \ldots, k$ and the set V^{**}, the machines $(M_{\downarrow V^{**}}, (s', \vec{v}_{i\downarrow V^{**}}))$ are identical. Let $(M_{\downarrow V^{**}}, (s', \vec{v}'_{\downarrow V^{**}}))$ be the obtained projection.

A sequence that confirms the configuration $(s, \vec{v})$ in the set C may be found in the distinguishing machine $[(M,(s, \vec{v}))\text{-}(M_{\downarrow V^{**}},(s', \vec{v}'_{\downarrow V^{**}}))]$, as stated in the following.

Proposition 3.12. If a parameterized input sequence α takes the distinguishing machine $[(M,(s, \vec{v}))\text{-}(M_{\downarrow V^{**}},(s', \vec{v}'_{\downarrow V^{**}}))]$ from the initial configuration to a fail-state then it confirms the configuration $(s, \vec{v})$ in the set of all configurations with state s' and any contexts $\vec{v}_i$ such that $\vec{v}_{i\downarrow V^{**}} = \vec{v}'_{\downarrow V^{**}}$, which includes the set C.

❑

The converse is not true. The method implied by this statement uses at most $|S|(|S|+1)$ states in the distinguishing machine instead of $|S|(|S|+1)^k$ states, the upper bound for the number of states in a machine of Proposition 3.8. To further reduce the state number, one may confine the construction of this machine to its l-fragment, as discussed before.

Example. Assume that the configuration (1; 0,0) of the EFSM M (Figure 1) has to be confirmed in all the configurations with state 2 with a sequence of at most two inputs. In this case, the set V^* is empty, its closed subset V^{**} is also an empty set. The V^{**}-projection of the EFSM M is an FSM obtained from M by erasing all predicates, variables, and operations on them. The obtained machine $(M_{\downarrow V^{**}}, 2)$ has the graph of the EFSM M (Figure 1) and we do not reproduce it here again. Figure 3 presents the l-fragment of the distinguishing machine $[(M, (1; 0,0))\text{-}(M_{\downarrow V^{**}}, 2)]$ for l=2. The initial configuration is (12; u=0, w=0). We have only two three paths to the fail-state, among which only one, labeled with *ab*, is executable. Thus, the input sequence *ab* confirms the configuration (1; 0,0) in the set of all configurations with state 2.

Replace now state 2 by 4. The corresponding l-fragment of the distinguishing machine [(M, (1; 0,0))-($M_{\downarrow V^{**}}$, 2)] for l=2 contains, in fact, only two transitions from state 14 to the fail-state shown in Figure 3. The transition with a is not enabled, while the one with b is. It means that the input sequence b confirms the configuration (1; 0,0) in the set of all configurations with state 4.

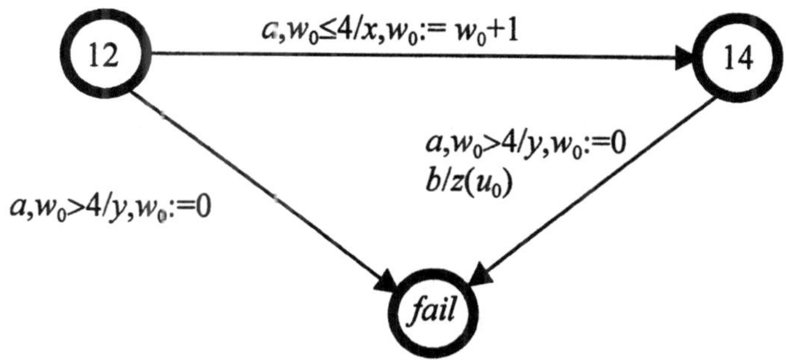

Figure 3. The l-fragment of the distinguishing machine [(M, (1; 0,0))-($M_{\downarrow V^{**}}$, 2)] for l=2

Finally, we illustrate projections and their use when the set V^* is not empty. Assume that the configuration (1; 0,0) of the EFSM M (Figure 1) has to be confirmed in all the configurations with state 3 and context variable u=1 with a sequence of at most two inputs. We have $V^*=\{u\}$. The closed subset V^{**} of this set is $\{u\}$, as the context variable u does not depend on w. The V^{**}-projection $M_{\downarrow V^{**}}$ is obtained from M (Figure 1) by erasing all statements and operations with the variable w in two transitions from state 1. We present here only the l-fragment of the distinguishing machine [(M, (1; 0,0))-($M_{\downarrow V^{**}}$, (3, u=1)] for l=2 in Figure 4.

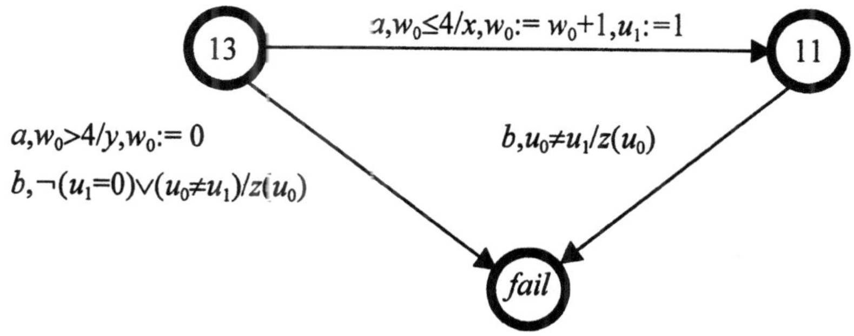

Figure 4. The l-fragment of [(M, (1; 0,0))-($M_{\downarrow V^{**}}$, (3, u=1)] for l=2

The obtained machine has the context variables u_0, w_0, and u_1. There are two executable paths from the initial configuration (13; u_0=0, w_0=0, u_1=1) to the fail-state, namely, b and ab. In other words, sequences b and ab confirm the configuration (1; 0,0) in all the configurations with state 3 and context variable u=1.

❑

The approach based on projections allows us to address the problem of CCS derivation even in a more general setting.

3.4 CCS for configurations of several states

Let $(s,\vec{v})$ be a configuration to be confirmed in a finite set $\mathfrak{I} = \{(s_1,\vec{v}_1^*), \dots, (s_k,\vec{v}_k^*)\}$, where $\vec{v}_i^*$ is a vector of $|V_i^*|$ components and V_i^* is a given subset of V. Vector $\vec{v}_i^*$ is called a *partial* context if $V^* \neq V$. A partial context $\vec{v}_i^*$ represents, in fact, a set of contexts $\{\vec{v} \in D_V \mid \vec{v}_{\downarrow V_i^*} = \vec{v}_i^*\}$, so the set $\mathfrak{I}$ is a set of configuration sets. States $s_1, \dots, s_k$ are not necessarily different. If this list includes all states in the given EFSM M and, in addition, $V_i^* = \varnothing$ for all i=1,…,k, then $\mathfrak{I}$ represents all the possible configurations of M. A parameterized input sequence α confirms c in the set $\mathfrak{I}$ if α confirms c in every set of $\mathfrak{I}$.

A pair $(M, (s_i, \vec{v}_i^*))$ is said to be a *weakly* initialized EFSM (as opposed to strongly initialized machines, defined in Section 2). Let V_i^{**} be the maximal closed subset of the set V_i^*. For each machine $(M, (s_i, \vec{v}_i^*))$ we determine its V_i^{**}-projection $(M_{\downarrow V_i^{**}}, (s_i, \vec{v}^*_{i\downarrow V_i^{**}}))$. Now, for each machine $(M_{\downarrow V_i^{**}}, (s_i, \vec{v}^*_{i\downarrow V_i^{**}}))$, we can verify whether the configuration $(s,\vec{v})$ of M can be separated from the configuration $(s_i, \vec{v}^*_{i\downarrow V_i^{**}})$ of $M_{\downarrow V_i^{**}}$. If so, there exists a sequence that confirms the configuration $(s,\vec{v})$ in the set of configurations represented by state s_i and partial context $\vec{v}_i^*$. The distinguishing machine $[(M, (s,\vec{v}))\text{-}(M_{\downarrow V_i^{**}}, (s_i, \vec{v}^*_{i\downarrow V_i^{**}}))]$ can be used to solve the problem. Hereinafter, we assume, with no loss of generality, that the configuration is distinguishable from the initial configuration of each machine in the set $\{(M_{\downarrow V_1^{**}}, (s_1, \vec{v}^*_{i\downarrow V_1^{**}})), \dots, (M_{\downarrow V_k^{**}}, (s_k, \vec{v}^*_{i\downarrow V_k^{**}}))\} = \aleph$.

The distinguishing machine of $(M, (s,\vec{v}))$ for the set of machines $\aleph$ possesses the following property.

Proposition 3.13. Let $\mathcal{J} = \{(s_1, \vec{v}_1^*), \ldots, (s_k, \vec{v}_k^*)\}$ be a finite set of configuration sets such that there exists a sequence that confirms $(s, \vec{v})$ in each configuration set represented by $(s_i, \vec{v}_i^*)$, and let $(s', s'_1, \ldots, s'_k)$ be a state, where a sequence α takes the distinguishing machine $\Delta[(M,(s, \vec{v})), \aleph]$ from the initial configuration. If $s'_i = fail$ for all $1 \le i \le k$ then α confirms $(s, \vec{v})$ in the set $\mathcal{J}$.

❑

If the distinguishing machine has no state with k fail-substates, one may try to determine a sequence that takes the distinguishing machine into a state that has a maximal number of fail-substates. Once found, such a sequence confirms a given configuration in a maximal number of sets in the given set $\mathcal{J}$. In fact, all the strategies discussed in Section 3.2 are applicable here as well.

Example. Assume that the configuration (1; 0,0) of the EFSM M (Figure 1) has to be confirmed in the set of sets of configurations {(1; 0,5), (3; u=1), (2), (4)} with a sequence of at most two inputs. In Section 3.1, we have determined that (1; 0,0) and (1; 0,5) are distinguishable; input a separates them. In Section 3.3, it is established that there are input sequences that confirm (1; 0,0) in each of the configuration sets (3; u=1), (2), and (4). Now we want to determine whether there exists a single input sequence of at most two inputs that confirms (1; 0,0) in at least three sets among (1; 0,5), (3; u=1), (2), and (4). To this end, we construct the l-fragment of the distinguishing machine of (M, (1; 0,0)) for the four machines, considered in Section 3.1 and 3.2. Since we are interested in finding a sequence that confirms the given configuration in at least three sets, we further exclude from the l-fragment each path in the graph that leads to a node with two or fewer fail-substates. The obtained fragment is presented in Figure 5.

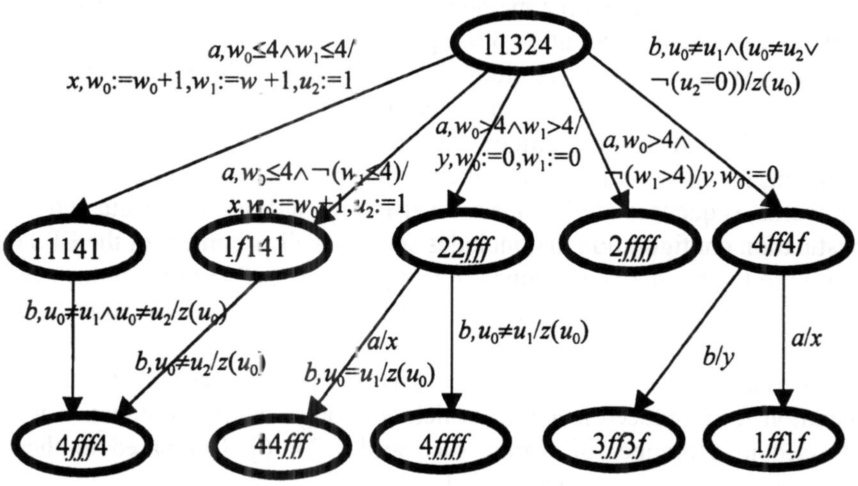

Figure 5. A fragment of the distinguishing machine of (M, (1; 0,0)) for four machines

Here the variables u_0, w_0 belong to the original machine *M*; u_1, w_1 belong to the first machine with the initial configuration (1; 0,5); and u_2 to the third machine with the initial configuration (3; u=1). Thus, (11324; u_0=0, w_0=0, u_1=0, w_1=5, u_2=1) is the initial configuration of the distinguishing machine. In the graph, there are eight nodes with three or four fail-substates and to solve the problem, it is sufficient to find an executable sequence that takes the machine from the initial configuration to one of these nodes.

By direct inspection, one may see that in Figure 5, there is only one executable path to the node (4*fff*4) caused by the input sequence *ab*. It means that *ab* confirms the configuration (1; 0,0) of the EFSM *M* in the set of sets of configurations {(1; 0,5), (3; u=1), (2)}.

4. COMPARISON WITH RELATED WORK

The list of papers addressing the issue of test derivation from EFSM specifications constantly grows in spite of the fact that all problems easily become intractable for non-trivial EFSMs. Most research concentrates on finding executable test cases that attempt to cover both the control flow and data flow aspects of an EFSM [10-21]. Symbolic execution and constraint solving appear as standard techniques for dealing with a general type of machines, though some researchers look for special EFSM classes, where the problems become somewhat easier and could be solved by other techniques.

Few attempts have been reported so far regarding the problem of state/configuration confirming sequence generation from an EFSM specification. Many papers mention using state verification sequences, usually in form of UIO, for checking the tail state of transitions. However, a formal definition along with a method of computation of these sequences for an EFSM can rarely be found in those papers. To the best of out knowledge, [22] and [23] are the few exceptions. [22] confines itself to a restricted class of EFSMs, while [23] addresses a wider class of EFSMs. [23] defines a so-called context independent unique sequence and presents an algorithm for computing it, while [22] attempts at finding a more general type of UIO. In both works, sequences are intended to verify only a destination state and not a destination configuration that includes state as well as context of the EFSM, opposed to the approach developed in this paper. Note that our approach can be adjusted to solve the problems considered in [22] and [23], due to its flexibility.

Note that in the realm of pure FSMs, the problem reduces to SIS or UIO generation for a reduced machine, which is a long-standing problem. It was addressed in a number of works, such as [7, 24, 6]. Compared to these techniques, our method, when applied to the FSM model, is similar to Gill's successor trees. Our notion of distinguishing machine is, in turn, pretty close to the FSM product used in many papers, mainly related to model checking and test derivation [25-28]. A distinguishing machine in this paper can be composed of several machines, whose behavior is compared to that of the specification machine, as in context of model checking. However, we further develop this notion to deal with the EFSM model.

Model checking techniques have already been tried for test derivation by many researchers, see, e.g., [29-32]. In these works, a product of two machines (one of them is a specification machine) has also been used to derive a test case that either fulfils a given test purpose (modeled by a second machine) or reveals a certain fault modeled by a mutated specification machine (a second machine). Even when a product machine is not explicitly constructed, as in, e.g., [19], it provides a theoretical foundation for the results. We further extend the concept of a product to a so-called distinguishing machine of a specification machine for a few machines that are obtained from the specification machine by projections and assigning proper initial configurations. It allows us to address a new (for the EFSM model) problem of determining the most "powerful" configuration-confirming sequence and apply widely used traversal methods, just as all the previous attempts did, to find a solution.

5. CONCLUSIONS

We investigated the problem of construction of a configuration confirming sequence for the EFSM model, specifically, given a configuration and an arbitrary set of configurations, determine an input sequence such that the EFSM in the given configuration produces an output sequence different from that of the configurations in the given set or at least in a maximal proper subset. The problem was also generalized to consider a set of configuration sets. We demonstrated that the problem can be reduced to the EFSM traversal problem, so that the existing methods and tools developed in the context of model checking become applicable. The theoretical framework for determining configuration-confirming sequences based on projections and products of EFSMs was presented. Based on this framework, a number of test derivation strategies with various heuristics could be elaborated and we indicated a few of them in this paper.

The approach to confirming sequence generation developed in this paper can be used to improve any existing test derivation tool that typically uses a model checker mainly to derive executable preambles and postambles. A most "powerful" confirming sequence satisfying user constraints could be generated and appended to the test body of a test case if desired. We believe that the techniques, based on projection and product elaborated in this paper, facilitate precise specification of requirements to configuration confirming sequences. The test designer may thus be provided with a new facility of incorporating test or fault hypotheses into the test derivation process. The added value would be in improved fault coverage of tests, according to the needs of the test designer. An attractive feature of our approach is its compatibility with a current practice of deriving tests from test purposes.

Our current work involves experiments with SDL specifications and a model checker to demonstrate the applicability of our approach to real systems.

ACKNOWLEDGMENTS

This work was supported by research contract 981B112 from France Télécom.

REFERENCES

1. A. Ek, J. Grabowski, D. Hogrefe, R. Jerome, B. Koch, M. Schmitt, *Towards the industrial use of validation techniques and automatic test generation methods for SDL specifications*, Proceedings of SDL'97 : Time for Testing - SDL, MSC and Trends, Elsevier, 1997.
2. A. Kerbrat, T. Jéron, R. Groz, *Automated test generation from SDL specifications*, Proceedings of SDL'99, Elsevier, 1999.
3. A. Petrenko, G. v. Bochmann, and M. Yao, *On fault coverage of tests for finite state specifications*, Computer Networks and ISDN Systems, 29, December 1996, pp. 81-106.
4. D. Lee, M. Yannakakis, *Principles and methods of testing finite state machines, a survey*, Proceedings of the IEEE, vol. 84, 8, 1996, pp. 1090-1123.
5. E. P. Hsieh, *Checking experiments for sequential machines*, IEEE Transactions on Computers, Vol. C-20, 10, 1971, pp. 1152-1166.
6. D. Lee, M. Yannakakis, *Testing finite-state machines: state identification and verification*, IEEE Transactions on Computers, Vol.43, 3, 1994, pp. 306-320.
7. A. Gill, *Introduction to the theory of finite-state machines*, Mc Graw-Hill, New York, 1962.
8. A. Petrenko, N. Yevtushenko, *Test suite generation for a given type of implementation errors*, Proceedings of the IFIP XII International Conference on Protocol Specification, Testing and Verification, 1992, pp. 229-243.
9. C.-J. Wang and M. T. Liu, *Generating test cases for EFSM with given fault model*, Proceedings of the IEEE INFOCOM'93, 1993, pp. 774-781.

10. H. Ural, A. Williams, *Test generation by exposing control and data dependencies within system specifications in SDL*, Proceedings of the IFIP Conference on Formal Description Techniques, 1993.
11. H. Ural, *Test sequence selection based on static data flow analysis*, Computer Communications, Vol. 10, No. 5, 1987.
12. R. E. Miller, and S. Paul, *Generating conformance test sequences for combined control and data flow of communication protocols*, Proceedings of IFIP XII International Conference on Protocol Specification, Testing and Verification, 1992, pp. 13-27.
13. K.-T. Cheng, A. S. Krishnakumar, *Automatic functional test generation using the extended finite state machine model*, Proceedings of the 30th Design Automation Conference, 1993, pp. 86-91.
14. W. Chun, P. D. Amer, *Test case generation for protocols specified in Estelle*, Proceedings of the IFIP Conference on Formal Description Techniques, III, 1991, pp. 191-206.
15. P. Qixiang, C. Shiduan, J. Yuchui, *Protocol conformance test suite generation*, Proceedings of ICCT'96, International Conference on Communication Technology, Vol.1, 1996, pp. 218-222.
16. R. E. Miller, Y. Xue, *Bridging the gap between formal specification and analysis of communication protocols*, Proceedings of the 1996 IEEE Fifteenth Annual International Phoenix Conference on Computers and Communications, pp. 225-231.
17. C.-M. Huang, M.-S. Chiang, *Protocol test sequence generation for EFSM-specified protocols*, Proceedings of the International Computer Symposium Conference, vol.2, 1994, pp. 842-847.
18. L.-S. Koh, M. T. Liu, *Test path selection based on effective domains*, Proceedings of ICNP, International Conference on Network Protocols, 1994, pp. 64-71.
19. C.-J. Wang, L.-S. Koh, M. T. Liu, *Protocol validation tools as test case generators*, Proceedings of the 7th International Workshop on Protocol Test Systems, 1994, pp. 155-170.
20. S. T. Chanson, J. Zhu, *Automatic protocol test suite derivation*, Proceedings of INFOCOM'94, Conference on Computer Communications, vol.2, 1994, pp. 792-799.
21. S. T. Chanson, J. Zhu, *A unified approach to protocol test sequence generation*, Proceedings of INFOCOM'93, vol.1, 1993, pp. 106-114.
22. X. Li, T. Higashino, M. Higuchi, K. Taniguchi, *Automatic generation of extended UIO sequences for communication protocols in an EFSM model*, Proceedings of the 7th International Workshop on Protocol Test Systems, 1994, pp. 225-240.
23. T. Ramalingom, A. Das, K. Thulasiraman, *A unified test case generation method for the EFSM model using context independent unique sequences*, Proceedings of the IFIP 8th International Workshop on Protocol Test Systems, 1995.

24. W. Chun, P. D. Amer, *Improvements on UIO sequence generation and partial UIO sequences*, Proceedings of the IFIP XII International Conference on Protocol Specification, Testing and Verification, 1992, pp. 245-260.
25. P. Camurati, M. Gilli, P. Prinetto, and M. S. Reorda, *Model checking and graph theory in sequential ATPG*, Proceedings of CAV'90, DIMACS Series in Discrete Mathematics and Theoretical Computer Science, Vol. 3, 1991, pp. 505-517.
26. G. Cabodi, P. Camurati, F. Corno, P. Prinetto, and M. S. Reorda, *The general product machine: a new model for symbolic FSM traversal*, Formal Methods in System Design, 12, 1998, pp. 267-289.
27. A. Petrenko, N. Yevtushenko, and G. v. Bochmann, *Testing deterministic implementations from their nondeterministic specifications*, Proceedings of the IFIP 9th International Workshop on Testing Communicating Systems, 1996, pp. 125-140.
28. R. Alur, C. Courcoubetis, and M. Yannakakis, *Distinguishing tests for nondeterministic and probabilistic machines*, Proceedings of the 27th ACM Symposium on Theory of Computing, 1995, pp. 363-372.
29. H. Cho, G. Hachtel, S.-W. Jeong, B. Plessier, E. Schwarz, F. Somenzi, *Results on the interface between formal verification and ATPG*, Proceedings of CAV'90, DIMACS Series in Discrete Mathematics and Theoretical Computer Science, Vol. 3, 1991, pp. 615-627.
30. M. Clatin, R. Groz, M. Phalippou, R. Thummel, *Two approaches linking a test generation tool with verification techniques*, Proceedings of the IFIP 8th International Workshop on Protocol Test Systems, 1995.
31. S. Huang, D. Lee, M. Staskauskas, *Validation-based test sequence generation for networks of extended finite state machines*, in Proceedings of the IFIP Conference on Formal Description Techniques, IX, 1996, pp.403-418.
32. J.-C. Fernandez, C. Jard, T. Jéron, C. Viho, *An experiment in automatic generation of test suites for protocols with verification technology*, Science of Computer Programming, 1996.

Remote testing can be as powerful as local testing*

Claude Jard, Thierry Jéron, Lénaïck Tanguy & César Viho
IRISA - Campus de Beaulieu, F35042 Rennes, France
e-mail: {jard, jeron, ltanguy, viho}@irisa.fr

* *This work is partially supported by the French Army (CELAR).*
C. Jard is member of CNRS, T. Jéron of INRIA,
L. Tanguy and C. Viho of IFSIC/University of Rennes I

Abstract

Designing test cases for remote asynchronous testing is error-prone. This is due to the difficulty to foresee all the disorders on the observations collected by the tester as well as the possible collisions between stimuli and observations. Designing correct synchronous test cases is easier, but transforming them into correct asynchronous ones is a difficult task. Moreover, it is difficult to compare remote testing and local testing as in general sets of conformant implementations are not comparable.

In this paper, we prove that by the use of logical stamps, remote testing can gain the same power as local testing: the conformant implementations in an asynchronous environment are exactly the same ones as in a synchronous environment. We give an operational method to derive the correct test cases for remote testing with this testing power. Furthermore, we show that test cases designed for a synchronous environment can test synchronous conformance in an asynchronous environment. This is achieved by the implementation of a test driver executed at runtime on the tester.

Keywords

Conformance Testing, Test Generation, Local and Remote Testing, Asynchronism, Stamp

1 INTRODUCTION

We consider the context of black box conformance testing in which an implementation under test (IUT for short) is tested in order to obtain the conviction that its behaviour conforms with its specification. The tester stimulates the IUT by sending messages on points of control and observation (PCOs) and observes on these same PCOs the reactions of the IUT (see figure 1, part A). Within sight of the reactions, a verdict (Fail, Pass or Inconclusive) is emitted. The underlying concepts have been formalized since the last years leading to the so called testing theory which identifies the notion of formal conformance

Figure 1 Synchronous and asynchronous testing

relation and gives a precise meaning of the verdicts (see [11] for example). Originally, the theory considered a synchronous interaction between the tester and the IUT. This made the implicit assumption that an IUT can refuse an event and that the tester can observe the refusal [1]. In practice however, one cannot always avoid taking into account the test environment intercalated between the tester and the IUT. As advocated in [4], conformance must then be defined as the conformance of the IUT in its test environment with the specification in a model of this environment. The most frequent example is that of remote testing architecture in which the tester reaches the IUT through a network connection. In this case, PCOs can be seen as composed of two FIFO queues, one for each direction of the interaction. The communication between the tester and the IUT is then asynchronous as illustrated in figure 1, part B.

The asynchronous nature of the PCOs poses some difficulties to design correct test cases. The possibility of disorder on the observations collected on different PCOs as well as the possible collision on a PCO between stimuli and observations have to be taken into account. This should be achieved by the inclusion of this asynchronous behaviour in the specification. But during the examination of existing test suites, one realizes (see for instance [2]) that this is not always done and it is the main reason of non-validity of some tests.

Another problem is the fundamental difference between conformance in synchronous or asynchronous environment. Precisely, for a particular specification, the sets of conformant IUTs in synchronous or asynchronous environment are in general not comparable. This explains why transforming a correct synchronous test case into a correct asynchronous test case is a difficult task.

It appears nevertheless that under certain conditions, the asynchronous deformation is foreseeable and even invertible. This topic is investigated in this paper and our main results are shortly described here:

1. We show that the **ioconf** conformance relation [12] has a simple characterization in terms of trace inclusion for input complete specifications. Moreover a specification can always be completed without modification of its set of conformant implementations.
2. We deduce that for complete specifications, synchronous conformance is preserved by the asynchronous environment. This means that a conformant IUT in synchronous environment is still conformant in an asynchronous environment. But asynchronous conformance is still more permissive: some

non conformant IUTs in synchronous environment can be conformant in an asynchronous environment.
3. Under the assumption of one PCO linking the IUT and its environment, and by using stamps (a simple counting mechanism), we prove that the permissiveness of remote asynchronous testing can be suppressed: conformant IUTs in synchronous environment are exactly the same ones as in asynchronous environment. From the preceding results, we deduce a method for the automatic generation of tests with synchronous testing power.
4. Furthermore, still in the case of one PCO, synchronous test cases can be used for asynchronous testing. This is achieved by a specific driver which inverts the asynchronous transformation using stamps. We think that this new mechanism can be of a great utility. It makes possible to conceive tests in a synchronous way while carrying out them in an asynchronous environment.

The presentation is organized as follows. We start by the definition of models. Then, we recall the local synchronous approach and the need to complete specifications to prepare their use in an asynchronous environment. Secondly, we present the distortions induced by the PCO queues and deduce the asynchronous conformance relation. Then we define the counting mechanism in the case of one PCO. We deduce a transformation of the specification allowing to generate automatically test cases for asynchronous testing with the power of synchronous testing. The next part is devoted to a possible implementation of a test driver which controls a synchronous test case in an asynchronous environment with the information acquired dynamically from the IUT. We finish by some prospects of generalization in the case of multiple PCOs.

2 LOCAL SYNCHRONOUS TESTING

Because of the asymmetrical nature of the testing activity, the models have to differentiate input and output actions. In this paper, we will use the model of IOLTS (Input-Output Labeled Transition Systems) to describe the different objects involved in the conformance testing.

2.1 Models

Definition 21 *An IOLTS is a tuple* $M=(Q^M, P_I^M, P_O^M, A_I^M, A_O^M, T^M, q_{init}^M)$ *where* Q^M *is a set of states,* $q_{init}^M \in Q^M$ *is the initial state,* P_I^M *and* P_O^M *are finite sets of input and output ports.* A_I^M *and* A_O^M *respectively are finite input and output alphabets.* $A^M \subseteq P_I^M \times \{?\} \times A_I^M \cup P_O^M \times \{!\} \times A_O^M$ *is the alphabet of observable actions constructed from the sets of input-output ports and input-output alphabets.* $\tau \notin A^M$ *denotes an internal action.* $T^M \subseteq Q^M \times A^M \cup \{\tau\} \times Q^M$ *is the transition relation, we note* $p \xrightarrow{\alpha}_M q$ *for* $(p, \alpha, q) \in T^M$.

Let $\alpha_i \in A^M$, $\mu_i \in A^M \cup \{\tau\}$, $\sigma \in (A^M)^*$, $q, q', q_i \in Q^M$:
• $q \xrightarrow{\mu_1 \cdots \mu_n} q' =_{def} \exists q_0 = q, q_1 \ldots, q_n = q', \forall i \in [1, n], q_{i-1} \xrightarrow{\mu_i} q_i$,

• $q \xrightarrow{\mu_1 \cdots \mu_n} =_{def} \exists q', q \xrightarrow{\mu_1 \cdots \mu_n} q'$ and $q \overset{\mu_1 \cdots \mu_n}{\not\longrightarrow} q' =_{def} \neg(q \xrightarrow{\mu_1 \cdots \mu_n})$,
• $q \overset{\epsilon}{\Rightarrow} q' =_{def} q = q'$ or $q \xrightarrow{\tau \cdots \tau} q'$ and $q \overset{\alpha}{\Rightarrow} q' =_{def} \exists q_1, q_2, q \overset{\epsilon}{\Rightarrow} q_1 \overset{\alpha}{\rightarrow} q_2 \overset{\epsilon}{\Rightarrow} q'$,
• $q \overset{\alpha_1 \cdots \alpha_n}{\Longrightarrow} q' =_{def} \exists q_0 = q, q_1 \ldots, q_n = q', \forall i \in [1, n], q_{i-1} \overset{\alpha_i}{\Rightarrow} q_i$,
• $enable(q) =_{def} \{\alpha \in A^{\mathrm{M}} \mid \exists q'$ and $q \overset{\alpha}{\Rightarrow}_{\mathrm{M}} q'\}$ is the set of observable actions possible in q, $In(q) =_{def} \{a \in A_I^{\mathrm{M}} \mid \exists p \in P_I^{\mathrm{M}}, p?a \in enable(q)\}$ is the set of possible inputs in q, and $Out(q) =_{def} \{a \in A_O^{\mathrm{M}} \mid \exists p \in P_O^{\mathrm{M}}, p!a \in enable(q)\}$ is the set of possible outputs in q,
• q *after* $\sigma =_{def} \{q \in Q^{\mathrm{M}} \mid q \overset{\sigma}{\Rightarrow}_{\mathrm{M}} q\}$ is the set of reachable states from q by the sequence of observable actions σ,
• $Traces(q) =_{def} \{\sigma \in (A^{\mathrm{M}})^* \mid q \text{ after } \sigma \neq \emptyset\}$,
• if $\alpha \in A^{\mathrm{M}}$ is an observable action, we note $\overline{\alpha}$ its mirror action: if $\alpha = p!a$ then $\overline{\alpha} = p?a$ else $\overline{\alpha} = p!a$. This can be extended to sequences of actions.

Definition 22 *An IOLTS M is said*
• **deterministic** *if* $\forall \sigma \in (A^{\mathrm{M}})^*, |M \text{ after } \sigma| \leq 1$ *where* $|X|$ *is the cardinality of the set* X,
• **controllable** *if in each state of M, either only one output is enabled or all inputs are enabled: for any sequence* $\sigma \in (A^{\mathrm{M}})^*$,
$In(M \text{ after } \sigma) = A_I^{\mathrm{M}}$ *or* $(In(M \text{ after } \sigma) = \emptyset \wedge |Out(M \text{ after } \sigma)| \leq 1)$,
• **input-complete** *if any input is possible after each trace:* $\forall \sigma \in (A^{\mathrm{M}})^*$, $In(M \text{ after } \sigma) = A_I^{\mathrm{M}}$.

As usual [10, 12], a **specification** of a system S will be modeled by an IOLTS $S = (Q^{\mathrm{S}}, P_I^{\mathrm{S}}, P_O^{\mathrm{S}}, A_I^{\mathrm{S}}, A_O^{\mathrm{S}}, T^{\mathrm{S}}, q_{\mathrm{init}}^{\mathrm{S}})$ and an **implementation** by a deterministic input-complete IOLTS $I = (Q^{\mathrm{I}}, P_I^{\mathrm{I}}, P_O^{\mathrm{I}}, A_I^{\mathrm{I}}, A_O^{\mathrm{I}}, T^{\mathrm{I}}, q_{\mathrm{init}}^{\mathrm{I}})$, with $P_I^{\mathrm{I}} = P_I^{\mathrm{S}}$, $P_O^{\mathrm{I}} = P_O^{\mathrm{S}}$, $A_I^{\mathrm{S}} \subseteq A_I^{\mathrm{I}}$, and $A_O^{\mathrm{S}} \subseteq A_O^{\mathrm{I}}$. A **test case** is a set of sequences of actions describing all the interactions occurring between an IUT and a tester which wants to verify that an IUT conforms with its specification. A test case is modeled by a deterministic IOLTS $T = (Q^{\mathrm{T}}, P_I^{\mathrm{T}}, P_O^{\mathrm{T}}, A_I^{\mathrm{T}}, A_O^{\mathrm{T}}, T^{\mathrm{T}}, q_{\mathrm{init}}^{\mathrm{T}})$ such that: $A_I^{\mathrm{T}} = A_O^{\mathrm{I}}$ (every possible output of the IUT must be considered as an input of the test case), $A_O^{\mathrm{T}} = A_I^{\mathrm{S}}$ (a test case should only send outputs that are waited by the specification), $\{pass, fail\} \in Q^{\mathrm{T}}$ with $enable(pass) = enable(fail) = \emptyset$, and the last transition leading to *fail* is an input. In general, it is assumed that a test case is controllable and input-complete.

Remark: In practice $A_I^{\mathrm{T}} = A_O^{\mathrm{I}}$ is unknown. Thus, only inputs not leading to *fail* can be denoted, the other inputs are implicitly leading to fail or are denoted by "? otherwise Fail", like in TTCN (see [9]-part 3). ⋄

2.2 Conformance

Formalizing conformance testing [4] necessitates to define a conformance relation relating models of IUTs to specifications. We will consider here a conformance relation which states that outputs produced by an IUT after a trace of the specification are foreseen by the specification [10, 12].

Definition 23 *(Conformance relation) Let S be the IOLTS describing the specification and I an (input-complete) IOLTS describing an implementation:* $I\ ioconf\ S \iff \forall\sigma \in Traces(S), Out(I\ after\ \sigma) \subseteq Out(S\ after\ \sigma)$.

Definition 24 *(Synchronous Testing) The synchronous application of a test case to an IUT is defined as a parallel composition* $\|$ *of the test case T and the IUT I:* $\forall\alpha \in A_I^T \cup A_O^T, \frac{T \xrightarrow{\alpha} T' \quad I \xrightarrow{\bar{\alpha}} I'}{T\|I \xrightarrow{\alpha} T'\|I'}$.

Definition 25 *(Test failure and unbiased test case)*
$T\ fails\ I =_{def} \exists I', \exists\sigma, T\|I \stackrel{\sigma}{\Rightarrow} fail \| I'$. *A test case T is unbiased with respect to S if and only if* $\forall I, T\ fails\ I \Rightarrow not(I\ ioconf\ S)$.

The definition of *ioconf* authorizes IUTs to diverge from the specification starting from unspecified inputs: the specification implicitly authorizes any behaviour in the IUT after an unspecified input. We may need to make these behaviours explicit by considering input-complete specification. Moreover for input complete specifications, *ioconf* has a very simple characterization as stated by the following proposition.

Proposition 21 *Let S and I be two input-complete IOLTS.*
We have $I\ ioconf\ S \iff Traces(I) \subseteq Traces(S)$

Proof: Suppose $I\ ioconf\ S$ and let $w \in Traces(I)$. Suppose $w \notin Traces(S)$, then w can be split in two sequences $w = w_1.w_2$ where w_1 is the maximal prefix of $w \in Traces(S)$. Let α be the first action in w_2. If α is an input, as S is input complete, $u_1.\alpha \in Traces(S)$. If α is an output, as $w_1 \in Traces(S)$ and $I\ ioconf\ S$, $\alpha \in Out(S\ after\ w_1)$, and $w_1.\alpha \in Traces(S)$ which contradicts the hypothesis. Thus, $Traces(I) \subseteq Traces(S)$. Thus in both cases w_1 is not maximal and proves that $w \in Traces(S)$.

The converse i.e. $Traces(I) \subseteq Traces(S) \Rightarrow I\ ioconf\ S$ is evident even for a non input-complete specification S. ■

As we may have to deal with specifications which are not input complete, we may want to complete them in order to benefit from the preceding result. But this completion should not alterate the set of conformant IUTs. This is achieved by using the following completion similar to the notion of trap-state [13] and illustrated by figure 2.

Definition 26 *(Completion)*
Let S be an IOLTS. Comp(S) is defined as an IOLTS such that:
$Traces(Comp(S)) = Traces(S) \bigcup (\cup_{w \in Traces(S), a \in U(w)} w.?a.(\{?\} \times A_I \cup \{!\} \times A_O)^*)$
where $U(w) = A_I \setminus In(S\ after\ w)$ *is the set of unspecified inputs of the state of S reached after the trace w.*

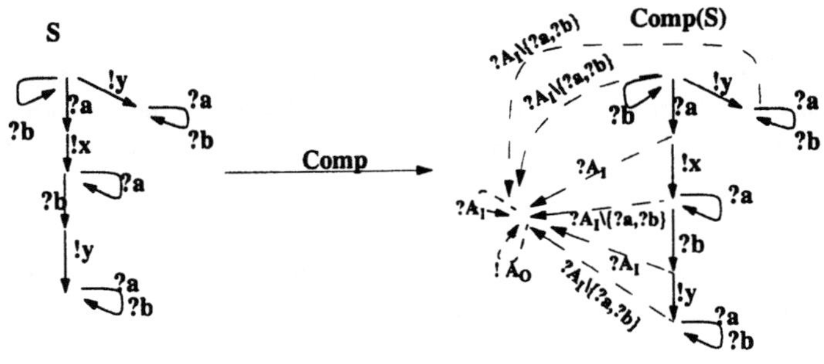

Figure 2 Example illustrating the completion (dotted lines represent transitions added by the transformation)

Proposition 22 *Let S be an IOLTS and $Comp(S)$ its input completion. For all input complete IOLTS I, I ioconf $S \Longleftrightarrow I$ ioconf $Comp(S)$.*

Proof: Suppose $\neg(I \textit{ ioconf } S)$. By definition of *ioconf*, this means $\exists \sigma \in Traces(S), \exists z \in Out(I \textit{ after } \sigma)$, and $z \notin Out(S \textit{ after } \sigma)$. By definition of *Comp*, we have $Traces(S) \subseteq Traces(Comp(S))$, so $\sigma \in Traces(Comp(S))$. But $\sigma.z \notin Traces(Comp(S))$ as $z \in A_O$ and the input completion only adds $a \in A_I \backslash In(S \textit{ after } \sigma)$. This implies $\neg(I \textit{ ioconf } Comp(S))$.
Suppose now $\neg(I \textit{ ioconf } Comp(S))$, this means $\exists \sigma \in Traces(Comp(S))$, $\exists z \in Out(I \textit{ after } \sigma)$, and $z \notin Out(Comp(S) \textit{ after } \sigma)$. If $\sigma \in Traces(S)$ then $z \notin Out(Comp(S) \textit{ after } \sigma)$ induces $z \notin Out(S \textit{ after } \sigma)$ as $Traces(S) \subseteq Traces(Comp(S))$. Otherwise $\sigma \in Traces(Comp(S)) \backslash Traces(S)$ is of the form $\sigma_1.a.\sigma_2$ with $\sigma_1 \in Traces(S)$ and $a \in A_I$. In this case, $\sigma_1.a.\sigma_2.z \in Traces(Comp(S))$, i.e. $z \in Out(Comp(S) \textit{ after } \sigma)$ which contradicts the hypothesis. ■

Remark: The validity of the proposition 22 heavily depends on the notion of completion used. For example completing with loops on unspecified inputs does not work.

There also exist different notions of input completion. For instance in SDL, input completion is implicit in the case of unspecified inputs. This means that in any control state, if an unspecified input occurs it is consumed. But as transitions are atomic, this happens only in any control (stable) states. Thus in terms of transition systems, only some states are input complete. The other states should be completed with *Comp* in order to apply 22.

Notice also that the notion of completeness of Mealy machines (see [8] for example) corresponds more to this implicit completeness of SDL than to the notion of input-completeness of definition 26. ◇

Propositions 21 and 22 lead to the following proposition 23 which gives a very simple characterization of *ioconf*.

Proposition 23 *I ioconf $S \Longleftrightarrow Traces(I) \subseteq Traces(Comp(S))$.*

3 REMOTE ASYNCHRONOUS TESTING

In practice, the testing activity is generally done through an environment intercalated between the tester and the IUT. As advocated in [4] the conformance of an IUT with respect to a specification in a context should be defined as the conformance of the implementation in its context with respect to the specification in a model of the context. For example, in the context of a remote testing architecture, the tester reaches the IUT through a network. In this case, the interaction between the tester and the IUT is asynchronous and PCOs can be seen as composed of two FIFO queues.

We first define the asynchronous transformation $\mathcal{A}$ on IOLTS which describes the impact of FIFO queues on observable behaviours. We use $\mathcal{A}$ to define the conformance relation in an asynchronous environment. We notice how this affects conformance: the set of conformant IUTs in synchronous or asynchronous architectures are not comparable in general. However, we show that for input-complete specifications, synchronous conformance implies asynchronous conformance.

3.1 Asynchronous testing

As already done in [14], we define the asynchronous transformation $\mathcal{A}$ as follows:

Definition 31 *Let* $M = (Q^M, P_I^M, P_O^M, A_I^M, A_O^M, T^M, q_{init}^M)$ *be an IOLTS.* $\mathcal{A}(M) = (Q^{A(M)}, P_I^{A(M)}, P_O^{A(M)}, A_I^{A(M)}, A_O^{A(M)}, T^{A(M)}, q_{init}^{A(M)})$ *with:*
- $Q^{A(M)} = Q^M \times \prod_{p \in P_I^M} A_I^{M*} \times \prod_{p \in P_O^M} A_O^{M*}$ *and* $q_{init}^{A(M)} = < M, (\epsilon \cdots \epsilon), (\epsilon \cdots \epsilon) >$
- $P_I^{A(M)} = P_I^M$ *and* $P_O^{A(M)} = P_O^A$, $A_I^{A(M)} = A_I^M$ *and* $A_O^{A(M)} = A_O^M$,
- $T^{A(M)}$ *is described by the following operational rules defined for all* $q, q' \in Q^M, a \in A_I^M, b \in A_O^M, p_{I_k} \in P_I^M, p_{O_l} \in P_O^M$:

R1 (inputs of $\mathcal{A}(M)$ from Env)

$$< q, (p_{I_1} \cdots p_{I_k} = w \cdots), (p_{O_1} \cdots) > \overset{p_{I_k}?a}{\rightarrow}_{A(M)} < q, (p_{I_1} \cdots p'_{I_k} = w.a \cdots), (p_{O_1} \cdots) >$$

R2 (outputs of $\mathcal{A}(M)$ to Env)

$$< q, (p_{I_1} \cdots), (p_{O_1} \cdots p_{O_l} = b.w \cdots) > \overset{p_{O_l}!b}{\rightarrow}_{A(M)} < q, (p_{I_1} \cdots), (p_{O_1} \cdots p_{O_l} = w \cdots) >$$

R3 (internal actions)

$$\frac{q \overset{\tau}{\rightarrow}_M q'}{< q, (p_{I_1} \cdots), (p_{O_1} \cdots) > \overset{\tau}{\rightarrow}_{A(M)} < q', (p_{I_1} \cdots), (p_{O_1} \cdots) >}$$

R4(inputs of M from queues)

$$\frac{q \overset{p_{I_k}?a}{\rightarrow}_M q'}{< q, (p_{I_1} \cdots p_{I_k} = a.w \cdots), (p_{O_1} \cdots) > \overset{\tau}{\rightarrow}_{A(M)} < q', (p_{I_1} \cdots p_{I_k} = w \cdots), (p_{O_1} \cdots) >}$$

R5 (outputs of M to queues)

$$\frac{q \overset{p_{O_l}!b}{\rightarrow}_M q'}{< q, (p_{I_1} \cdots), (p_{O_1} \cdots p_{O_l} = w \cdots) > \overset{\tau}{\rightarrow}_{A(M)} < q', (p_{I_1} \cdots), (p_{O_1} \cdots p_{O_l} = w.b \cdots) >}$$

Remark: $\forall$ *IOLTS M, Traces(M)* $\subseteq$ *Traces*($\mathcal{A}(M)$). $\diamond$

For the asynchronous context the conformance relation now becomes:

Definition 32 *(Conformance in an asynchronous environment) Let I and S be two IOLTS with I input-complete. I* $ioconf_A$ *S* $=_{def}$ $\mathcal{A}(I)$ *ioconf* $\mathcal{A}(S)$.

The notion of test failure and unbias in an asynchronous environment are then straightforward.

3.2 Problems in asynchronous testing

Let us consider the specification S described on the left part of figure 3. For sake of clarity, we will suppose that all the indicated observable actions occur on the same PCO. Notice also that S is not input-complete. The right

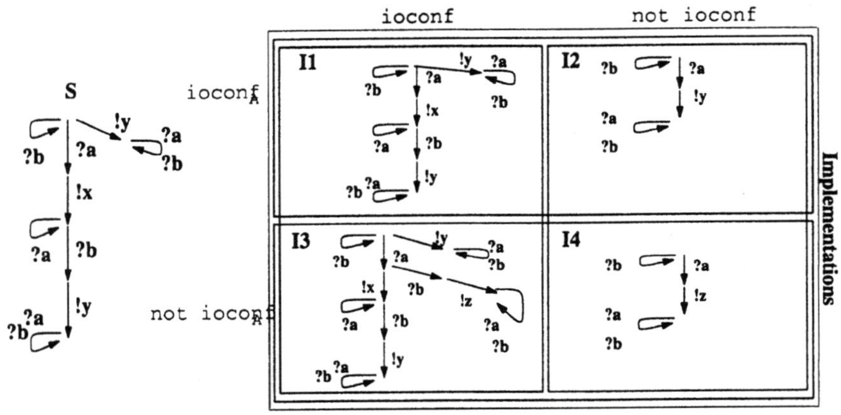

Figure 3 Difference between synchronous and asynchronous testing

part of figure 3 contains different implementations which show that testing synchronously or asynchronously is not comparable. More precisely, they show two main problems when testing in an asynchronous environment.

Permissiveness: The IUT I_2 shows that asynchronous testing is more permissive than synchronous testing: $\neg(I_2$ *ioconf* $S)$ but $(I_2$ $ioconf_A$ $S)$. We have $\neg(I_2$ *ioconf* $S)$ because $Out(I_2 \text{ after } ?a) = \{y\} \not\subseteq Out(S \text{ after } ?a) = \{x\}$. In an asynchronous environment, outputs can be delayed. Thus the trace $!y.?a$ of S can be observed as $?a.!y$ in $\mathcal{A}(S)$. Thus, we have $(I_2$ $ioconf_A$ $S)$ as $Out(\mathcal{A}(I_2) \text{ after } ?a) = \{y\} \subseteq Out(\mathcal{A}(S) \text{ after } ?a) = \{x, y\}$.

Non preservation of conformance: This problem is brought to light by the IUT I_3: I_3 *ioconf* S but $\neg(I_3$ $ioconf_A$ $S))$. In fact, I_3 *ioconf* S even though $?a.?b.!z$ is not a trace of S. This is because *ioconf* authorizes divergence from the specification starting from an unspecified input. The sequence

$?a.?b$ is in $Traces(\mathcal{A}(S))$, and we have $Out(\mathcal{A}(S)$ *after* $?a.?b) = \{x, y\}$ but $Out(\mathcal{A}(I_3)$ *after* $?a.?b) = \{x, y, z\}$. Thus $\neg(I_3$ *ioconf*$_A$ $S)$.

Notice that these problems are more complex in the context of several PCOs as inputs and outputs orders are not preserved. Similar remarks concerning the non preservation of conformance in asynchronous environment have been done in [14]. But these were made regarding the synchronous conformance relation *conf* (which does not distinguish inputs and outputs) and an asynchronous conformance relation similar to *ioconf*$_A$.

The problem of permissiveness is inherent to the transformation by a context. But the non preservation of conformance is due to the fact that S is not input-complete. Lemma 31 states that this can be avoided input-complete specifications and **monotonic** transformations by contexts. A transformation $\mathcal{T}$ on IOLTS is **monotonic** if it preserves trace inclusion:
$Traces(M) \subseteq Traces(M') \Rightarrow Traces(\mathcal{T}(M)) \subseteq Traces(\mathcal{T}(M'))$.

Lemma 31 *Let S and I be input complete IOLTS. If $\mathcal{T}$ is a monotonic transformation, I ioconf $S \Rightarrow \mathcal{T}(I)$ ioconf $\mathcal{T}(S)$.*

Proof: Suppose I *ioconf* S and S and I are input complete. By proposition 21 we have $Traces(I) \subseteq Traces(S)$. By monotonicity of $\mathcal{T}$, this implies $Traces(\mathcal{T}(I)) \subseteq Traces(\mathcal{T}(S))$ which implies $\mathcal{T}(I)$ *ioconf* $\mathcal{T}(S)$. ∎

The asynchronous transformation $\mathcal{A}$ is monotonic because it is applied on traces independently of others. Application of lemma 31 gives the corollary:

Corollary 31 *For input complete IOLTS S and I, I ioconf $S \Rightarrow I$ ioconf$_A$ S.*

As a consequence, the only difference between *ioconf* and *ioconf*$_A$ is permissiveness. We show in the next section that this can be avoided if the order of occurrence of events on the IUT is captured by an appropriate stamp mechanism.

4 STAMPED ASYNCHRONOUS TESTING

The idea is to instrument the IUT so that each output from the IUT to the tester (via the environment) can bring to the tester an additional information on the real order in which the IUT has produced the events. Linking time stamp techniques used in observing distributed systems to the problem of generating tests has been already advocated in different contexts like in [6, 7].

This instrumentation can be defined by the synchronous parallel composition of the IUT with a stamp process ST (as illustrated in figure 4). The role of the stamp process ST is to code the history of occurrence of events on the IUT and to transmit it to the environment by piggybacking each output.

We will consider the particular case of one PCO in which history is coded by an integer counter.

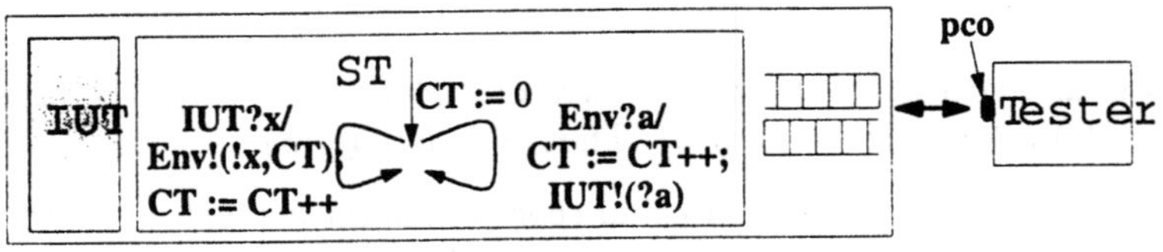

Figure 4 Implementation of the counting mechanism *st*

This stamping process ST implements a function *st* defined on traces as follows.

Definition 41 $st : (A_I \cup A_O)^* \rightarrow (A_I \cup (A_O \times \mathbb{N}))^*$, $\forall \sigma \in (A_I \cup A_O)^*, a \in A_I$, *and* $z \in A_O$: $st(\epsilon) = \epsilon$, $st(\sigma.!z) = st(\sigma).(!z, length(\sigma))$, *and* $st(\sigma.?a) = st(\sigma).?a$.

By extension, one can regard the transformation *st* as being carried out on the IOLTS by an on-the-fly traversal of the graph. Let us consider the example of an interaction between an IUT and its environment illustrated on figure 5. From the point of view of events produced on the IUT, one has: $st(?x.!s.?y.?t.?u.!z) = ?x.(!s,1).?y.?t.?u.(!z,5)$.

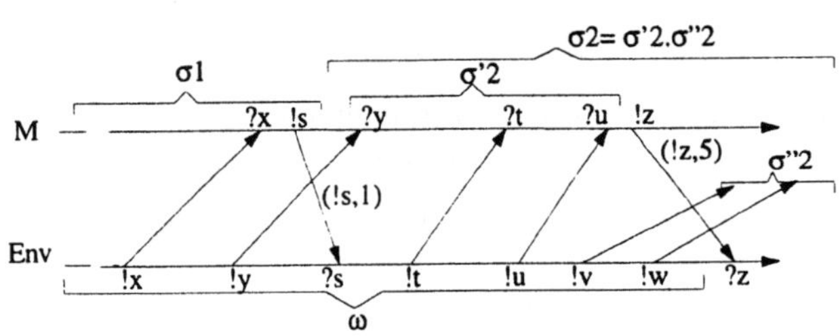

Figure 5 Illustration of the computation of st and $\overline{st}$

This counting information is intended to be decoded by the tester as described in section 5. We thus define an opposite transformation $\overline{st}$ (illustrated in figure 5), which orders events of a system M in the environment. From a sequence of events ω in the environment *Env*, $\overline{st}$ computes two sequences of events $\overline{st}(\omega)[1]$ and $\overline{st}(\omega)[2]$ using the counting information associated to outputs of the system M:

- $\overline{st}(\omega)[1]$ is the sequence of events of the system M which precede the last output z of M received by the environment and including this output.
- $\overline{st}(\omega)[2]$ is the sequence of inputs of M (corresponding to outputs of *Env*) which were not received before the sending by M of z. As we assume a FIFO channel between *Env* and M, this sequence will be received in this order but M may have some intercalated outputs of M.

Notice that $\overline{st}$ can be computed on-the-fly by *Env*. When *Env* makes an output, the corresponding input is added to the tail of $\overline{st}(\omega)[2]$. When *Env* makes an input z with the counting information, $\overline{st}(\omega)[1]$ and $\overline{st}(\omega)[2]$ are updated accordingly.

Definition 42 *$\overline{st} : (A_I \cup (A_O \times \mathbb{N}))^* \rightarrow (A_I \cup A_O)^* \times A_I^*$ is defined recursively as: $\overline{st}(\epsilon) = (\epsilon, \epsilon)$, and $\forall \omega \in (A_I \cup (A_O \times \mathbb{N}))^*$, if $\overline{st}(\omega) = (\overline{st}(\omega)[1], \overline{st}(\omega)[2])$ then $\overline{st}(\omega.?a) = (\overline{st}(\omega)[1], \overline{st}(\omega)[2].?a)$ and $\overline{st}(\omega.(!z, i)) = (\overline{st}(\omega)[1].\sigma_2'.!z, \sigma_2'')$ where σ_2' and σ_2'' are sequences such that $\overline{st}(\omega)[2] = \sigma_2'.\sigma_2''$ with $length(\sigma_2') = i - length(\overline{st}(\omega)[1])$.*

For example, let us consider the sequence of events occurring on the tester in figure 5 (denoted by their corresponding names on the IUT). Then we have $\overline{st}(?x.?y.(!s,1).?t.?u.?v.?w) = (?x.!s,\ ?y.?t.?u.?v.?w)$, and $\overline{st}(?x.?y.(!s,1).?t.?u.?v.?w.(!z,5)) = (?x.!s.?y.?t.?u.!z,\ ?v.?w)$. In this last case, this means that upon reception of z, the tester knows that $?x.!s.?y.?t.?u.!z$ has occurred in this order on the IUT, but it does not know yet what will be the order in the future including the receptions of v and w. The counting information gives the index for inserting the output in the sequence.

The following proposition states that the composition of st with the asynchronous transformation $st \circ \mathcal{A}$ is invertible: for a trace σ, the application of $\overline{st}$ on any trace of $st \circ \mathcal{A}(\sigma) = \mathcal{A}(st(\sigma))$ can reconstruct σ.

Proposition 41 *$\forall M \in IOLTS, \forall \sigma \in Traces(M), \forall \omega \in Traces(\mathcal{A}(st(\sigma)))$, we have: $\overline{st}(\omega) = (\sigma_1, \sigma_2)$ with $\sigma = \sigma_1.\sigma_2$.*

Proof: *st* associates with each output the index of this output in the sequence. This information can be used by $\overline{st}$ to recover the order since the transformation $\mathcal{A}$ (in the case of only one FIFO queue in each direction) preserves the relative order of inputs and the relative order of outputs. Notice that outputs do not need to be FIFO ordered as they are numbered by *st*. ■

We can now prove the main theorem which says that for input complete IOLTS communicating asynchronously with their environment using one FIFO in each direction, remote asynchronous testing using stamps has the same testing power as local synchronous testing.

Theorem 41 *Let S and I be two input complete IOLTS communicating asynchronously with their environment using one FIFO in each direction.*
We have: $I\ ioconf\ S \iff st(I)\ ioconf_{\mathcal{A}}\ st(S)$

Proof: We first prove that $I\ ioconf S \Rightarrow \mathcal{A}(st(I))\ ioconf\,\mathcal{A}(st(S))$. Since $\mathcal{A}$ and *st* are monotonic (*st* is defined on traces), $st \circ \mathcal{A}$ is also monotonic and lemma 31 applies.

Let us prove the converse: if $\neg(I\ ioconf S)$ then $\neg(\mathcal{A}(st(I))\ ioconf\,\mathcal{A}(st(S)))$. By definition of *ioconf*, $\neg(I\ ioconf S) \Rightarrow \exists \sigma \in Traces(S), \exists z \in Out(I after \sigma)$ such that $z \notin Out(S after \sigma)$. Let $\omega = st(\sigma)$. For any $IOLTS\ M, \sigma \in Traces(M)$ implies $st(\sigma) \in Traces(st(M))$ and $Traces(M) \subseteq Traces(\mathcal{A}(M))$. Thus, we

have: $\omega \in Traces(\mathcal{A}(st(S)))$ and $\omega.(z, length(\sigma)) \in Traces(\mathcal{A}(st(I)))$. We want to prove that $\omega.(z, length(\sigma)) \notin Traces(\mathcal{A}(st(S)))$.
Since $z \notin Out(S after \sigma)$, we must show that the output $(z, length(\sigma))$ after ω cannot be created by the $st \circ \mathcal{A}$ transformation. First, note that the traces of $\mathcal{A}(\sigma)$ are produced from σ by the following semi-commutation $!x.?y \rightarrow ?y.!x$ which can only delay the outputs with respect to inputs. Let us suppose that z belongs to a trace σ' of S. We show that σ' is identical to σ. $\forall \sigma' \in Traces(S), \exists \omega \in \mathcal{A}(st(\sigma'))$ and $\omega = st(\sigma) \Rightarrow \sigma = \sigma'$. This is based on the properties of $\mathcal{A}$: a) $length(\sigma) = length(\sigma')$, b) σ and σ' are on the same alphabet, c) the outputs are numbered by the same stamps, d) the inputs are not changed by the transformation.
Thus, $\exists \omega \in Traces(\mathcal{A}(st(S))), \omega.(z, length(\sigma)) \notin Traces(\mathcal{A}(st(S)))$ which implies $\exists \omega \in Traces(\mathcal{A}(st(S))), (z, length(\sigma)) \in Out(\mathcal{A}(st(I)\ after\ \omega)$, and $(z, length(\sigma)) \notin Out(\mathcal{A}(st(S)\ after\ \omega)$. ■

Stamped asynchronous tests generation Theorem 41 induces a method for the generation of test cases for asynchronous testing with the power of synchronous testing.

Proposition 22 says that for any specification S its input completion $Comp(S)$ has the same set of conformant implementations. Theorem 41 says that if we assume a communication with the environment with one FIFO in each direction, an IUT I is conformant to $Comp(S)$ if and only if $\mathcal{A}(st(I))$ is conformant with $\mathcal{A}(st(Comp(S)))$. Now, there exists test generation algorithms implemented in tools [3, 15]. Applied to a specification S, they produce unbiased test cases with respect to S and *ioconf*. These methods are (theoretically) exhaustive: assuming bounded fairness of implementations, all non conformant implementations may be rejected by a test case. Using these algorithms on $\mathcal{A}(st(Comp(S)))$ will thus produce a test suite which is unbiased and exhaustive with respect to $\mathcal{A}(st(Comp(S)))$ and *ioconf*. Moreover, by theorem 41 this test suite has exactly the same testing power as the synchronous test suite.

But, generating tests from $\mathcal{A}(st(Comp(S)))$ has some drawbacks. The first one is the state space explosion due to the asynchronous transformation and the unfolding caused by st. A second one is the relevance of test cases. In the example of figure 3, $?a.?b.(!x, 1)$ and $?a.(!x, 1).?b$ are both sequences of $\mathcal{A}(st(Comp(S)))$. So one could generate a test case with a sequence $!a.!b.(?x, 1)$. But this would be artificial because as S sends x after the input a, it is preferable to wait for x before sending b.

So, when testing the conformance of I with respect to S in a remote testing architecture, we are mainly interested in traces of S but not in all traces of $\mathcal{A}(st(Comp(S)))$, even if we have to consider them as possible traces. One way to limit these drawbacks is thus to use test purposes which accept traces in $Traces(S)$ in order to select test cases from $\mathcal{A}(st(Comp(S)))$ and to generate test cases on-the-fly as in [3]. Another completely different and more efficient solution is investigated in the following section.

5 REMOTE ASYNCHRONOUS TESTING WITH SYNCHRONOUS TEST CASES

The idea is that from a sequence of events of $\mathcal{A}(st(I))$, $\overline{st}$ can reconstruct, with a certain latency due to the asynchronous communication, the sequence of events which occurred on I. Thus this sequence can be checked on test cases produced from S to check the conformance of I with respect to S and *ioconf* as in a local synchronous testing architecture. Thus local synchronous test cases are generated from $Comp(S)$. Then, test cases are decorated with counters using the transformation st. They can then be played in a remote asynchronous testing architecture with a kind of test driver implementing $\overline{st}$. This is illustrated in figure 6.

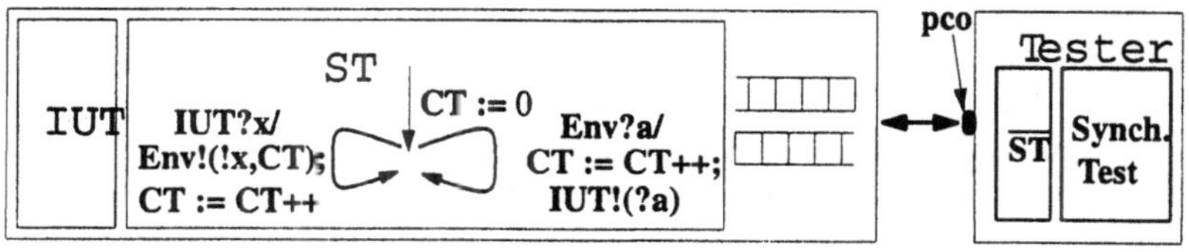

Figure 6 Architecture for "remote local" testing

The mechanism necessitates some attention. In fact the difference between synchronous local testing and asynchronous testing is that the control of the tester on the IUT is weakened. Inputs of the IUT cannot be completely controlled by the tester. This can be illustrated on the example of figure 3. According to the specification, the tester may choose an output (say a) and computes $\overline{st}(?a) = (\epsilon, ?a)$. But between the output $!a$ by the tester and its corresponding input $?a$ by the IUT, the IUT may decide to perform the output $!y$ associated with the counter 0. When receiving $(y, 0)$, $\overline{st}(?a.(!y, 0)) = (!y, ?a)$. Thus the tester knows that the IUT has performed $!y$ first and will later receive a. The IUT has thus chosen a different behaviour from the one chosen by the tester. Nevertheless, the tester must evolve according to the behavior of the IUT but also to its own past behavior. In particular, a has been sent and this cannot be cancelled. Thus the tester will have to wait for a new input from the IUT or choose a new output according to the sequence $!y.?a$. All this information is contained in $\overline{st}(?a.(!y, 0))$.

The tester thus computes $\overline{st}(\omega)$ on-the-fly. Only $\overline{st}(\omega)[1]$ should be used for verdicts because it is a sequence of I. The information on the sequence of inputs of the IUT of $\overline{st}(\omega)[2]$ is not complete as outputs can be intercalated in the sequence. The tester may sometimes choose to wait for an output of the IUT to complete its information. But this is not always possible because the IUT may also wait for an input. So outputs of the tester must be chosen and this is done according to $\overline{st}(\omega)[1].\overline{st}(\omega)[2]$.

A fundamental difference with synchronous local testing is that test cases are controllable in synchronous local testing: a test case never has the choice between two outputs or between an input and an output. This comes from the fact that the tester controls the inputs of the IUT. This is not anymore the case when synchronous test cases are played in an asynchronous environment. In this case, in any state all possible outputs of the IUT must be taken into account. The controllability property thus has to be relaxed into **semi-controllability**. An IOLTS is said **semi-controllable** if in any state at most one output is possible and all possible inputs are considered:
$\forall \sigma \in (A^{M})^*,\ In(M\ after\ \sigma) = A_I^M \wedge |Out(M\ after\ \sigma)| \leq 1$.

Also specifications have to be input complete or completed by *Comp*. In fact if $\overline{st}(\omega)[1] = \sigma.!z \in Traces(I)$, either $z \notin Traces(S\ after\ \sigma)$ and in this case $!z$ should produce a fail verdict, or $z \in Traces(S\ after\ \sigma)$ and $\overline{st}(\omega)[1].\overline{st}(\omega)[2]$ must be in $Traces(S)$. As $\overline{st}(\omega)[2]$ is composed of inputs, input completeness always ensures this.

Notice that the first component of $\overline{st}$ is monotonic for the prefix ordering i.e. $\omega \leq \omega'$ for the prefix ordering implies $\overline{st}(\omega)[1] \leq \overline{st}(\omega')[1]$.

The algorithm which has to be performed on the test driver is described below. It uses two sequences σ_1 and σ_2 which respectively contain the two components of the sequence of events ω performed by the tester: $\sigma_1 = \overline{st}(\omega)[1]$ and $\sigma_2 = \overline{st}(\omega)[2]$. The variable ω is only used to describe the invariant.

■ **Input:** TC: test case. Output: verdict
Invariant $\sigma_1 = \overline{st}(\omega)[1]$ and $\sigma_2 = \overline{st}(\omega)[2]$
Initialization: $\omega := \epsilon$; $\sigma_1 := \epsilon$; $\sigma_2 := \epsilon$; verdict:= nil
(* $\overline{st}(\epsilon) = (\epsilon, \epsilon)$ *)
while verdict = nil **do**
  non-deterministic choice
    if $Out(TC\ after\ \sigma_1.\sigma_2) \neq \emptyset$
      send(a) where $\{a\} = Out(TC\ after\ \sigma_1.\sigma_2)$
     (* a is unique as test cases are semi-controllable *)
      $\omega := \omega.\overline{?a}$; $\sigma_2 := \sigma_2.?a$; (* i.e. $\overline{st}(\omega.?a) = (\overline{st}(\omega)[1], \overline{st}(\omega)[2].?a$*)
    if input queue is not empty
      receive((z, i)); $\omega = \omega.(!z, i)$
      let σ_2', σ_2'' s.t. $length(\sigma_1.\sigma_2') = i$ and $\sigma_2 = \sigma_2'.\sigma_2''$
      $\sigma_1 := \sigma_1.\sigma_2'.!z$; $\sigma_2 := \sigma_2''$; (* $\overline{st}(\omega.(!z, i)) = (\overline{st}(\omega)[1].\sigma_2'.!z, \sigma_2'')$ *)
      verdict := *verdict assigned to the state* $(TC\ after\ \sigma_1)$
end

6 CONCLUSION AND PERSPECTIVES

We have described how remote asynchronous testing with counters could have the same testing power as local synchronous testing in the case where the communication between the IUT and the environment is done through one PCO. This can be generalized to multiple PCOs but needs more sophisticated mechanisms. The general idea is that the tester needs to reorder events of the IUT in order to reach the same testing power as in the synchronous case. It is natural to suppose that the IUT can only transmit information to the tester by piggybacking its outputs. In order to emit a correct verdict on an output of the IUT, the tester needs to know the whole sequence of events of the IUT which precedes this output (see figure 7). Thus a theoretical instrumentation consists in associating to each output of the IUT the sequence of its predecessors in the IUT.

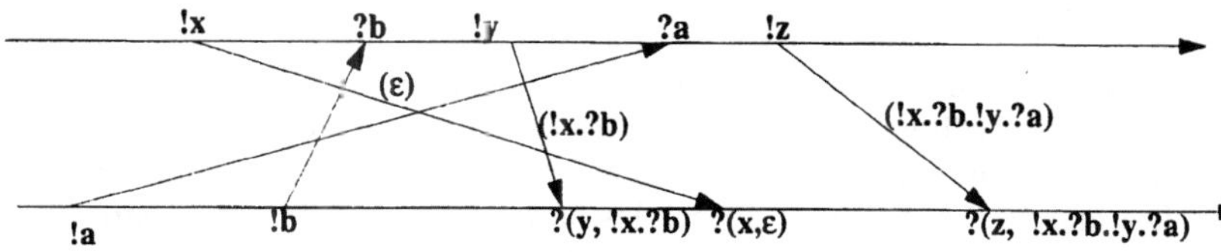

Figure 7 Theoretical instrumentation in the general case

This instrumentation is not realistic because this induces redundant information and the piggybacked sequences continuously grow. A non redundant instrumentation consists in segmenting the information into sub-sequences between two consecutive outputs. But as an output can be overtaken by a following one, the order of this output in the sequence should be piggybacked as in the case of one PCO. The functions st and $\overline{st}$ must be extended in the case of multiple PCOs.

Another prospect is timers management. Deadlocks and output quiescence are considered as outputs in the testing theory [12]. In our framework of asynchronous instrumented testing, inputs of test cases (outputs of the IUT) carry counters. This must be generalized to timeouts but suggests that timers should be managed by the instrumentation of the IUT. Moreover starting and cancelling a timer must be done by the instrumentation but initiated by the tester with messages.

REFERENCES

[1] E. Brinksma. A theory for the derivation of tests. In S. Aggarwal and K. Sabnani, editors, *Protocol Specification, Testing and Verification*, North Holland, 1988. IFIP Transactions.

[2] L. Doldi, V. Encontre, J.-C. Fernandez, T. Jéron, S. Le Bricquir, N. Texier, and M. Phalippou. Assessment of automatic generation methods of conformance test suites in an industrial context. In B. Baumgarten, H.-J. Burkhardt,

and A. Giessler, editors, *IFIP TC6 9th International Workshop on Testing of Communicating Systems.* Chapman & Hall, p. 346–361, September 1996.

[3] J.-C. Fernandez, C. Jard, T. Jéron, and C. Viho. Using on-the-fly verification techniques for the generation of test suites. In A. Alur and T. Henzinger, editors, *Conference on Computer-Aided Verification (CAV '96), New Brunswick, New Jersey, USA*, LNCS 1102. Springer, July 1996.

[4] FMCT ISO/IEC JTC1/SC21 WG7, ITU-T. SG 10/Q.8. Information Retrieval, Transfer ad Management for OSI; Framework: Formal Methods in Conformance Testing. Committee Draft CD 13245-1, ITU-T proposed recommendation Z 500. ISO - ITU-T, Geneve, 1996.

[5] ITU-T Z.100. Specification and Description Language (SDL). Recommendation ITU Z.100, 1996.

[6] C. Jard. How to observe interoperability at the service layer of protocols. In T. Mizuno, T Higashino, and N. Shiratori, editors, *IFIP TC6 7th International Workshop on Protocol Test Systems.* Chapman & Hall, p. 259–270, November 1994.

[7] M. Kim, S.T. Chanson, S. Kang, and J. Shin. An approach for testing asynchronous communicating systems. In B. Baumgarten, H.-J. Burkhardt, and A. Giessler, editors, *IFIP TC6 9th International Workshop on Testing of Communicating Systems.* Chapman & Hall, p. 141–155, September 1996.

[8] D. Lee and M. Yannakakis. Principles and Methods of Testing Finite State Machines - A survey. In Proceedings of the IEEE, vol. 84(8), p. 1090–1123, August 1996.

[9] OSI Open Systems Interconnection. Information Technology - Open Systems Interconnection Conformance Testing Methodology and Framework - Part 1 : General Concept - part 2 : Abstract Test Suite Specification - part 3 : The Tree and Tabular Combined Notation (TTCN). *International Standard ISO/IEC 9646-1/2/3*, 1996.

[10] M. Phalippou. *Relations d'implantations et Hypothèses de test sur les automates à entrées et sorties.* PhD thesis, Université de Bordeaux, 1994.

[11] J. Tretmans. *A formal approach to conformance testing.* PhD thesis, University of Twente, Enschede, The Netherlands, 1992.

[12] J. Tretmans. Test generation with inputs, outputs and repetitive quiescence. *Software - Concepts and Tools.* Springer-Verlag, vol. 17, No. 3, p. 103–120, 1996.

[13] S. H. Unger. *Asynchronous sequential switching circuits.* John Wiley and Sons, 1969.

[14] L. Verhaard, J. Tretmans, P. Kars, and E. Brinksma. On Asynchronous Testing. In G. Von Bochman, R. Dssouli, and A. Das, editors, *Fifth International Workshop on Protocol Test Systems*, North Holland, 1993. IFIP Transactions.

[15] R.G. de Vries, and J. Tretmans. On-the-Fly Conformance Testing using SPIN. In Holzmann, G. and Najm, E. and Serhrouchni, A., editors, *Fourth Workshop on Automata Theoretic Verification with the SPIN Model Checker.* ENST 98 S 002, p. 115-128, November 1998, Paris, France.

HIT-OR-JUMP: AN ALGORITHM FOR EMBEDDED TESTING WITH APPLICATIONS TO IN SERVICES

Ana Cavalli *, David Lee ° Christian Rinderknecht *, Fatiha Zaïdi *
* *Institut National des Telecommunications*
9 rue Charles Fourier
F-91011 Evry Cedex
{Ana.Cavalli, Christian.Rinderknecht, Fatiha.Zaidi}@int-evry.fr
° *Bell Laboratories, Lucent Technologies*
600 Mountain Avenue
Murray Hill, NJ 07974, USA
lee@research.bell-labs.com

Abstract This paper presents a new algorithm, Hit-or-Jump, for embedded testing of components of communication systems that can be modeled by communicating extended finite state machines. It constructs test sequences efficiently with a high fault coverage. It does not have state space explosion, as is often encountered in exhaustive search, and it quickly covers the system components under test without being "trapped", as is experienced by random walks. Furthermore, it is a generalization and unification of both exhaustive search and random walks; both are special cases of Hit-or-Jump. The algorithm has been implemented and applied to embedded testing of telephone services in an Intelligent Network (IN) architecture, including the Basic Call Service and five supplementary services.

Keywords: conformance testing, embedded testing, communicating extended finite state machines, IN.

1. INTRODUCTION

With the advanced computer technology and the increasing demand from the users for sophisticated services, communication protocol systems are becoming more complex yet less reliable. Conformance testing, which ensures correct protocol implementations, has become indispensable for the development of reliable communication systems. Traditional testing methods tend to test these systems as a whole or to test their components in isolation. Testing these systems as a whole becomes difficult due to their formidable size. On the other hand, testing system components in isolation may not be always feasible due to the interactions among the system components. Embedded testing or testing in context has become one of the main focuses of conformance testing research in recent years. The goal of embedded testing is to test whether an implementation of a system component conforms to its specification in the context of other components. It is generally assumed that the tester does not have a direct access to the component under test; the access is obtained through other components of the system. According to the standard: "if control and observation are applied through one or more implementations which are above the protocol to be tested, the testing methods are called embedded" [9].

Different approaches for embedded testing have been proposed in the published literature. They are based on fault models [15], on reducing the problem to testing of components in isolation [16], on test suite minimization [11, 13, 18, 1], on fault coverage [19], on the test of systems with semicontrollable and uncontrollable interfaces [5], or on-the-fly [6]. Most of these approaches resort to reachability graphs to model the joint behaviors of all the system components, and are exposed to the well-known state space explosion.

Our goal is to test pre-specified parts of a system component that is embedded in a complex communication system. The pre-specified parts are determined by practical needs or by system certification requirements. For instance, for a given system component, we may want to test all the transitions or certain boundary values of system variables. We can first construct a reachability graph, which is the Cartesian product of all the system components involved, and then derive a test that covers all the pre-specified parts of the component under test. Unfortunately, this exhaustive search technique is often impractical; it is impossible to construct a reachability graph for practical systems due to the state space explosion. To avoid this problem random walks have been proposed; at any moment we only keep track of the current states of all the components and determine the next step of test at random. This approach indeed avoids the state space explosion but it may repeatedly test covered parts and take a long time to move on to the untested parts.

We propose a new technique: Hit-or-Jump. It is a generalization and unification of both the exhaustive search technique and random walks, yet it does not have the drawbacks of the two approaches. The essence of our approach is as follows. At any moment we conduct a local search from the current state in a neighborhood of the reachability graph. If an untested part is found (a Hit), we test that part and continue the process from there. Otherwise, we move randomly to the frontier of the neighborhood searched (Jump), and continue the process from there. This procedure avoids the construction of a complete system reachability graph. As a matter of fact, the space required is determined by the user - the local search, and it is independent of the systems under consideration. On the other hand, a random walk may get "trapped" at certain part of the component under test [11]. Our algorithm is designed to "jump" out of the "trap" and pursue the exploration further.

The Hit-or-Jump algorithm has been applied to the embedded testing of services on a telephone network. With the aid of ObjectGEODE tool, this case study is on a real system that has been specified using the SDL language. It describes telephone services in an Intelligent Network (IN) architecture. In addition to the Basic Call Services (BCSs), five other services are included: Originating Call Screening (OCS), Terminating Call Screening (TCS), Call Forward Unconditional (CFU), Call Forward on Busy Line (CBL) and Automatic Call Back (ACB).

The paper is organized as follows. Section 2 introduces the basic concepts and testability of embedded components. Section 3 describes the test generation algorithm Hit-or-Jump for embedded components. Section 4 discusses the implementations. Section 5 reports the experimental results on IN. Section 6 concludes the paper with remarks on the generalization and variations of the algorithm and on related issues.

2. BASICS

In this work we use extended finite state machines to model system components: the environment, the components under test and their implementations. It is only for the convenience of presentation; our technique can be adapted to other mathematical models, such as Transition Systems [14], Petri Nets [17] and Labeled Transition Systems [2].

Definition 1. An *extended finite state machine* (EFSM) is a quintuple $M = (I, O, S, \vec{x}, T)$ where I, O, S, $\vec{x}$, and T are finite sets of input symbols, output symbols, states, variables, and transitions, respectively. Each transition t in the set T is a 6-tuple:

$$t = (s_t, q_t, a_t, o_t, P_t, A_t)$$

where s_t, q_t, a_t, and o_t are the start (current) state, end (next) state, input, and output, respectively. $P_t(\vec{x})$ is a predicate on the current variable values and $A_t(\vec{x})$ defines an action on variable values.

Initially, the machine is in an initial state $s^{(0)} \in S$ with initial variable values $\vec{x}^{(0)}$. Suppose that the machine is at state s_t with the current variable values $\vec{x}$. Upon input a_t, if $\vec{x}$ is valid for P_t, i.e., $P_t(\vec{x}) = \text{TRUE}$, then the machine follows the transition t, outputs o_t, changes the current variable values by action $\vec{x} := A_t(\vec{x})$, and moves to state q_t.

We model the environment and component under test by EFSM's. During an execution the states and variable values can be determined as in the construction of reachability graphs for EFSM's [8, 12]. From now on we use the following notation : C is the environment EFSM, A is the specification EFSM under test, and B is the implementation of A. Machine C and $A(B)$ communicate synchronously. We represent A in the context of C by the following notation : $C \times A$. We want to test the conformance of B to A in the context of C where C and A are known and B is a "black-box". It should be noted that $C \times A$ may not be minimized or strongly connected even if C and A are. Also they can be partially (incompletely) specified.

In general, it is not always possible to test for isomorphism of embedded components, even in the case of FSM's. Assume that A and B are FSM's. Denote machine isomorphism by $A \cong B$. Then we have:

Proposition 1. $B \cong A$ implies $C \times B \cong C \times A$. However, the converse is not true in general.

The first part of the proposition is trivial. We show the second part by an example.

Example 1.

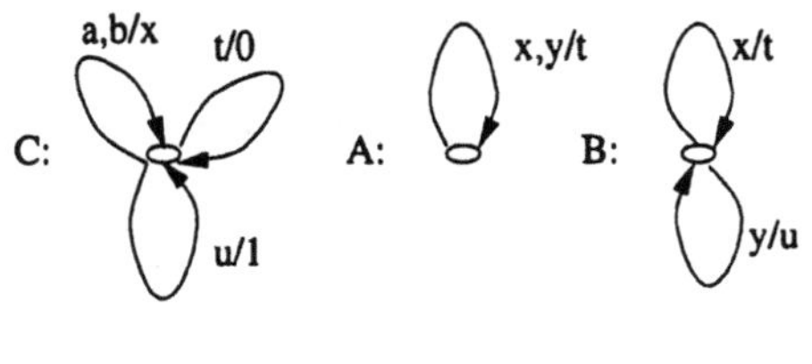

Figure 1.1

In Figure 1.1, a and b are external inputs, 0 and 1 are external outputs, and x, y, t, u are internal input/outputs. Obviously $C \times B \cong C \times A$. But $B \ncong A$. Therefore, it is impossible to test $A \cong B$ in the context of C.

In practice, what we want is that B "behaves correctly" in the context of C. That is, $C \times B \cong C \times A$. Therefore, the problem is reduced to testing if $C \times B \cong C \times A$. However the real goal is to test the component A, assuming that the environment machine C is correctly implemented. Suppose that we test A in isolation. Then we may want to test all the transitions of A. That is, we want to obtain a testing sequence such that all the transitions of A are exercised. Similarly, in embedded testing we want to obtain test sequences (with external inputs) such that all the transitions of the component A are exercised. Specifically, we want to derive tests for $C \times A$ such that all the transitions of A are tested. We may want different coverage of A than testing all the transitions. For instance, we often want to test the boundary values of the variables. In general, we want to obtain a test sequence for $C \times A$, i.e., for testing the component A in the context of C, such that the component machine A is covered according to a pre-specified criterion. This criterion can be specified by assigning a distinct color to each entity (transition or variable value, for instance) to be tested and by covering all the assigned colors by test sequences generated. On the other hand, we do not worry about the coverage of C, since it is assumed to be correctly implemented. For clarity, we assume that the system under test does not have deadlock nor livelock, which are well studied topics in validation [8].

3. EMBEDDED TESTING

We first briefly survey three commonly used and related methods and then present our Hit-or-Jump algorithm, which is a generalization and unification of all these three methods. For clarity, we describe a procedure that covers all the transitions of the component machine under test. This is a commonly used criterion in practice. As indicated earlier, other coverage criteria can be reduced to color covering of the component under test, and our procedure can be easily adapted to generating tests for the color covering; it is only a marking process. We shall further elaborate on this issue when describing the algorithm.

3.1 A STRUCTURED ALGORITHM

From the initial state we want to generate a test sequence such that all the transitions of component machine A are covered at least once. The algorithm includes three steps: (1) Assign a distinct color to each transition of A; (2) Construct a reachability graph of $C \times A$ where each edge of $C \times A$ is marked with a color from A if it is derived from that transition of A; (3) From the initial node of $C \times A$, find a path of minimal length such that all the colors are covered at least once. This approach requires a construction of the reachability graph of $C \times A$, which leads to the well-known state explosion. Even if we could obtain such a graph the problem is still NP-hard. Consequently, unstructured algorithms such as random walks are considered, which do not require the construction of reachability graphs.

3.2 RANDOM WALK

Starting from the initial node $(s_C^{(0)}, s_A^{(0)}, \vec{x}^{(0)})$ where $s_C^{(0)}, s_A^{(0)}$ and $\vec{x}^{(0)}$ are the initial state of C and A and initial variable values, respectively. Among all the possible outgoing edges in the reachability graph from the initial node, we select one uniformly at random, and follow that edge to the next node in the reachability graph. Suppose that after k steps we arrive at a node $(s_C^{(k)}, s_A^{(k)}, \vec{x}^{(k)})$. We examine all the outgoing edges from this node and select one uniformly at random to follow. Meanwhile, if there are colors associated with the chosen edges that have not been marked (exercised), we mark them off. We repeat the process until all the colors are marked off. During the walk, we only keep track of: (1) The current node $(s_C^{(k)}, s_A^{(k)}, \vec{x}^{(k)})$; (2) The colors that have not been marked off; and (3) The edges that have been walked through with the associated external I/O sequence, and that is the test sequence obtained from this walk. Obviously, there is no need to construct a whole reachability graph of $C \times A$.

3.3 GUIDED RANDOM WALK

The procedure is the same as the random walk except for the following. When we arrived at a node $(s_C^{(k)}, s_A^{(k)}, \vec{x}^{(k)})$ among all the possible outgoing edges, we classify them: (1) With transitions of A involved, some of which are not marked; (2) With transitions of A involved and all of them are marked; and (3) Without any transitions of A involved. If the set (1) is not empty, we select one uniformly at random and follow that edge; else if (2) is not empty, we select one uniformly at random and follow that edge; and, finally, if none of the above is true, (3) must be non-empty, and we select one uniformly at random and follow that edge. Guided random walks [11] favor transitions of the embedded component under test, and among them give first priority to the transitions that have not been tested.

3.4 HIT-OR-JUMP ALGORITHM

The problems with random walks are: (1) To be "trapped" in a small neighborhood; (2) With a low probability to cross a "narrow bridge" to test the parts beyond the bridge; and (3) To miss the unmarked transitions of A even if they are nearby (more than one step from the current node). The Hit-or-Jump algorithm is designed to avoid these problems yet without the construction of a reachability graph. It does not require a construction of a reachability graph of $C \times A$ either, and performs better than pure random walks. Local search is used in the procedure, and it can be Depth-first or Breadth-first search.

ALGORITHM HIT-OR-JUMP

initial condition. The environment machine C is in an initial state $s_C^{(0)}$, the component machine under test A is in an initial state $s_A^{(0)}$, and the system variables have initial values $\vec{x}^{(0)}$.

termination. The algorithm terminates when all the colors (transitions) of A have been marked off.

execution.

1. **HIT** From the current node $(s_C^{(k)}, s_A^{(k)}, \vec{x}^{(k)})$ conduct a search in $C \times A$ until:
   (a) Reach an edge which is associated with unmarked colors of the component machine A: a Hit. Then :
      i. Include the path from the current node to the edge (inclusive) in the test sequence under construction;
      ii. Mark off the newly exercised colors of A;
      iii. Arrive at a node $(s_C^{(k+1)}, s_A^{(k+1)}, \vec{x}^{(k+1)})$;
      iv. Erase the searched graph;
      v. Repeat from 1.

   or

   (b) Reach a search depth or space limit without hitting any unmarked colors of A. Then move to 2.

2. **JUMP**
   (a) We have constructed a search tree, rooted at $(s_C^{(k)}, s_A^{(k)}, \vec{x}^{(k)})$.
   (b) Examine all the leaf nodes of the tree, and select one uniformly at random.
   (c) Include the path from the root to the selected leaf node in the test sequence.
   (d) We arrive at the selected leaf node $(s_C^{(k+1)}, s_A^{(k+1)}, \vec{x}^{(k+1)})$: a Jump.
   (e) Repeat from 1.

Suppose that the local search depth is set to one. Then, obviously, Hit-or-Jump becomes a Random Walk. If we enforce priorities then it becomes a Guided Random Walk. On the other hand, if we do not set any bound on the local search depth then we construct a reachability graph in the worse case; Hit-or-Jump becomes a structured algorithm. Therefore, Hit-or-Jump is a generalization of Random and Guided Random Walks and also the structured algorithm. Furthermore, this technique unifies these three apparently quite different approaches.

4. IMPLEMENTATIONS OF HIT-OR-JUMP

In this section we describe the implementation of Hit-or-Jump. It was not possible to use the ObjectGEODE tool (Verilog) alone because our algorithm needed special handling of features of the tool that are not widely used (and hence poorly documented). For our purpose, we developed a software tool which drives the ObjectGEODE simulator. Due to space limit we do not include the details here, and the interested readers are referred to [3].

Our goal is to get a test sequence in the fully deployed automaton, corresponding to a path starting at the initial state, that contains all the transitions of the embedded component under test yet without constructing the fully deployed automaton.

Interface. We supply our tool with the following informations. The result is a file containing the test sequence for the embedded component. The sequence is a series of pairs of inputs and outputs. (1) A disjunctive stop condition modeling the set of the embedded system component transitions. It defines the embedded system. (2) A positive integer denoting a depth limit that will be passed along to the simulator; we stop when a search (DFS, BFS or B-DFS) reaches it. (3) The inputs that the simulator can fire from the environment in order to stimulate the whole system (here opposed to the *embedded* system). (4) The protocol-dependent variable for the simulator ("let" clauses, like number of users, actions allowed, maximum number of actions per user, service subscriptions etc.). (5) An initial scenario that starts all the processes and put then in their initial state.

Configuration. The first step of our tool is to configure and produce three start-up files that will be used to drive the simulator. (1) main.startup. It loads the inputs that the simulator can fire from the environment, the protocol-dependent variables, the current scenario, and specifies the exhaustive and B-DFS mode. Then it runs the simulator. (2) stop_search.startup. It is devoted to the identification of the stop condition in the disjunction that actually interrupted the simulation. (3)

final.startup. It loads the inputs that the simulator can fire from the environment and replays the final scenario we got after hitting all the colors (transitions) of the embedded component. As a result we get an ObjectGEODE log file from which we extract the test sequence.

Simulation. We start the simulation with the main.startup file. There are two possible situations: (1) *The simulator outputs a scenario.* This file is output if and only if we Hit an uncovered transition of the embedded system. Two cases can occur: (A) *There was only one stop conditions remaining.* Then we have successfully completed the algorithm. We run the simulator with the final.startup and extract from the log file the test sequence. (B) *There were at least two stop conditions remaining.* Then we run again the simulator with the stop_search.startup file in order to identify the stop condition that corresponds to the hit transition among the current disjunction (i.e., the set of uncovered transitions). This start-up prints the stop condition statuses and identifies the one assessed to true. (2) *The simulator does not output a scenario.* This means that the simulator stopped after reaching the depth limit- a Jump is to be made. In other words, it did not find any transition that satisfies one of the stop conditions in the disjunction. We nevertheless got a file, containing the partially deployed automaton, as a result of the interrupted simulation, but we know neither the current state in the EFSM (SDL specification), nor the path from the initial state. Thus we parse the deployed automaton and conduct a DFS on it. We choose uniformly at random a leaf node and find a (shortest) path for the current state to the selected leaf node. We append the path at the end of the constructed scenario and resume the simulation.

5. CASE STUDY: IN TELEPHONE SERVICES

In this section we report experimental results of applying the Hit-or-Jump test generation technique to Intelligent Network (IN) telephone services. The service integrates the supplementary services: Originate Call Screening (OCS), Terminal Call Screening (TCS), Call Forward Unconditional (CFU), Call Forward on Busy Line (CBL) and Automatic Call Back (ACB). The system has been described using the SDL language [10] as far as call treatment, service invocation and user management are concerned [4]. It is located at the Global Functional Plane (GFP), taking some concepts of the Distributed Functional Plane (DFP). It consists of different functional entities that are represented by the Network block. The Network block is composed by two blocks: the Basic Service, which represents the Basic Call Service (BCS) and a Features

Block (FB) that represents the services. The BCS block contains three processes: the Call Manager (deals with the management of a call); the Call Handler (which takes in charge the call itself) and the Feature Handler (which allows to access to services). The FB block is composed of five processes that represent the services: Black List which is instantiated twice in order to obtain a black list on calls start, the OCS service, and a black list at calls arrival, the TCS. The other services are CFU, CBL and ACB as mentioned above. This block includes also a process: Feature Manager (which establishes a link between the Feature Handler and the services). The architecture of this specification is depicted in Figure 1.2.

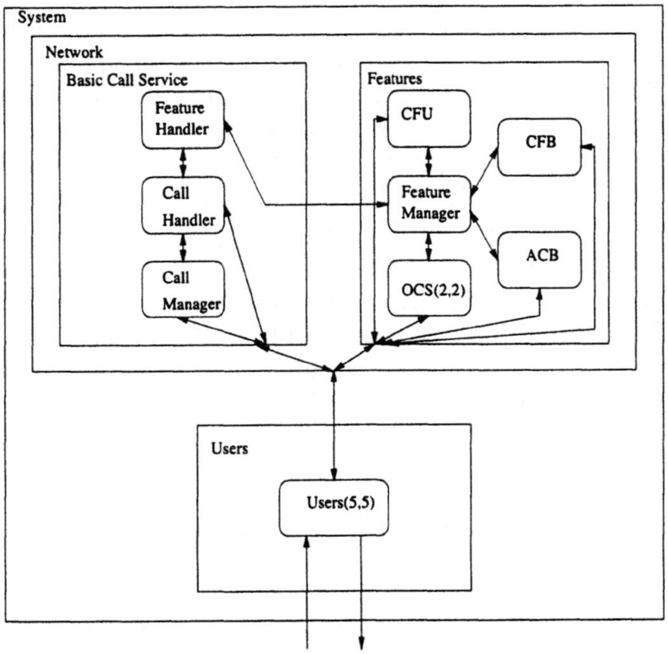

Figure 1.2 Global architecture

The model is described in such a way that it allows the execution of different calls in parallel and also calls initiated by the network.

The environment sends messages to the process Users, that are modeled as SDL process instances that composed the Users Block. The user process represents a combination of a phone line, a terminal and a user. It is relatively complete with respect to the service-usage life-cycle, with user-activations, deactivations, updates and invocations all modeled.

In order to provide a general idea of the complexity of the SDL system specifications, we present in Figure 1.2 the global architecture of the system and in Figure 1.3 some relevant metrics. The global system was simulated using exhaustive simulation in a mood to obtain the complete reachability graph. Figure 1.4 gives some information concerning the numbers of states, transitions, etc, obtained after a manual stop of the exhaustive search/simulation. It is impossible to construct the whole reachability graph due to the formidable state space requirement.

Lines	3,098
Blocks	4
Process	9
Procedures	12
States	88
Signals	50
Macro definition	12
Timers	0

Figure 1.3 Metrics of the service specification

#states	674,814
#transitions	2,878,800
Max depth reached	28
Duration	43mn 49s
Transition coverage	46.07%
States coverage	70.37%

Figure 1.4 Partial simulation of the complete specification

We now report detailed results on test generation of OCS and CFU services modules. It is a system component that is embedded in the Features block and does not possesses any link with the environment. For the embedded testing of this module, we want to traverse at least once each of its branches, which is depicted in figure 1.5. Stop conditions are used to represent the characteristics of each branch. To distinguish each branch of the component, we hand-crafted the stop conditions. Figure 1.6 illustrates the stop conditions of OCS module.

In order to perform the simulation of the system we configure a startup, that initialize some variables and some services: the subscribers that invocate the services and actions each subscribers can do (eg. hangups, activations, disactivations, normal dialing). For this case study, and the results obtained, we have set these variables around 80 actions for each users.

The results are shown in Figure 1.7. The line "Number of transitions" indicates the number of fired transitions at each simulation (i.e., the size of the deployed automaton). Hence it is aimed to be a measure of resource consumption, not of the size of the corresponding test

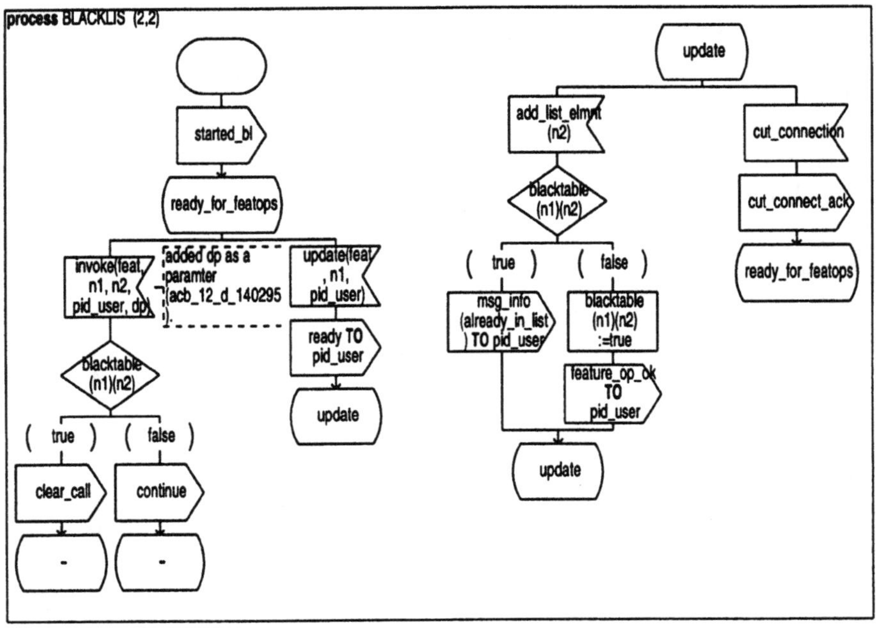

Figure 1.5 Blacklist process

stop if	output continue from blacklist		#1
or	output clear_call from blacklist		#2
or	output ready from blacklist		#3
or	and	input add_list_elmnt to blacklist output feature_op_ok from blacklist	#4
or	trans blacklist : from_update_input_cut_connection		#5
or	and	input add_list_elmnt to blacklist output msg_info from blacklist	#6

Figure 1.6 Stop conditions of the Blacklist process

subsequence, which is a path in the deployed automaton. Note that in the worst case, when finding the stop condition input add_list_elmnt to blacklist and output msg_info from blacklist (stop #6), the simulator only passed through 103 transitions. It clearly shows that Hit-or-Jump algorithm effectively finds untested transitions without constructing the reachability graph.

Furthermore, the total test sequence is short. Note that the time corresponds to the CPU real user time (Sun Sparc Ultra-1).

Stops	#1	#2	#3	#4	#5	#6
Number of transitions	11	65	95	3	4	103
Max depth reached	11	50	50	3	4	50
Duration (seconds)	2.5	4.4	6.7	8.6	10.4	11.1

Figure 1.7 Simulation results for each stop condition

Once all the transitions of the embedded component OCS module have been traversed, we obtain a single test sequence, which corresponds to the total path that has been traversed from the environment to the last transition of the module. The obtained sequence is of length 150; we only need to take 150 transitions to cover the whole OCS module in the context.

We have exercised a Random Walk (see section 3.2) and got a test sequence of 1.402 transitions. It is clear that Hit-or-Jump produces a test sequence with a same fault coverage as a Random Walk but is an order of magnitude shorter.

We have also performed experiments on the embedded testing of the service CFU. Figure 1.8 illustrates the results obtained for OCS and CFU services. Moreover we have also applied the Hit-or-Jump algorithm to the process Responder of the INRES protocol [7]. The results for the module Responder of the INRES protocol are relevant, in fact we obtained a sequence of length 44 in a BFS mode. We have also obtained various test sequence lengths with hit-or-jump algorithm in different modes of search and a Random Walk (RW).

Modules	OCS				CFU			
Modes	DFS	BFS	B-DFS	RW	DFS	BFS	B-DFS	RW
Depths	50	50	50	-	100	100	100	-
#Stops	6	6	6	6	6	6	6	6
Sequences	834	150	167	1402	deadlock	137	261	586
#Jumps	18	1	0	-	< 70	0	0	-

Figure 1.8 Results of the Modules OCS and CFU

6. CONCLUSION

We have presented a new algorithm Hit-or-Jump to perform testing of components that are embedded in a complex communication system. It is a generalization and unification of Random and Guided Random Walks and also the structured algorithm. Yet it does not have the state space explosion problem as is encountered by the structured algorithms, and it generates high coverage test sequences that are much shorter than that from random walks.

Hit-or-Jump is a new technique for system state search. We have applied it for embedded testing. It can also be used for verification and validation, which depend on system state search, and our method could help dealing with the state explosion problem there.

For clarity we have presented a straightforward version of Hit-or-Jump algorithm. It has a number of variations and generalizations, and their implementations are simple modifications of the version presented. We briefly describe a few here. For a Jump we select uniformly at random a leaf node of the locally searched graph (tree) and proceed from there. Instead, we can enforce certain priorities in selecting the leaf nodes as in a Guided Random Walk [11], and then conduct a "Guided Jump" according to the leaf node priorities as in a Guided Random Walk. Another variation is: if there has been no Hit for a large number of Jumps, one might "backtrack" to the previous Hit, and Jump to a different node to proceed with testing. The idea behind is: get back when one has gone "astray". Even though in our experiments with IN we have not encountered such problem, it might not be a surprise for testing components that are embedded in a complex system. Also when constructing a search tree on-line, we can compress internal transitions of $C \times A$ [13] to further save space.

We have been focused on covering all the transitions of A. The algorithm can be easily extended to: (A) Covering some (not necessarily all) transitions of A, which are specified by users or the testers; (B) Covering some states of A; and (C) Covering some transitions and states of A along with specified variable values such as boundary values. We can assign a distinct color to each entity to be covered, and run Hit-or-Jump until all the colors are covered. Several approaches [16], [15] use fault models. We have a general procedure, independent of any fault models. However, we can assign colors to the entities of the component machine under test for the coverage from fault models, and our procedure can be used for test generation associated with fault models.

We have not specified the depth of local search for a Jump in case there is no Hit. For IN we tested on a few depth values, i.e., 50 and 100. Intuitively, a larger depth value increases the probability of hitting an uncovered part of the component under test. However, it requires more space and time for each step. Furthermore, a long "Jump" implies a longer subsequence in the test for this step. We believe that it depends on the system under test to choose a good depth value. As indicated earlier, one can always choose a depth value that is within the limit of affordable memory space. As local search for a Hit-or-Jump, we have tested both Breadth-first-search and Depth-first-search. Breadth-first-search seems to perform better; it is "unbiased" and makes an "equi-distance" random Jump.

References

[1] C. Bourhfir, R. Dssouli, E. Aboulhamid, and N. Rico. A guided incremental test case generation procedure for conformance testing for CEFSM specified protocols. In *IWTCS'98*, Tomsk, Russia, August 1998.

[2] E. Brinksma. A theory for the derivation of tests. In *Proc. IFIP WG6.1 8th Int. Symp. on Protocol Specification, Testing and Verification*. North-Holland, 1988.

[3] A. Cavalli, D. Lee, C. Rinderknecht, and Fatiha Zaidi. Hit-or-jump: An algorithm for embedded testing with applications to in services. In *Tech. Memo, Bell Laboratories*, May 1999.

[4] P. Combes and B. Renard. Service validation, tutorial. In *SDL Forum'97*, France, 1997.

[5] M. A. Fecko, U. Uyar, A. S. Sethi, and P. Amer. Issues in conformance testing: Multiple semicontrollable interfaces. In *Proceedings of FORTE/PSTV'98*, Paris, France, November 1998.

[6] J.-C. Fernandez, C. Jard, T. Jeron, and C. Viho. Using on-the-fly verification techniques for the generation of test suites. In *CAV, LINCS 1102*, USA, July 1996.

[7] D. Hogrefe. Osi formal specification case study: the inres protocol and service, revised. Technical report, Institut für Informatik Universität Bern, may 1992.

[8] G. J. Holzmann. *Design and Validation of Computer Protocols.* Prentice Hall, New Jersey, 1991.

[9] ISO. *Information Technology, Open Systems Interconnection, Conformance Testing Methodology and Framework, International Standard IS-9646*, 1991.

[10] ITU. *Recommendation Z.100 : CCITT Specification and Description Language (SDL)*, 1992.

[11] D. Lee, K. Sabnani, D. Kristol, and S. Paul. Conformance testing of protocols specified as communicating finite state machines - a guided random walk based approach. In *IEEE Transactions on Communications*, volume 44, No.5, May 1996.

[12] D. Lee and M. Yannakakis. Principles and methods of testing finite state machines - a survey. *Proc. of the IEEE*, 84(8):1090–1123, August 1996.

[13] L. P. Lima and A. Cavalli. A pragmatic approach to generating test sequences for embedded systems. In *Proceedings of IWTCS'97*, Cheju Island, Korea, September 1997.

[14] R. Milner. *Communication and Concurrency.* Prentice-Hall, Englewood Cliffs, New Jersey, 1989.

[15] A. Petrenko, N. Yevtushenko, and G. V. Bochmann. Fault models for testing in context. In *Proceeding of FORTE/PSTV'96*, Kaisersläutern, Germany, October 1996.

[16] A. Petrenko, N. Yevtushenko, and G. V. Bochmann. Testing faults in embedded components. In *Proceedings of IWTCS'97*, Cheju Island, Korea, September 1997.

[17] A. A. Petri. *Kommunikation mit Automaten.* Ph. D. thesis, Universitat Bonn, 1962.

[18] N. Yevtushenko, A. Cavalli, and L. P. Lima. Test suite minimization for testing in context. In *IWTCS'98*, Tomsk, Russia, August 1998.

[19] J. Zhu and S. T. Vuong. Evaluation of test coverage for embedded system testing. In *IWTCS'98*, Tomsk, Russia, August 1998.

PART TWO

FDTs I

Scalable and Object Oriented SDL State(chart)s

Birger Møller-Pedersen[1], and Dagbjørn Nogva[2]
[1]) NorARC - Applied Research Center, Ericsson Norway: [2]) Telox, Norway

Key words: Statecharts, SDL, UML, state type, state type inheritance, virtual state

Abstract:

The notion of composite state in SDL is introduced. The features include entry/exit points to enforce encapsulation, type/subtypes of any composite state, virtual states in order to control the redefinition of states when defining state subtypes, and parameterized state types in order to allow maximum reuse of the same state type in different contexts. This notion of composite states will be part of SDL-2000. The same mechanism can, however, as well be introduced in other languages with a state machine concept.

1. INTRODUCTION

For a family of applications it is important to describe objects by means of state machines. Specification languages support this by the notions of either simple finite state machines, extended (with variables) finite state machines, or variations of Harel's Statecharts [9]).

The most important reason for using state machines is that it ensures that all possible states of the object and its reaction to all possible events are covered. State machines may, however, become large and complex, and their specification correspondingly large and complex. The notion of hierarchical states (sometimes also called states with nested states or composite states) introduced by Harel's Statecharts is a recognized solution to this problem, and it has been adopted by many notations and specification languages.

SDL as of 1996 [6] supports extended finite state machines with one level of states. In addition, procedures may be described by state machines, but the states of a procedure are entered when the calling state machine is in a transition.

The emerging proposal for SDL-2000 includes support for composite states. This paper presents the most important aspects of these composite states:

- *Types and subtypes of states* for states in general, and not only for the outermost state as in Statecharts, combined with the introduction of *state diagrams* with *state connection points* facilitating encapsulation and scalability of state definitions.
- The distinction between *virtual state types* (of a supertype state) that can be *redefined* in a subtype state, and states that can *not* be redefined (in the spirit of SDL's finalized types/transitions and Java's final methods).
- *Parameterized state types* in line with SDL types with context parameters and UML parameterized object classes, enabling the definition of composite state types that can be used in different state machines within the same system and even in different systems.

2. STATE MACHINES, COMPOSITE STATES

Figure 1 is a partial state machine for an ATM as supported by UML statecharts [8].

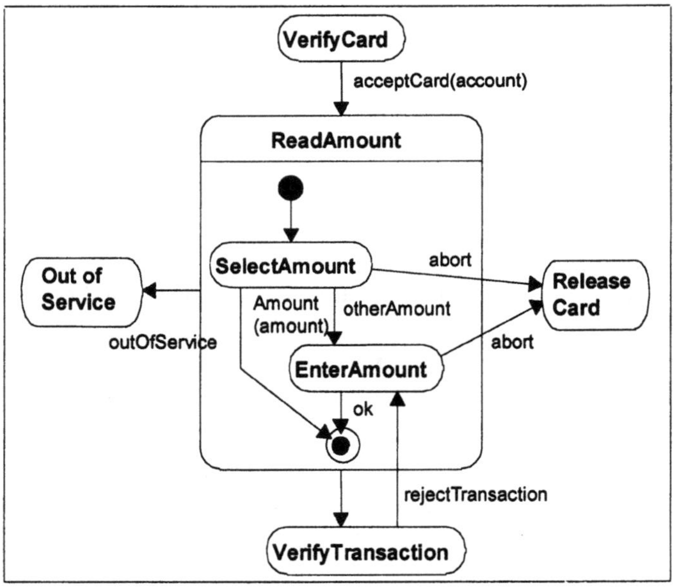

The example covers the part of an ATM that reads the desired amount: the user either selects the amount (the choices given by the ATM), or enters the amount. One of the benefits of Statecharts is illustrated by the outOfService event: this applies to all states being part of the composite state ReadAmount.

Figure 1. Statechart (UML)

Figure 1 provides the overview of which states are involved and the transitions between these. As specified here, the transitions are denoted by

the events that trigger the transitions. In general it is possible to specify more than this. Actions being part of transitions are specified textually, in separate diagrams or editor windows, and just associated with the transition arcs.
The same partial state machine is specified in SDL-2000 in *Figure 2* and *Figure 3*.

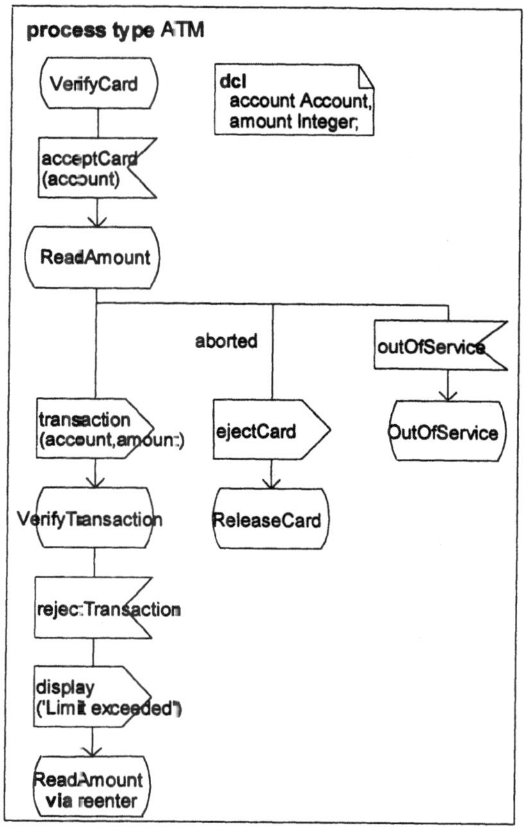

Figure 2. SDL State Machine with Composite State

The most apparent difference between statecharts and SDL composite states is that the content of the composite state ReadAmount is described in a separate diagram in SDL. *Figure 2* is a partial process type diagram with a state machine specification that *references* a state diagram.

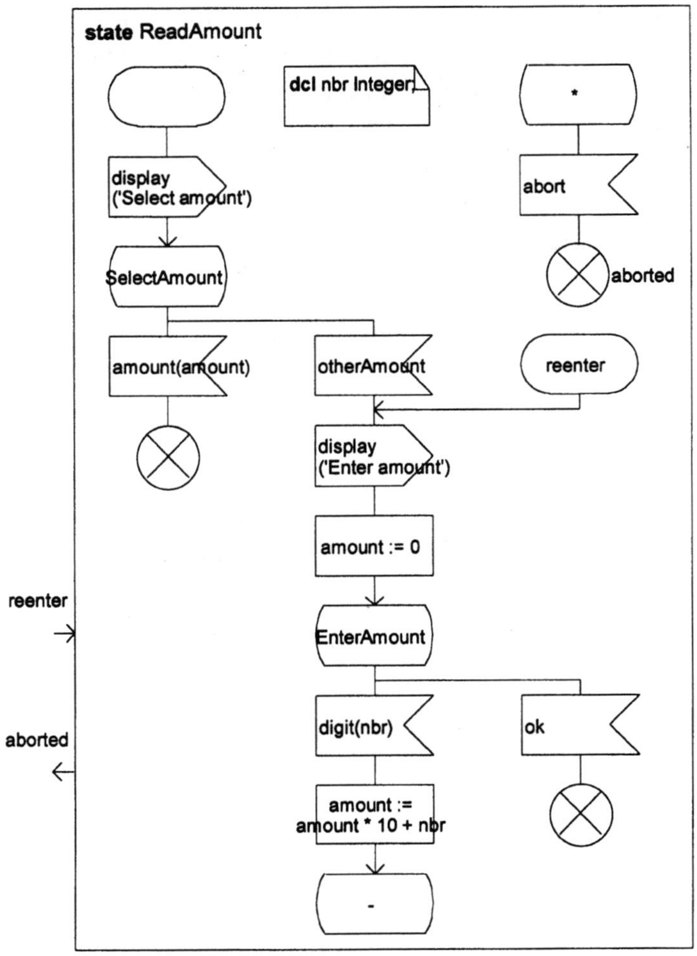

Figure 3. Composite State Diagram

A composite state is entered and exited by means of connection points defined as part of the interface of the composite state. Within the state diagram for the composite state, an entry connection point is represented by a special *start symbol* with the name of the entry points (reenter in *Figure 3*), An exit connection point is represented by an *exit symbol* with the name of the point associated (aborted in *Figure 3*).

The unlabeled start and exit have the following correspondence to Statecharts: The initial state of Statecharts corresponds to the state following the unlabeled start symbol, and the final state corresponds to the exit symbol without any exit point associated.

Input transitions associated with a composite state apply to all substates. As an example, in process type ATM in *Figure 2*, the input transition with signal outOfService applies to all substates of the composite state ReadAmount. A transition in SDL is the input of a signal and the following sequence of actions.

As a summary of the SDL composite as introduced above, they differ from Statecharts in the following respects:

- SDL composite states scale up to handle real, large systems by having composite states described in *separate state diagrams*.
- Entering and exiting a composite state is *not* specified by crossing state boundaries as in Statecharts, but by means of *connections* to *state connection points*, thereby providing encapsulation. The internal specification of a composite state can be changed without effecting the enclosing state machine.
- While Harel statecharts appear *state-oriented*, SDL state machines are specified in a *transition-oriented* way. While state orientation provides overview, some developers prefer the transition orientation when it comes to the specification of the transitions. It is usually also regarded as a benefit to see both transitions caused by incoming signals *and* outputs of signals.

3. ENTERING AND EXITING COMPOSITE STATES

Entering a composite state at a specific point is specified by the name of the composite state followed by via <entry point name>. In *Figure 2*, if the ATM rejects the transaction, it reenters the composite state ReadAmount via the reenter connection point, and in the state diagram for ReadAmount it is specified where the state is reentered (by a start symbol with the label reenter.

Entering a composite state in the "initial state" is a special case of this; specified by the name of the composite state, without any via <entry point name>. The composite state is then entered at the unlabelled start symbol. In *Figure 2* this happens when the ATM has got the signal acceptCard (in state VerifyCard). The composite state is entered at the unlabelled start symbol in the left uppermost part of state ReadAmount.

The composite state may exit in two different ways:

- it has got the amount signal in state SelectAmount or the ok signal in state EnterAmount, and simply terminates, in that case the process type ATM continues at the unlabelled connection line after ReadAmount, or
- it aborts; in that case the ATM continues at the connection line labeled abort following ReadAmount.

While a Statecharts transition crossing a state boundary is *one* transition, specified as part of the whole statechart, a corresponding SDL-2000 transition is a combination of a inner sub-transition defined in the composite state and an outer sub-transition defined in the enclosing state diagram. In *Figure 4* it is illustrated what that would mean in UML Statecharts. An SDL-2000 transition being just one sequence of actions is just a special case, where the outer or inner transition is left empty.

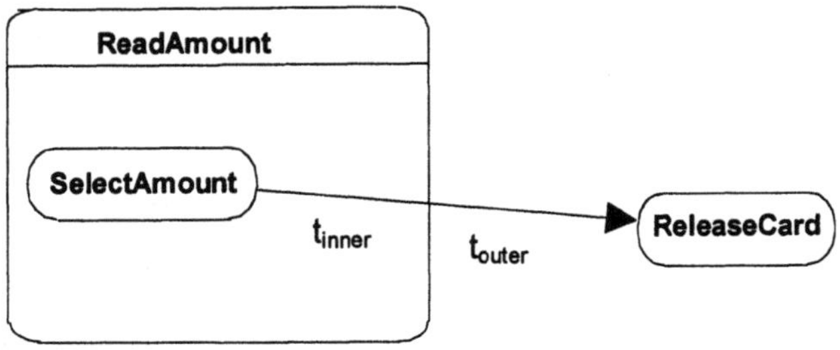

Figure 4. Transitions as combination of outer and inner transitions

Entry/exit actions are specified by means of textual procedures defined locally in the state diagram. The procedures, named *entry* and *exit*, are called implicitly when a state is entered and exited, respectively. The names *entry* and *exit* are predefined names.

The entry/exit actions may not take parameters, but may access variables defined in the state diagram and in enclosing entities according to the usual scope rules.

Note that there is no notational difference between a composite state and a non-composite. One reason for this is that states with entry/exit actions (but not necessarily any contained states) are described by separate state diagrams. As described below, a state may also have just internal transitions. Finally, the decision of whether a state is a composite state or not, may change during development of the complete state machine.

4. SPECIAL FEATURES

The SDL composite state example also illustrates a special SDL feature: the *asterisk state* with input of the signal abort implies that this transition applies to *all* states. In general, a transition can be specified to be applicable to

- all states,
- all states, except a specified list of states,
- a list of states.

This feature provides an alternative way of grouping states than provided by composite states. An input transition specified for a composite state applies to all substates, while an input transition specified for a list of states applies to a set of states that are not necessarily substates. SDL-2000 with composite states will provide both alternatives and their combination.

5. INTERNAL TRANSITIONS

An internal transition is a transition that remains in a state. An internal transition is not equivalent to a self-transition from a state back to the same state. A self-transition causes the exit and entry actions on the state to be executed, whereas an internal transition does not.

Internal transitions in SDL-2000 are part of composite states, even if they do not have substates. In a state with no substates, the asterisk state is merely a syntactical way of referring to the composite state itself, see *Figure 5*.

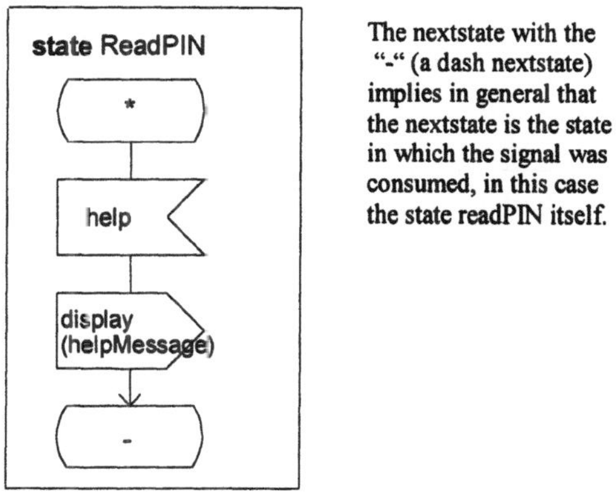

Figure 5. Internal transition in SDL

6. STATE OVERVIEW DIAGRAMS

A composite state can as a special case have transitions only (and not just internal transitions). This can be used to obtain state overview diagrams very similar to Statecharts. The state diagram of ATM in *Figure 6* is an example of a state overview diagram, where only the connections between the states are specified. The transitions corresponding to these connections are specified in separate state diagrams. These are sketched in *Figure 7*.

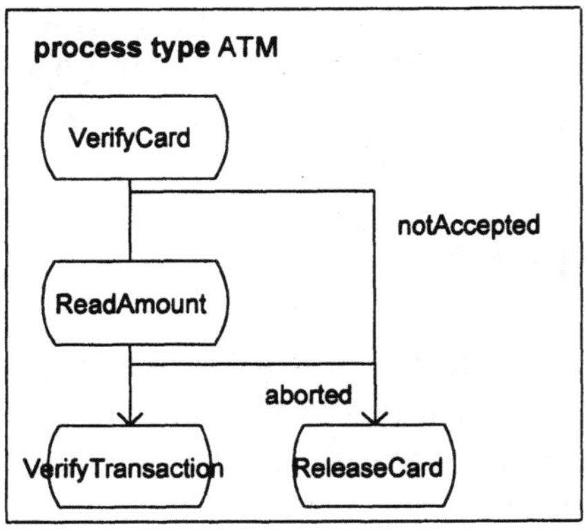

This state overview diagram relies on the fact that the composite states VerifyCard and ReadAmonut define the transitions. In this diagram it is only specified that there may be two transitions from VerifyCard, and two transitions from ReadAmount. The details about these transitions, e.g. which signals that trigger them and what actions are performed as part of the transitions, are found in the separate state diagrams.

Figure 6. State overview diagram

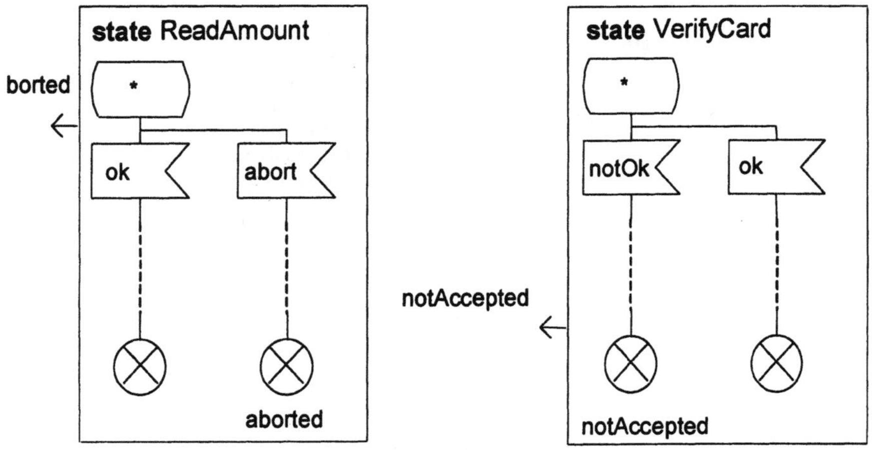

Figure 7. Composite State Diagrams covering all Transitions from the State Overview Diagram

7. STATE TYPES AND SPECIALISATION

If a composite state shall be used in different situations, a *composite state type* is defined, for short called a *state type*. A state with the properties of a state type is called a *type-based state*. The syntax is simple. The type name is specified after the name of the state, separated by a colon: "<state name>:<state type name>". This is by the way the standard syntax for type-based instances in SDL.

State types are useful in cases where the properties of composite state are applicable in many systems in the same domain, or even in different parts of the same system. Several larger state machines may use composite state types describing the states and transitions needed for e.g. starting/stopping, providing status or handling general management signals.

A state type may be defined as a *specialisation* of another (super) state type, *inheriting states and transitions* of the supertype and *redefining virtual states* and *transitions*. This is the same mechanism as already present in SDL. States of the supertype cannot be deleted, but virtual states can be redefined. States and transitions can be added in a subtype.

Entry/exit procedures may also be defined as virtual procedures that can be redefined.

Figure 8 shows an example of a state type that inherits a general type and adds abort functionality. Adding this functionality includes adding a state exit point (aborted). The asterisk state in the subtype specifies that the abort input applies to all states, including the inherited states (in this case SelectAmount).

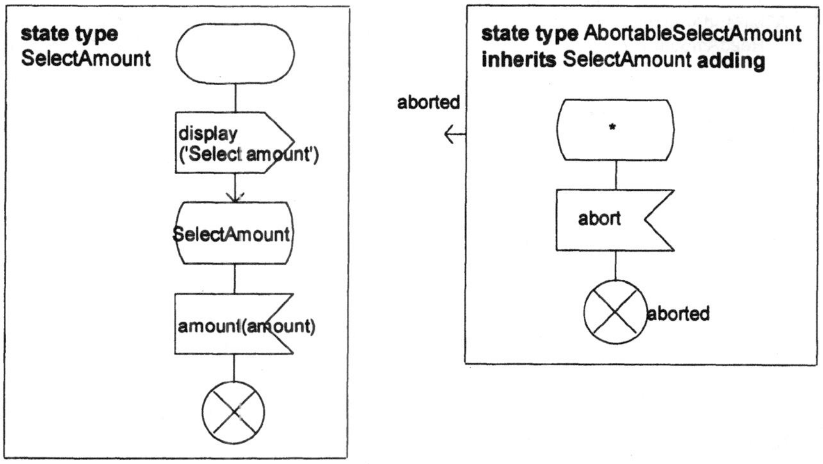

Figure 8. State Type Inheritance

In SDL-2000, any composite state type at any level can be defined as a subtype of another state type. For Statecharts associated with object modeling notations, the outermost state is associated with a class, and inheritance for states follows the class inheritance. The implication of this is that inheritance only applies to the outermost state.

8. VIRTUAL STATES

A state type may be defined to be a virtual type in order to allow redefinition, when the enclosing type is subtyped. *Figure 9* and *Figure 10* provide an example of the redefinition of a virtual state type when the enclosing process type is subtyped. The example also shows how an exit transition in the enclosing process type is defined as virtual and redefined in the subtype.

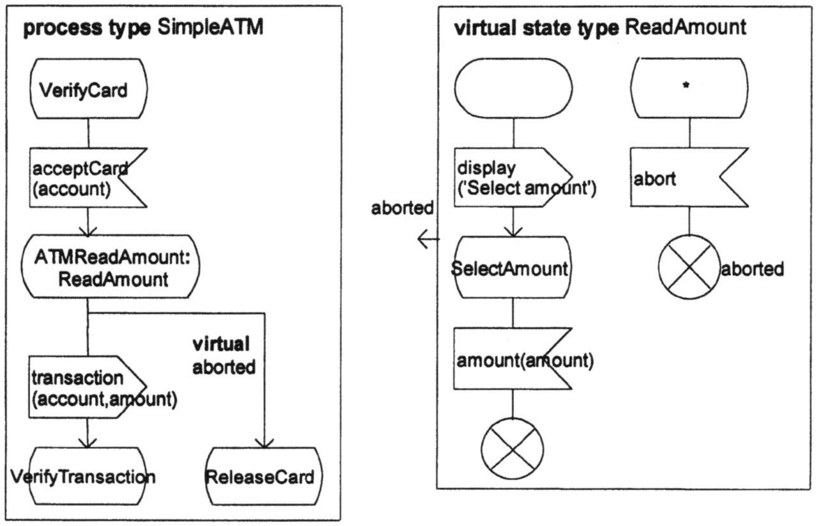

Figure 9. Virtual states/redefinition of states (1)

The state type ReadAmount is defined as a virtual state type and used for defining a state, by the expression

"ATMReadAmount: ReadAmount"

The definition of the contents of the virtual state type will in all redefinitions be inherited, that is only virtual properties of the virtual state type can be redefined.

In *Figure 10* the redefined state type inherits the corresponding state type definition in the super process type, and extends it with the state EnterAmount and with three transitions. The state SelectAmount in the redefinition is the inherited state with the same name defined in the virtual state type.

The virtual exit transition (aborted) is a transition defined as part of SimpleATM and not as part of ReadAmount. It is therefore redefined as part of the enclosing process subtype ATM.

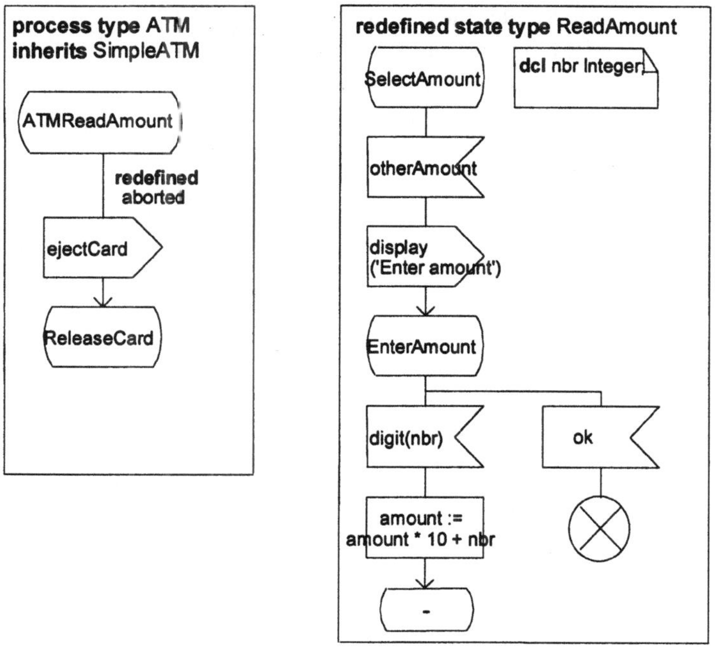

Figure 10. Virtual States/redefinition of States (2)

A virtual state can not just be redefined to any state with the same name. A virtual type has a constraint, and redefinitions have to be extensions (by inheritance) of the constraint. A constraint is either another, more general state type, or it is the virtual type itself. In the last case, the redefinition is simply an extension of the virtual state.

9. PARAMETERIZED STATE TYPES

In order for state types to be defined in *packages* and used in different systems with different local definitions, state types can have type parameters, in the same way as provided for classes (template parameters) in many other languages.

Virtual types in SDL (the correspondence would be virtual nested classes in UML and virtual inner classes in Java) provides one way of expressing that a type has type parameters, in that a virtual type can be redefined to any (visible) type that satisfies the constraint on the virtual type. For a further description of this in SDL, see [5]. For its application to Java, [4] is recommended. Virtual state types have been described above.

SDL has, in addition to virtual types, the notion of context parameters in general for type definitions, and this mechanism is in SDL-2000 also applied to state types. Context parameters are a little more general than just type parameters: a context parameter can be of any kind of entity that may be defined in the context of the type. Possible context parameters for state types are data types, signals, procedures, variables and other state types. *Figure 11* gives an example, where the parameter is a variable.

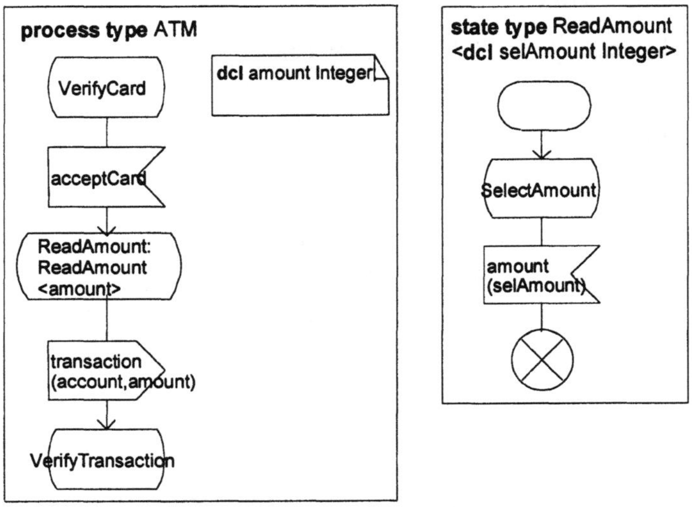

Figure 11. Parameterised State Type

The state type ReadAmount is defined to have an Integer typed variable as parameter. Binding this variable to a local variable in the process type ATM implies that all accesses to selAmount in the state type will become accesses to the process local variable amount.

Parameterised types are necessary in order to facilitate reuse in real system development, and together with virtual types it allows the definition of frameworks by means of abstract (super) types that are specialized to actual applications. In [5] it is demonstrated how virtual types can be used to make frameworks. It covers only the virtuality of types like system-, block-, and process types, and virtual procedures. With virtual state types it is possible to extend the notion of frameworks to cover also the state machine description of process types.

10. FORMAL DEFINITION

In the current stage of standardization, composite states are defined by transformation to basic SDL, which in turn has a formal definition. This transformation is part of the Z.100 recommendation. This scheme may change with the new approach to the formal definition of SDL, using Abstract State Machines (ASMs), [11].

11. RELATED WORK

UML state machines have a notion of *submachinestate*, which resembles the SDL-2000 notion of either a composite state that *references* a separate state diagram, or a type-based composite state. According to the UML Semantics, this mechanism should "facilitate reuse and modularity", but submachinestate is not a Classifier and not a GeneralizableElement, so inheritance is not supported. It is also said in the UML Semantics that it is a shorthand "that implies a macro-like expansion". Graphical macros have been removed from SDL-2000, and state diagrams in SDL-2000 are *not* macros, but real scope units. This implies that a state diagram forms its local name space, i.e. names of entities defined in enclosing scope units are visible in the same way as for other kinds of scope units in SDL. Entities of the composite state scope unit have to be specified as part of its interface in order to be visible from outside.

A *stubstate* in UML state machines corresponds to a *nextstate via an entry point* in SDL-2000. The difference is that a stubstate can reach any substate to any depth, while a nextstate via an entry point can only reach a first level starting point within the composite state. In addition, this starting point has to be defined as an entry point as part of the *interface* of the composite state. The SDL-2000 notion of nextstate as part of a composite state is therefore less flexible than stubstate, but it supports encapsulation and thereby independent development. The enclosing state machine only has to know the entry/exit points of a composite state, and the content of a composite state can be changed without changing the use of it in the enclosing state machine.

As mentioned above, the notion of type/subtype of composite states is in UML state machines only supported for the topmost state associated with a class. But even for this topmost state, the UML Semantics does not say anything definite about inheritance of states and transitions. A note describes some alternatives that may be considered.

SDL-2000 has first of all the notion of type/subtype for composite state at any level, not only the topmost. Secondly, it is well defined how inheritance works. *Virtual states and transitions* as parts of a state type (or directly of a process type or procedure) can be *redefined* in state subtypes (or process subtypes or subprocedures). If they are finalized (like Java's final), they can not be redefined in further subtypes. States and transitions can also be added to the states and transition inherited from the supertype.

Even though UML does not define inheritance for state machines, various methods for using UML recommend approaches that are similar to the one described here. In [10] it is recommended to follow a.o. these rules when inheriting a state machine type (in [10] called a "state model"):

- New states and transitions may be added
- States and transitions defined by the parent cannot be deleted
- Action and activity lists may be changed
- Actions and activities may be specialised.

Other similar notations have an opposite view on this. In [3] it is argued that strict inheritance (as the above approach is called in [3]) does not guarantee anything, so in ROOM it is possible to delete states and transition. It is true that strict inheritance is a kind of syntactic restriction on inheritance (that cannot guarantee anything in terms of the actual application), but it is still better than complete freedom in deletion of properties.

No other approaches to inheritance of state machines have the notion of constraints on redefinition of inherited, virtual states. The "strict inheritance" in SDL-2000 is extended with the notion of a virtual state type *constraint* (in terms of a more general state type or the virtual state type itself). A virtual state type with constraint can only be redefined to a state type that is a subtype of the constraint This assures that all redefinitions are extensions of this constraint. If no constraint is specified, the virtual state is itself the constraint; the implication is that redefinitions are extensions of the virtual state. The last case is covered by the rules in [10]. For SDL-2000 it is also covered by the language, and thereby checked by tool.

The notion of constraints on virtual state types is not a mechanism special for SDL. It first appeared in [2], and recent proposals for adding type parameters to Java ([4] being one of these) also provide constraints on type parameters. This is needed in order for Java to be a strongly typed language. The parameterised class can use the type parameters according to the constraint. The same is the case for virtual states in SDL-2000: the enclosing state machine can use the constraint of a virtual state, e.g. entry/exit connection points.

12. CONCLUSION

It has been demonstrated that it is possible to obtain the best of two "worlds": the transition-orientation of SDL state machines *and* the state-orientation of Statecharts. The transition-orientation has focus on the where the functionality is specified and therefore provides designers with the right mechanism during detailed design. The state-orientation is superior for providing an overview of large and complex state machines. The notion of composite state of SDL-2000 provides this combination, but it does more than this. It scales up and handles large state machines by state references and by separate state diagrams. It supports encapsulation by defining interfaces of composite states, instead of allowing any substate of a composite state to be entered directly. Finally it has been demonstrated that it is possible to have real object oriented composite states. Any composite state can be subject to the type/subtype mechanism; state types can inherit states and transitions, and redefine virtual states; state types can be generalised by type parameters in order to be used in different contexts.

ACKNOWLEDGEMENTS

Although the authors are responsible for the idea behind composite states in SDL, for the first versions of their definition, and of course for the contents of this paper, the notion of composite states is the result of the joint effort of the working group ITU-T Study Group 10 Question 6.

REFERENCES

[1] D. Harel & E Gery: *Executable Object Models with Statecharts*. IEEE Computer, 1997.

[2] O. Lehrmann Madsen & B. Møller-Pedersen: *Virtual Classes - a Powerful Mechanism in Object Oriented Programming*. OOPSLA'89, Sigplan Notices, Vol 24 Number 10, 1989.

[3] B. Selic: *An efficient Object-Oriented Variation of the Statecharts Formalism for Distributed Real-Time Systems*. http://www.objectime.com/otl/technical/efficient.html.

[4] K. Thorup: *Genericity in Java with Virtual Types*. European Conference on Object-Oriented Programming, LNCS 1241 Springer-Verlag, 1997.

[5] Bræk & B. Møller-Pedersen: *Frameworks by means of virtual types*. FORTE XI/PSTV XVIII'98, Paris November 1998.

[6] ITU Z.100 Specification and Description Language SDL, 1996.

[7] A. Olsen et al: *Systems Engineering Using SDL-92*. North-Holland 1994.

[8] OMG *Unified Modeling Language Specification* (draft, Version 1.3, 1999).

[9] D. Harel: *Statecharts: a visual formalism or complex systems*. Science Computer Program Vol. 8, 1987.

[10] B. P. Douglass: Real-Time UML – Developing Efficient Objects for Embedded Systems. Addison-Wesley 1998.

[11] R. Gotzheim, B. Geppert, F. Rössler, and P. Schaible: Towards a new formal SDL semantics. In Proc. Of the 1st Workshop of the SDL Forum Society and SDL and MSC, Berlin June 98.

Prototyping Framework for SDL with Evolving Semantics

Peter Csurgay
Department of Telematics, Norwegian University of Science and Technology, c/o Traffic Analysis and Network Performance Laboratory, Ericsson Telecommunications Ltd, Address: Laborc utca 1, Budapest 1037, Hungary, Phone: (+36 20) 979 3049
E-mail: csurgay@item.ntnu.no

Key words: SDL, prototyping, formal semantics, extension of FDTs, Java technology

Abstract: The definition of the ITU-T Specification and Description Language (SDL) evolves, its semantic interpretation and modelling features are subject to modifications and extensions to meet demands of new application domains and new modelling contexts. Tools are slow in following this evolution, although automated support is vital in applying formal methods for the design and evaluation of telecommunications protocols. This paper introduces a prototyping framework that supports traditional and new tool functions for specifications in SDL with evolving semantics, thus eases the evolution process as well. The framework includes automatic parser generation and abstract grammar definition from Z.100-style EBNF descriptions, and an easily extensible object-oriented formal definition of the semantic interpretation model. Java technology has been used to implement an open framework of distributed visualisation, control, and evaluation components. Early implementation of the framework was applied to prototyping systems with extended SDL semantics in the performance evaluation context.

1. INTRODUCTION

Initially, SDL was meant for unambiguous specification and description of telecommunications systems. As Z.100 states: 'specification' of a system is the description of its required behaviour, and 'description' is the description of its actual behaviour [5]. Recently, however, SDL has been used in various other contexts, like performance evaluation or automatic implementation of formally specified systems. Behaviour of an SDL system is of non-deterministic nature, not just because it has language constructs to express non-determinism on process level, but because it is composed of asynchronously communicating concurrent extended finite state machines (CEFSMs). This non-determinism is the basis of modelling concurrent entities of the real world, and also of specifying and validating a set of possible traces in a concise manner. Time and ordering of internal events plays key role in SDL's non-determinism. Due to the vague time semantics of SDL (signal transactions and transitions may or may not take time), tool builders and users of SDL choose legal, but different interpretations of time semantics when simulating or interpreting SDL systems. A legal interpretation for example (most probably inspired by the way some leading tools handle advancing of time) is that time passes only if there are no armed transitions (no valid input signals in queues) of processes. This interpretation is legal, even reduces the set of possible traces simplifying e.g. validation, but not appropriate for e.g. performance modelling [12]. A prototyping framework should support generation of possible traces of system behaviour based on different interpretation models. This is achieved by separation of different interpretation functions in a modular object-oriented model of formal semantics. Due to the unsatisfying formal semantic definition of Z.100 for new application domains various modifications and extensions to SDL's semantics [10,13], and even new formal semantic interpretations emerge [11]. (These improvements will influence the recommendation, and eventually, a new version SDL-2000 might include modifications [7].) Current tools are not open in the sense that they employ a specific language definition and specific semantic interpretation with little room for experimentation with them. A prototyping framework should support the evolution process and protocol engineering with evolved SDL semantics by taking formal semantics and interpretation models along with syntax and system specification as input to the prototyping process.

The ITU-T Recommendation Z.100 defines the Specification and Description Language (SDL) with (1) its concrete grammars SDL/PR and SDL/GR, (2) an abstract syntax AS1 equivalent to the concrete grammars, (3) rules to transform a system specification into abstract syntax, and (4) the SDL-machine semantics to interpret a specification given in terms of the abstract syntax. The structure of the definition that eventually leads to the behaviour interpretation of SDL systems and the effect of the evolution of SDL semantics on it is illustrated on Figure 1. (1) is given by an extended Backus-Naur Form (EBNF) notation. (2) uses a subset of Meta IV. (3) and

(4) is given both informally and formally in the Annex F to Z.100 as the static and dynamic semantics respectively. On over 500 pages, it defines a set of Meta IV (a formal language based on synchronously communicating concurrent processes) programs for static correctness analysis, transformation of concrete syntax into abstract syntax with removal/substitution of several language constructs, and interpretation of behaviour the SDL-machine. The Annex F formal model does not help human understanding of SDL's semantics simply because its size, and Z.100 gives precedence to the informal definition.

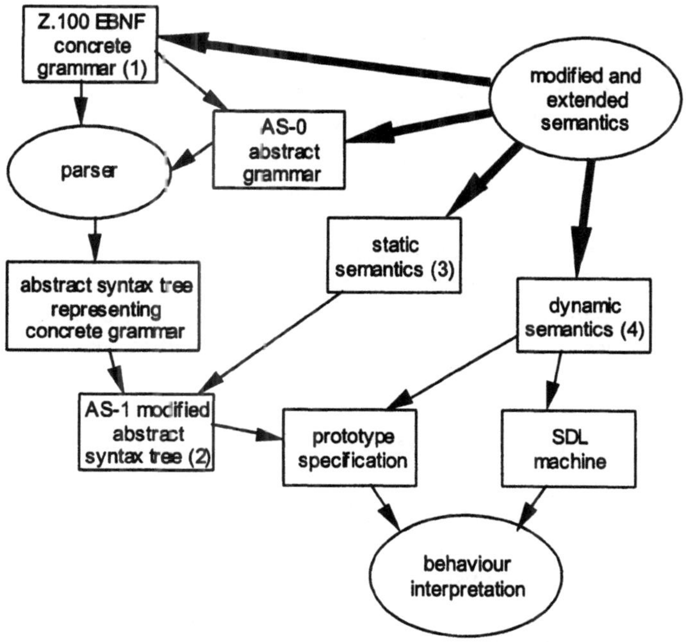

Figure 1. Effect of evolving semantics on Z.100 SDL definition

Because of the evolutionary nature of SDL, a prototyping framework shall support semi-automatic processing for (1), support for formal definition of semantic actions and semi-automatic generation of (2), a concise and flexible formal definition of (3), an extensible formal interpretation model for (4), and finally, a portable, open interpretation platform for visualisation, control and evaluation of systems specified in SDL with extended semantics (Figure 2).

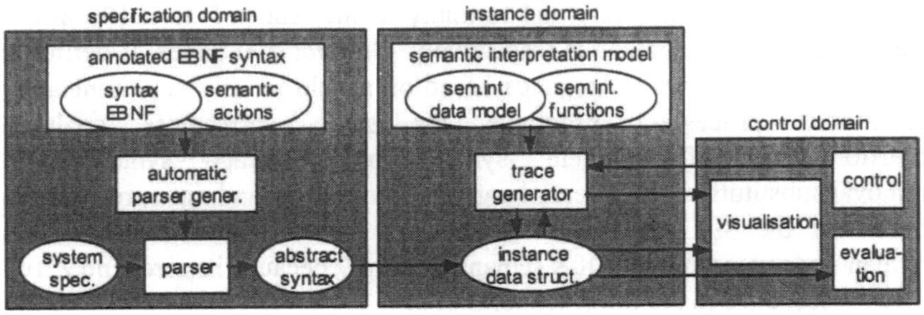

Figure 2. Functions and data structures of the prototyping framework

Section 2 describes the automatic parser and abstract syntax tree generation process. Section 3 shows the new approaches in the object-oriented semantic interpretation model with examples from the specification and instance domains. Section 4 illustrates the application of the framework to specifications in SDL with extended semantics in the performance evaluation context. Two different packet transfer protocol alternatives were modelled in PerfSDL (a semantic extension to SDL [10]), and they were compared by performance evaluation based on performance traces produced by a prototype interpretation model. The paper concludes in Section 5.

2. AUTOMATIC PARSER GENERATION

The traditional or most understood ways of translating a sentence of a formal language to a sentence of another follow the analysis-synthesis model of compilation. The analysis part breaks up the source program into constituent pieces (tokenising or lexical analysis), and creates an intermediate representation (parsing or syntax analysis). The synthesis part constructs the target language sentence from this intermediate representation. While the analysis part has well settled methods and algorithms [8], the synthesis part requires specialised techniques. Lexical analysers can be easily built from the lexical rules of a language, and it is most often done following rather intuitive than automated or formal procedures. The aim of syntax analysis is to generate the intermediate representation (often an abstract syntax tree) of the source sentence from its tokenised form. This Section will focus on the syntax analysis part, while Section 3 and 4 illustrates the synthesis part through examples.

2.1 Parsing the BNF grammar of Z.100

The recommendation contains the specification of both the graphical (SDL-GR) and textual (SDL-PR) form of the language in a mixed dual fashion having <sdl specification> the common root symbol. The project presented in this paper only processes the textual form, i.e. the grammar is taken to be the set of rules of the SDL-PR definition without the phrases referring to graphical elements of SDL-GR. To be able to automatically process the BNF definition given by Z.100, first a tokenizer, parser, and abstract syntax tree have to be built. For reasons mentioned above, an automated parser generation approach was chosen based on the grammar's BNF definition directly derived from the text of the Z.100 recommendation. The recommendation uses an extension of the original Backus Naur Form notation, however, it does not follow the "ISO/IEC 14977 Extended BNF" standard for the extensions. The following meta-BNF grammar (using the very same style BNF) can be derived from the syntactic meta-rules given by Z.100 informally. Semantics of the notation is given in Table 1.

```
<rule>    ::=  <<nonterminal>> ::= <production>
<production>    ::=  <alternative> { | <alternative> }*
<alternative>   ::=  { <concatenation> }+
<concatenation>      ::=  <terminal> |
           <<nonterminal>> |
           [<production>] |
           {<production>} [ * | + ]
```

Table 1. Semantics of BNF extensions in Z.100

::=	Defining symbol of rules
< >	Enclosing nonterminals
|	Separating alternatives
[ ]	Enclosing optionals; zero or one occurrence
{ }	Enclosing groups; one occurrence
{ }*	Enclosing repetition; zero or more occurrences
{ }+	Enclosing repetition; one or more occurrences

This small grammar will be the basis for all further automatic parser generation issues. Note, that for each rule of this grammar the first terminal symbol (lookahead) unambiguously determines which branch to parse on. Therefore a simple predictive parser (that is a recursive descent parser with no backtracking) can syntax analyse its language (i.e. the BNF rules of Z.100).

2.2 Abstract syntax tree for Z.100

According to the BNF grammar above and the semantics of the meta-description in Z.100, node types for the abstract syntax tree was chosen. These node types and their relation for the valid branching topologies are illustrated in Figure 3. For each node type there is a Java class that is capable of predictive parsing the Z.100 BNF input stream, and building on the abstract syntax tree according to the fired rules.

Given the complete abstract syntax tree of the Z.100 BNF definition, several interesting observations can be made. For instance by producing a pretty print of all rules, and comparing them to the original definition, meta-information can be derived about the recommendations and the algorithm. The Z.100 document Addendum 1 gives of list of changes and corrections to the recommendation. Contact the author for a list of bugs not covered by the Addendum, that were made during formal processing of the Z.100 recommendation.

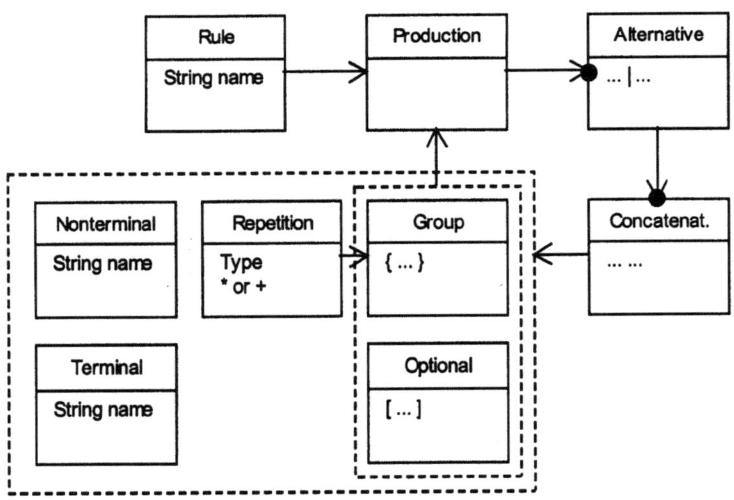

Figure 3. Z.100 BNF abstract syntax tree node types and their branching

2.3 Recursive descent parsing with backtracking

The most common method to syntax analyse formal languages given by context-free grammars is to pre-process the grammar first (e.g. convert it to Greibach Normal Form), and construct a parser of linear complexity (e.g. predictive parser). In this project however, to preserve local integrity of rules of our original BNF grammar for reasons explained earlier, we decided to generate a recursive descent parser for SDL. Therefore this method is shortly introduced here. In the family of top-down parsers, the most intuitive (or more importantly for our reasons, the closest following a context free grammar in structure) is the recursive descent parser. Note, that not all context free grammars can be analysed with recursive descent parsers (left recursive grammars leads them into infinite looping), but the grammar in Z.100 is in a proper form in this sense. In recursive descent parsers a logical value returning procedure is associated with each nonterminal symbol. These procedures return true if the input matches their terminal symbols and the procedures associated with their nonterminal symbols also returned true, and false in any other case. These procedures might call each other recursively, implementing an implicit stack automaton needed to analyse context free languages in classical theory of formal languages. Table 2 explains the mapping from BNF constructs and operation of logical values generated by terminal and nonterminal symbols of a rule.

Table 2. Logical operations for BNF constructs in recursive descent parsing

Alternative	x | y	If x or y then true else false
Concatenation	x y	If x and y then true else false
Optional	[x]	If x; true
Group	{ }	Parentheses in logical expressions
Repetition	{x}*	While x; true
Repetition	{x}+	If x then while x; true else false

Beside the advantages of this way of parsing (i.e. no modification to the BNF rules are needed and semantic constraints can easily be applied to the generated abstract syntax tree), there is one serious inconvenience. There is no problem until all procedures (implementing rules of the original BNF) returns true. That would mean that the first input symbol (lookahead) unambiguously determines which branch to follow, and we get back recursive parsing with no backtracking (i.e. predictive parsing) that we used for the BNF definition earlier. Note however, that by the time a procedure returns false, it might have consumed a number of symbols from the input stream. Therefore the input stream pointer must also be stacked along the recursive procedure calls, and each time a rule implemented by its procedure is recognised not to be applicable, this pointer has to be restored. This is called backtracking, and there are only a few known parsers using this technique [8], because it's rarely needed for preprocessed context free grammars. But this is the price we pay at the trade off to keep our parser closely follow the untouched standardised set of BNF rules in structure.

2.4 Synthesis of the SDL parser

Node types of the abstract syntax tree are generated from the BNF definition of SDL according to decisions explained earlier. Roughly stated, there is a node type for each nonterminal symbol of the BNF definition. For an example, Figure 4 shows the simplified node type "textual system definition" of the abstract syntax tree, that was generated from rule <textual system definition> of the original BNF definiton.

```
<textual system definition>  ::=
     { <package reference clause> }*
     {    system <system name> <end>
          {<entity in system>}+
          endsystem [<system name> ] <end>
          | <textual typebased system definition> }
```

Modelling rules in Z.100 define semantic equivalence of possibly infinite sets of syntax units. Modelling rules and semantic constraints in Z.100 are given with local scope on syntax units defined by individual BNF rules. Therefore it is straightforward to check and enforce them based on an abstract syntax tree of our kind as opposed to one that the "Abstract grammar" definitions in Z.100 would imply. Besides constructing the tree builder functionality from SDL-PR at the synthesis of the SDL parser, these constraints must be implemented manually (remember that they are given informally) as so-called semantic subroutines [2] and put into the synthesis process. Semantic subroutines manipulate on the abstract syntax tree, that will be the basis of Java code generation according to concepts of the Specification domain explained in Section 3.

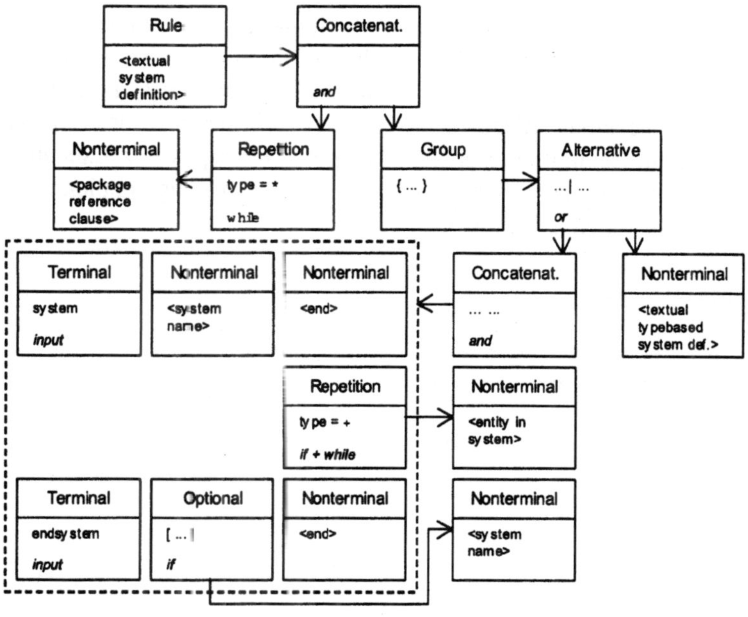

Figure 4. Generated node type for rule "textual system definition"

Whenever a procedure associated with a nonterminal is called, a new instance of the corresponding node type is created, linked to the abstract syntax tree, and all its branches is attempted to parse in recursively. The problem of backtracking of course also applies to these branches of the growing tree. After successful parsing, the derivation tree for the SDL-PR source is given by either the sequence of processes returning true or the abstract syntax tree. The same does not stand for unsuccessful parsing. If the input is not a valid sentence of the SDL language, from one moment in the parsing process all procedures will return false until the last, the one corresponding to the root symbol of the grammar, will also return false. We get the information that the source is syntactically incorrect, but the input stream and the tree were backtracked during the false procedure returns. To get the information how far the parsing process could successfully reach into the input stream, a special algorithm was developed to track the longest possible prefix of the sentence believed to be valid prefix of a valid sentence. Note, that the problem is that by the time the parser gets known about the invalidity of the sentence, the erroneous position, lexical unit (token), and the tree is rewound already by the backtracking. Therefore the algorithm delays the rewinding of the stream with some remembered backtracking parameters.

Let 'maxpos' be the end marker of the longest prefix believed so far and 'revertpos' be the earliest position in the stream until we had to rewind the input during backtracks. 't' will denote time, and pos(t) is our current position of the input stream in the backtracking recursive descent parsing process. In the beginning of parsing Maxpos = revertpos = 0. Executing the following instructions after each step of the parsing implements the algorithm illustrated in Figure 5.

```
if ( pos(t) > maxpos )
{
     if ( revertpos < maxpos ) { delayed_rewind(); }
     revertpos = maxpos = pos(t);
}
if ( pos(t) < revertpos ) { revertpos = pos(t); }
```

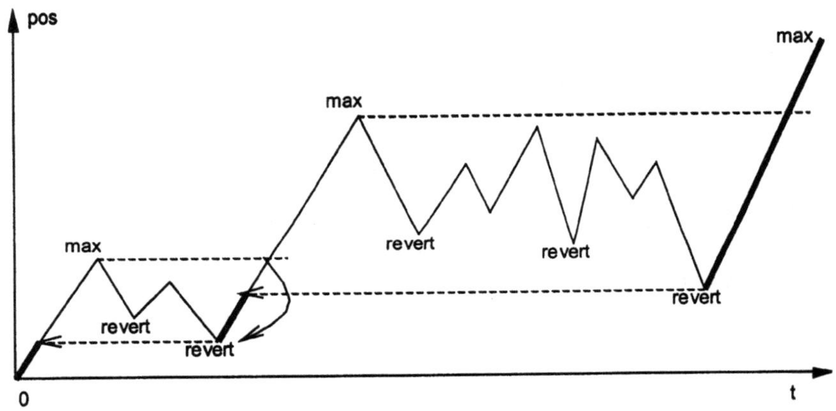

Figure 5. Backtracking parameters for delayed rewinding

3. SEMANTIC INTERPRETATION MODEL

In the specification domain Z.100 introduces the concept of named entities. A named entity is either a type or an instance. Processes and signals are examples of types, system, blocks, and variables are examples of instances. The difference is illustrated in [6] as "In general, we can identify instances by the fact that they are bound to a specific place within the system, whereas a type may be used in more than one place". System here clearly relates to the specification domain. By the distinction between specification and instance domains, a more precise definition can be formulated. The difference can roughly be defined as the relation between the entity that is given name in the specification domain, and the existence it possibly corresponds to in the instance domain. If this relation is one-to-many, than the named entity is a type, if one-to-one, than it is an instance.

Decisions had to be made concerning the selection of Java constructs for the specification domain (how the original SDL specification phrases are translated), and the instance domain (how the dynamic semantics of SDL comes to life). The decisions for the two domains are of course not independent, and priority was given to the specification domain mapping to achieve an ideal one-to-one readability of the SDL phrases in the generated Java source code. This approach is new and does not follow the transformation rules given by Annex F of Z.100, as will be explained shortly. However, it would enable easy manipulation of the Java code according to SDL terms, therefore this approach fits research oriented, experimenting, prototyping applications most. The accepted trade-off is, that this decision implies hardest requirements on the Java solutions for the instance domain, since all features that could have been handled by manipulation of the abstract syntax tree (or in the specification domain), must be implemented there. The overall process of the prototyping of SDL specifications is illustrated in Figure 1.

3.1 Formal definition of SDL semantics

Z.100 gives formal definition of SDL semantics in Annex F. F.1 is introduction to the Meta-IV language that is used in F.2 and F.3 describing the static and dynamic semantics respectively. Their relation is illustrated by the top segment of Figure 6 along with the AS0 and AS1 syntaxes introduced by Z.100. AS0 is not formally defined in the recommendation, but can be derived from the BNF syntax. A constructive example was given to this as a generated abstract syntax tree in Section 2. AS1 is given in Z.100 as the Abstract Grammar. However, with the F.2 transformation from the abstract syntax (representing concrete textual syntax) to AS1 (input to the dynamic behaviour definition of SDL), specification domain information is lost. Such information would regard type inheritance hierarchies or virtuality. Losing this information causes no problem when interpreting SDL, because it is object oriented only in the specification domain, i.e. does not feature polymorphism or means of late binding.

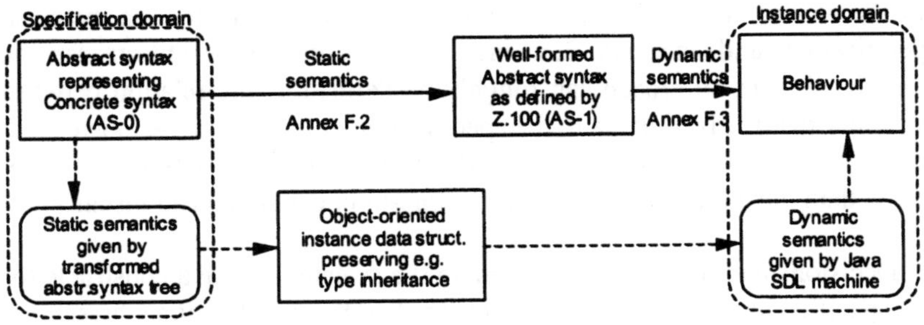

Figure 6. Alternative to Annex F derivation of SDL behaviour

For the same reason one should have wide knowledge in principle about these transformations to produce quality specifications reflecting exactly his/her intentions. For example even though transitions of asterisk states in virtual process types are transformed into transitions for each states of the process type, its asterisk semantics shall also apply to brand new states introduced by redefined process types down the inheritance hierarchy.

In other words, there is a gap between specification domain concepts and instance domain representation, which the designer should be aware of. Our framework attempts to narrow this gap on the price of preserving specification domain concepts for an interpretation with heavier runtime overhead, e.g. using delegation to resolve added and redefined virtual properties down the type inheritance hierarchy. The bottom segment of Figure 6 illustrates this alternative way to derive dynamic behaviour of an SDL specification with the representation forming the basis of behaviour interpretation being closer to the specification domain.

3.2 Process types and process instances

Basically each named entity in SDL is mapped to its own Java class. Classes for SDL instances will be inner classes of their corresponding SDL type's Java class. In this manner local management of instance life-cycle, and clear structure of Java specification is achieved.

Taken SDL processes as an example, process Monitor of block Game in Figure 7 will be mapped to the following Java construct:

```
class SDLprocessType_Monitor extends SDLprocessType {
  Pool instancePool = new Pool();
  public SDLprocessType_Monitor( int initNo, int max ) {
    while( initNo-- > 0 ) create( null );
```

```
    }
    public SDLprocessInstance create( SDLprocessInstance parent) {
      return(instancePool.add(new SDLprocessInstance_Monitor( parent )))
    }
    public void rcvSignal(Signal s) {
      instancePool.rcvSignal(s);
    }
    class SDLprocessInstance_Monitor extends SDLprocessInstance {
      Queue signalQueue = new Queue();
      public SDLprocessInstance_Monitor( SDLprocessInstance parent ) {..}
      public void rcvSignal(Signal s) {
        signalQueue.rcvSignal(s);
      }
      public void start() { ... }
    }
}
```

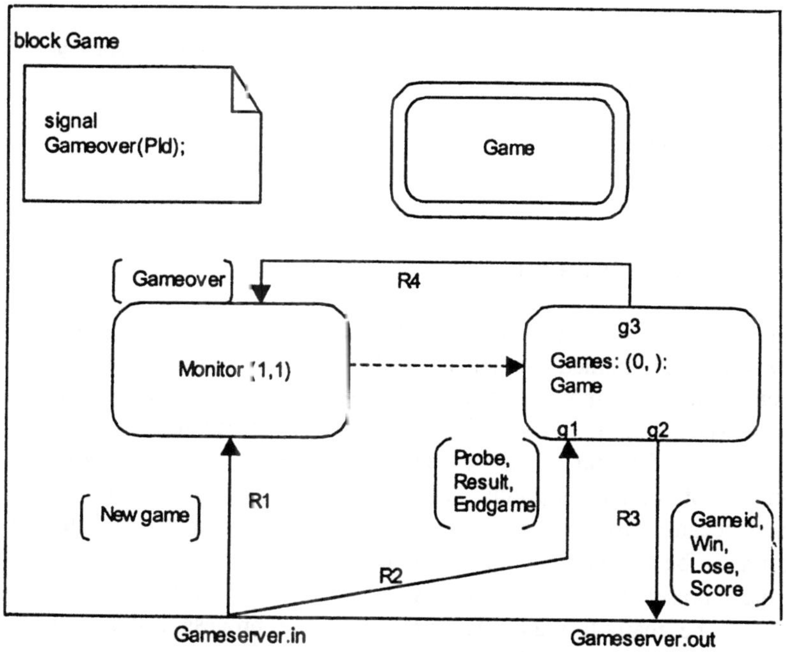

Figure 7. Example SDL block definition

Process types will create instances of their own type and will deliver incoming signals in case of output-via type signal sending. New to the mapping applied by this project that it allows generic context parameterisation and type inheritance to appear in the instance data structure of the instance domain.

3.3 Signals

Signals are mapped to an abstract class having a pointer to an array of objects as the list of parameters they can carry. Each signal in the SDL specification will inherit from this abstract class. Methods handling signals identify them by their class type that is given by the Java RTTI (runtime type identification) construct Object.class. Classes responsible for input part of transitions will have a field containing the class type of the input signal they act upon. This mapping enables signal types (i.e. signals having formal context parameters) to be handled the same way signal instances are handled.

```
abstract class SDLsignal {
     SDLprocessInstance sender;
     Object[] par;
     public SDLsignal(...)
{...}
}
```

```
class Newgame extends SDLsignal
{
     public Newgame(...) {...}
}
```

3.4 Instance domain

Figure 8 shows the most important classes and their inheritance relations representing SDL named elements for the instance domain. Path is the abstraction of the common properties of named elements that can carry signals in SDL. Plug represents connection points of these paths, typically on the bounderies of blocks, processes, services, and corresponding types. The restriction that connection points on boundaries can only be 1-to-many was removed because with the introduction of types, gates and specialization it was found to be too restrictive. Because the interpretation model applied by the project preserves types for the instance domain, connection points are allowed to be many-to-many. Process types and process instance types reflect the structure explained above.

The following example illustrates the Java skeleton of the runtime system (RTS) with a simple scheduling algorithm. With this algorithm it is possible to fire one armed transition at a time, including spontaneous transitions, continuous and priority signals, and expired timers. Responsibility of checking the armed property of transitions is delegated down the hierarchy of simulation environment, the RTS, processes, their states, and transitions of any kind of these states. This mechanism enables parameterised types to be brought from the abstract syntax of the specification domain into the instance data structure of the instance domain. Another advantage is the scheduler's global information about the armed state of the SDL machine, i.e. the disjunction of armed states of all process instances in the system instance. Different interpretation models of e.g. time semantics of SDL can be implemented with specialisation of this RTS class in the object oriented semantic interpretation model.

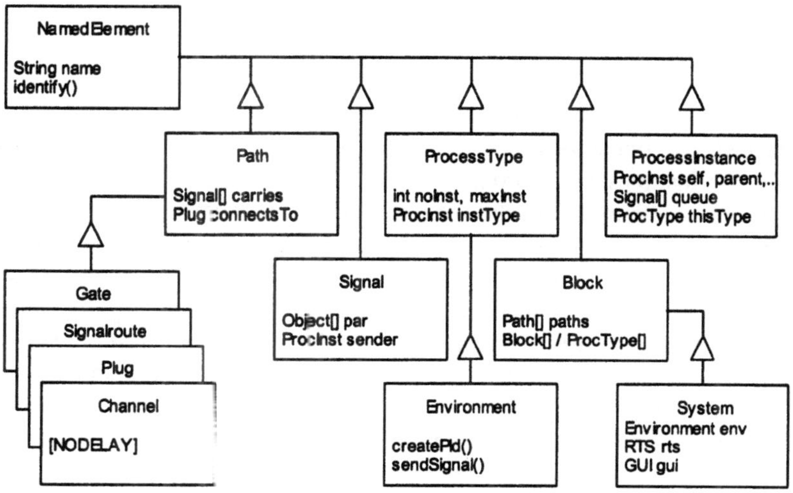

Figure 8. Classes and inheritance relations of the instance domain

The scheduler below fires the first armed transition from a list of active processes, and schedules the fired process into the end of the list. An interpretation generating a larger set of possible traces can be implemented by using 'any' instead of 'first' in the algorithm below.

```
class RTS {
  Vector instancePool = new Vector();
  public void start() {
    for (int i=0;i<instancePool.size();i++)
    ((SDLprocessInstance)instancePool.elementAt(i)).start();
  }
  boolean step() {
    boolean couldStep = false;
    for (int i=0;(!couldStep)&&(i<instancePool.size());i++) {
      int result=((SDLprocessInstance)instancePool.elementAt(i)).step()
      if (result!=0) couldStep = true;
```

```
      if (result==1) {
        SDLprocessInstance aux =
          (SDLprocessInstance)instancePool.elementAt(i);
        instancePool.removeElementAt(i);
        addProcessInstance( aux );
      }
    }
    return(couldStep);
  }
}
```

3.5 A new approach to time semantics in SDL

Due to the current time semantics of SDL, time may or may not be advanced by transitions and transactions during system interpretation. Figure 9 illustrates an approach where the chosen interpretation of time semantics is orthogonal to the scheduler algorithm above. This solution is compatible to different interpretations of time semantics with respect to the generated set of possible traces. Whenever a process instance encounters setting of a timer, it is interpreted as sending a signal to a global timer process with parameters of signal name, sender id, and expiration time. The timer process's queue is sorted by these expiration time values, and handled by the scheduler the same way other processes are handled. Upon interpreting the reception of a signal by the timer process, it sets the global simulation time to the specified expiration time parameter, and sends a signal with the specified signal name parameter to the specified sender parameter. If the timer process is always scheduled last among all processes, the interpretation will generate a trace set equivalent to the "time may not be advanced..." interpretation. If the scheduler randomly selects armed processes including this timer process, the resulting set of possible traces will be equivalent to the "may not be advanced..." interpretation. Note, that with this separation and orthogonal modelling of semantic interpretation functions, different concepts can be experimented with by prototyping in the modular object-oriented framework of semantic interpretation models.

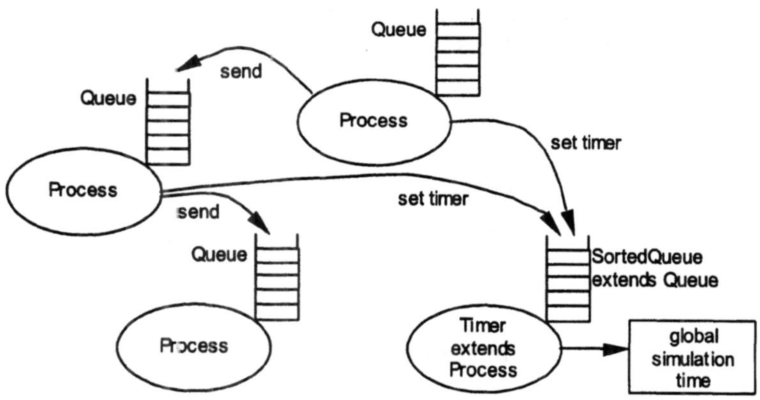

Figure 9. Time semantics interpreted with global timer process approach

4. APPLICATION TO PERFORMANCE EVALUATION

4.1 PerfSDL, protocol performance modelling in SDL

Covering timing properties and identifying performance bottlenecks in testing of proprietary protocols is one method to relate protocol performance to system performance. Because an SDL system is not limited by processing resources (i.e. assumed to be fast enough to process any load it is offered), it would be cumbersome to assess performance measures by means of modifying protocol specifications with available SDL constructs. Instead, PerfSDL [10] describes protocol performance integrated into system performance, where new SDL constructs provide appropriate correlation and feedback. These new constructs, illustrated in Figure 10, imply modifications to the BNF definition of the concrete grammar, to the semantic actions on the abstract syntax tree, and to the Java implementation of the meta-process "SDL-machine". In the concrete grammar, the production rule for <action> has been extended with <timed transition> and <probe>, and production rules for these have been defined. After automatic generation of the parser and AS0 abstract syntax tree, the semantic actions are systematically included. Implementation of the dynamic behaviour of the constructs involves a new layer to the SDL-machine that manages system time and collects information during simulation. This information can be analysed off-line, and results can be fed back to subsequent iterations of the protocol development process.

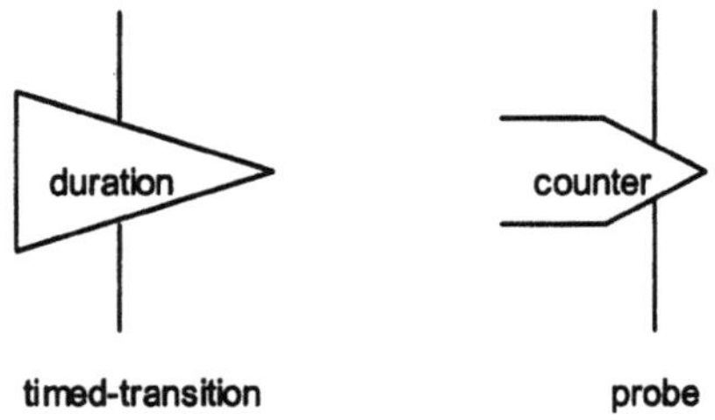

Figure 10. Introduction of new constructs for performance testing in [10]

4.2 First experiences applying the framework to a small example

Two simple packet transfer protocol alternatives were specified in PerfSDL using the new constructs, and two special SDL processes. One generated system stimuli as packets to be sent according to different stochastic arrival processes. Information relevant to protocol performance was sent to the other, an observer process, as signal parameters. These signals were monitored and logged, and later evaluated. Based on the evaluation, performance of the protocol alternatives was compared under different workload conditions.

5. CONCLUSIONS

A prototyping framework for SDL protocol specifications with extended and modified SDL semantics was introduced based on automated parser and abstract syntax tree generation and an open object-oriented semantic interpretation model using Java technology. The framework supports experimenting with SDL specifications of extended semantic foundations. A systematic approach was applied for treating the effects of semantic extensions on different phases of the prototyping process. A new general interpretation model of time semantics was described that is orthogonal to other semantic interpretation functions and can be specialised for modelling different actual interpretations of SDL's time semantics. Application of early implementation of the prototyping framework was shown in the field of SDL-based protocol performance engineering.

6. ACKNOWLEDGEMENTS

The author is thankful to Miklós Boda for the support of Traffic Lab at Ericsson Telecommunications, where the project has been carried out. At the Technical University of Budapest Professor Gyula Csopaki and Professor Iván Bach gave support and helpful instructions. Thanks to Professor Finn Arve Aagesen for commenting the paper and for the supervision throughout the years at the Department of Telematics, Norwegian University of Science and Technology.

7. REFERENCES

1. Ferenc Belina, Dieter Hogrefe, Amardeo Sarma, *SDL with Applications from Protocol Specification*, Prentice Hall International, 1991
2. Ivan Bach, *Computer Linguistics*, lecture notes, 1991
3. Rolv Bræk, Øystein Haugen, *Engineering Real Time Systems*, Prentice Hall International, 1993
4. Anders Olsen, Ove Færgemand, Birger Møller-Pedersen, Rick Reed, J.R.W. Smith, *Systems Engineering Using SDL-92*, Elsevier, 1994
5. ITU-T Recommendation Z.100, *CCITT Specification and Description Language (SDL)* and Annex F. *Static and Dynamic Semantics of SDL,* 03/93, International Telecommunication Union, 1995
6. Jan Ellsberger, Dieter Hogrefe, Amardeo Sarma, *SDL – Formal Object-oriented Language for Communicating Systems*, Prentice Hall Europe, 1997
7. Øystein Haugen, Birger Møller-Pedersen, *SDL – a platform for telecom system development also in the future?*, Ericsson and Telelogic Joint User Group Conference 1998
8. Alfred V. Aho, Ravi Sethi, Jeffrey D.Ullman, *Compilers - Principles, Techniques, and Tools*, Addison Wesley, March 1998
9. Olivier Dedieu, *A Dynamically Extensible Web Proxy*, NoTeRe colloquium, 20-23 October 1998.

10. Mazen Malek, *PerfSDL: an Interface to Protocol Performance Analysis by means of Simulation*, Accepted paper, 9th SDL Forum, Montreal, June 1999
11. R.Gotzhein, B.Geppert, F.Rossler, P.Schaible, *Towards a New Formal SDL Semantics*, SAM98, 1st SDL and MSC workshop of the SDL Forum Society, Berlin, July 1998
12. Andreas Mitschele-Thiel, *Performance Evaluation of SDL Systems*, SAM98, 1st SDL and MSC workshop of the SDL Forum Society, Berlin, July 1998
13. E.Holz, J.Fischer, *Introduction of Gate Types into SDL*, SAM98, 1st SDL and MSC workshop of the SDL Forum Society, Berlin, July 1998
14. Andreas Mitschele-Thiel, *Results of the discussion*, Workshop on Performance and Time in SDL and MSC, Erlangen, February 1998

Framework for Automatic SDL to C++ Translation

Dirk Trossen, Christian Cseh, and Roman Kogan
Department of Computer Science IV
Technical University of Aachen
Ahornstr. 55
D-52074 Aachen, Germany
{dirk,cseh,kogan}@i4.informatik.rwth-aachen.de

Key words: SDL, Automatic SDL Translation, Lazy Evaluation

Abstract: Due to the increasing complexity of distributed systems the need for a formal description of these systems arises. Specification description languages like SDL enable the formal specification of distributed systems for verification purposes of the system. For simplification of the development process the need for an automatic translation from the specification language to a programming language like C or C++ arises. High performance and a good readability of the translated code are crucial for the usability of an automatic translation. This paper introduces a framework for the automatic translation of an SDL specification to C++ providing high performance together with good readability of the resulting code. For that we present an object model together with an efficient implementation of the finite state machine with constant costs only using polymorphism. Additionally we present mechanisms to reduce copy and memory allocation operations. A performance evaluation for these mechanisms is also presented.

1. INTRODUCTION

Due to the increasing complexity of distributed systems the need for a formal description of these systems arises. Specification description languages like *SDL* [1][4] enable the formal specification of distributed systems with the means of finite state automatas which allow together with testing languages like *TTCN* [3] to verify the system on an abstract level without implementing it. For simplification of the development process the need for an automatic translation from the specification language to a programming language like C or C++ arises. Especially an object-oriented language like C++ with its features for component software like *polymorphism* and *templates* [8] forms an ideal platform for specification-based implementations of distributed systems. For the usability of an automatic translation there are mainly two criterias which have to be fulfiled. The translated system should provide high performance which mostly depends on the implementation of the finite state machine and on the number of copy operations which may occur when signaling between different processes due to missing pointers in languages like SDL. For code maintenance and perhaps manual improvements a good readability of the translated code should also be provided.

This paper introduces a framework for the automatic translation of an SDL specification to C++ program code providing high performance together with good readability of the resulting code. For that we present an object model to map the SDL instances to appropriate C++ objects. An efficient implementation of the finite state machine with constant costs only is outlined using polymorphism. Additionally we present a mechanism to reduce copy operations called lazy evaluation and a memory pool approach to avoid costly memory allocation and deallocation for fixed-sized objects like SDL signals.

The remainder of the paper is organized as follows. Chapter 2 gives an overview of related work in the area of automatic SDL to C/C++ translation. In the third chapter the framework is presented by outlining the object and activity model of our approach as well as mechanisms to reduce copy operations and memory allocations in the translated system. In chapter 4 a performance evaluation of the mechanisms is presented before concluding the paper.

2. RELATED WORK

In this chapter two existing techniques for the mapping of SDL entities to a programming environment, the server model and the activity thread model, are introduced. Their mode of operation and main features are presented, taking into account the problems of high performance and good readability.

Another aspect related to the automatic translation from SDL to C++ is the issue of memory management. The efficiency of the generated code depends on the number of copy operations that must be performed. Two mechanisms will be introduced: the *integrated packet framing* (IPF) and a linked list of buffers (*mbufs*).

The following discussion will prove that none of the existing approaches sufficiently fulfil the requirements of an efficient automatic SDL to C++ translation with respect to the translation model or the memory management.

2.1 Server Model

In the *server model* [5] active entities of the SDL specification, like a process, a timer or a channel, are implemented by threads or processes of a programming language or operating system. For this reason the operating system usually takes over the scheduling of active threads, for example in a round robin fashion. SDL signals are implemented as passive objects, not as threads. The code is generated with the help of a runtime library, where each SDL entity is mapped to a corresponding class. A communication channel is implemented as a thread with a FIFO queue, able to send and receive signals to and from other processes. If a process receives a signal, a virtual function transition() is called. Based on the incoming event and nested if-statements or a switch-case construct, a corresponding code segment is executed. A possible output is copied to the input queue of another process or of a communication channel. The *finite state machine* (FSM) of an SDL process is therefore realized by these conditional statements of the transition() function. The readability of the generated code is supported by the straight forward translation of SDL entities to object classes, which allows also to adopt the naming of the SDL specification. The readability of the FSM implementation within the process depends on the chosen technique. Mostly if-else or switch-case constructs are used to choose the appropriate transition which results in bad readability of the translated code (see chapter 4.1).

The performance of the server model is limited due to the direct coupling between active SDL entities and corresponding threads. A large number of active SDL objects, like processes, timers and channels, leads to a large task switching overhead and poor performance. The communication relationships between the active components may be analyzed at system start-up and are made available to all classes involved in the communication process. This information is accessed via a hash table which allows an efficient exchange of signals. Nevertheless for each output a signal instance is created by the new operator, which has a negative effect on the performance. Because the structure of the communication relations may be analyzed beforehand, it is possible to extend the server model in the way that a group of SDL processes that communicate with each other is mapped to a single thread. This optimization improves the performance of the server model to a certain degree, but still does not provide convincing results.

2.2 Activity Thread Model

In the *activity thread model* [2] SDL processes are implemented as procedures with their own private data space. The data part contains the variables and state information of the process instance. The FSM of the SDL process is realized through these procedures with signals being the active part. An incoming signal triggers the corresponding procedure, which handles the event. A possible output activates the next procedure, resulting in a chain of procedure calls, the so called activity thread. This thread is not related to a thread executed in an operating system, like in the server model described above. However the synchronous communication by procedure calls violates the semantics of SDL. In order to overcome this problem, the output signal should be buffered until the transaction is completed. For this reason two approaches have been proposed: the *basic activity model* (BAT) and the *enhanced activity model* (EAT).

The BAT implements a ring buffer, where the signals are stored. An incoming signal is removed from the head of the list and a possible output signal is enqueued at the tail. As a consequence this output signal is processed only after all preceding signals are consumed. The BAT is able to handle only a single external signal, if the ring buffer is empty and contains no more waiting signals. The processing of all other external signals is deferred, until the handling of the current external signal is completed. The EAT allows the processing of several external signals in parallel by creating an operating system thread for each of the external signals.

Due to the use of the *call/return* mechanism, the performance of the BAT model is considerably better than the one of the server model. The implementation of the call/return function is much more efficient than the task switching used in the server model. The EAT falls in between the server and the BAT model. The task switching overhead, associated with the threads of the EAT model, depends on the number of external signals and thwarts the advantages of the BAT model.

The readability of the generated code is poor, since all tasks and states of an SDL process are translated to separate procedures, which are mutually invoked. To the knowledge of the authors no automatic SDL translation tool based on the activity thread model is available up to now.

2.3 Integrated Packet Framing (IPF)

A common task inside a protocol layer n is the extension of a *service data unit* (SDU) received from the upper layer n+1 by the *protocol control information* (PCI), forming a *protocol data unit* (PDU). This PDU of layer n will then become the SDU of the lower layer n-1, and so forth. Usually copy operations are performed in each layer appending the PCI to the SDU. The main idea of the *IPF approach* [9] is to allocate from the beginning a memory area large enough to accommodate all PCIs of all layers. This avoids the copying of SDUs to PDUs and allows all protocol layers to access the same memory area for changes of the data. A disadvantage of IPF is the requirement that no data manipulations must be performed on the data after an output operation, since this would violate the SDL semantics. Another drawback is the fact that IPF can not be used, if the protocol has to perform SDU fragmentation, due to a maximum message size. In this case the SDU has to be divided into several parts, where the PCI is appended to each of the parts, corrupting the idea of a continuous, linear built up of the final PDU. As many protocols provide a maximum transfer unit, fragmentation is very common in protocol design, making IPF an improper choice.

2.4 mbufs

In the implementation of the TCP/IP protocol another approach for efficient memory operations was used [10]. With this method a data instance is represented by a linked list of memory buffers (*mbufs*) of fixed size, called the mbuf chain. Each list element contains a header and a payload field. The header consists of a reference to its predecessor and successor, a length and type field as well as a pointer to the data. The data can be stored either in the payload section of the mbufs element or, if a larger area is needed, in a cluster allocated separately. The memory operations performed by the protocol, like adding a pre- or suffix or cutting out data, are realized by simple pointer manipulations inside the linked list, without any copying of data. Only fixed sized mbuf elements are handled, allowing the use of a memory pool for the efficient generation and deallocation of mbuf structures. Another advantage of the mbuf technique is the efficient transport of mbuf chains. If a data object, like a PDU, has to transferred to another process, a pointer reference to the beginning of the mbuf chain is sufficient. However, problems arise, when the semantics of SDL must be implemented with mbufs. Like with IPF, the access of data after an output operation will lead to an incorrect manipulation, since the reference to a single mbuf chain is shared by several processes. In order to comply to the SDL semantics, the sending process must provide a copy of the mbuf chain as an output. This copy operation is more costly for the linked list of an mbuf chain, than for a continuous memory area. Another drawback of the mbufs technique is the fact that the programmer has to provide functions for freeing the mbuf chain, which introduces another source for errors. Furthermore the automatic code generation process must be able to identify the proper locations for the deallocation of mbuf chains inside the protocol specification, which is not a trivial task.

3. FRAMEWORK FOR SDL TO C++ TRANSLATION

This chapter introduces the framework for an automatic translation of an SDL specification to C++. For that, we present the object model of the framework followed by the activity model describing the realization of the finite state machine. Additionally we describe a mechanism to minimize the copy operations within the system to improve the performance of the translated code and a memory pool approach to improve memory allocation of fixed size elements like SDL signals.

3.1 Object Model

The object model of our framework is adapted from the hierarchy of an SDL system. The SDL components system, block, process, state, signalroute and channel are mapped to appropriate C++ objects. A generic *Block* is the central entity of the object model which is inherited by the *System* object for the representation of an SDL system.

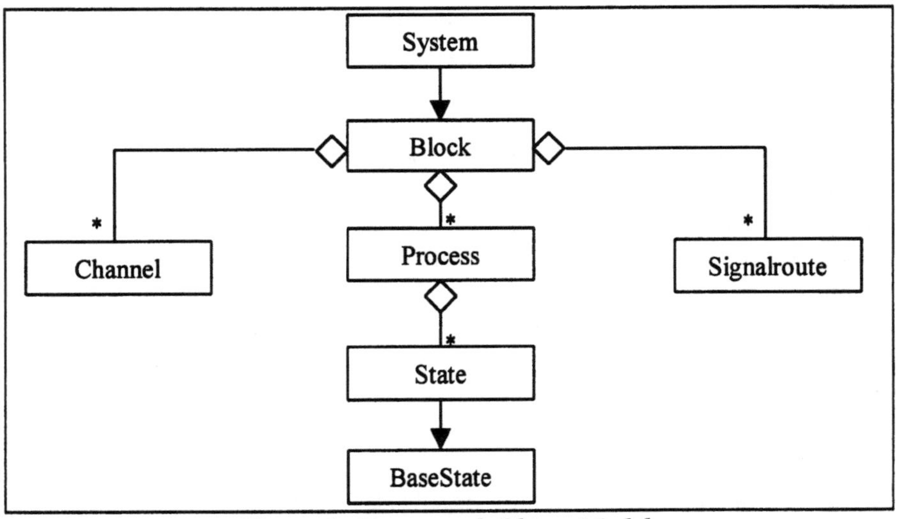

Figure 1: Framework Object Model

There may be different System objects in the code as well as several Block objects. Within each block there are several *Process* objects consisting of several states represented by *State* objects. The states extend the *BaseState* object for an efficient implementation of the finite state machine (see next section). *Channel* and *Signalroute* objects complete the object model which is shown in figure 1 given in UML notation [7]. In the remainder of this paper we will concentrate on the realization of the finite state machine involving the Block, Process, State, and BaseState object. Due to space restricitions the Channel and Signalroute objects are not within the scope of this paper.

3.2 Activity Model

The activity model of the automatic translation is very important for the performance of the translated code. The presentation of the activity model is divided in two parts. In the first one the mapping of the SDL onto the operation system entities is shown before presenting the realization of the finite state machine in each block.

3.2.1 Mapping onto the Operation System

As presented in the related work there are several approaches to map SDL components like SDL system, SDL block and SDL process to the underlying operation system.

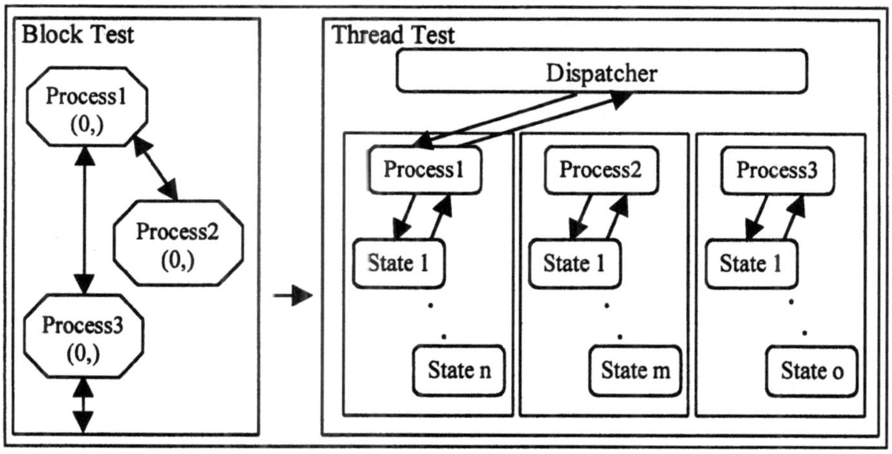

Figure 2: Framework Activity Model

In our framework the SDL system is created as a process while the SDL blocks are mapped to threads in the sense that each SDL block has its own thread. Within the thread the dispatcher of that SDL block is implemented using the call/return model. The dispatcher determines the addressed SDL process from the actual incoming signal and calls the appropriate state transition method within that SDL process. With this model the number of threads is equal to the number of SDL blocks in the system, so it is constant which avoids costly thread creation like in the Activity Thread model [2].

3.2.2 Implementation of the Finite State Machine

Crucial for the performance of the translated system is the implementation of the finite state machine. In our framework we use the call/return method, which means that the appropriate state transition function is called and the dispatcher waits for completion of the function (return). Therefore there is a mapping between states, incoming signals, and appropriate functions to handle the state transitions. Common approaches use state functions in which all possible incoming signals are handled either by switch-case or if-constructs. Both approaches are poor in performance (linear dependence on the number of signals) as well as poor in readibility of the translated code.

In our approach we define a *BaseState* class for each process with a default transition for all possible signals which may be delivered to the process. Default transition means to consume the signal without any further action. Each incoming signal has its own virtual function with the default transition. For each possible state in the system a subclass is defined overloading the appropriate methods of class *BaseState* according to the possible incoming signals in this state. To demonstrate that, we assume a process with three possible incoming signals *e1*, *e2* and *e3* and three possible states *s1*, *s2* and *s3*. Signal *e2* is possible in all states while *e1* and *e3* may be incoming in state *s3* only. So the class definition looks like

```
class BaseState {
    virtual e1() {};
    virtual e2() {};
    virtual e3() {};
};
class s1 : public BaseState {
    virtual e2() {actions};
};
class s2 : public BaseState {
    virtual e2() {actions};
};
class s3 : public BaseState {
    virtual e1() {actions};
    virtual e2() {actions};
    virtual e3() {actions};
};
```

There are two advantages of that class structure. The first one is the readability of the code, because each state has its own class definition consisting of methods for all incoming signals within this state. The second advantage is the efficiency of dispatching which is explained next.

During initialization of the dispatcher, for each process in the block, an array of method offsets is built using the offsets of the *BaseState* class of that process. These offsets are used in the dispatcher to call the appropriate method of the appropriate subclass. This is possible because the subclass consists of overloaded methods of the *BaseState* class only, so the offsets are the same. The offset array is indexed by the signal which is stored as an enumeration type. The actual state is stored as the pointer *ActualState* of type *BaseState* to the appropriate class and the method is called by jumping to the appropriate offset in the virtual table function of the actual state. If there is an illegal signal in a state, the dispatcher calls the default method of the *BaseState* class and consumes the signal (additionally some debug or exception code may be integrated in the default transition).

For our simple example the variable *OffsetArray* is filled during the process initialization with the offsets of the *BaseState* object by the instructions

```
OffsetArray[e1] = &BaseState::e1();
OffsetArray[e2] = &BaseState::e2();
OffsetArray[e3] = &BaseState::e3();
```

During runtime of the system, incoming signals are dispatched by using these offsets to call the method of the appropriate state. For instance, if there is an incoming signal *e2* and the current state is *s3*, the method is called by the instruction

```
(ActualState->*(OffsetArray[e2]))(Parameters)
```

with ActualState containing the pointer to the instance of s3 and signal a enumeration type of all signals. The dispatching is shown in figure 3. It can be seen that method calls of not implemented (illegal) signals fall back to the default transition of the *BaseState* class.

If there is a state transition in SDL, this is simply inplemented by changing the *ActualState* pointer after calling the transition method. This state transition approach is shown in figure 4. In the first step, the appropriate state method is determined and called. In the second step, at the end of the method, the *NextState* pointer is set to the new state. After returning from the state method, the dispatcher sets the *ActualState* to the *NextState* pointer and the state transition is completed.

To access process-wide variables we propose a *template-based* [6] implementation of the state objects. The parameter of the template is a pointer to the parent process to which the signal instance belongs. Using this pointer, process-wide variables like mapped SDL variables as well as pointers to the other state instances (needed for state transition) can be accessed within each state method.

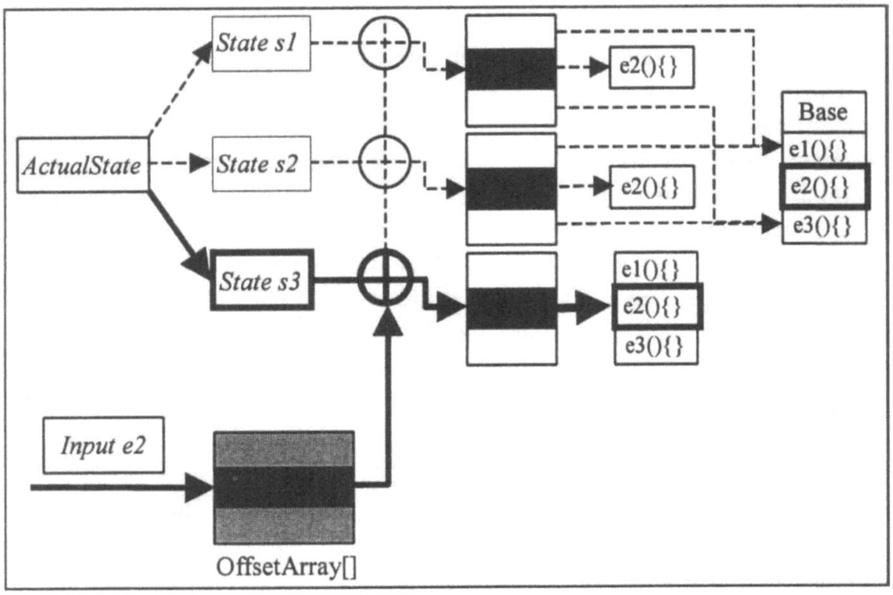

Figure 3: Dispatching Input Signal e2 in State s3

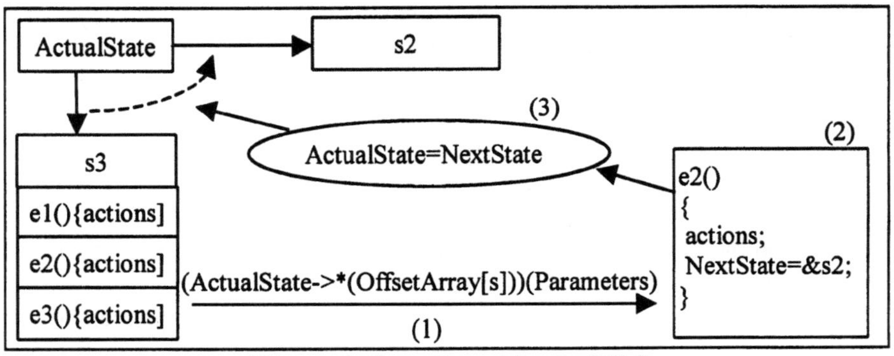

Figure 4: State Transition in FSM

With this dispatching method we are able to obtain constant instead of linear costs (see performance evaluation in chapter 4.1) for the dispatching which improves the overall performance of the state machine implementation. Additionally the readability of the translated code is improved by providing separate objects for each state with separate methods for each incoming signal for this state.

3.3 Lazy Evaluation

Beside the activity model and the finite state machine implementation of the translation, the functions for data manipulation are one of the main performance factors of the automatic translation from SDL to C++. Data manipulation in protocol specifications may occur either when sending SDL signals from one SDL process to another or by accessing the memory for operations like checksum calculation, adding prefixes/suffixes or cutting specific fields from the data. Especially the processing of protocol data units causes a high overhead during runtime when accessing and copying the memory [2].

For the optimization of those operations we define a specific abstract SDL data type, called *TransferBuffer*, which extends the mbufs principle [10] to any operation which may occur to buffers together with the *SmartPointer* approach [6] for reference counting of memory. This approach allows an optimization of copy actions within an SDL translation together with a high readability of the translated code.

The basic principle of the approach is very similar to the mbuf approach. Instead of determining the new buffer content after each operation (copy, cut or forwarding), a linked list of operations is built. The linked list is evaluated when building an output buffer (sending the buffer to the environment for instance) by calling the appropriate operations stored in the linked list. When signaling a TransferBuffer to another SDL process the pointer to the linked list is forwarded instead of copying the entire structure. Thus, operations are delayed until they are needed. This principle is also used in functional and imperative programming languages [6]. It is also possible to optimize the copy operations by deleting operations in the linked list when they are not necessary. When there are several instances of the memory (logical copies of instances) during runtime, each instance gets its own branch of operations which is unique for building the buffer at output of the data. Combined with the reference counting of the data, copying the memory for the operations (the data buffer) is avoided which is the main difference compared to the mbuf approach, where each logical copy of an instance gets its own memory copy.

As we said before, a new abstract data type is defined for this type of optimization. The abstract data type provides standard operations like assignments as well as special buffer operations like AddPrefix, AddSuffix or Cut. The translation is done by defining a C++ object called *TransferBuffer*. The object-oriented approach allows us to implement the operations of the abstract SDL data type directly as object methods either by overloading operations (like assignment) or defining appropriate methods like `AddPrefix()`. Thus the readability of the translated code is improved which can be seen in the simple example in table 1. Two layers (implemented as SDL processes) are invoking certain operations on the data buffers which they get as an input signal. In the n-th layer some protocol control information (PCI) is added as a prefix to the service data unit (SDU) and the resulting protocol data unit (PDU) is signaled to the next ((n-1)-th) layer. In this layer, a specific area of the input SDU is cutted and a suffix is added before signaling this PDU to the next layer. Beside the FSM code (see chapter 3.2.2) the operations are directly translated in the appropriate C++ code which is very similar to the SDL code. The last `Out()` method of the specific signal instance (`Instance_c` in the example) to the environment is responsible for building the output buffer by evaluating the operation tree of the signal buffer and executing the operations.

	SDL	C++
Layer(n)	dcl pdu TransferBuffer;	TransferBuffer pdu;
	dcl pci TransferBuffer;	TransferBuffer pci;
	...	...
	input a(sdu);	void State_X::Input_a(TransferBuffer sdu){
	...	...
	task pdu:=AddPrefix(sdu, pci);	pdu=AddPrefix(sdu, pci);
	...	...
	output b(pdu);	Instance_b->Out(pdu);
		}
Layer(n-1)	dcl pdu TransferBuffer;	TransferBuffer pdu;
	dcl pci TransferBuffer;	TransferBuffer pci;
	...	...
	input b(sdu);	void State_Y::Input_b(TransferBuffer sdu){
	...	...
	task pdu:=Cut(sdu, from, size);	pdu=Cut(sdu, from, size);
	task pdu:=AddSuffix(pdu,pci);	pdu=AddSuffix(pdu, pci);
	...	...
	output c(pdu);	Instance_c->Out(pdu);
		}

Table 1: Example for Lazy Evaluation

In figure 5 the resulting linked list of operations can be seen. The data of the n-th SDU is protected using the reference counting technique. The linked list consists of an operation pointer and a parameter structure for that operation. Implementing this as a union of possible parameters allows a linked list of fixed-size elements, so a *memory pool* approach (see chapter 3.4) can be used for further improvement by avoiding costly new/delete operations for memory allocation of the linked list elements.

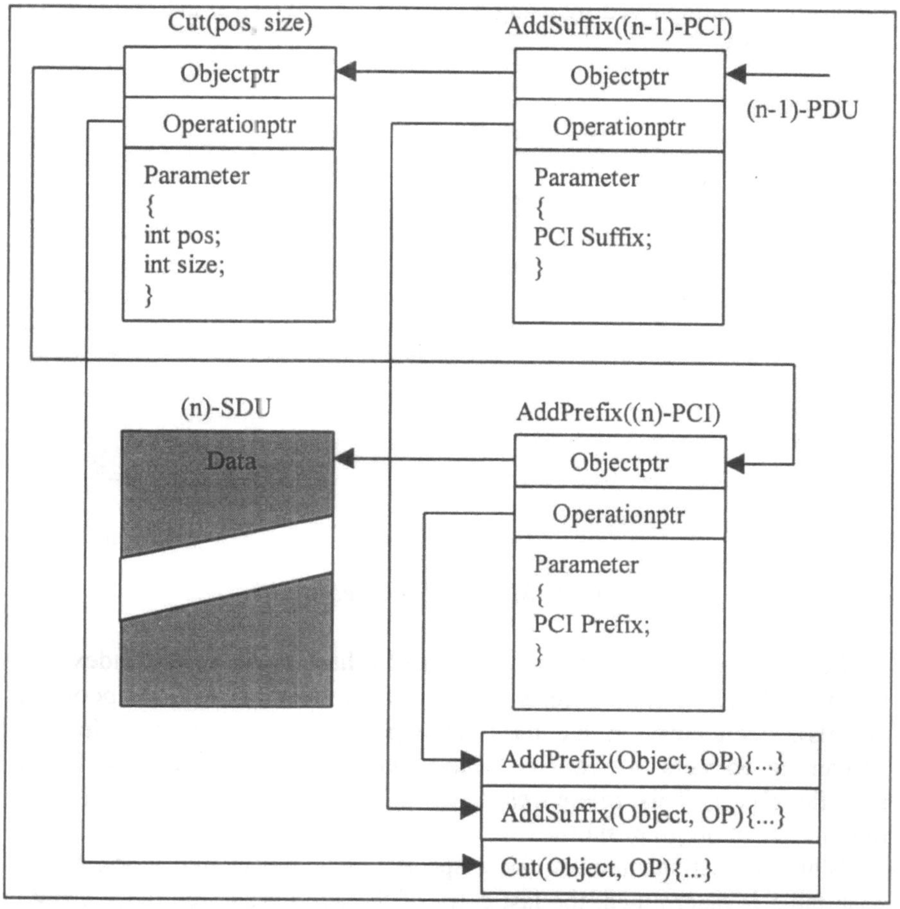

Figure 5: Example Linked List of Operations for Lazy Evaluation

The data part in figure 5 may also be implemented as a simple pointer or an mbuf structure as well instead of using the reference counting technique. But in that case, the programmer is responsible for the protection of the data when accessing several copies of it, e.g. by copying the data after each assignment.

3.4 Memory Pool

Another crucial performance issue for the translation of SDL to C++ is the memory allocation. Standard new/delete mappings lack in poor performance, because that sequence of allocation/deallocation has to be done dynamically during runtime at each creation and deletion of an SDL object like signals or data objects. Especially the creation of fixed-sized objects like signals may be improved by using a memory pool which is also proposed in our framework.

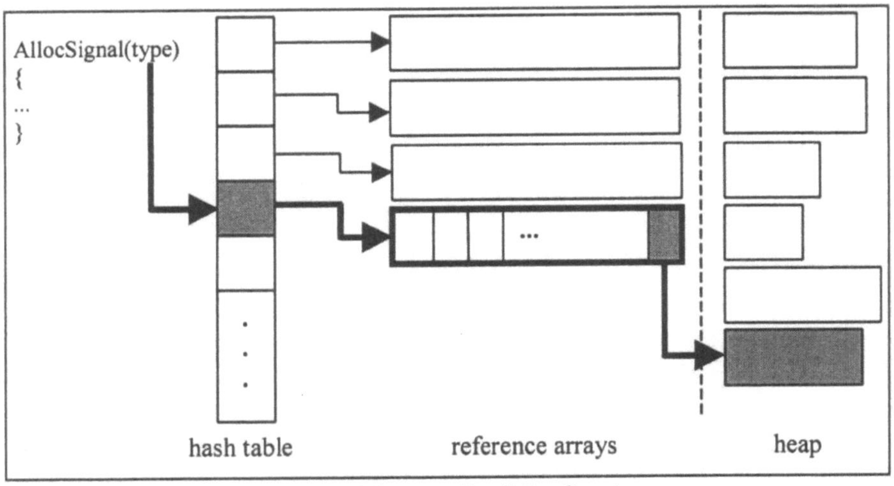

Figure 6: Memory Pool Realization

For the realization of the memory pool a hash table is built indexed by the type of the different objects we want to administrate with the pool (e.g. signals). Each entry in the hash table points to an array of references containing pointers to available heap-based memory of fixed size for the specific object. This references array is allocated according to a given default size at program startup.

When allocating memory for a specific SDL object, the index for the hash table is determined, the last entry of the appropriate reference array is returned, and the entry in the reference array is deleted. If the reference array is empty, a new object is allocated from the heap and returned. Alternatively, a certain amount of new objects may be created and added to the reference array to reduce heap allocation time.

When deallocating memory, the entry is inserted at the end of the appropriate reference array after determining the index from the hash table. The access to the reference arrays is coordinated using a simple mutex for protection.

The default size of the memory pool is very important for the performance. On the one hand, large default pools avoid the allocation of memory for resizing the pool using the heap during runtime. On the other hand, small default pools reduce the total amount of memory of the runtime environment at startup. So, there is a trade-off between performance and workset size when adjusting the default pool. Strategies like the expected or maximum pool size may be used to set the pool size at the startup of the program.

4. PERFORMANCE EVALUATION

The main focus of our framework for the automatic translation of SDL to C++ is the performance of the translated code. Nevertheless the readability of the translated code is another big issue for maintenance and manual adjustment of the resulting code. The latter issue is fulfilled by the object-oriented design approach we used for the translation. In the following we concentrate on the performance evaluation of our framework compared to related work analyzing the main parts of our approach, namely realization of the finite state machine, the lazy evaluation, and the memory pool. All results were obtained by measurements on a Pentium PC running with Windows NT4.0. In the diagrams the average values are shown.

4.1 Finite State Machine

Crucial for the performance of a finite state machine realization is the implementation of a state transition. While common approaches often use a simple if-else or switch-case model our framework uses a lookup in a virtual table indexed by the incoming signal. The goal of the performance evaluation is to look at the dependence of the state transition performance on the number of possible signals in a state. For that we compare our framework with the if-else and switch-case model.

The type of the signals received by the process, the number of states and the task to run in a transition does not influence the performance of the chosen dispatching method. Thus we use a very simple state machine for the evaluation consisting of one state only remaining in the same state after each transition. Additionally there are no tasks to be executed during a state transition and there are no parameters for the different signals. Thus the state machine for the evaluation is given by

```
STATE s1
    INPUT e1
        NEXTSTATE s1
...
    INPUT en
        NEXTSTATE s1
```

We generate a random sequence (of length 100000) of input signals for the process and run the same sequence for the if-else/switch-case model and our approach. Figure 7 shows the results for the comparision of if-else/switch-case and our approach with a varying number of possible signals.

As expected the values for the if-else method are increasing more than linearly with the number of possible signals. The values for our approach are constant due to the constant overhead for the table lookup of the appropriate signal method. Surprisingly the switch-case method also has constant costs for the state transition. Those tests were performed with an implementation compiled with the Microsoft Visual C++ compiler which implements a switch-case sequence with linear numbers (the signals are enumerated as integers) as a table lookup as well, instead of executing consecutive if-else sequences.

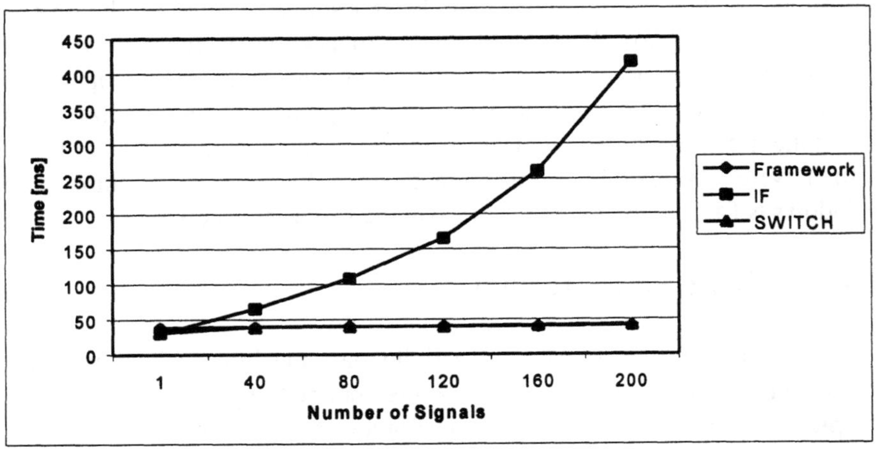

Figure 7: Performance of Finite State Machine

But this depends highly on the chosen compiler and it is not guaranteed that all compilers optimize the compiled code in that way. Using our framework leads to constant costs independent of the chosen compiler. It can also be seen from the diagram that the performance tradeoff between the if-else method and our framework is around six signals but the total performance difference (with one signal per process only) is lower than five percent. Considering the higher readability of the translated code we propose the usage of our framework independent of the number of signals.

4.2 Lazy Evaluation

The lazy evaluation approach, provided by the TransferBuffer data type together with the reference counting approach, minimizes copy operations of transfered data like SDL signals. Additionally a high readability of the translated code is provided by encapsulating the functionality in a C++ object.

In this section a comparison is presented between the lazy evaluation and the pure reference counting approach invoking operations directly instead of delaying them. The reference counting approach is used instead of a simple pointer implementation because SmartPointers (implemented with the reference counting technique) are common used in C++ for shared data [6].

Additionally the SmartPointer approach complies to the SDL semantics in contrast to the simple pointer approach. Two evaluation scenarios are performed. In the first one, the dependence on the number of processes traversed by a signal and the size of transfered data is shown while the second one illustrates a more realistic segmentation/reassembly example.

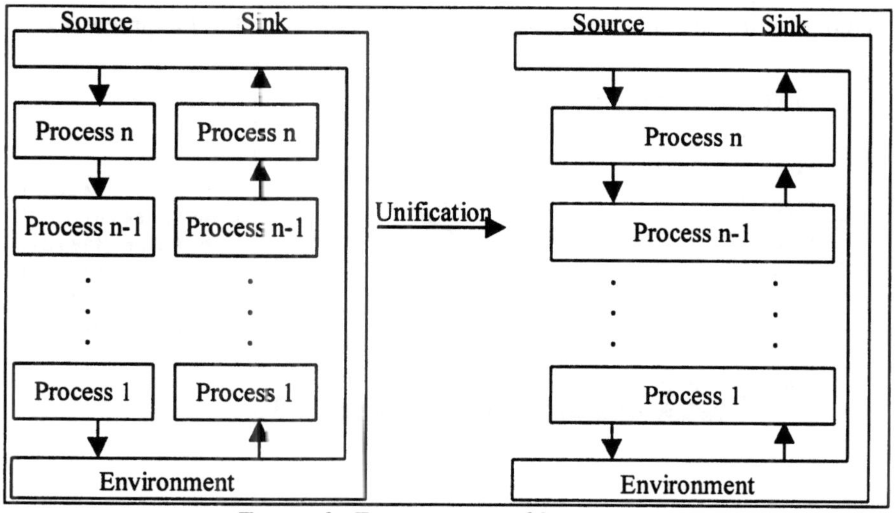

Figure 8: Environment of Scenario 1

Copy operations often occur when signaling data buffers from one SDL process to another. For the comparison of both approaches we use a layered process environment consisting of n processes which varies during the measurement. This results in building a 'vertical' chain of a certain number of processes transfering and processing signal data within the SDL specification before calculating an output buffer when sending the signal to the environment. The data is transfered from the top most process of the source to the environment that in turn sends the data to the sink which transfers the data up to the top most process again. At last the data is sent to the environment for output. For simplification, the source and sink are implemented in a unified environment as shown in figure 8. The total amount of time is measured for the transfer of 50 signals between starting the transfer for the first until the output of the last is finished at the sink. At each process of the source an `AddPrefix()` operation is invoked while the sink cuts the prefix from the data buffer using the `Cut()` operation. The amount of transfered data in each signal is set to 128 bytes, 64kBytes, and 128kBytes. The fixed-sized list elements of the lazy evaluation are allocated using the memory pool technique.

Figure 9 shows the results of the comparison. The term LE stands for lazy evaluation and denotes our approach. It can be seen that with increasing buffer size and number of layers the total amount of time for the transfer can be dramatically reduced using the lazy evaluation technique. In the 128 byte case, the maximum performance gain is about 50 percent, while the transfer of 128kByte data is improved by a factor of more than 20 when using our approach. There is a small dependence on the number of processes only when using the lazy evaluation method. That can be explained with the avoidance of copy and memory allocation operations within the processes.

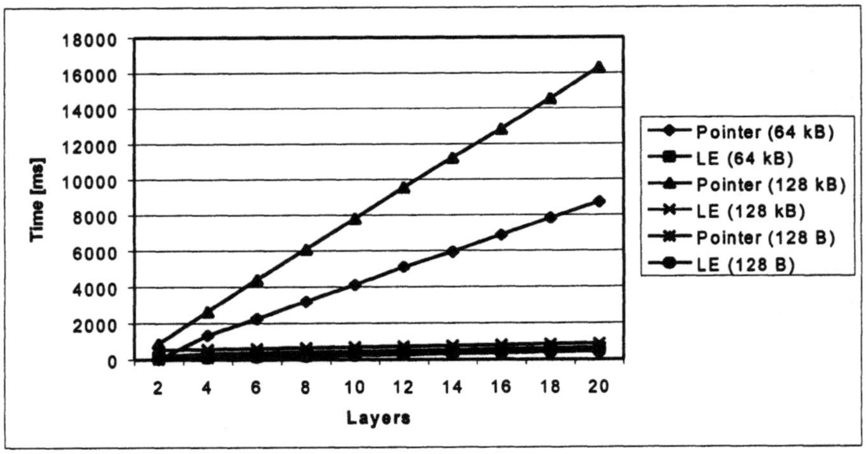

Figure 9: Performance of Scenario 1 for Lazy Evaluation

In the second evaluation scenario, a more realistic *fragmentation-reassembly* example is performed. There are four processes in the system performing operations like `AddPrefix()`, `Cut()` and `Union()` on the signal data (see figure 10). The data size is set to 64kByte and 128kByte while the segment size varies from 128 to 4096 bytes (the segments are built in the Split process and rebuilt in the Union process). The total amount of time for 30 data transfers is measured.

In figure 11 the comparison of both techniques can be seen. The performance gain is not as high as in the first scenario, because the length of the chain is much smaller (only five). The maximum improvement is about 80 percent in the 128kByte case and 45 percent for the smaller data units. With higher segment sizes the performance gain decreases due to the decreasing number of operations to be performed.

It can be seen from both scenarios that the lazy evaluation technique leads to a significant performance improvement depending on the chain length of processes until the data is sent to the environment and the size of data transfered with the signal.

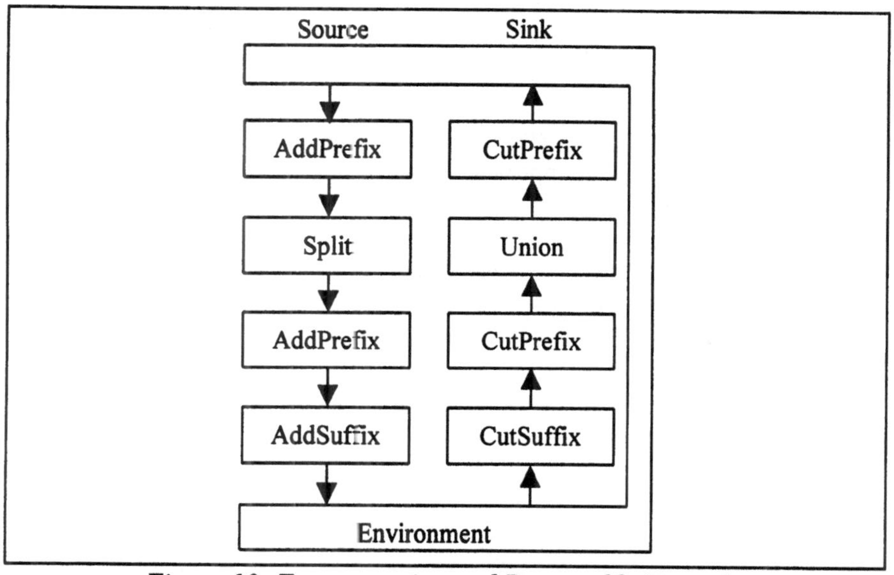

Figure 10: Fragmentation and Reassembly Example

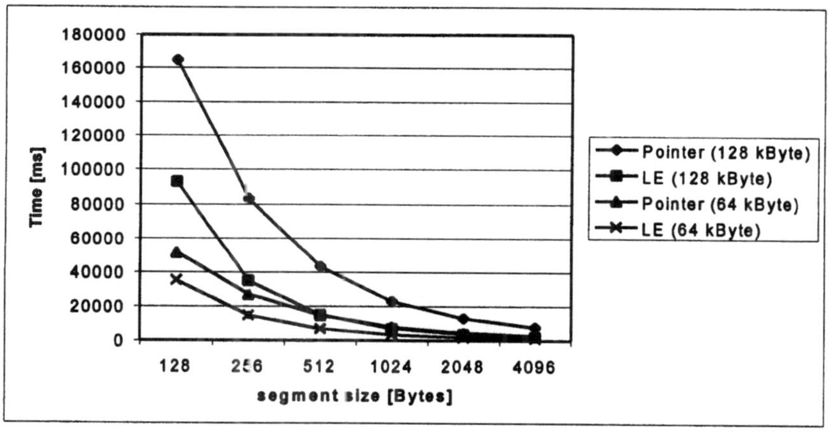

Figure 11: Performance of Scenario 2 for Lazy Evaluation

4.3 Memory Pool

Our framework proposes a memory pool for the optimization of memory allocation and deallocation for fixed-sized objects like SDL signals. Common approaches use the memory management functions of the operating or runtime system for allocation and deallocation. The performance evaluation of the memory pool procedure compares these approaches by generating a random sequence of object allocation and deallocation requests of length m. The randomly generated number of objects n_i to allocate in request i is less than the memory pool size and the total number of objects N is defined as the sum over all n_i.

The objects are allocated and directly deallocated using the appropriate method (new/delete or pool approach). The time between the start of the first object allocation and the end of the last deallocation is measured. The size of the allocated objects in terms of user data is crucial for the overall performance. A user data size of zero bytes means that there is no specific data stored in the object except of a fixed size for administration of the object. In our measurements a user data part of zero and 104 bytes is used.

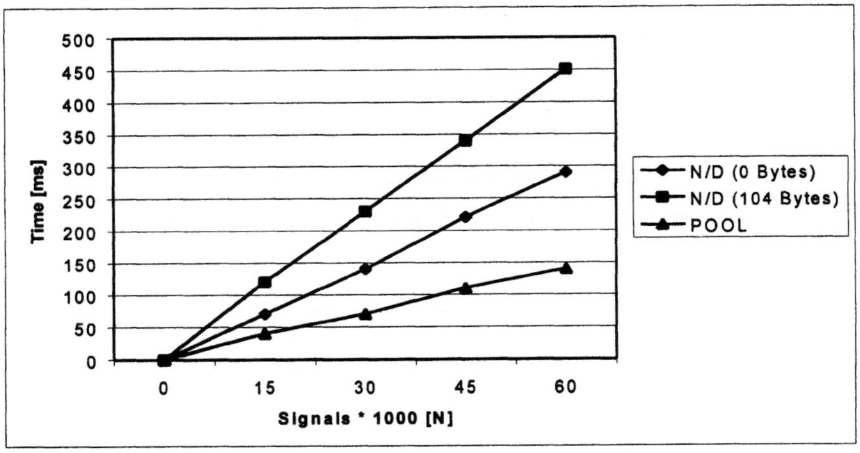

Figure 12: Performance of Memory Pool

Figure 12 shows the overall time used by the performance evaluation for the memory pool approach and the new/delete operation. It can be seen that for the zero byte case a performance gain of about 100 percent is obtained which is increased to 200 percent when allocating 104 bytes. The memory pool performance is independent of the size of the allocated object, because the hash table lookup does not depend on the size of the object.

5. CONCLUSION

Many distributed systems are specified using specification languages like SDL. For simplification of the development process the need for an automatic translation from the specification language to a programming language like C++ arises.

In this paper we presented a framework for the automatic translation of an SDL specification in C++ program code. For that, the main components were outlined and described which are crucial for the performance and readability of the translated system. For the activity model of the framework we proposed a mapping of SDL blocks to operation system threads dispatching the incoming signals in each SDL process of the block. The dispatcher in each block was realized by using a hierarchy of subclasses derived from a default base class accessing the appropriate state transition method by a simple hash table. We also presented a mechanism to minimize buffer operations by delaying the evaluation of the operations until the buffer is sent to the output of the system. The framework was completed by a pool approach for fixed-sized objects to avoid costly memory allocation and deallocation operations. Our performance evaluation of the framework showed that the proposed mechanisms perform very well compared to other approaches.

It can be summarized that our framework for automatic SDL to C++ translation provides mechanisms with high performance and good readability of the translated code using the features of an object-oriented language like polymorphism.

In our future work we will concentrate on the comparison to other object-oriented languages like Java. Additionally we are going to implement the framework for the automatic translation from SDL to C++.

6. REFERENCES

[1] F. Belina, D. Hogrefe, A. Sarma: SDL with Applications from Specification, Prentice Hall, 1991

[2] R. Henke, P. Langendörfer, A. Mitschele-Thiel: Effiziente Implementierung formal spezifizierter Protokollarchitekturen (in german), Otto-von-Guericke-Universität, Fakultät Informatik, Technical Report IRB005/96, 1996

[3] ITU: Recommendation X.292 - OSI conformance testing methodology and framework for protocol recommendations for CCITT applications - The Tree and Tabular Combined Notation (TTCN), 1992

[4] ITU: Recommendation Z.100 - Specification and Description Language (SDL), March 1993

[5] N. Kikuchi,Y. Shigeta, K. Miyake, W. Tanaka, M. Nabeta: An Integrated System Development Method and Support System based on SDL and C++, SDL'89: The Language at Work, Elsevier Science Publishers, p.135-143, 1989

[6] S. Meyers: More Effective C++, 35 New Ways to Improve your Programs and Design, Addison-Wesley, 1997

[7] T. Quatrani: Visual Modeling with Rational Rose and UML, Object Technology Series, Addison-Wesley, 1997

[8] B. Straustrup: The C++ Programming Language, Addision-Wesley, 1998

[9] L. Svoboda: Implementing OSI Systems, IEEE Journal on Selected Areas in Communications 7(1989), p.1115-1130, 1989

[10] G. R. Wright, W. R. Stevens: TCP/IP Illustrated Volume2, The Implementation, Addison-Wesley, 1995

PART THREE

Test II

Stress Testing of Distributed Multimedia Software Systems

Jian Zhang[1], Shing-Chi Cheung[2] and Samuel T. Chanson[2]

[1]*Lab. of Computer Sci., Inst. of Software, Chinese Academy of Sciences*
[2]*Dept. of Computer Sci., Hong Kong Univ. of Sci. and Tech.*

Key words: Multimedia applications, stress testing, test case generation, real-time systems, resource modeling, symbolic execution, constraint solving

Abstract:

With the advance in network technologies, the use of distributed multimedia data has become very popular. Distributed multimedia systems are complex and error prone. While there are well established testing techniques for the functionalities of such systems, the technique of stress testing to improve test efficiency has not received much attention. It has been argued that faults are more likely to occur in situations where there is a fierce competition for resources. When the system is heavily loaded, errors like mismanagement of buffers can be easily identified. In this paper, we present several criteria for selecting test cases, and describe two methods for generating test cases which maximize system resource usage. Our approach is based on symbolic execution and constraint solving. An example is given to illustrate the concepts and techniques.

1. INTRODUCTION

Multimedia software systems systems are becoming popular in recent years. In multimedia applications, various kinds of information media are involved, including video, audio, image and text. The architecture of interactive multimedia applications can be roughly divided into two classes: client-server (like remote learning, video-on-demand and product presentations), or peer-to-peer (like video phone and video conferencing).

Distributed multimedia software systems (DMSSs) are complicated and error-prone. They have to be responsive to user events, and they often require timely exchange of large amounts of media data. Furthermore, some of the media objects have to be synchronized (e.g., a piece of text and a speaker's voice reading the text). There can also be very complex user interaction scenarios, as in the case of virtual shopping.

To facilitate the design and implementation of DMSSs, a Distributed Multimedia System Environment (DMSE) was developed at the Hong Kong University of Science and Technology [W97]. In this environment, the programmer can specify the architecture of a DMSS in an object-oriented, textual language. He/she may also describe different views of the system using some graphical notations (including a variant of Petri net and an extended time-line diagram).

A framework for testing DMSS is described in [MCC98], where only generic testing methods are discussed. It is impractical and often infeasible to test a complex software system exhaustively. A system is more likely to misbehave when its resources are heavily utilized. In this paper, we focus on the derivation of test data which drive the system to critical states where almost all available resources are consumed. Experience tells us that in such a situation the system is more vulnerable and some types of errors (like mismanagement of buffers) can be easily identified. Testing is more effective if we concentrate on these cases.

In this paper, we give some criteria for selecting test cases such that the system is tested under more stressful conditions. We also describe two methods for generating such test cases. The basic idea is to execute the system's specification symbolically, and then use constraint solving and optimization techniques to find the timings of events which maximize system resource usage. The paper is organized as follows. In the next section, we briefly describe our specification formalism, with emphasis on a variant of Petri net. Then in Section 3, we discuss test case selection criteria for stress testing of multimedia applications, and present two methods for automatic test case generation. Finally, our work is compared with other related work, and some observations are made.

2. MODELING OF DMSS

A model for specifying DMSS architectures is given in [W97]. It allows users to author a DMSS by describing different aspects of the system. A major part of the model is an extension of the Petri net, called Temporal Event Petri Nets (TEPNs). An event can denote some action of the user (e.g., EXIT_CLICKED) or a signal from some media object (e.g., EOF which means the end of an object). A TEPN offers a visual representation of system activities and user interactions.

TEPN adopts a discrete time model. That is, the time domain consists of time points which are non-negative integers. We shall use $nt(pl,t)$ to denote the number of tokens in place pl at time t. Place pl is active at time t if $nt(pl,t) > 0$. A transition is enabled at time t if every input place is active at that time. When it fires, the number of tokens at each input place decreases by 1 and the number of tokens at each output place increases by 1. But if transition v forms a self-loop (i.e., goes from place P to itself), we assume that when v fires, the number of tokens in P first decreases by 1 and then increases by 1 to get back to the same value.

A TEPN is an extension of a classical Petri net. A transition may be marked with a triggering condition (i.e., the occurrence of some user interaction). It is firable if it is enabled and the condition is met. For simplicity, we assume that no two user interactions occur at the same time instant. Otherwise, the behavior of the system can be nondeterministic. A transition can also have a timeout value t_{max}. Then the transition fires if it has been enabled for t_{max} time units, or if the triggering condition is met before that time.

A place in a TEPN may be associated with a presentation (e.g., displaying a picture and then some text), which takes a finite amount of time. In such a case, we shall use the same name for a place and its presentation. Each presentation may consist of several media objects (like a paragraph of text, an image and an audio stream). The designer can specify the temporal relationships between the objects using an extended time-line diagram or using some temporal logic formulas (e.g. *A* `ends_with` *B*). A presentation is executed when the corresponding place changes from inactive to active; and it is stopped when the place becomes inactive, even though the presentation is not over. A transition can be associated with some actions (like the migration of objects). The associated actions are executed when the transition fires.

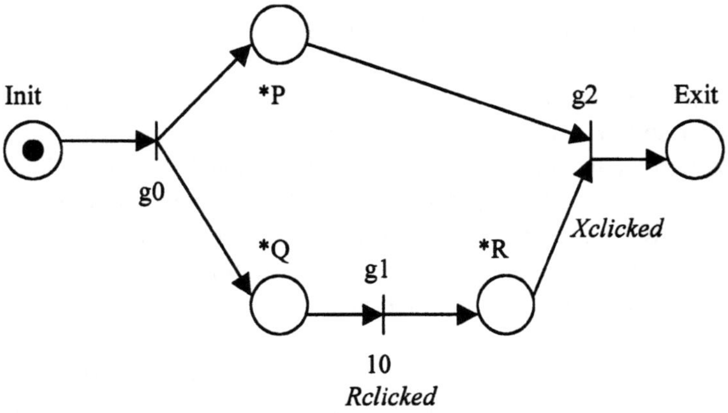

Figure 1. Example TEPN

Throughout this paper, we shall use the simple example in Figure 1 to illustrate the concepts and techniques.

The TEPN has 5 places: *Init*, *P*, *Q*, *R*, *Exit*. Among them, *P*, *Q* and *R* are associated with presentations (indicated by a star beside the name). Suppose there is a media object *A* in *P*, another object *B* in *Q*, two objects *C* and *D* in *R*, and the durations of *A*, *B* and *D* are 5, 12, and 8, respectively. In addition, there is a temporal constraint: *C* during *D*. This means that *C* can only be active when *D* is. The duration of *C* is undetermined. Transition g_1 has a timeout value of 10.

2.1 A Discrete Time Model of TEPN Executions

The TEPN and temporal specifications provide a basis for specifying, analyzing and testing distributed multimedia applications. For timed systems, many analysis problems are undecidable [BD91], [GMMP91]. The reason is that unbounded time values can make the state space infinite. But for testing purposes, we may consider the system's behavior within a finite period of time.

Given a specification, an *execution trace* is defined to be a set of timed events, i.e., $\{ (ev_i, t_i) \}$. Here each t_i is a non-negative integer denoting a time point, and ev_i denotes an event in the general sense. An event is defined to be the beginning or ending of a presentation, the activation or ending of a media object, or a user interaction. For a presentation P, we use $P\uparrow$ and $P\downarrow$ to denote its beginning and ending events, respectively, and similarly for media objects. At any time t ($t > 0$), event $P\uparrow$ happens if $nt(P,t-1) = 0$ and $nt(P,t) > 0$; and event $P\downarrow$ happens if $nt(P,t-1) > 0$ and $nt(P,t) = 0$.

In an execution trace, more than one events can happen at the same time point, and an event can happen many times. An execution trace should be well-defined. For instance, there should be an event $P\downarrow$ between two consecutive $P\uparrow$ events.

For the TEPN in Figure 1, a possible execution trace is:

$$\{ (P\uparrow, 1), (Q\uparrow, 1), (A\uparrow, 1), (B\uparrow, 1), (A\downarrow, 6), (Q\downarrow, 11), (B\downarrow, 11), (R\uparrow, 11), (D\uparrow, 11), (C\uparrow, 13), (Xclicked, 15), (P\downarrow, 15), (R\downarrow, 15), (C\downarrow, 15), (D\downarrow, 15) \}.$$

The special places *Init* and *Exit* are omitted for the sake of brevity. In this trace, g_0 fires at time 1, which activates presentations P and Q and their media objects A and B. A ends at time 6 because its duration is 5, and so on. At time 15, the user clicks on X, and g_2 fires. This terminates the presentation R (as well as its media objects C and D). Place P also becomes inactive.

3. STRESS TESTING

Specification-based testing of real-time systems and multimedia applications has been studied by some researchers (see for example, [MMM95], [AS96]). These approaches are based on syntactic features of the specification, such as the structures of formulas.

In what follows, we shall study stress testing [B84], i.e., exercising the system under strenuous conditions. The basic idea is that the system is more susceptible to errors or performance degradation when much resource is being used. For a DMSS, the three most critical resources are CPU cycles, memory and network bandwidth. Heavy communication may cause buffer overflow and packet loss. Errors may also result from insufficient CPU power or programming mistakes. From our experience, mismanagement of buffers is one of the common types of errors in developing multimedia systems. Horn and Girod also reported that when time consuming applications are running simultaneously, buffer overflow and packet loss may happen due to limited CPU capacity [HG97].

3.1 Resource Consumption

The amount of resources consumed by an active media object depends on several factors, like the supporting software/hardware, and the required quality of service (QoS). Based on our observations, we assume that for a particular media type and some fixed QoS, the amount of resources consumed is constant during the object's lifecycle, except for the following two cases:

(1) Immediately after an object is activated, much more resources are needed. This transient period may last for less than a second to a few seconds.

(2) For some media types, random peaks may occur from time to time. For example, when a singer raises her voice in a recital. Such peaks may occur at any time instant. If they occur frequently, we consider the maximum amount of resource usage; otherwise, we consider the average amount.

3.2 Test Case Selection Criteria

As mentioned earlier, we adopt the discrete time model in our analysis. For the purpose of testing, it is reasonable to restrict the time domain to be finite. Let us denote by *TT* the total testing time. Like an execution trace, a test case is also a (well-defined) set of timed events.

Given a set of execution traces, a trace will be selected as a test case if one of the following conditions is met:

(1) its total resource usage during the period [0..*TT*] is maximum;
(2) it has the largest number of object beginning events;
(3) there is a large peak resource usage at some time point t ($0 < t \leq TT$).

The second condition is necessary since there is usually a significant increase in the amount of CPU usage when a media object starts or when some user interaction occurs. However the exact amount is unpredictable.

Generally speaking, it is not necessary or practical to produce test data that are strictly optimal in the above sense. Nevertheless, appropriate techniques and tools can help us find good test cases.

We shall be mainly interested in the amount of data required by the active media objects. Usually, the more data a client needs, the more resources the system consumes. Our analyses will be based on the data rates of the media objects.

For an object with duration *d*, we assume that the amount of required data is *w(m,q)* * *d*, where *w* is the average or maximum amount of data needed per unit time. The parameters *m* and *q* denote the media type and the quality of services, respectively. In the following, we simply write *wA* for the value of the function in case of object *A*. For example, if *A* is an audio stream of telephone quality, then *wA* is about 56 Kbps. If the data are compressed before communication, and we are interested in maximizing the amount of data transfer, then we should divide *wA* by the compression ratio. For example, instead of 56 Kbps, we use 16 Kbps.

3.3 Maximizing Total Resource Usage

Obviously, if there are many time points, the number of possible execution traces can be very large. A better way is to use *symbolic execution traces*. Such a trace is also a set of timed events $\{(ev_i, t_i)\}$. But the t_i's can be either integer variables or specific integers. To avoid confusion, we shall call an execution trace a *concrete execution trace* if every t_i is a constant.

As mentioned earlier, a place can be active more than once during an execution. In a given execution trace, we use bP_i and eP_i to denote the *i*'th beginning and ending times of place *P*, respectively. Moreover, we define $fP_i = \min(eP_i, TT)$, where *TT* is the total testing time. (The letter *i* can be omitted if each place is active only once during the execution.) In the example in Section 2, we have, $bQ = 1$, $eQ = 11$ and so on. If $TT = 10$, then $fQ = 10$.

Given a specification, we can derive a set of finite symbolic execution traces. This is done by choosing sequences of transition firings. For each sequence, we record its associated conditions, and then determine the values of the variables such that some resource usage criterion is satisfied.

Let us illustrate the ideas using the previous example.

For the TEPN in Figure 1, we can see that there is only one firing sequence, namely, g_0 fires first, then g_1 and then g_2. We assume that the timings of the firings are as follows:

$$1 = bP = bQ; \quad 1 + x = eQ = bR; \quad 1 + x + y = eP = eR.$$

where $1 \leq x \leq 10$ and $1 \leq y$.

For a media object O and its j'th activation, let bO_j and eO_j denote the *scheduled* beginning and ending times, respectively. Again, the subscript j can be omitted if the object is active only once. In the present example, we can derive the following equations and inequalities from the temporal specification of the presentations:

$$eA = bA + 5, \;\; ..., \;\; bC > bD, \;\; eC < eD.$$

Note that the object may end earlier than the scheduled time if the place becomes inactive (due to some user events) before that time. It is also possible that some objects are not activated at all. The actual running time of object A, denoted by rA, can be defined as:

$$rA = \begin{cases} 0, & \text{if } fP \le bA; \\ fP\text{-}bA, & \text{if } bA < fP \le eA; \\ eA\text{-}bA, & \text{if } fP > eA. \end{cases}$$

The variables rB, rC and rD can be defined in a similar way. Let wA (wB, wC, wD) denote the amount of data needed for object A (B, C, D) per unit time. Our goal is to maximize ($wA * rA + wB * rB + wC * rC + wD * rD$) under the temporal constraints. This represents maximal resource utilization and contention.

Given the above constraints and objective function, we can find solutions using constraint solving and optimization techniques. Suppose $TT = 15$, $wA = 100$, $wB = 1400$, $wC = 1800$, $wD = 64$. The optimal solution is: $x = 7$, $y = 7$. That is, object B runs for 7 time units, then the event *Rclicked* happens, and then place R is active for 7 time units. If we change the value of wC to 800, then the solution becomes: $x = 10$, $y = 4$. In this case, object B runs for 10 time units when transition g_1 times out. The rationale is to select the timing of events in a way which makes communication intensive objects run as long as possible, so as to maximize the total amount of data usage while satisfying the timing constraints of the application.

If we adopt the second criterion in Section 3.2, we need only change the objective function to the number of all object beginning events (before TT). This can be obtained by summing up the following quantities (over all activations of each object A):

$$nA_i = 1, \text{ if } bA_i < fP_i; \;\; 0, \text{ otherwise.}$$

Here P is the presentation which contains A.

We can see that even though the TEPN's control structure is very simple, several different test sequences are possible, depending on the resource usage information and on the total testing time. For more complicated systems, it is hard to determine the best (or even near optimal) test data without automated tool support. We have implemented a tool for generating the constraints. The tool NCL [Zh98] is used to obtain the solutions.

3.4 Maximizing Peak Usage

A multimedia software system can be considered to be in a critical state if several media objects are active simultaneously and they require a large amount of data. To find a time point at which the resource utilization is maximized, we first analyze the object instances to see which of them can be active at the same time. A pair of instances A_i and B_j can not be active simultaneously, if $eB_j \leq bA_i$ or $eA_i \leq bB_j$. Obviously, for any pair of integers i, j ($i \neq j$), it is impossible that A_i and A_j are active simultaneously.

The above information can be depicted as a graph, which we call the *object instance relation graph*. Each vertex represents an active instance of some media object. A vertex is associated with a weight, which represents the amount of resource consumption per unit time. An edge connects two vertices if the two object instances can be active at the same time. What needs to be solved is an extended maximal clique problem. (Some additional constraints have to be satisfied.) That is, find a subset of the vertices such that every pair of vertices are connected and the sum of the weights is maximum.

To represent the clique, we associate a set of binary variables with the vertices. For example, $xA_i = 1$ means that A_i is included in the clique. We study the satisfiability of the following constraints:

(1) $(xA_i = 0)$ or $(xB_j = 0)$, if there is no edge between vertex A_i and B_j;
(2) $(xA_i = 1) \rightarrow (bA_i < t < eA_i)$;
(3) $(xA_i = 1) \rightarrow (t < eP_i)$;
(4) $t < TT$;
(5) the constraints characterizing the symbolic execution trace.

Here P is the presentation which contains A. The objective function to be maximized is the summation of $(xA_i * wA)$ over all object instances A_i. If the problem has a solution, we obtain a concrete execution trace which uses maximal amount of data at time point t.

Let us consider again the TEPN in Figure 1. Suppose that presentation *P* consists of 3 objects *A*, *E* and *F*, with the constraints *A* `begins_with` *F*, and *E* `ends_with` *F*. The durations of the objects and resource consumption weights are given in Table 1. We note that the objects *A* and *E* can not be active simultaneously, neither can the presentations *Q* and *R*. So we get the object instance relation graph as shown in Figure 2.

Table 1. Durations and Weights of Objects in the Example

	A	*B*	*C*	*D*	*E*	*F*
duration	5	12	6	8	6	13
weight	100	1400	1800	64	150	56

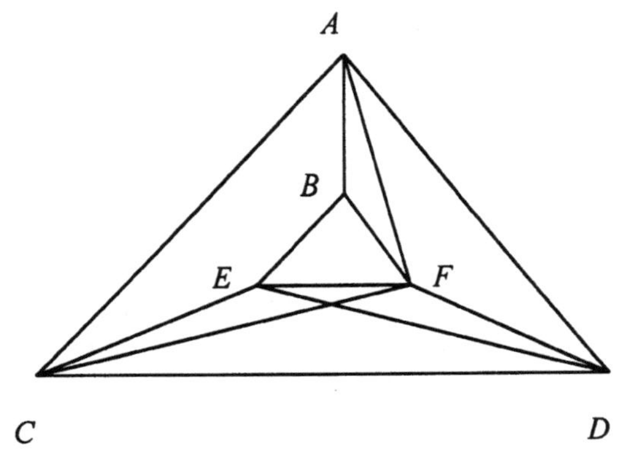

Figure 2. Object Instance Relation Graph of the Example

This optimization problem has a solution, which can be found by NCL in less than 1 second on a SPARCstation 5. The maximal clique consists of the media objects *C*, *D*, *E* and *F*. Their beginning/ending times are: $bC = 3$, $eC = 9$, $bD = 2$, $eD = 10$, $bE = 8$, $eE = 14$, $bF = 1$, $eF = 14$. At time point $t = 8$, the total amount of data usage is maximized at 2070.

Summary Given a specification and resource consumption information about each media object, we can prepare test data for stress testing in one of the following ways:

(1) Arbitrarily generate many concrete execution traces. Then pick from them the best ones using the test selection criteria given in Section 3.2.
(2) Generate some symbolic execution traces, and then use constraint satisfaction and optimization techniques to determine concrete traces with the largest amount of data usage.

The total testing time *TT* can be varied as needed.

4. RELATED WORK

Specification and Testing of DMSS

Several authors have extended Petri nets to specify distributed multimedia applications. Most notably, Little and Ghafoor proposed Object Composition Petri Nets (OCPN) [LG90]. It was extended later into XOCPN [WQG94]. Sénac *et al.* proposed the so-called Hierarchical Time Stream Petri Nets (HTSPN) [SSW95].

Considerable amount of work exists on conformance testing of protocols and distributed systems (see for example [S93]). Most of the test data generation methods are based on graph models of protocols. Testing of multimedia systems is more complicated because the exact timings of events are very important.

Some researchers have attempted to adapt protocol testing methods to multimedia system testing. Ates and Sarikaya [AS96] discussed an approach based on timed automata. A finite state machine (FSM) is derived from the automata, and then timed traces are obtained from the FSM. Each timed trace serves as a test case, which is a sequence of transitions from the initial state to some final state. A complex system may have a large number of timed traces. In our approach, only the most significant test cases are considered. These test cases drive the system to states where potential bugs can be revealed with high probability. It is possible to add some criteria to the approach of [AS96], for example, each transition is visited only once. But this can lead to inaccuracy. Sometimes a transition fires more than once before the system enters a critical state.

Modeling and Analysis of Real-time Systems

Real-time systems have been extensively studied. Various formalisms have been proposed to model and analyze them. They include, among others, time Petri nets [BD91], timed automata [AD94], I/O automata[LSGL95], timed CSP [ABSS96], Duration Calculus [Zcc93], Modechart [JM94].

Roughly speaking, there are three classes of techniques for studying the behaviors of real-time systems based on formal specifications:

(1) Simulation. The specification is executed. Typical executions may reveal certain properties of the system. But in most cases, there are many possible executions.

(2) Searching. This includes reachability analysis of timed Petri nets [BD91], [GMP94] and temporal logic model checking [LPW96], [YMW97]. Search-based tools are attractive in that they are automatic. But usually they have difficulty with infinite or very large state spaces. While model checking procedures are quite effective for untimed systems, verifying properties of timed systems is much more difficult in general.
(3) Theorem proving. The user develops a theory about the system based on some logic, like first-order logic or temporal logic. System properties are expressed as formulas and their truth is to be proved in the logic. Constructing nontrivial proofs tends to be arduous, although some steps can be mechanized as done in provers like the Larch Prover [LSGL95] and PVS [SS94], [AH96].

While reachability analysis and symbolic execution are well-known techniques for time Petri nets, they have not been used with resource consumption information.

The Petri net is a natural formalism for describing asynchronous systems. In contrast, many model checking tools are intended for synchronous systems. For example, the UPPAAL tool [LPW96] is designed to verify networks of timed automata. In such a model, the basic communication mechanism is through send/receive primitives which synchronize *two* automata. Yang *et al.* [YMW97] introduce a logic called SREL for specifying synchronous systems, and described an extension of McMillan's SMV tool to verify Modecharts. These approaches still have some limitations. It is hard to specify a requirement such as the following: The meeting begins when all the 3 participants are ready.

Test Data Generation and Constraint Solving

In general, automatic test data generation is an undecidable problem. However, in certain cases, constraint solving methods can be used effectively in protocol testing [CA90], [CZ93]. A test case derivation algorithm was given in [HB94] for a subset of LOTOS called P-LOTOS. The only data types in this subset are integers and Booleans, and the operations on integers are restricted to addition, subtraction and comparison. None of them considered timing or resource consumption.

Mandrioli *et al.* [MMM95] described how to generate functional test cases for real time systems given their temporal logic specifications. The test generation criteria are based on the structure of formulas in the specification. They used an interactive tool. In contrast, constraint solving tools are automatic and highly efficient. In this paper, we are mainly interested in the resources consumed by the media objects. We believe that for distributed multimedia systems, it is more effective to focus on those system states where large amounts of resources are being used.

Resource Modeling and Load Testing

The traditional method of representing resource consumption in Petri nets is by way of the number of tokens [M89], [YSS94]. When a transition fires, n ($n \geq 1$) tokens may be deleted from or added to the related places, where the value of n depends on how much resource is needed. This is not adequate for multimedia applications since a presentation (which is associated with a place in the net) consists of several types of media objects, and different objects may consume different amounts of resources.

Avritzer and Weyuker [AW94] discussed load testing of software based on some operational profile. They used Markov chains, and their purpose is to examine the *most probable* system states more thoroughly. Yang and Pollock [YP96] described an algorithm for analyzing sequential programs to identify load sensitive parts. But no method was given to find input data to exercise these parts.

5. CONCLUDING REMARKS

In this paper, stress testing of distributed multimedia systems is studied. A multimedia application is specified by a temporal event Petri net (TEPN) together with some time-line diagrams (or the corresponding temporal formulas). We define execution traces of a specification assuming a discrete time model. From such a trace one can derive a test case of the application. When the media objects' resource consumption data are available, we may use some criteria to select the test cases which drive the application to stressful states.

The focus of the paper is on automatic generation of test cases for stress testing. Essentially our approach is to generate some symbolic execution traces first, and then use constraint satisfaction and optimization techniques to determine the timings of events. We give two methods for generating the constraints, which correspond to different test case selection criteria. One appealing feature of our approach is that the method is easy to automate. Another is that the generated test cases are executable. Preliminary experimental results are quite encouraging. Using a state-of-the-art constraint solver, we are able to determine the timings of user interactions such that the test criteria are satisfied. On the examples we tried, the test case generation times are reasonable.

Of course, some assumptions are needed in our work, and our model is only an approximation of the real world. For instance, in the model, it is assumed that the firing of every transition takes the same amount of time (i.e., 1 time unit). In reality the transition times may be different from each other, depending on such factors as the distance between a client and a server. Moreover, we assume that the amount of data transfer is constant for a given media object. In practice the bit rates of audio and video objects may vary for different compression techniques. Nevertheless, the results of our analysis can serve as a guideline for test engineers. In most cases, audio and video streams consume much more resources than other media objects.

ACKNOWLEDGEMENTS

This work is partially supported by RGC grant HKUST6088/97E. The paper was written during the first author's visit to the Hong Kong University of Science and Technology, sponsored by the Croucher Foundation. He is also supported in part by the NSF of China.

REFERENCES

[AD94] R. Alur and D.L. Dill. A theory of timed automata. *Theoretical Computer Science* 126(2): 183-235, 1994.

[AH96] M. Archer and C. Heitmeyer. Mechanical verification of timed automata: a case study. *Proc. IEEE Real-time Technology and Applications Symp.*, 192-203, 1996.

[ABSS96] A.F. Ates *et al.* Using timed CSP for specification, verification and simulation of multimedia synchronization. *IEEE J. on Selected Areas in Comm.* 14(1): 126-137, 1996.

[AS96] A.F. Ates and B. Sarikaya. Test sequence generation and timed testing. *Computer Networks and ISDN Systems* 29(1): 107-131, 1996.

[AW94] A. Avritzer and E.J. Weyuker. Generating test suites for software load testing. *Proc. Int'l Symp. on Software Testing and Analysis (ISSTA)*, 44-57, 1994.

[B84] B. Beizer. *Software System Testing and Quality Assurance*, Van Nostrand Reinhold, New York, 1984.

[BD91] B. Berthomieu and M. Diaz. Modeling and verification of time dependent systems using Time Petri Nets. *IEEE Trans. on Softw. Eng.* 17(3): 259-273, 1991.

[CZ93] S.T. Chanson and J. Zhu. A unified approach to protocol test sequence generation. *Proc. IEEE INFOCOM'93*, 106-114, 1993.

[CA90] W. Chun and P.D. Amer. Test case generation for protocols specified in Estelle. *Formal Description Techniques III (FORTE'90)*, 191-206.

[GMMP91] C. Ghezzi *et al.* A unified high-level Petri net formalism for time-critical systems. *IEEE Trans. on Softw. Eng.* 17(2): 160-172, 1991.

[GMP94] C. Ghezzi, S. Morasca and M. Pezze. Validating timing requirements for time basic net specifications. *J. of Systems and Software* 27(2): 97-117, 1994.

[HB94] T. Higashino and G.v. Bochmann. Automatic analysis and test case derivation for a restricted class of LOTOS expressions with data parameters. *IEEE Trans. on Softw. Eng.* 20(1): 29-42, 1994.

[HG97] U. Horn and B. Girod. Scalable video transmission for the Internet. *Computer Networks and ISDN Systems* 29, 1833-1842, 1997.

[JM94] F. Jahanian and A.K. Mok. Modechart: a specification language for real-time systems. *IEEE Trans. on Software Engineering*, 20(12): 933-947, 1994.

[LPW96] K.G. Larsen, P. Pettersson and Wang Yi. Diagnostic model-checking for real-time systems. *Hybrid Systems III*, LNCS Vol. 1066, Springer, Berlin, 575-586, 1996.

[LG90] T.D.C. Little and A. Ghafoor. Synchronization and storage models for multimedia objects. *IEEE J. on Selected Areas in Comm.* 8(3): 413-427, 1990.

[LSGL95] V. Luchangco *et al.* Verifying timing properties of concurrent algorithms. *Formal Description Techniques VII (FORTE'94)*, 259-273.

[MMM95] D. Mandrioli *et al.* Generating test cases for real-time systems from logic specifications. *ACM Trans. on Computer Systems*, 13(4): 365-398, 1995.

[MCC98] V.B. Misic, S.T. Chanson and S.-C. Cheung. Towards a framework for testing distributed multimedia software systems. *Proc. Int'l Symp. on Software Engineering for Parallel and Distributed Systems*, Kyoto, Japan, 72-81, 1998.

[M89] T. Murata. Petri nets: Properties, analysis and applications. *Proceedings of the IEEE*, 77(4): 541-580, 1989.

[S93] B. Sarikaya. *Principles of Protocol Engineering and Conformance Testing.* Ellis Horwood, New York, NY, USA, 1993.

[SSW95] P. Sénac, P. de Saqui-Sannes and R. Willrich. Hierarchical time stream Petri net: a model for hypermedia systems. *Proc. 16th Int'l Conf. on Application and Theory of Petri Nets*, 451-470, 1995.

[SS94] J. Skakkebaek and N. Shankar. Towards a duration calculus proof assistant in PVS. *Proc. Int'l Symp. on Formal Tech. in Real-Time and Fault-Tolerant Syst.*, 660-679, 1994.

[W97] K.-K. Wong. *Distributed Multimedia Authoring Using Application Structures.* Master Thesis, Hong Kong Univ. of Sci. and Tech. 1997.

[WQG94] M. Woo, N.U. Qazi and A. Ghafoor. A synchronization framework for commu-nication of pre-orchestrated multimedia information. *IEEE Network* 8(1): 52-61, 1994.

[YP96] C.-S.D. Yang and L.L. Pollock. Towards a structural load testing tool. *Proc. Int'l Symp. on Software Testing and Analysis (ISSTA)*, 201-208, 1996.

[YMW97] J. Yang, A.K. Mok and F. Wang. Symbolic model checking for event-driven real-time systems. *ACM Trans. on Prog. Lang. and Syst.* 19(2): 386-412, 1997.

[YSS94] J.C. Yee *et al.* Resource synchronization specification and modeling based on timed stream Petri net. *Proc. of the IASTED/ISMM Int'l Conf. on Distributed Multimedia Systems and Applications*, 171-175, 1994.

[Zcc93] Zhou Chaochen. Duration calculi: an overview. *Proc. Formal Methods in Programming and Their Applications*, 256-266, 1993.

[Zh98] J. Zhou. *The user manual of NCL 1.0.* Loria and Inria-Lorraine, France, 1998.

FAIR TESTING THROUGH PROBABILISTIC TESTING

Manuel Núñez and David Rupérez
Dept. de Sistemas Informáticos y Programación
Universidad Complutense de Madrid, Spain
{manuelnu,ruperezd}@eucmax.sim.ucm.es.

Abstract In this paper we define a probabilistic testing semantics which can be used to alternatively characterize *fair* testing. The key idea is to define a probabilistic semantics in such a way that two non-probabilistic processes are fair equivalent iff any *probabilistic version* of both processes are equivalent in our probabilistic testing semantics. In order to get this result we define a simple probabilistic *must* semantics by saying that a probabilistic process *must* pass a test iff the probability with which the process passes the test equals 1. Finally, we present an algorithm for deciding whether the probability with which a finite-state process passes a finite-state test equals 1. Alternatively, this algorithm can be used for computing whether a finite-state process *fairly* passes a finite-state test.

Keywords: Testing semantics, fair testing, probabilistic processes.

1. INTRODUCTION

Formal models of concurrency have been proved to be very useful to properly specify concurrent and distributed systems. In order to verify the properties held by these specifications some kind of semantics must be given to these syntactic processes. Several semantic models have been proposed in the literature, for example, failures model, bisimulations, testing semantics, axiomatic semantics, etc.

In our opinion, among the previously cited semantic models, the closest to *reality* is testing. Testing semantics was defined and thoroughly studied in [de Nicola and Hennessy, 1984, Hennessy, 1988], and this will be the approach considered here. In their model tests are just processes but they may contain a special action indicating the *successful* termination of the testing procedure. Then, two testing interpretations are defined: a process *may* pass a test if the composition of process and test has a successful computation; a process *must* pass a test if every computation of the composition is successful. From these interpretations three testing semantics can be defined: a process is greater in the may (resp. must) sense than another one if every test that the latter process may (resp. must) pass is also may (resp. must) passed by the former. Besides, a may–must interpretation can be defined, but given the fact that for non-divergent processes must equivalence and may–must equivalence coincide, this interpretation does not add too much information. While may semantics identifies too much processes, because it takes into account only the possible traces of processes, must semantics distinguishes enough processes to be considered a useful relation: it is stronger than may semantics, weaker than weak bisimulation, and equivalent to failures semantics (actually, failures semantics and bisimulation have been characterized in terms of tests [Milner, 1981, Brookes et al., 1984]). Although must semantics is very suitable for giving meaning to processes, further research has shown some weakness in the original definition. For example, there are problems dealing with *conformance* and *fairness* (see [Tretmans, 1996, Francez, 1986] for good overviews on the subjects).

Intuitively, under a conformance assumption, a process is considered *better* than another (or implements another) if the former can execute more sequences of actions than the latter, while presenting less nondeterminism. For example, using a CCS notation, $a + b$ should be better than a, since the former process can execute both a and b deterministically, while the latter cannot execute b. But this does not hold in the must testing semantics: these processes are not related at all. The extension of [Hennessy, 1988] to deal with conformance has been treated in [de Frutos-Escrig et al., 1997, de Frutos-Escrig et al., 1998].

The lack of *fairness* in the *classical testing theory* is the main purpose of this paper. A usual justification for fairness is that *unfair* behavior is unlikely. We will illustrate this problem by showing two simple examples. Consider the processes P_0 and P_1 depicted in Figure 1.1. In the must testing semantics these processes are not equivalent: P_0 may execute an infinite sequence of internal actions τ, so it is considered to be divergent (i.e. somehow *catastrophic*). On the contrary, we can observe that the action a is offered infinitely often (actually, it is offered continuously), so it is reasonable to think that this action should be finally executed, or in other words, that P_0 *should* pass T_0 (this insight was presented in [Koomen, 1985] and treated formally in [Baeten et al., 1987]). Thus, P_0 and P_1 should be considered equivalent under a *fairness* assumption.

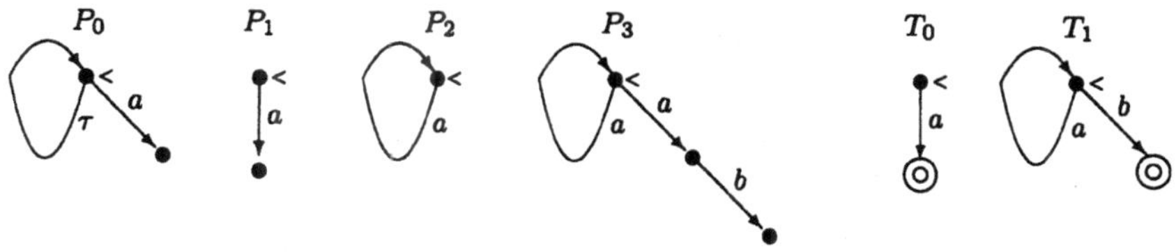

Figure 1.1 Examples of processes and tests.

Consider now the processes P_2 and P_3. Even though P_3 could execute the sequence $a\,;b$, this sequence might not be executed. But considering a fair interpretation, P_3 presents the capability of executing the sequence $a\,;b$, and so, if this sequence is infinitely often offered, P_3 should be able to execute it, while P_2 is not. For instance, P_3 *should* pass T_1 while P_2 should not.

In the previous decade, there were several proposals for including different kinds of fairness into process algebras (e.g. [Hennessy, 1987, Bergstra et al., 1987, Costa and Stirling, 1987, Brinksma, 1988]). Almost simultaneously, two groups of authors [Natarajan and Cleaveland, 1995, Brinksma et al., 1995] have come up with what is called *fair testing.* They proposed two equivalent testing semantics with the property of abstracting from certain divergences, like weak bisimulation, in contrast to the classical must testing. The idea is to modify the classical definition in order to consider as successful those computations such that, under a fairness assumption, success can always be reached. Both papers present alternative characterizations of the new fair testing semantics. In a later work [Brinksma et al., 1996], the framework described in [Brinksma et al., 1995] is extended with a set of sound axioms and more examples showing the usefulness of this fair testing semantics.

In this paper we probabilistically characterize fair testing by showing the connection between probabilities and fairness. This point has been motivated by Koomen's Fair Abstraction Rule (KFAR) [Koomen, 1985], that has been intuitively explained with probabilistic arguments in the sense of "No matter how small the probability of success, if you try often enough you will eventually succeed". Taking into account that fair testing is fully abstract with respect to this rule, it must be possible to justify fair testing in probabilistic terms. This is exactly what we will do in the present paper. As the probabilistic intuitions motivating KFAR have never been formalized before, this paper consolidates these intuitions.

Let us consider the previous examples. When composing the process P_3 and the test T_1 we would like that the process must pass the test because the sequence $a;b$ is available an infinite number of times. Let us add a probability for choosing between both actions a in the process P_3. So, let π be the probability of executing the left hand side a transition (leading to the initial state), and let $1-\pi$ the probability associated with the other a transition (leading to a state where only b is available). It is clear that the probability of executing the trace $\langle a,b\rangle$, by the composition of process and test, equals $1-\pi$; the probability of executing $\langle a,a,b\rangle$ equals $\pi\cdot(1-\pi)$, etcetera; in general, the probability of executing the trace $\langle a^n,b\rangle$ equals $\pi^{n-1}\cdot(1-\pi)$. These traces lead to a successful state of the test, and the sum of these probabilities is given by:

$$\sum_{i=1}^{\infty}(\pi^{i-1}\cdot(1-\pi)) = (1-\pi)\cdot\sum_{i=1}^{\infty}\pi^{i-1} = (1-\pi)\cdot\frac{1}{1-\pi} = 1$$

This example illustrates our technique. We consider non-probabilistic processes and non-probabilistic tests. Then, we assign arbitrary (randomly) probabilities within every choice of the process and the test, and if the probability with which the process passes the test equals 1, then we say that the test is passed, otherwise we say that it is not passed. This final answer is returned for the composition of the (non-probabilistic) process and test. If we consider finite-state processes and tests, as in the previous example, the probability of passing the test can be computed by solving a linear equations system, which is built from the states of the interaction system. Even though there could be more efficient algorithms for checking fair testing (by directly checking the induced graph) this algorithm can be used in more general probabilistic testing semantics (e.g. [Christoff, 1990, Cleaveland et al., 1992, Núñez and de Frutos, 1995]).

It could be thought that we add too much complexity to afford the definition of this semantics but we claim that this is not the case. The definitions of processes, tests, interaction systems, and probabilistic testing are standard. Moreover, tests could be described without any probabilistic information because in our setting probabilities do not add any discriminatory power to tests. Nevertheless, in order to keep compatibility with the classical framework (both in the probabilistic and non-probabilistic cases) we consider that tests are *like* processes (i.e. probabilistic). Note that in general probabilistic frameworks, probabilities add distinguishing power to tests (see, for example, [Christoff, 1990, Núñez and de Frutos, 1995] where different probabilistic semantics are defined according to different sets of tests).

There are also previous papers relating probabilistic and classical testing. For instance, [Cleaveland et al., 1992] compares in some settings satisfaction with probability 1 to ordinary must testing. In [Núñez and de Frutos, 1995], failing a test with probability 0 is related to the classical must testing. In [Kumar et al., 1998] some notions of non-probabilistic and probabilistic testing are related. Regarding fair testing, our proposal extends the last one given the fact that we deal with infinite and divergent processes and our characterization of fair testing works in both ways, that is, from our probabilistic must testing semantics to fair testing and back.

The rest of the paper is structured as follows. In Section 2 we present the probabilistic model: probabilistic processes and tests, interaction systems and the definition of the probabilistic must testing semantics. In this semantics we consider that a process *must* pass a test if the test is passed with a probability 1. In Section 3 we probabilistically characterize fair testing. We compare fair satisfaction to satisfaction with probability 1. We recover the *fair* testing passing of tests (and so the fair testing preorder) from our probabilistic must semantics. In Section 4, we give an algorithm which effectively computes the probability with which a finite-state process passes a finite-state test (or equivalently, it decides whether a process *fairly* passes a test). Finally, in Section 5 we present our conclusions and some lines for future work.

The results of this paper are an extension of those appearing in the second author's Master Thesis [Rupérez, 1997], where only finite-state processes were considered.

2. THE PROBABILISTIC MODEL

In this section, we present the probabilistic model. Probabilistic processes will be modeled by using *probabilistic labeled transitions systems*, which are an extension of usual labeled transition systems by labeling the arcs by an action and a probability. Later, we will define the notion of *probabilistic test*. Then, we will give a notion of a process passing a test with a certain probability, as well as the definition of the *probabilistic must testing semantics*.

We suppose that we are working with a fixed and finite set of actions Act, and we assume the existence of a special action $\tau \notin \mathrm{Act}$, which represents an internal action. We denote by Act_τ the set $\mathrm{Act} \cup \{\tau\}$; a will denote a generic action in Act, while e will denote a generic action in Act_τ.

Definition 1 A *probabilistic process* P is a triple (S, μ, s_0) where S is a (possibly infinite) *set of states*, ranged over $s_0, s_1, \ldots$; $s_0 \in S$ is the *initial state*, and $\mu : S \times \mathrm{Act}_\tau \times S \longrightarrow [0,1]$ is the *transition probability distribution function*, which satisfies that for all $s \in S$, $\sum_{e,s'} \mu(s,e,s')$ equals 1, or 0 (blocked state). □

We will use the following conventions:

- $s \xrightarrow{e}_p s'$ stands for $\mu(s,e,s') = p$.
- $s \xrightarrow{e} s'$ stands for $\mu(s,e,s') > 0$.
- $s \longrightarrow$ if $\exists e \in \mathrm{Act}_\tau, s' \in S: \ s \xrightarrow{e} s'$.
- $s \not\longrightarrow$ if $\nexists e \in \mathrm{Act}_\tau, s' \in S: \ s \xrightarrow{e} s'$.
- $\mu(s,e) = \sum_{s'} \mu(s,e,s')$.

In the following definition, we present some process algebraic operators. These operators will be used in the last section of this paper. Specifically, a prefix operator $a\,;P$, a *CCS*-like probabilistic choice operator $P +_\pi Q$ and a stop operator.

Definition 2 Let $P = (S, \mu, s_0)$ be a probabilistic process and $e_0 \in \mathrm{Act}_\tau$. We define the *prefixing* of P by e_0, denoted by $e_0\,;P$, as the probabilistic process $P' = (S \cup \{s_0'\}, \mu', s_0')$ where $s_0' \notin S$, and the function μ' is defined as

$$\mu'(s,e,s') = \begin{cases} \mu(s,e,s') & \text{if } s,s' \in S \\ 1 & \text{if } s = s_0' \ \wedge \ s' = s_0 \ \wedge \ e = e_0 \\ 0 & \text{otherwise} \end{cases}$$

Given n probabilistic processes $P_i = (S_i, \mu_i, s_i)$ such that if $i \neq j$ then $S_i \cap S_j = \emptyset$, and $p_1, p_2, \ldots, p_n > 0$ such that $\sum p_i = 1$, we define the *probabilistic choice* of P_i with associated probabilities p_i, denoted by $\sum_{i=1}^{n} [p_i] P_i$, as the process $P = (\bigcup S_i \cup \{s_0\}, \mu, s_0)$ where $s_0 \notin \bigcup S_i$, and the function μ is defined as:

$$\mu(s,e,s') = \begin{cases} \mu_i(s,e,s') & \text{if } s,s' \in S_i \\ \frac{p_i \cdot \mu_i(s_i,e,s')}{R} & \text{if } s = s_0 \ \wedge \ s' \in S_i \ \wedge \ R \neq 0 \\ 0 & \text{otherwise} \end{cases}$$

where $R = \sum_{i=1}^{n} p_i \cdot \delta_i$, δ_i being equal to 1 if $s_i \longrightarrow$, and 0 otherwise.

We consider stop as the probabilistic process without transitions, that is, the probabilistic process $(\{s\}, \mu, s)$ such that $\mu(s,e,s) = 0$ for any $e \in \mathrm{Act}_\tau$. □

In the definition of $\sum [p_i] P_i$ we use the factor R in order to avoid *sub-stochastic* processes in the case that some of the P_i's are deadlocked (i.e. equivalent to stop). Also note that we keep the transitions from the initial states of the original processes for keeping the possible *loops* of the former processes.

A probabilistic test is just a probabilistic process (*plts*) extended with a set of successful states. As in [Cleaveland et al., 1992, Yuen et al., 1994] we slightly differ from the standard formulation of tests given in [Hennessy, 1988] because successful states, instead of a distinguished acceptance event, are used to denote the (successful) passing of tests. Anyway, these successful states can be easily simulated by adding a new action *ok* to the set of actions.

Definition 3 A *probabilistic test* is a 4-tuple (S, μ, s_0, Suc) such that (S, μ, s_0) is a probabilistic process and $Suc \subseteq S$ is the set of *successful states.*

We consider *ok* as the probabilistic test without transitions, and with a unique successful state, that is, the probabilistic test $(\{t\}, \mu, t, \{t\})$ such that $\mu(t, e, t) = 0$ for any $e \in \text{Act}_\tau$. □

As we commented in the introduction of this paper, in order to define our probabilistic must semantics the probabilities appearing in the tests are completely irrelevant, so we could be working with non-probabilistic tests. We have preferred to keep probabilities in tests in order to follow the classical testing theory where processes and tests are (almost) the same. In the rest of this section, we usually omit the adjective "probabilistic" when referring to probabilistic processes (or probabilistic tests) and no confusion arises.

Now we will follow the process used, for example, in [Cleaveland et al., 1996]. First, we define the total probability of the successful computations of a test. Afterwards, we define the composition of a process and a test, resulting into a test. Then, in order to compute the probability with which a process passes a test we just study the corresponding composition as a test.

Definition 4 Let $T = (S, \mu, s_0, Suc)$ be a test, and let $s_1 \in S$. We call *computations from* s_1 the (possibly infinite) sequences of transitions:

$$C = s_1 \xrightarrow{e_1} s_2 \xrightarrow{e_2} \cdots s_n \xrightarrow{e_n} \cdots$$

We say that the computation C is *successful* if there exists n such that $s_n \in Suc$ and such that $\forall\, 1 \leq i < n: \; s_i \notin Suc$. We will denote by $\tilde{C}_{s_1}$ the *set* of successful computations from s_1.

We say that the state $s_i \in S$ is *non-successful* in C, denoted by $ns(s_i, C)$, if there does not exist any successful computation extending the partial computation $C_1 = s_1 \xrightarrow{e_1} s_2 \xrightarrow{e_2} \cdots s_i$. In this case, we say that C_1 is a *Non-Successful Partial* computation, in short NSP, if we have that $ns(s_i, C)$ holds, $ns(s_j, C)$ does not hold for any $j < i$, and $s_1, s_2, \ldots, s_i \notin Suc$.

Successful and NSP computations are defined as the *maximal* computations. For finite computations we say that $length(C) = n - 1$. Given a finite computation C, we define the *probability of executing* C as:

$$Pr(C) = \prod_{i=1}^{length(C)} \mu(s_i, e_i, s_{i+1})$$

Finally, we define the probability of *reaching* a successful state from s_1, denoted by $Pr(T, s_1)$, as $\sum_{C \in \tilde{C}_{s_1}} Pr(C)$. □

Note that what we call *maximal* computations are always finite. Let us remark that finite computations have positive probability, while infinite computations may have zero probability. We will see later that the composition of processes and tests will produce tests such that no further action can be executed from a successful state. Thus, the condition imposing that no previous state in the computation belongs to *Suc* in the definition of successful computations would be redundant.

Next we define the way a process and a test interact by using *interaction systems*. Intuitively, an interaction system will have a transition labeled by a visible action from a state iff the corresponding states of process and test have a transition labeled by this visible action.

Definition 5 Let $P = (S, \mu_P, s_0)$ be a process and $T = (S', \mu_T, s'_0, Suc)$ be a test. The *interaction system* for P and T, denoted $P \mid T$, is a test $I = (S \times S', \mu_I, (s_0, s'_0), S \times Suc)$. In order to define μ_I, we need a *normalization function* ν_I. This function computes the sum of the probabilities associated with the actions that the interaction system can execute, and is formally defined as:

$$\nu_I(s, s') = \sum_{a \in \mathrm{Act}} (\mu_P(s, a) \cdot \mu_T(s', a)) \;+ \mu_P(s, \tau) \;+\; \mu_T(s', \tau)$$

The function μ_I is defined as:

$$\mu_I((s_1, s'_1), e, (s_2, s'_2)) = \begin{cases} \frac{\mu_P(s_1, e, s_2)}{\nu_I(s_1, s'_1)} & \text{if } s'_1 = s'_2 \wedge e = \tau \\ & \quad \wedge\, s'_1 \notin Suc \\ \frac{\mu_T(s'_1, e, s'_2)}{\nu_I(s_1, s'_1)} & \text{if } s_1 = s_2 \wedge e = \tau \\ & \quad \wedge\, s'_1 \notin Suc \\ \frac{\mu_P(s_1, e, s_2) \cdot \mu_T(s'_1, e, s'_2)}{\nu_I(s_1, s'_1)} & \text{if } e \neq \tau \wedge s'_1 \notin Suc \\ 0 & \text{otherwise} \end{cases}$$

In the first three cases of the definition of μ_I we assume $\nu_I(s_1, s'_1) > 0$ (otherwise the result is 0). We say that P *passes* T with a probability p

if $Pr(I, (s_0, s'_0)) = p$. that is, if the probability of reaching a successful state from the initial state (s_0, s'_0) equals p. □

A normalization function ν_I is used in most probabilistic models having a notion of parallel composition (i.e. a parallel operator, or the parallel composition of a process and a test; e.g. [Christoff, 1990, Cleaveland et al., 1992, Núñez and de Frutos, 1995, D'Argenio et al., 1999, Núñez, 1999]). Given that we do not allow the synchronization in τ, our normalization factor is similar to that in [Núñez and de Frutos, 1995]. Let us comment that our definition of interaction systems is slightly different from that in classical testing. Following [Natarajan and Cleaveland, 1995] we avoid useless computations after the interaction system has reached a successful state (that is why the clause $s_1 \notin Suc$ appears in the definition of μ_I). Thus, once the test reaches a successful state, no more transitions are possible Note that this decision does not change the rest of the semantic treatment. Finally, we define the desired relation between processes.

Definition 6 Let $P = (S, \mu_P, s_0)$ be a process and $T = (S', \mu_T, s'_0, Suc)$ be a test. We say P must T if $Pr(I, (s_0, s'_0)) = 1$, where $I = P \mid T$. Moreover, given two processes P, P' we write $P \sqsubseteq_{\text{must}} P'$ if for any test T we have P must T implies P' must T. □

Next we give two results about the computations of interaction systems. They will be used in order to prove that the previously defined testing semantics characterizes, in a way that we will specify later, fair testing. Whereas the results hold for finite-state systems, they are not correct for general infinite-state systems. In the latter case we have to add two *regularity* conditions which will ensure that the considered infinite-state systems do not have *strange* behaviors.

Definition 7 Let P be a process, T be a test, and consider $P \mid T = (S, \mu, s_0, Suc)$. We say the composition of P and T is *regular* if the two following conditions hold:

- There exists a non-zero probability π_m such that for any $s, s' \in S$, $e \in$ Act we have $\mu(s, e, s') > 0$ implies $\mu(s, e, s') > \pi_m$.
- Let us consider the digraph induced by S and μ where probabilities are discarded. There exists a value $M > 0$, such that for any $s \in S$, if there exists a path from s to a successful state then there exists $s' \in Suc$ such that a path from s to s' has length less than M.

□

From now on we will assume that the considered compositions are always regular. Let us note that finite-state systems are trivially regular.

Also note that the first condition implies that systems are finite branching (any state has, at most, $\frac{1}{\pi_m}$ transitions). Finally, let us remark that regularity conditions could have been defined on probabilistic transitions systems, instead of using our formulation on interaction systems.

The first condition indicates that the probabilities appearing in interaction systems have a lower bound, or equivalently that probabilities less than one have an upper bound. This condition is necessary in order to avoid *strange* successions of probabilities. By strange we mean an infinite succession of probabilities such that all of its members are different from one, but the product of the components is different from zero. For example, in the succession $a_n = (\frac{1}{2})^{\frac{1}{n!}}$ we have $\prod_{n=0}^{\infty} a_n = (\frac{1}{2})^e$. This is formalized in the following lemma which states that if an infinite computation is executed with a probability greater than zero then there exists a point in the computation such that from this point on all the probabilities associated with the transitions are equal to 1, and so, a prefix of the computation is already a NSP computation. The proof is not difficult and follows from the existence of the previously commented lower bound.

Lemma 1 Let $P = (S, \mu_P, s_0)$ be a process, $T = (S', \mu_T, s'_0)$ be a test, and $I = P \mid T$ be the corresponding interaction system. If I is regular and there exists an infinite computation of I

$$C = (s_0, s'_0) \xrightarrow{e_1} (s_1, s'_1) \xrightarrow{e_2} (s_2, s'_2) \cdots \xrightarrow{e_n} (s_n, s'_n) \cdots \qquad [Pr(C) > 0]$$

then there exists n_0 such that $\mu((s_{n-1}, s'_{n-1}), e_n, (s_n, s'_n)) = 1$ for all $n > n_0$.

The second regularity condition is necessary in order to ensure that we cannot get a set of infinite computations having positive measure, and such that they do not have NSP prefixes. Note that we need this condition because the first regularity condition ensures that there will not be what we called strange computations (previous lemma) but dealing with infinite systems we can define a set of computations behaving like one of this *undesirable* computations that we want to avoid. In particular, this extra condition allows us to formulate the following result. The proof is highly technical and it is based on the idea that from prefixes of non-NSP computations we can always find a successful state whose associated probability is not arbitrarily small.

Lemma 2 Let P be a process and T be a test T such that $P|T$ is regular. We have that the sum of the probabilities of the *maximal* computations (successful and NSP computations) of $P \mid T$ is equal to 1.

3. COMPARING FAIR AND PROBABILISTIC TESTING

In this section we probabilistically characterize the fair testing semantics given in [Natarajan and Cleaveland, 1995] by comparing fair satisfaction to satisfaction with probability 1. First, we formally define the concept *made from*, that is, we obtain a probabilistic process by putting arbitrary nonzero probabilities on the transitions of a non-probabilistic process. Then, we show that the specific probabilities associated with a process do not change its probabilistic must semantics. We will assume again that all the interaction systems appearing in this section are regular.

Definition 8 A *non-probabilistic process* P is a triple $(S, \rightarrow, s_0)$ where S is a *set of states*, s_0 is the *initial state*, and $\rightarrow \subseteq S \times \text{Act}_\tau \times S$ is the *transition relation*. We say that the probabilistic process $P_\pi = (S', \mu, s'_0)$ is *made from* P if there exists a bijection $f : S \longrightarrow S'$ s.t. $f(s_0) = s'_0$ and $\forall s_1, s_2 \in S, e \in \text{Act}_\tau : (s_1, e, s_2) \in \rightarrow$ iff $\mu(f(s_1), e, f(s_2)) > 0$. □

Let us note that P_π is a probabilistic process, so for any $s' \in S'$ we have $\sum_{e,s''} \mu(s', e, s'')$ equals 1, or 0 if no transitions are possible. The key of the proof of the following results relays on the equivalence: $P_1 \mid T$ has a NSP-computation iff $P_2 \mid T$ has one.

Proposition 1 Let P_1, P_2 be two probabilistic processes made from a non-probabilistic process P and let T be a probabilistic test. We have P_1 must T iff P_2 must T.

Proposition 2 Let P be a probabilistic process and T_1, T_2 be two probabilistic tests made from a non-probabilistic test T. We have P must T_1 iff P must T_2.

These results allow us to define a testing semantics for non-probabilistic processes, just by assigning whatever positive probabilities to the transitions of the labeled transition system (with the restriction that the sum of these probabilities equals one for any state having transitions) and getting its probabilistic must semantics, which by Propositions 1 and 2 does not depend on the assigned probabilities. That is, in order to decide if a process passes a test (both non-probabilistic), we assign positive probabilities to process and test, and check whether the associated probabilistic process passes the test. This reasoning leads us to the next definition.

Definition 9 Let P, T be a process and a test, both non-probabilistic. Let P_{π_1} be a probabilistic process made from P, and let T_{π_2} be a probabilistic test made from T. We write P must* T iff P_{π_1} must T_{π_2}. □

Now we will relate our testing equivalence with *fair testing* as described in [Natarajan and Cleaveland, 1995].

Definition 10 Let P, T be a process and a test, both non-probabilistic, and let us consider $I = P \mid T$. We say that a state of I is *potentially successful* if there exists a path leading to a successful state. A computation is *potentially successful* if all the states along the computation are potentially successful. A test is passed by a process, denoted by $P \text{ must}_{NC}\ T$, if all the computations of $P \mid T$ are potentially successful. Given two processes P, P' we write $P \sqsubseteq_{\text{must}_{NC}} P'$ if for all the tests we have $P \text{ must}_{NC}\ T$ implies $P' \text{ must}_{NC}\ T$ □

Now, we can formally give the result we were looking for: The recovering of fair testing from our *non-probabilistic* must semantics.

Theorem 1 Let P, T be a process and a test, both non-probabilistic. We have $P \text{ must}^* T$ iff $P \text{ must}_{NC}\ T$.
Proof: For the right to left implication, suppose $P \text{ must}^* T$ does not hold. Then, by Lemma 2, there exists a maximal computation C of $P \mid T$ being NSP. If C is NSP then every computation that extends C is not potentially successful, so $P \text{ must}_{NC}\ T$ does not hold.

For the left to right implication, suppose $P \text{ must}^* T$ does hold. Then, the successful computations of $P \mid T$ sum up 1. So, again by Lemma 2, there does not exist any NSP computation. Let us consider an arbitrary computation of $P \mid T$, $C = s_1 \xrightarrow{e_1} s_2 \xrightarrow{e_2} \cdots s_n \xrightarrow{e_n} \cdots$. Suppose that C is not potentially successful and we get a contradiction. If C were not potentially successful then there would exist a state s_i in the computation such that there is no path from s_i leading to a successful state. Then, there exists $j \le i$ such that $C' = s_1 \xrightarrow{e_1} s_2 \xrightarrow{e_2} \cdots s_j$ is a NSP computation, which it is a contradiction. □

Corollary 1 Let P, Q be non-probabilistic processes and let P_{π_1}, Q_{π_2} be probabilistic processes made from P and Q, respectively. Then, $P \sqsubseteq_{\text{must}_{NC}} Q$ iff $P_{\pi_1} \sqsubseteq_{\text{must}} Q_{\pi_2}$.

4. COMPUTING MUST TESTING FOR FINITE-STATE PROCESSES

In this section, we restrict our study to finite-state processes. We present an algorithm to decide whether a finite-state process passes a finite-state test in our must semantics, or equivalently if a finite-state process fairly passes a finite-state test. Finally, we show some examples about how this algorithm works.

In order to compute passing of tests we give an algorithm for computing the probability which a process passes a test with. This algorithm is given by solving a linear equations system. Note that we will have a finite number of equations because in this section we work with finite-state processes. Specifically, given a process P and a test T, let $I = P \,|\, T$ be the associated interaction system $I = (S, \mu, s_0, Suc)$, where $S = \{s_0, s_1, \ldots, s_n\}$. We denote by q_i the probability with which P passes T from s_i (i.e. $Pr(I, s_i)$) and by a_{ij} the sum $\sum_e \mu(s_i, e, s_j)$. The probability with which a process passes a test is calculated by solving the next linear equations system consisting of $n + 1$ equations:

$$\forall i = 0..n: \; q_i = \sum_{j=0}^{n} a_{ij} \cdot q_j$$

We also have the following restrictions:

$$q_i = 1 \text{ if } s_i \in Succ \;\; \wedge \;\; \forall i = 0..n: \; \sum_{j=0}^{n} a_{ij} \in \{0, 1\}$$

which are imposed by the definition of passing tests (meaning that a successful state has been reached) and by the restriction that the sum of the probabilities associated with the transitions from one state is either 1 or 0 (the last case corresponds to a blocked state). Once the system is solved (for instance by using Gauss method) we can obtain relations over the variables with an infinite number of solutions, that is, relations of the form

$$q_i = \sum_{k \in K \subseteq \{0..n\}} (a'_k \cdot q_k) \; + \; \pi$$

In this case, we assign 0 to all the values in K, and so $q_i = \pi$, given the fact that such a relation indicates that no successful state can be reached from the states in K.

Finally, in order to decide whether a process passes a test in the probabilistic must semantics, it is enough to check if q_0 is 1. Due to Theorem 1 this method can be applied to decide if a non-probabilistic process passes a test in the *fair* sense.

Let us remark that this algorithm can be trivially adapted for checking properties about ongoing execution behavior. For instance to check if a process will execute an action (for example the action a in the process P_0 in Figure 1.1) or a sequence of actions (for example the process P_3 in Figure 1.1), etcetera, with a probability greater than π, just by controlling the value of q_0.

Example 1 Let $P_0 = (\tau \,;\, P_0) +_\pi \; a$ and $P_1 = a$. Consider the test $T_0 = a \,;\, ok$ and the interaction system $I = P_0 \mid T_0$. The associated system has got 2 equations (I has 2 states), and it is:

$$\begin{cases} q_0 = \pi \cdot q_0 + (1-\pi) \cdot q_1 \\ q_1 = 1 \end{cases}$$

where $I_{s_1} = \text{stop} \mid ok$, so $s_1 \in Suc$. In order to solve the system we just have the equation $q_0 = \pi \cdot q_0 + (1-\pi)$, that is, $q_0 \cdot (1-\pi) = (1-\pi)$. Given that $0 < \pi < 1$, the only solution is $q_0 = 1$. So, P_0 must T_0.

Consider the interaction system $I = P_1 \mid T_0$. The associated system of 2 equations is:

$$\begin{cases} q_0 = 1 \cdot q_1 \\ q_1 = 1 \end{cases}$$

where $I_{s_1} = \text{stop} \mid ok$, so $s_1 \in Suc$. We have $q_0 = 1$, and so P_1 must T_0.

Example 2 Let $P_2 = a\,; P_2$ and $P_3 = (a\,; P_3) +_\pi (a\,; b)$. Consider the test $T_1 = (a\,; T_1) +_{\pi'} (b\,; ok)$ and the interaction system $I = P_3 \mid T_1$. The associated equations system is:

$$\begin{cases} q_0 = \pi \cdot q_0 + (1-\pi) \cdot q_1 \\ q_1 = 1 \cdot q_2 \\ q_2 = 1 \end{cases}$$

where $I_{s_1} = b \mid T_1$ and $I_{s_2} = \text{stop} \mid ok$, so $s_2 \in Suc$. In order to solve the system we get the same equation of the previous example, $q_0 = \pi \cdot q_0 + (1-\pi)$, and so P_3 must T_1.

Consider the interaction system $I = P_2 \mid T_1$. The associated equations system has got only one equation $q_0 = 1 \cdot q_0$, where $q_0 \notin Suc$, so $q_0 = 0$, that is, P_2 must T_1 does not hold.

5. CONCLUSIONS AND FUTURE WORK

In this paper we have studied *fair* testing by using an intermediate probabilistic semantics. We have defined a probabilistic *must* testing semantics, and we have shown that two probabilistic processes are related by our probabilistic semantics iff the corresponding non-probabilistic processes are equivalent under *fair* testing. Finally, we have given an algorithm that computes whether a finite-state probabilistic process *must* pass a finite-state probabilistic test.

As future work we plan to define a denotational characterization of fair testing based on a probabilistic model (e.g. [Núñez et al., 1995, Gregorio and Núñez, 1999]). We also would like to study the extension of *fair* testing to timed models. Following a probabilistic testing approach, we will use a probabilistic-timed model like those given in [Cleaveland et al., 1996, Gregorio and Núñez, 1996, Gregorio et al., 1997]. Finally, it would be interesting to study how the framework could be adapted to deal with other notions of fairness.

Acknowledgments. This research has been partially supported by the CICYT project TIC 97-0669-C03-01. We would like to thank David de Frutos-Escrig for fruitful discussions on the topic. We also would like to thank a referee of a previous version of this paper for the careful reading and for pointing out a mistake in the characterization.

References

[Baeten et al., 1987] Baeten, J., Bergstra, J., and Klop, J. (1987). On the consistency of Koomen's fair abstraction rule. *Theoretical Computer Science*, 51(1-2):129–176.

[Bergstra et al., 1987] Bergstra, J. J.W.Klop, and Olderog, E. (1987). Failures without chaos: A new process semantics for fair abstraction. In *Formal Description of Programming Concepts III*, pages 77–103. Elsevier.

[Brinksma, 1988] Brinksma, E. (1988). A theory for the derivation of tests. In *Protocol Specification, Testing and Verification VIII*, pages 63–74. North Holland.

[Brinksma et al., 1995] Brinksma, E., Rensink, A., and Vogler, W. (1995). Fair testing. In *CONCUR'95, LNCS 962*, pages 313–327. Springer.

[Brinksma et al., 1996] Brinksma, E., Rensink, A., and Vogler, W. (1996). Applications of fair testing. In *Formal Description Techniques for Distributed Systems and Communication Protocols (IX), and Protocol Specification, Testing, and Verification (XVI)*, pages 145–160. Chapman & Hall.

[Brookes et al., 1984] Brookes, S., Hoare, C., and Roscoe, A. (1984). A theory of communicating sequential processes. *Journal of the ACM*, 31(3):560–599.

[Christoff, 1990] Christoff, I. (1990). Testing equivalences and fully abstract models for probabilistic processes. In *CONCUR'90, LNCS 458*, pages 126–140. Springer.

[Cleaveland et al., 1996] Cleaveland, R., Lee, I., Lewis, P., and Smolka, S. (1996). A theory of testing for soft real-time processes. In *8th International Conference on Software Engineering and Knowledge Engineering*.

[Cleaveland et al., 1992] Cleaveland, R., Smolka, S., and Zwarico, A. (1992). Testing preorders for probabilistic processes. In *19th ICALP, LNCS 623*, pages 708–719. Springer.

[Costa and Stirling, 1987] Costa, G. and Stirling, C. (1987). Weak and strong fairness in CCS. *Information and Computation*, 87:207–244.

[D'Argenio et al., 1999] D'Argenio, P. R., Hermanns, H., and Katoen, J.-P. (1999). On generative parallel composition. In *PROBMIV'98, Electronics Notes in Theoretical Computer Science 21*. Elsevier.

[de Frutos-Escrig et al., 1997] de Frutos-Escrig, D., Llana-Díaz, L., and Núñez, M. (1997). Friendly testing as a conformance relation. In *Formal Description Techniques for Distributed Systems and Communication Protocols (X), and Protocol Specification, Testing, and Verification (XVII)*, pages 283–298. Chapman & Hall.

[de Frutos-Escrig et al., 1998] de Frutos-Escrig, D., Llana-Díaz, L., and Núñez, M. (1998). An invitation to friendly testing. *Journal of Computer Science and Technology*, 13(6):531–545.

[de Nicola and Hennessy, 1984] de Nicola, R. and Hennessy, M. (1984). Testing equivalences for processes. *Theoretical Computer Science*, 34:83–133.

[Francez, 1986] Francez, N. (1986). *Fairness*. Springer.

[Gregorio et al., 1997] Gregorio, C., Llana, L., Núñez, M., and Palao, P. (1997). Testing semantics for a probabilistic-timed process algebra. In *4th International AMAST Workshop on Real-Time Systems, Concurrent, and Distributed Software, LNCS 1231*, pages 353–367. Springer.

[Gregorio and Núñez, 1996] Gregorio, C. and Núñez, M. (1996). Specifying and verifying the Alternating Bit Protocol with Probabilistic-Timed LOTOS. In *COST 247 International Workshop on Applied Formal Methods in System Design*, pages 38–50.

[Gregorio and Núñez, 1999] Gregorio, C. and Núñez, M. (1999). Denotational semantics for probabilistic refusal testing. In *PROBMIV'98, Electronics Notes in Theoretical Computer Science 21*. Elsevier.

[Hennessy, 1987] Hennessy, M. (1987). An algebraic theory of fair asynchronous communicating processes. *Theoretical Computer Science*, 49:121–143.

[Hennessy, 1988] Hennessy, M. (1988). *Algebraic Theory of Processes*. MIT Press.

[Koomen, 1985] Koomen, C. (1985). Algebraic specification and verification of communications protocols. *Science of Computer Programming*, 5:1–36.

[Kumar et al., 1998] Kumar, K. N., Cleaveland, R., and Smolka, S. (1998). Infinite probabilistic and nonprobabilistic testing. In *18th Conference on Foundations of Software Technology and Theoretical Computer Science, LNCS 1530*, pages 209–220.

[Milner, 1981] Milner, R. (1981). A modal characterization of observable machine-behaviour. In *6th CAAP, LNCS 112*, pages 25–34. Springer.

[Natarajan and Cleaveland, 1995] Natarajan, V. and Cleaveland, R. (1995). Divergence and fair testing. In *22nd ICALP, LNCS 944*, pages 648–659. Springer.

[Núñez, 1999] Núñez, M. (1999). An axiomatization of probabilistic testing. In *5th AMAST Workshop on Real-Time and Probabilistic Systems, LNCS 1601*, pages 130–150. Springer.

[Núñez and de Frutos, 1995] Núñez, M. and de Frutos, D. (1995). Testing semantics for probabilistic LOTOS. In *Formal Description Techniques VIII*, pages 365–380. Chapman & Hall.

[Núñez et al., 1995] Núñez, M., de Frutos, D., and Llana, L. (1995). Acceptance trees for probabilistic processes. In *CONCUR'95, LNCS 962*, pages 249–263. Springer.

[Rupérez, 1997] Rupérez, D. (1997). Una semántica de pruebas *fair* mediante un modelo probabilístico. Master Thesis. Dept. Sistemas Informáticos y Programación. Universidad Complutense de Madrid.

[Tretmans, 1996] Tretmans, J. (1996). Conformance testing with labelled transition systems: Implementation relations and test generation. *Computer Networks and ISDN Systems*, 29:49–79.

[Yuen et al., 1994] Yuen, S., Cleaveland, R., Dayar, Z., and Smolka, S. (1994). Fully abstract characterizations of testing preorders for probabilistic processes. In *CONCUR'94, LNCS 836*, pages 497–512. Springer.

A Formal Approach to Conformance Testing of Distributed Routing Protocols [*]

Jun Bi and Jianping Wu
Dept. of Computer Science, Tsinghua University, Beijing 100084, P. R. China

Abstract: This paper proposes a formal approach to conformance testing for OSPF v2, a widely used distributed routing protocol in Internet. Routing is performed in a distributed way, and, consequently, testing is more involved. For routing function testing concurrent TTCN handles concurrent test behaviors and is suitable for our needs. We first discuss a concurrent TTCN based OSPF test architecture. Then we present a specification model CEBE to specify an OSPF entity and to generate (combine the data and control flow) test suite. Finally, we design and implement a concurrent TTCN based test system to test OSPF implementations.

Key words: Conformance Testing, Concurrent TTCN, Routing Protocols

1. INTRODUCTION

Protocol conformance testing involves checking correctness of protocol implementations with respect to its (formal or informal) specification or requirements. Conformance testing is indispensable for the interoperability and performance of the protocol implementations. OSPF (Open Shortest Path First) is an important routing protocol for Internet. The network performance and reliability depend on the correct implementation of routing protocols of the routers. On the other hand, routing in the Internet is performed in a distributed fashion among different Autonomous Systems (AS) and in different areas (not in only a single router). How to formally specify and test concurrent behaviors of OSPF is a challenge to both

[*] This work has been supported by National Natural Foudation of China under Grant No. 69682002 and No. 69725003.

In this paper, we propose a formal approach to test OSPF protocols based on Concurrent Tree and Tabular Combined Notation (C-TTCN). A router running OSPF could be regarded as an entity in a distributed system [1]. An entity in a distributed system is associated with a set of interaction points and a set of concurrent behaviors at these interaction points, and it is more complex than the traditional peer-to-peer protocol entity [2][3]. To generate C-TTCN test suite, we present an entity specification model called Concurrent External Behavior Expression (CEBE). CEBE could directly specify concurrent external behaviors (both the I/O actions and data) at different interaction points of an entity in a distributed system and reduce internal states, actions and data. Therefore it is an ideal model to formally specify the entity in a complex distributed system and to generate (combine the data and control flow) C-TTCN test suite. The above methods have been implemented in an integrated test system PITS.

The rest of this paper is organized as follows. In section 2, we describe a C-TTCN based OSPF test architecture. Then we present the test suite generation method in section 3. Section 4 discusses the OSPF test suites and test system design. Section 5 concludes this paper.

2. TEST ARCHITECTURE

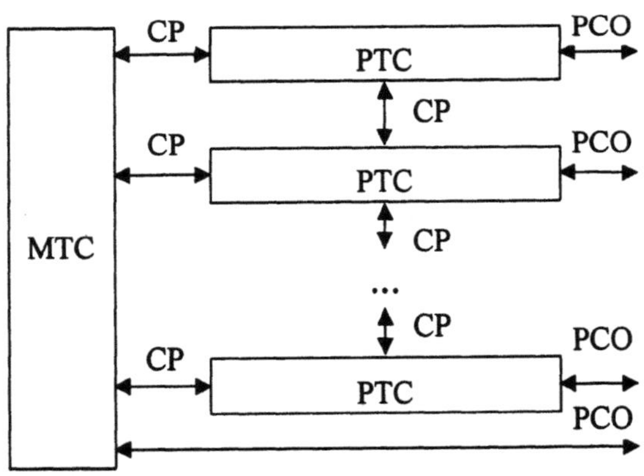

Figure 1. TC model of concurrent TTCN

OSPF v2, herein abbreviated simply as OSPF, is an IPv4 routing protocol documented in RFC 2328 [4]. It is a link-state routing protocol and all routers run the exact same algorithm in parallel. From the network topology database, each router constructs a tree of shortest paths with itself as the root. This tree defines a route to each destination in the AS. Externally derived routing information appears in the tree as leaves. Each individual piece of database is a particular router's local state (e.g., the usable interfaces and reachable neighbors). The router distributes its local state throughout the AS by flooding. OSPF adjusts quickly to network topology changes, utilizing a minimum of routing protocol traffic.

TTCN is recommended by ISO to describe abstract test suite. C-TTCN is an extension of TTCN, and it allows a test system to execute test cases by several test components (TCs) running in parallel. As figure 1 shows, a tester consists of exactly one main test component (MTC) and any number of parallel test components (PTCs). TCs are linked by coordination points (CPs) to convey coordination messages (CMs). Communications between TCs with environment are through points of control and observation (PCOs). C-TTCN is suitable for specifying systems that consist of several parallel testers or lower testers, which is common in router testing.

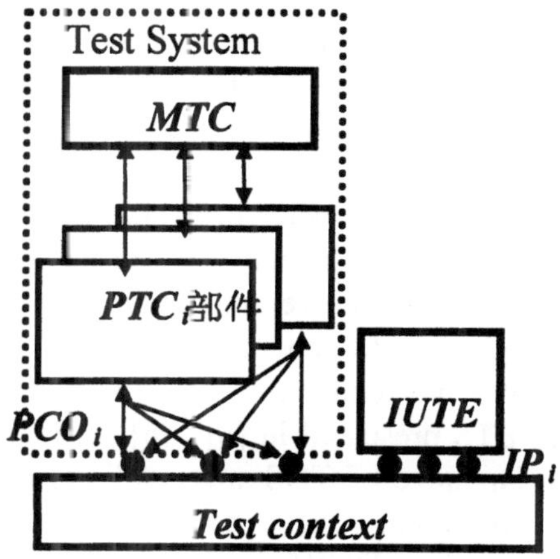

Figure 2. Test architecture

Test architecture is a description of the environment in which an IUT is tested. For a router entity *e*, there are k concurrent behaviors happening at interaction points. The router entity *e* conforms to the specification if and only if all these behaviors conform. Therefore, OSPF test architecture must deal with these concurrent behaviors at several interaction points.

We use the test architecture shown in figure 2. It consists of a test system, an entity implementation under test (IUTE), a test context, several PCOs and several interaction points. Test system is an implementation of a test suite based on C-TTCN. It carries out experiments by executing test cases and observing test results. Interactions between IUT and its environment take place at interaction points (IPs). Test system communicates with IUT indirectly: through *test context*. The points where tester communicates with test context, and indirectly controls and observes IUT, are called Points of Control and Observation (PCOs). Execution of a test case starts with the execution of MTC. It is the concern of MTC to set up all PTCs, and to manage all PCOs and CPs to be connected to. PTCs can be created by MTC on demand. A '*create*' operation associates a PTC with a behavior tree. The newly created PTC starts execution of its assigned behavior tree concurrently with MTC. MTC may explicitly terminate a PTC by executing a '*down*' operation.

3. C-TTCN TEST GENERATION

3.1 Basics of CEBE

In this section, we describe an entity specification model CEBE. CEBE aims to improve specification abilities of concurrent behaviors (actions) and interaction points (gates). It is the extension of EBE presented in [5]. The essence of both EBE and CEBE is to describe data in addition to control behavior from the point of view of an external system. Then CEBE can directly specify concurrent external behaviors (both the I/O actions and data) at different interaction points of an entity in a distributed system and reduce internal states, actions and data. Therefore, it is an idea model to formally specify the complex object and to generate (combine data flow and control flow) C-TTCN test suite. A specification of CEBE can be obtained from ETS (the extension of ETS presented in [6]). [6] gives the translation from standard specification language LOTOS and Estelle to ETS. In [7], an ETS semantics for Z is presented. CEBE can be obtained from ETS by eliminating the internal actions. Therefore, the CEBE may be translated from the standard specification model. There are two phases in our test generation method: phase 1: Derive test sequence from CEBE; phase 2: Generate test suite in C-TTCN.

Definition 3.1 A *CEBE* is a 7-tuple $<S, G, \Sigma, A, R, s_0, v_0>$ where (1) S is a finite set of *external* states whose elements are pause states of interactions exchanged between entity and its external environment; (2) G is a set of gates over which a CEBE can communicate; (3) Σ is set of data on CEBE (see below); (4) A is a set of *external* actions on CEBE (see below); (5) R is a set of logic relations of transitions between external states (see below); (6) $s_0 \in S$ is the initial state of the entity, which represents initial external state; and (7) v_0 is a set of initial assignments (initial values of all variables and timers) of Σ.

Definition 3.2 $\Sigma = IO \cup VAR \cup CONS \cup TIMER \cup PARA$, is a set of data on CEBE, where (1) IO is a set of input or output events (i.e., the ASPs and PDUs); (2) *VAR* is a set of variables of various types; (3) *CONS* is a set of constants of various types; (4) *TIMER* is a set of timers and (5) *PARA* is a set of parameters of events in IO. For an event $e \in IO$, the parameters $p_1, p_2, \ldots, p_n$ are denoted as $e.p_1, e.p_2, \ldots, e.p_n$, e is denoted as $e(e.p_1, e.p_2, \ldots, e.p_n)$.

Definition 3.3 A behavior b is the elements of G with a list of input and output events and their value declaration. Then set of b is denoted as: $B=\{ <g_i, ?\, r_i\, (r_i.p_1, r_i.p_2, \ldots, r_i.p_n), !\, s_i\, (s_i.p_1, s_i.p_2, \ldots, s_i.p_n) > \mid g_i \in G, r_i, s_i \in IO, r_i.p_i, s_i.p_i \in PARA \}$, where symbol "!" indicates the output of an event s_i to external environment and symbol "?" indicates the input of an event r_i from the external environment. The absence of r_i or s_i could be denoted as symbol "–". For example $b= <g_i, ?\, r_i\, (r_i.p_1, r_i.p_2, \ldots, r_i.p_n), !\, -- >$. Here we give some notations: (1) $b_1 >> b_2 >> \ldots >> b_n$, the behaviors are sequential; (2) $b_1\, [\,]\, b_2\, [\,] \ldots [\,]\, b_n$, the behaviors are equally valid alternatives (choice); (3) $b_1 \,||\, b_2 \,||\, \ldots \,||\, b_n$, the behaviors are parallel.

Definition 3.4 An action a is set of concurrent behaviors. $A = \{ a_i \mid a_i \in A \}$, $a_i =_{def} b_j \in B \mid a_i\, [\,]\, a_i \mid a_i >> a_i \mid a_i \,||\, a_i \mid (a_i)$, where the operator precedence is '( )' > '[]' > '>>' > '>>'.

Definition 3.5 Transition relation $R \subseteq S \times A \times S \times P \times PowerSet\,(O) \times PowerSet(F)$ is the set of transition relations where: (1) A is a set of actions; (2) S is a set of states of CEBE; (3) $P(\Sigma)$: $PowerSet(\Sigma) \rightarrow \{ TRUE, FALSE \}$ is a set of predicates functions (=, ≥, ≤, >, <, ∧, ∨, ¬, Timeout, et.al.); (4) $O(\Sigma)$: $PowerSet(\Sigma) \rightarrow PowerSet(\Sigma)$ is a set of operation functions (:=, +, -, *, /, StartTimer, CancelTimer, et. al.); (5) $F(\Sigma)$: $PowerSet(\Sigma) \rightarrow PowerSet\,(\Sigma)$ is a set of functions (i.e. max(), min(), et.al).

The intuitive meaning of a transition $r \in R$ is that if a CEBE is in state s and the enabling action a is offered, then the enabling predicate p is evaluated on current assignment of variables. When p is true, the CEBE will go into a new state s' and the environment are updated by an operation o and a function f. The absence of predicate could be denoted as *TRUE*, the absence of operation or function is denoted as *NIL*.

3.2 Test sequence derivation

CEBE based test sequence derivation strategy is defined as two steps. First identifying all possible valid *IPs* (Interaction Path) from CEBE specification by imposing constraints set on IPs. Then generating *SIPs* (I/O

Subpath) from the above IPs through checking dependencies between transitions in each IP.

Step 1: An interaction path *IP* is the externally observable track on which a sequence of interactions between entity and its external environment occurs, starting from initial external state s_0 of the entity and ending in s_0. Any loop (or cycle) interaction in an *IP* is traveled only once. *IP* with loop is an interaction path in which there exists one or more loop interactions.

Step 2: An I/O subpath *SIP* is the externally observable track $r_1, \; . \; . \; . \; , r_k$ in an *IP*, where:

1. $r_1 =< s_{r1}, a_{r1}, s'_{r1}, p_{r1}, O_{r1}, F_{r1}>$ is the first transition relation when satisfies:

a) a predicate $p_{r1} (\Sigma_{pr1})$, and/or

b) a set of operation $\{ \ldots , o_{r1i} (\Sigma_{or1i}), \ldots \}$, and /or

c) a set of function $\{ \ldots , f_{r1i} (\Sigma_{fr1i}), \ldots \}$;

2. $r_k =< s_{rk}, a_{rk}, s'_{rk}, p_{rk}, O_{rk}, F_{rk}>$ is obtained when logical relations satisfy:

a) $(\forall i, \; (\Sigma_{prk} \cap o_{r1i} (\Sigma_{or1i})) \neq \emptyset) \vee (\forall i, \; (\Sigma_{prk} \cap f_{r1i} (\Sigma_{fr1i})) \neq \emptyset) \vee (\forall i, \forall j, \; (\Sigma_{orkj} \cap o_{r1i} (\Sigma_{or1i})) \neq \emptyset)) \vee (\forall i, \; \forall j, \; (\Sigma_{orkj} \cap f_{r1i} (\Sigma_{fr1i})) \neq \emptyset)) \vee (\forall i, \forall j, (\Sigma_{frkj} \cap o_{r1i}$
$(\Sigma_{or1i})) \neq \emptyset)) \vee (\forall i, \forall j, (\Sigma_{frkj} \cap f_{r1i} (\Sigma_{fr1i})) \neq \emptyset))$, and

b) a predicate $p_k (\Sigma_{prk})$, and/or

c) a set of operation $\{ \ldots , o_{kj} (\Sigma_{orkj}), \ldots \}$ and /or

d) a set of function $\{ \ldots , f_{kj} (\Sigma_{frkj}), \ldots \}$;

3.3 C-TTCN test suite generation

Based on SIP, a test tree with nondeterminsm, correctness, and defense branches can be derived, and three kinds of verdicts (PASS, FAIL and INCONC) are assigned to each leaf of the test tree. The test tree can be mapped to a test case in C-TTCN. Finally, these test cases are grouped according to some rules, and are mapped to a complete test suite in C-TTCN.

As discussed above, in the test case specified by C-TTCN, the behaviors of each test component are expressed with a *behavior tree (BT)* in a test case where BT is a tree-like presentation of the temporal relations between test events. In order to get formal definition of the semantics, we need to introduce the Test Behavior Expression (TBE) to express the behaviors of test components in algebraic form.

Definition 3.6 *The syntax of a TBE*

TBE B $=_{def}$ *stop* | *exit* | *id?a;B* | *id!y;B* | *B []B* | *B>>B*| *[q];B* | *(I:=val);B* | *(B)*| *start tid val;B* | *cancel tid; B* | *timeout tid;B* | *create id B* | *done id*

The operator precedence is *';'* > *'[]'* > *'>>'*. *'stop'* denotes termination of a running, *'exit'* denotes a successful quitting from an attached TBE. *'id?a;B'* and 'id!y;B' represent respectively a receiving and a sending event in PCO or CP identified by *'id'*. *';'* denotes sequential composition. *'[]'* denotes choice. *'q'* is the qualifier which is a boolean expression, its value is either *TRUE* or *FALSE*. *'I:=val'* is an assignment, where *'I'* is an identifier and *'val'* is a value. *'>>'* denotes attaching operation. *'start'* and *'cancel'* is a pair of timer operator, *'timeout'* represents the expiration of a timer. *'create id B'* creates a PTC named id whose behavior tree is B, while *'done id'* waits until the PTC named id has finished its running.

To understand the algorithm we use TBE to describe the rules for C-TTCN generation. In the real tool TUGEN, the test suite in C-TTCN MP is generated.

Rule 1: Each SIP is mapped to one C-TTCN test case;

Rule 2: In one SIP, if there are several relations $r_1, r_2, \ldots, r_n$, then attach the tree *i+1* to tree *i*.

TBE$_M$□r1>>r2>>...>>rn

Rule 3: In one action *a* of relation *r*, if $b_1 >> b_2 >> \ldots >> b_n$, then $b_1, b_2, \ldots, b_n$ are behaviors and it is successfully completed if b_{i+1} happens immediately after the successful completion of b_i $(i=1,2,\ldots,n-1)$:

TBE$_M$□create P1 TBE$_1$;start t1;(CP1?CM_1;down P1;stop) *[]* (timeout t1;stop))

TBE$_1$□b1>>b2>>...>>bn>>(CPi!CM_i;stop)

Rule 4: In one action *a* of relation *r*, if $b_1 [] b_2 [] \ldots [] b_n$, then an alternative is matched if the behavior b_i corresponding to the alternative happens first:

TBE$_M$□create P1 TBE$_1$;start t1;(CP1?CM_1;down P1;stop) *[]* (timeout t1;stop))

TBE$_1$□(b1 >> (CPi!CM_i;stop)) *[]* (b2 >> (CPi!CM_i;stop)) *[]...[]* (bn >> (CPi!CM_i;stop))

Rule 5: In one action *a* of relation *r*, if $b_1 \| b_2 \| \ldots \| b_n$, then $b_1, b_2, \ldots, b_n$ are started in parallel and the parallel tree completes only if $b_1, b_2, \ldots, b_n$ complete:

TBE$_M$□create P1 TBE$_1$; ... ;create Pn TBE$_n$;start t1;(CP1?CM_1;down P1;start t2;(...;(CPn?CM_n;donePn; stop)*[]* (timeout tn;stop));...)*[]* (timeout t2;stop)) *[]*... *[]* (CPn?CM_n;down Pn;start t2;(......)*[]* (timeout t2;stop))*[]* (timeout t1;stop)

TBE$_i$□bi>>(CPi!CM_i;stop)

Rule 6: In each behavior *b* of action *a* in a relation *r*, $<g_i, ?\, r_i\, (r_i.p_1, r_i.p_2, \ldots, r_i.p_n), !\, s_i\, (s_i.p_1, s_i.p_2, \ldots, s_i.p_n) >$, then
TBE$_i$: g_i! r$_i$;g_i? s$_i$;CP1!CM_1;stop

Rule 7: In the declaration part, the *timer, variable* and *constant* could be mapped from *TIMER, VAR, and CONS* with the initial value from v_0;

Rule 8: In the declaration part, the *ASP, PDU* can be mapped from *IO* with the parameters *PARA* for the events;

Rule 9: In the constrain part, the *timer, variable, ASP* and *PDU* can be mapped from each event $e_i\, (e_i.p_1, e_i.p_2, \ldots, e_i.p_n)$ in *b* with the value of parameters associated with the rules defined in section 3.1.

4. OSPF TESTING

4.1 Test Suite Design

We implement the above test method by an automatic tool TUGEN. The generated OSPF test suite has over 4000 SIPs when search depth is 6. We delete the redundant or useless ones and select 543 test cases that have clear test purposes. Table 1 shows the structure of OSPF test suite.

OSPFv2 Conformance Test Suite	
Test Group	Test Cases
DOWN	23
INIT	70
EXSTART	107
EXCHANGE	123
LOADING	68
FULL	152

Table 1 OSPFv2 Test Suite

In CEBE specification, we delete some internal states defined in the protocol, such as Attempt, 2-Way, and some interface state (Loop, Unloop) because these states are not observable externally. Moreover, we reduce some internal events defined in the protocol such as: Interface up/down and loop/unloop. The following events are replaced by the external input/output

PDUs (with different parameters) and timer events: Start, 2-WayReceived, NegotiationDone, ExchangeDone, BadLSReq, Loading Done, AdjOK?, SeqNumMismatch, 1-Way, KillNbr, InactivityTimer, and LLDown. We give examples of the C-TTCN based OSPF test suite. Here are two transition relations $r_{44(1)}$ and $r_{44(2)}$:

S4 [<PCO1, ?--, !DDP0 (Ver,Type,Len, D:=MyID, AID, CSM, AT, Auth, MTU, Option,I,M,MS, Seq:= YouSeq)> |
PREDICATE: TRUE;
OPERATION: HaveSendDDP:=1;
START RtxDDPTM;
*FUNCTION: NIL]*S4 {r4-4(1)}*
+ [<PCO1, ?LSU(...), !LSAck(...)> >> (<PCO2, ?--, !LSU(...)> || <PCO3, ?--, !LSU(...)> || ... || <PCOn, ?--, !LSU(...)>) |
PREDICATE: HaveSendDDP=1;
OPERATION: NIL;
*FUNCTION: NIL]*S4 {r4-4(2) }*

In $r_{44(1)}$, predicate is TRUE. In $r_{44(2)}$, predicate indicates that it needs "HaveSendDDP=1". It means that in state 4 (FLOOD), IUT must send last DDP first. The $r_{44(1)}$ is an associated transition. When it fires, associated function $f_{44(1)}$ and operation $o_{44(1)}$ is executed. $o_{44(1)}$ is to set HaveSendDDP:=1 and start retransmit timer RtxDDPTM. When OSPF-CEBE variables and timers are set, DDP (with parameters ID and Seq are assigned) are assembled (their parameters are set) and sent. Then $p_{44(2)}$ becomes TRUE and it can be executed. Therefore, when generating test case, $r_{44(1)}$ and $r_{44(2)}$ should be in one SIP (test case). In $r_{44(2)}$, if IUT receives an Update packet, it will send an Ack packet, then it updates its database and floods this new information in an Update packet to all the neighboring routers. This flooding process is concurrent and is followed with the receiving of Update (we use >> and ||). The generated C-TTCN test case is shown as follows.

```
TREE_M: behavior specification of M
  CREATE(P1, TREE_P1)
   START T1
    CP1?CM_1
     ?DONE(P1)
      CREAT(P1, TREE_P1');
       START T2;
        CP1?CM_1
         ?DONE(P1')
           CREAT(P2, TREE_P2)
```

```
                ...
                CREAT(Pn, TREE_Pn');
                  START T3;
                    CP2?CM_2
                      ?DONE(P2)
                        ...
                          CPn?CM_n
                            ?DONE(Pn)    PASS
                        ...
                      CPn?CM_n
                        ?DONE(Pn)
                          ...
                        ...                       FAIL
TREE_P1: behavior specification of P1
  PCO1?DDP0                    DDP_12
    CP1!CM_1
TREE_P1'
  PCO1?LSU                     LSU_10
    PCO1!LSAck                 LSAck_10
      CP1!CM_1
TREE_Pi: behavior of Pi
  PCOi?LSU                     LSU_i
    CPi!CM_I
```

4.2 A C-TTCN based test system

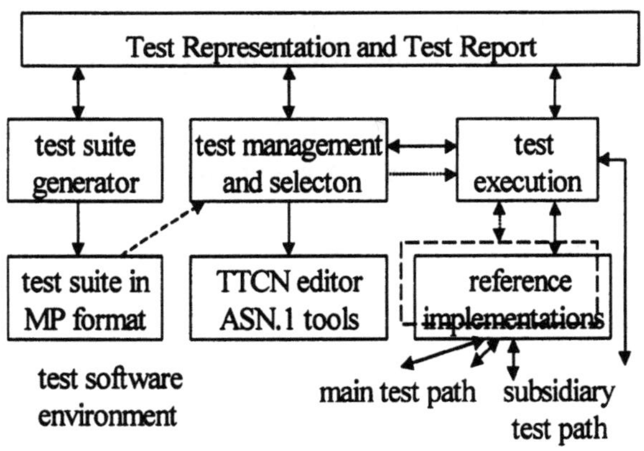

Figure 3. The architecture of PITS

We have implemented an integrated test system named PITS over SUN workstation and Solaris operating system. Figure 3 shows the architecture of

PITS. It provides an integrated platform to test for both protocols (OSI and TCP/IP) and entities (in distributed systems), based on C-TTCN (and its extension). This system conducts its testing process on the basis of TTCN/C-TTCN test suites, and parallel processing is explored to increase its efficiency. After test selection on the basis of PICS and PIXIT by TM (test management), test cases are interpreted and then executed by TE (test execution) step by step. TE is an engine that interacts with other components of PITS, and controls test process according to the content of TTCN or C-TTCN, and simultaneously generates all information required to produce test reports. Since SUT OSPF is an entity in the distributed routing area, there may be many PCOs and conceptual PTCs for these PCOs. By adjusting the reference implementation and test path, we can get flexible test configuration to meet the different need of entities in the distributed systems.

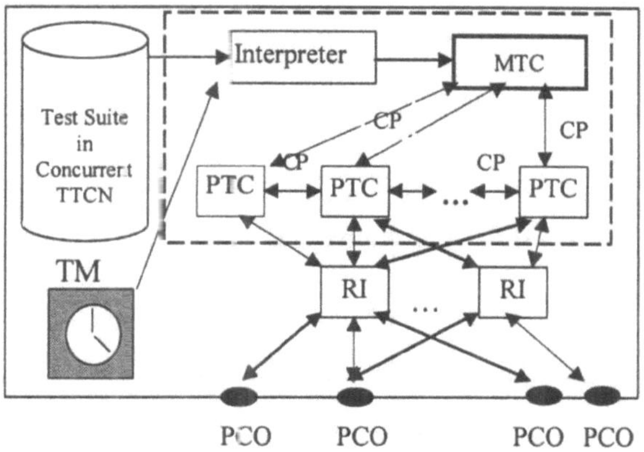

Figure 4. Architecture of test execution

TE can be regarded as an abstract TTCN/C-TTCN machine. It interprets and executes test cases strictly according to operational semantics of TTCN/C-TTCN. As shown in figure 4, TE is composed of an interpreter, one MTC, and zero or more PTCs. The interpreter gets the offset of a test case from TM, and then obtains an abstract test case from the test suite. The interpreter transfers an abstract test case to an executable test case by 3 steps. (1) *Syntax analysis:* check the syntax of an C-TTCN based test case; (2) *Static semantics analysis:* check the static semantics of the test case; and (3) *Transformation to an evaluation tree:* construct an evaluation tree by appending default behaviors, converting REPEAT constructs into GOTO constructs, and expanding attached trees.

MTC is the control part of test execution. It gets an executable test case from the interpreter and executes the test case by traversing a path in the evaluation tree of the test case. When meeting *'create'* event, MTC creates a

PTC and assigns the behavior tree defined by *'create'* event to PTC. Configuration declaration table determines the location of PTC. Initially, PTCs are demon processes listening to some special ports. After MTC executes *'create'* test event, the corresponding PTC will be active and run test behavior tree conveyed from MTC. Each PTC keeps its own test log during executing of a test case. When the execution of the test case completes, it reports log to MTC. Then MTC reports all logs to test result module. The test execution is performed by the method presented in [8].

Figure 5 shows the OSPF test architecture with PITS when IUT is the OSPF router entity. The RT is optional and depends on the interface of router under test. This test architecture has been used to test router products such as CISCO 4700. CISCO passed most test cases. However, we found some problems such as DDP interactions and MTU setting with CISCO version 11.

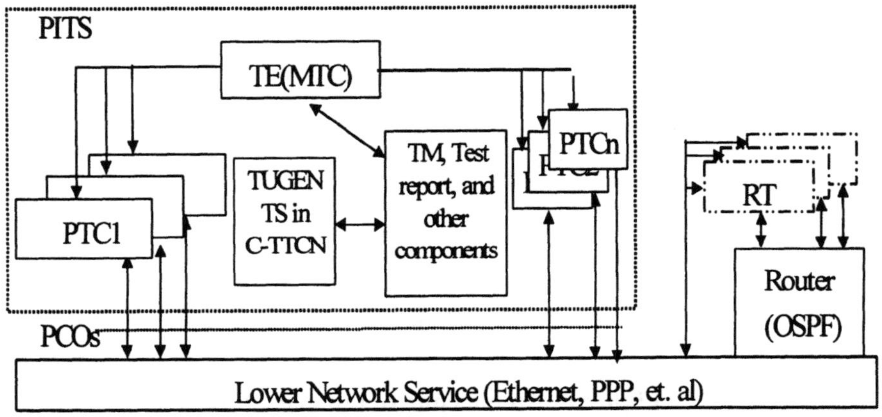

Figure 5 OSPF router entity test architecture with PITS

5. CONCLUSIONS

In this paper, we have applied C-TTCN for OSPF routing protocol testing. The real testing is accomplished by using a C-TTCN based test system PITS to test CISCO 4700 router. C-TTCN OSPF test suite is derived by a tool TUGEN based on CEBE. From our experience, when we write OSPF CEBE specifications, we find many internal states and actions that can be reduced, simplified, or replaced by external events in CEBE. Currently, we are involved in the work of testing the whole OSPF AS and Area as a black box by observing its external behaviors to stimuli provided by simulations of other Areas or ASs. Then it is necessary to specify CEBE of an entity Area or AS and to generate test suite in C-TTCN.

CEBE and C-TTCN based concepts and methods presented in this paper can also be used in the area of conformance testing in distributed systems. For instance, CEBE could be used to model communicating systems and entities of OSI, TCP/IP and ODP. Furthermore, we are studying CEBE model and C-TTCN generation algorithm to improve the test execution efficiency of the generated test suite. For instance, we want to combine TBEs of different PTCs in one test case to deduce the size of behavior tree.

REFERENCES

[1] A.Ulrich, L.Logrippo and J. Meer, "An approach to testing distributed software systems", *15th Int. Conference on Protocol Specification, Testing and Verification,* IFIP, 1995.

[2] A.A.F. Loureiro and S.T. Chanson, "Checking unstable properties in distributed testing of communication protocols", *10th Int. Workshop on Testing of Communicating Systems,* IFIP, 1996.

[3] A.C.Y. Wong, S.T.Chanson, S.C. Cheung, and H.Fuchs, "An framework for distributed object-oriented testing", *10th Int. Conference on Formal Description Techniques for Distributed Systems,* IFIP, 1997.

[4] J. Moy, OSPF version 2, RFC 2328, 1998.

[5] J. Wu and S.T.Chanson, "Testing sequence derivation based on external behavior expression", *2nd Int. Workshop on Protocol Test Systems*, IFIP, 1989.

[6] J. Wu and S.T.Chanson, "Translation from LOTOS and Estelle specifications to extended transition system and its verification", *2nd Int. Conference on Formal Description Techniques for Distributed Systems*, IFIP, 1989.

[7] J. Derrick, E. Boiten, H. Bowman and M. Steen, "Supporting ODP - translating LOTOS to Z", *1st Int. Workshop on Formal Methods for Open Objet-based Distributed Systems*, IFIP, 1996.

[8] R. Hao and J. Wu, "A formal approach to protocol interoperability testing", *6th Singapore International Conference on Networks*, IEEE, 1997.

Invited Talk II

PART FOUR

FDTs II

DERIVING ACTIVITY THREAD IMPLEMENTATIONS FROM FORMAL DESCRIPTIONS USING TRANSITION REORDERING

Peter Langendörfer and Hartmut König
Brandenburg University of Technology at Cottbus
Department of Computer Science
PF 101344, D-03013 Cottbus
Germany
{pl,koenig}@informatik.tu-cottbus.de

Abstract In this paper we show how formal descriptions can efficiently be mapped into activity thread implementations using transition reordering. This technique prevents semantic conflicts by reordering of statements at compile time. It considerably simplifies the derivation process and improves the efficiency of the generated code. The technique is implemented in the SDL compiler *COCOS*. We describe the approach as well as its implementation and show how semantic conflicts are resolved. Finally we present measurements which prove a considerable performance gain up to 120 per cent.

Keywords: Formal description techniques, SDL, automated protocol implementation, activity threads, transition reordering, configurable FDT compiler

1. MOTIVATION

The development of communication software is known to be an expensive and tedious process. Formal description techniques (FDTs) can help to considerably increase the quality of the protocols and telecommunication systems, and to shorten the time for their development. However, the benefits of FDTs are mainly used in the design, specification, performance prediction, verification, and testing phase. Automated implementation still represents a gap in this chain of development steps. Implementations are mostly carried out in traditional manner, i.e. manually. The main reason for this is that code automatically generated by an FDT compiler is mostly inadequate for applications in a real-life environment [Held95, Mans97]. It is more suited for prototyping rather than for final implementations. This shortage of automated protocol implementation cannot be compensated by other benefits such as a considerable reduction of the duration of the implementation process, independence of subjective implementation decisions, better conformance to the specification, and simplification of changes.

The inefficiency of automatically derived implementations is caused by the rigid implementation model, which determines the set of transformation rules, and by semantic constraints of the FDT. Further these compilers lack of means to adapt to a given implementation context, to efficiently handle the exchange of data packets, and to integrate several protocol layers. The semantics of the FDT can be simplly preserved by straight transforming it into the implementation. This mostly leads to a considerably performance overhead of the generated code caused by semantics-related computations and storing of data. Automatically generated implementations, however, will be only then accepted for practical applications if their efficiency comes close to manually coded implementations. Therefore, optimizations are required for the code generation process. In recent years several approaches were proposed to improve the efficiency of automated implementation techniques [Fisc95, Held95, Gotz96]. These approaches propose optimizations within the limits of the respective FDT semantics. The performance gain achieved is not adequate for final implementations. Other approaches such as a flexible runtime support system [Lang99] and the integration of efficient manual implementation techniques like activity threads [Henk97a, Henk97b] and integrated layer processing [Clar90, Abbo93, Brau96, Leue96] promise more optimistic results. The mapping strategies used in these techniques, however, do not permit it to straightly map the FDT semantics into the implementation. This is in part due to the integration of several layers which is usually not considered in formal descriptions. The compliance with the specification requires, therefore, to apply different transformation rules which can lead to conflicts with the FDT semantics as shown in [Henk97b].

In this paper we focus on the derivation of activity threads. In [Henk97a, Henk97b] we presented an approach to derive activity thread implementations from formal descriptions. As shown in these papers we could observe a considerably performance gain. We now present an improved approach for automatically deriving activity thread implementations. Unlike the approach presented in [Henk97a] it uses a new technique - the transition reordering - which allows it to handle almost all semantic conflicts at compile time. This simplifies the transformation procedure and remarkably improves the efficiency of the generated code. We present the approach for SDL [ITU93]. It can be also applied to other FDTs if the semantic constraints are taken into account as discussed in [Henk97b]. In Section 2, we introduce the activity thread implementation technique and discuss the semantic conflicts which may appear. Section 3 describes the transition reordering approach and its implementation in the SDL compiler *COCOS*. We also show how the semantic conflicts are resolved. In Section 4 we present measurements which show the effect of the approach. We further compare the received code with code generated by means of a commercial tool - the *Cadvanced* code generator of the SDT tool. In the concluding remarks we give an outlook on further research steps.

2. MAPPING OF SDL SPECIFICATIONS ONTO ACTIVITY THREADS

In this section we describe the principle of the transformation of SDL specifications onto activity threads as well as the possible semantic conflicts.

2.1 MAPPING PRINCIPLE

The activity thread technique [Clar85, Svob89] implements a protocol entity as a set of procedures where for each input event a corresponding procedure exists. The active elements in this model are the events. An incoming event activates the corresponding procedure which immediately handles the event and when producing an output calls the respective procedure of the next entity. The sequence of inputs and outputs (input→output→input ... input→output) results in a sequence of procedure calls called activity thread. Note that the term activity thread does not refer to an operating system thread. It denotes the execution path an event triggers in the protocol stack. Activity threads can run in both directions, upwards and downwards. The respective procedure calls are also denoted as upcalls and downcalls.

The activity thread technique provides very efficient implementations because it avoids the storing of events when calling the next protocol entity [Clar85]. Its semantic model (concurrent computation of different activity threads and synchronous communication), however, considerably differs from the semantic models of the standardized FDTs. The synchronous communication causes diverse semantic conflicts. They are discussed in the next subsection. Another constraint of this transformation is that spontaneous transitions cannot be handled [Henk97a,

Henk97b]. They must be removed by refining the specification before code is generated. Therefore, FDT compilers usually apply the server model which supports a straight mapping of the FDT semantics instead of the more complicated activity thread approach.

The derivation approach of [Henk97a] separates the behaviour and data part of the SDL processes. It maps each process instance into a reentrant procedure and the data part into an instance control block. The events are the signals. Each output-statement is mapped on a procedure call. The delivery of the signals is organized via an additional loop construction which forms the frame of the implementation. This loop construction is mapped on an activity thread scheduler which handles concurrent activity threads, the delivery of the signals and semantic conflicts. Details of this mapping strategy can be found in the referenced papers (see also Section 3.1).

2.2 SEMANTIC CONFLICTS

The semantic conflicts which may appear when directly mapping SDL specifications onto activity threads can be divided in two groups: interferences of transitions and overtaking of signals. In both cases they are caused by the appearance of cyclic process call sequences at runtime.

Interference of transitions. In the communication between two or more processes the activity thread implementation principle may cause situations in which a signal is consumed by the receiving process before the sender has finished its transition. This may lead to the following errors:

- discarded signals, and
- inverted order of state changes and variable assignments.

The discarding of signals may occur when a procedure call implementing an output -statement is executed before the assignment of next state. In this case, the process instance cannot change its state until the called procedure has returned control. Figure 1.1 gives an example for such a situation.

An implementation that is conform to the specification has to guarantee that process *P1* changes to state *update* after sending the signal *auth*. Then it can accept the signals *account* or *reject* by means of which process *P2* responds to *auth*. In contrast to the expected correct behaviour, the signals *account* and *reject* respectively, are always discarded by process *P1*, because *P1* remains in state *login* until the procedure call

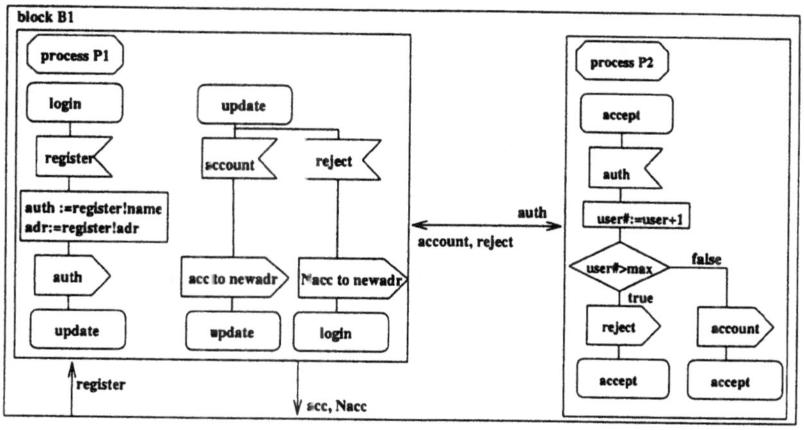

Figure 1.1 Example for discarding of a signal

for the output of *auth* is finished. Figure 1.2 shows the resulting cyclic process call sequence.

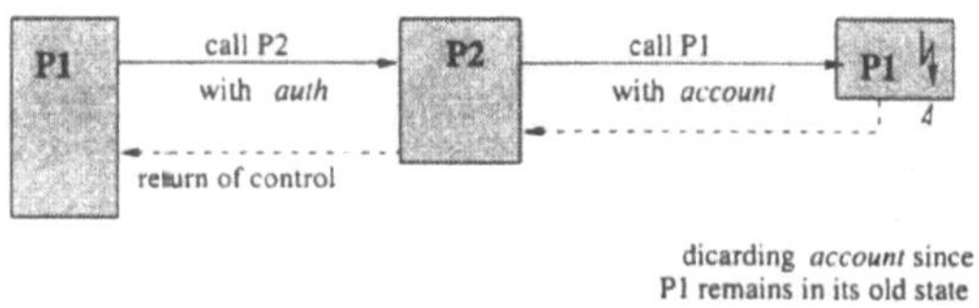

Figure 1.2 Process call sequence for example of figure 1.1

Overtaking of signals. SDL guarantees that the order of signals sent via a signal route or a channel is preserved at destination. If, for instance, a signal *s1* is sent at time *t* and a signal *s2* at time *t+x*, then *s1* has to be delivered before *s2*. When applying the activity thread approach signals may overtake each other.

This situation may happen if a transition contains several output-statements. The mapping of output-statements onto procedure calls causes that the second, third and any further procedure call is delayed until the first one and all successor calls have been finished. If one of these successors uses a signal route that is also used by one of the other output-statements of the transition with multiple outputs, the signals overtake. Figure 1.3 shows an SDL specification in which the preconditions for signal overtaking are fulfilled.

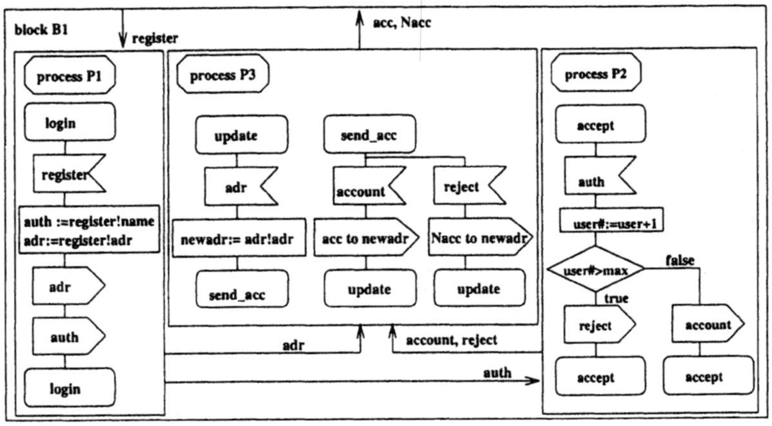

Figure 1.3 SDL System with cycles and the possibility of signal overtaking

In a straight forward generated activity thread implementation of the specification given in figure 1.3 the signal *register* would trigger the following execution. Process *P1* sends signal *auth* to process *P2* that replies with *reject* or *account*. *P1* immediately forwards these signals to process *P3* which responds with *acc* or *Nacc* to the address given in *newadr*. After that the control returns to *P1* which sends the signal *adr* to *P3*. In contrast to the specified behaviour, one of the signals *account* or *reject* now has overtaken *adr*. As consequence, *P3* sends the signal *acc* or *Nacc* to a wrong address. The resulting process call sequence is depicted in figure 1.4.

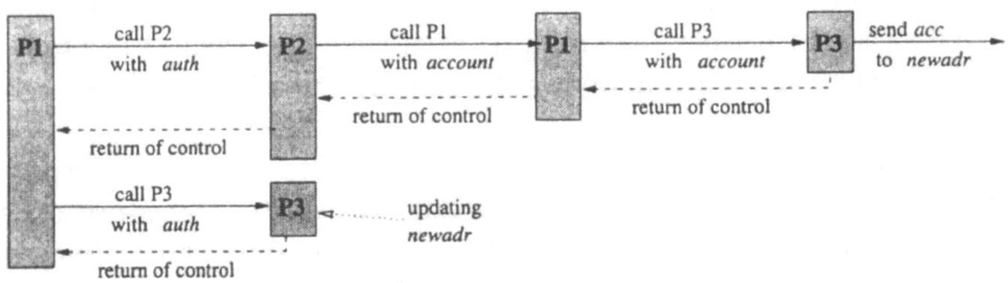

Figure 1.4 Process call sequence for example of figure 1.3

3. TRANSITION REORDERING

3.1 PRINCIPLE

In [Henk97a] the semantic conflicts discussed in section 2 are resolved at runtime. The solution consists in delaying the procedure calls for the output-statements until the end of the transition. This is done by the procedure represented in figure 1.5a. When executing an output-statement the signal and its receiver are stored in a signal list. The activity thread scheduler processes the signal list in a FIFO manner. It always activates the receiver process of the first entry. The receiver process consumes the signal and executes the respective transition. By this, further signals can be added by means of an append function to the signal list if the transition contains output-statements. After finishing the procedure the control is returned to the scheduler. The shortage of this mechanism is that the execution of the output-statement is delayed, i.e. it may be processed after output-statements of other transitions. This leads to an asynchronous communication and considerably increases the overhead of an activity-thread implementation.

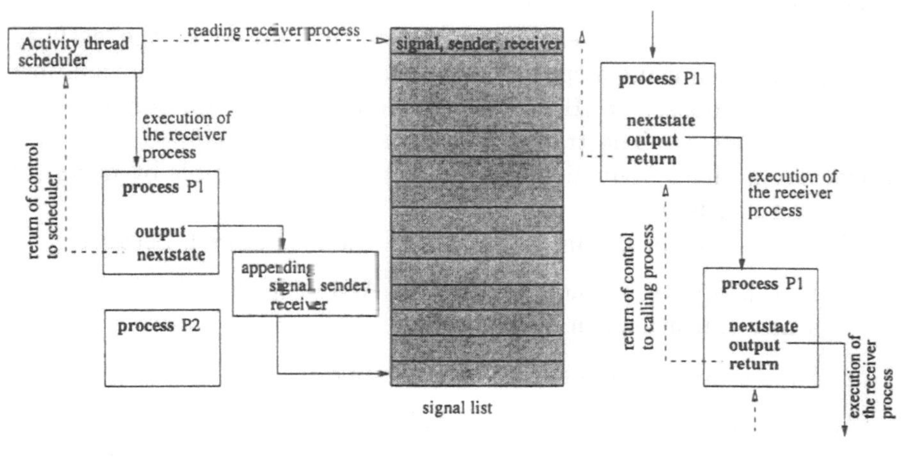

Figure 1.5 Mapping on activity threads

The transition reordering designates that the SDL statements of a transition are not implemented in the order as they are specified. They are reordered at compile time in such a way that the semantic conflicts described above cannot occur. The approach adapts the principle of code restruction which is applied for loop optimization in modern programming language compilers [Aho96].

The reordering only concerns the output-statements. They are implemented in two steps:

1. Replace the output-statement by an operation which stores the signal.
2. Append the output-statement with the stored signal to the end of the transition. If a decision-statement follows the output-statement the modified output-statement has to be appended to each branch of the decision-statement.

All other SDL statements are implemented in the order as they are specified, i.e. only the sending of the signals is delayed. This approach does not require any additional runtime support. The output-statements are only executed when the transition has reached the successor state.

Thus, according to the activity thread principle the receiver process is triggered by the sender process. The control only returns to the receiver process if no further output-statements have to be processed or if in connection with a server model implementation a signal is sent to an asynchronous (buffering) interface (see figure 1.5b).

Using transition reordering the indirect realization of the activity thread approach using a special loop construction and runtime support features as proposed in [Henk97a] can be avoided except some special cases (see below). The transitions can straight be mapped on procedures, i.e. a direct derivation of activity threads is possible by means of transition reordering.

The reordering of actions inside a transition is allowed, because:

1. SDL processes are nonpreemptive. A transition is finished before a new transition can be executed.
2. Signals cannot be modified after output.

These conditions guarantee that the shift of the output-statements to the end of the transition do not influence the compliance between specification and implementation. Condition (1) expresses that the output-statements are also executed if they are implemented at the end of the transition, because there is no statement in SDL which can interrupt the execution of the transition. Condition (2) ensures that the data transferred by a signal cannot be modified by a statement following the output-statement. Thus, the delay in the sending of the signals cannot lead to a transmission of falsified data.

In the next subsections we show how the semantic conflicts discussed above can be avoided by applying transition reordering.

3.2 AVOIDING TRANSITION INTERFERENCES

The reordering of transitions ensures that assignments to variables and state changes are carried out in the correct order even if cyclic process call sequences appear. It prevents the discarding of signals and the inversion of variable settings. The conformance between specification and implementation is preserved. Figure 1.6 shows the specification of the example from figure 1.1 after transition reordering.

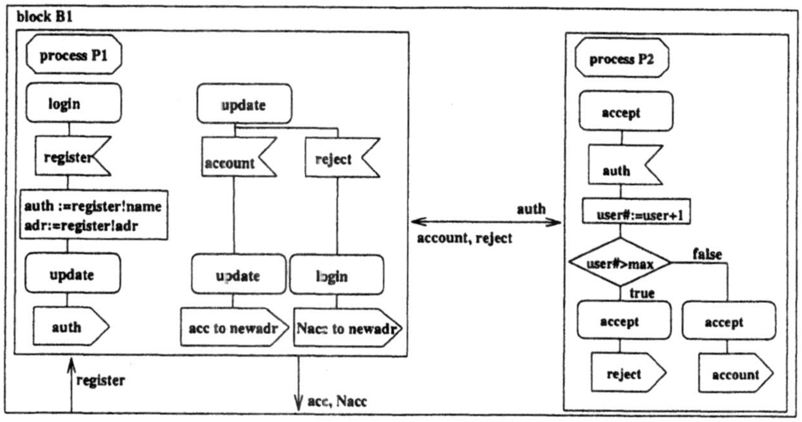

Figure 1.6 Example of figure 1.1 after transition reordering

Figure 1.7 depicts the respective process call sequence and indicates that the implementation is executed correctly.

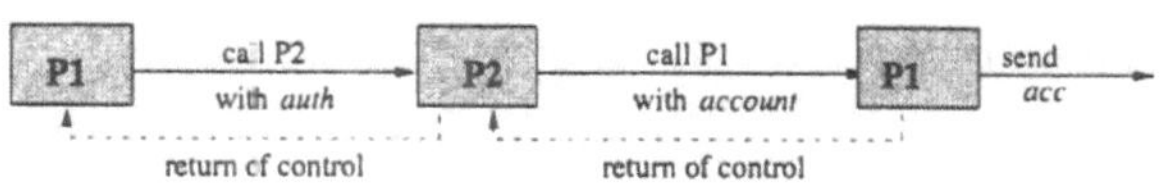

Figure 1.7 Process call sequence for example of figure 1.6

3.3 AVOIDING OVERTAKING OF SIGNALS

Since at compile time the structure of the SDL system as well as the logical structure of the implementation are known the compiler can determine whether there are transitions with multiple outputs which lead to an overtaking of signals. This is done by analysing the execution pathes and the state changes of transitions as described below. Based on this information it can determine in which order the procedure calls implementing output-statements have to be inserted into the final code to prevent overtaking of signals.

In the first step a data flow analysis is applied to detect possible cyclic process call sequences. For this purpose, for each signal sent in a transition with several output-statements the set of successor signals and of receiving processes is determined. The algorithm is recursively applied to the successor signals and terminates when the signals are either sent to the environment or to a process instance implemented according to the server model. For this analysis, it is assumed that each receiving process is in a state where it can accept the incoming signal and execute the corresponding transition. Thus, we get the execution path triggered by each output-statement. If possible cyclic behaviour is detected it is proved whether the first and the last transition of the call sequence use the same signal route. If this happens the following cases have to be distinguished:

1. the signal route is used for multiple output in the same transition,

2. the signal route is used by different transitions.

Case (1) allows it to implement the output-statements in the same order as they are specified. No additional measures have to be taken. In case (2) the output-statement triggering the cyclic process call sequence has to be executed last so that the signal which may be overtaken is sent before the other signals. These output-statements are appended to the end of the transition code.

Figure 1.8 shows the order in which the example given in figure 1.3 is implemented after reordering of the transitions.

Note that the outputs of process *P1* are not only delayed but that their order has also been changed. Figure 1.9 shows again the resulting process call sequence. In this implementation the signals *adr* and *account* arrive in the correct order at *P3*.

There are cases in which the transition reordering does not resolve the semantic conflicts. e.g. for multiple outputs on the same signal route or for cyclic process call sequences which lead to the same transition. In these cases the runtime solution proposed in [Henk97a] has to be applied as before.

3.4 IMPLEMENTATION OF THE TECHNIQUE

The transition reordering approach as described in Section 3.1 has been implemented for the activity thread technique in the SDL compiler

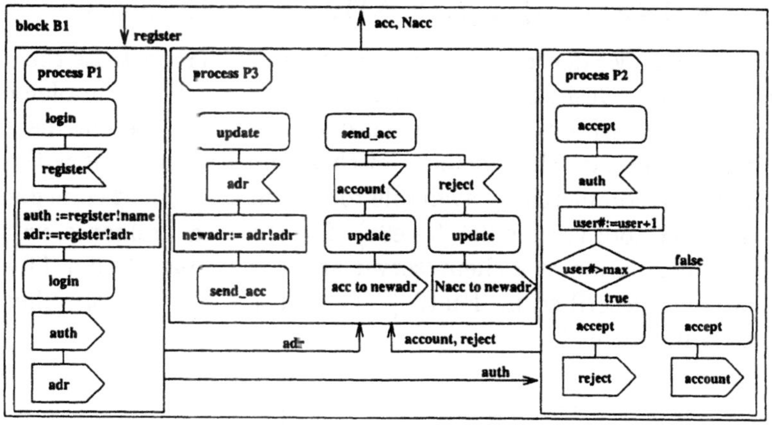

Figure 1.8 Figure 1.3 after transition reordering

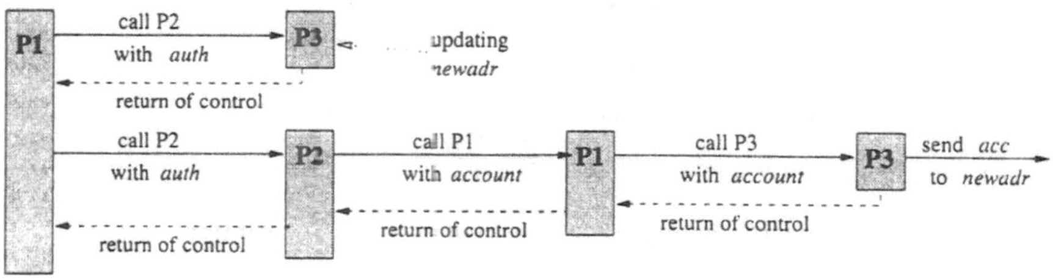

Figure 1.9 Process call sequence for example of figure 1.8

COCOS (Configurable Compiler for SDL) [Lang99]. *COCOS* is a configurable compiler that supports different implementation strategies. Currently these are the server model and the activity thread model. The integrated layer processing will be added soon. Another feature of *COCOS* is its ability to adjust to the given implementation context. This feature overcomes one of the main shortages of automated implementation: the rigid implementation model. Depending of the selected implementation strategy a tailored runtime system is generated which optimally adapts to the implementation environment. The configuration process is controlled by implementation-oriented annotation called *iSDL* which beside the selection of the mapping strategy integrates implementation-oriented information such as number of processors and kind of memory organization into the code generation process. A detailed description of *COCOS* and iSDL is contained in [Lang99].

4. PERFORMANCE MEASUREMENTS

In this section we describe measurements which demonstrate the effect of the approach. Besides measurements of activity thread implementations with and without transition reordering we also present measurements which compare the efficiency of these implementations with code generated by a commercially available tool - the *Cadvanced* code generator of the SDT tool version 3.4 [Tele98]. For the measurements, we used the SDL specification of a client/server application based on a TCP/IP protocol stack described in [Hint97]. The structure of the specification is depicted in figure 1.10.

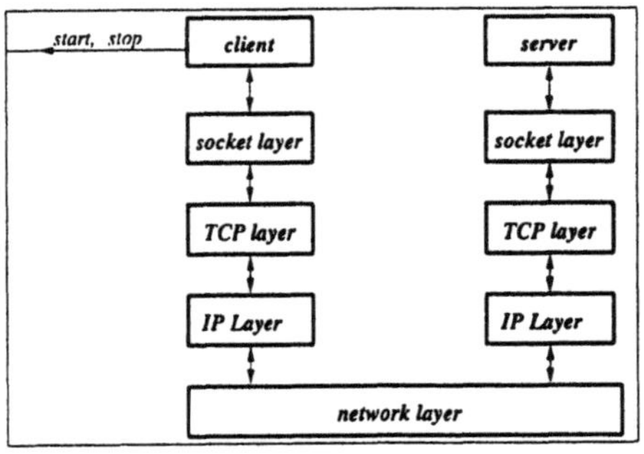

Figure 1.10 Structure of the applied SDL specification

The client process generates data which have to be transmitted to the server process and confirmed. Before the client can open a TCP connection it has to ask for a socket. The server process initiates a passive open to the socket layer. Then it listens. The socket process forwards the application data to the TCP process and vice versa. The TCP process contains the known functionality of the protocol: division of the application data into TCP segments, timer control of the transmission, discarding of duplicate IP packets, flow control, congestion control and error handling. The IP process has a simplified functionality in this specification. It only supplements the IP headers to TCP packets and assigns the incoming IP packets to TCP connections. Fragmentation is not included. The network process simulates the network and transfers the packets from the client to the server site.

We measured the time for the transmission of a sequence of packets starting from the transmission of the first packet until the reception of the last acknowledgement at client site. The process client sends the signal start to the environment when it forwards the first packet. The end of the transmission is indicated by the stop signal.

COCOS as well as *Cadvanced* implement the whole specification by a single operating system process. Thus, the interfaces between the generated code and the operating system do not influence the measurement results. We generated two activity thread implementations with *COCOS*: one without transition reordering and the other with applying this technique. In these implementations the processes socket, TCP, and IP were mapped on activity threads (as described above), while the processes client, server, and network were implemented according to the server model. (This mapping corresponds to the usually applied integration of activity thread implementations into the execution environment, because activity thread implementations require an asynchronous interface at the bottom and the top of the protocol stack.) The generated C code of all implementations was compiled with the gcc compiler without any optimizing options. The measurements were made on a Sun Sparc 20 workstation with four processors and Solaris 2.5 as operating system. We measured the time for the transmission of 1000, 10000 an 20000 packets with a packet size of 1024 byte. Each measurement was repeated 50 times. To exclude exceptional behaviour extreme good and extreme bad measured values were eliminated and then the arithmetic middle of the remaining measured values was calculated. The results are given in figure 1.11.

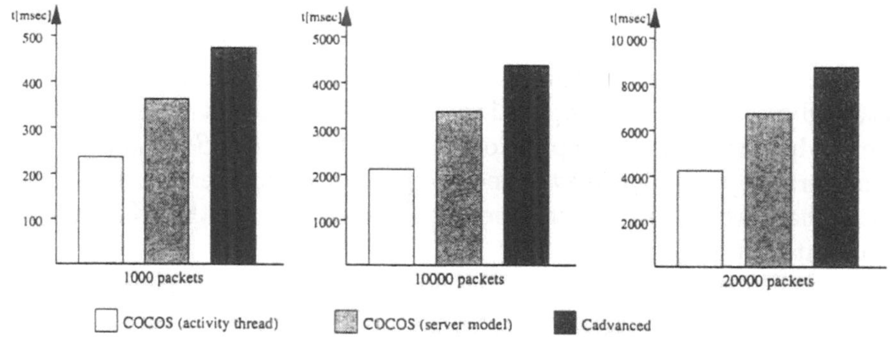

Figure 1.11 Measurements of different implementation strategies

The results show that activity thread implementations with transition reordering achieve a remarkable better performance (between 25 - 30 per cent) than those which apply the approach of [Henk97a]. This increase can be explained by the optimized mapping strategy which resolves semantic conflicts at compile time.

The comparison with the *Cadvanced* code generator shows that both activity thread implementations achieved an about 60 and 120 per cent, respectively, better performance. This proves that by using the activity thread technique a considerably better performance can be obtained. Note that both compilers use the same technique to reduce the operating system overhead. For each signal that is sent memory has to be allocated to store its data. This memory is usually allocated by the operating system at runtime. *Cadvanced* [Tele98] as well as *COCOS* allocate this memory during system initialization.

Further performance improvements are expected the avoidance of data copy operations [Lang99] and an optimized timer management.

5. CONCLUDING REMARKS

In this paper, we have presented with the transition reordering an approach which allows it to efficiently map formal descriptions in an EFSM-based FDT onto activity thread implementations. The transition reordering resolves almost entirely semantic conflicts at compile time. Additional runtime support is only needed for some exceptional cases. The approach has been integrated in the configurable SDL compiler *COCOS*. The measurements we have presented indicate a considerable increase in the efficiency of the generated code. For the application of the transition reordering, an increase of about 25 to 30 per cent was measured. Compared with the code generated by a commercially available tool like the *Cadvanced* compiler an increase up to 120 per cent could be observed. Further improvements are expected by optimizations such as avoidance of data copy operations and a more efficient timer management which are step by step introduced into the *COCOS* compiler. The considerable progress achieved and the mentioned further optimizations show that there is still a large potential for improving the efficiency of automated code generation.

Currently we are introducing the integrated layer processing as third implementation strategy in the configurable compiler. Thereafter, we plan to evaluate our tool and the different implementation techniques by comparing it with hand-coded implementations of the TCP/IP protocol stack or other appropriate protocols. The latter task seems simple, but there scarcely exist real-life protocol implementations of the TCP/IP protocol stack or other protocols which are derived from a formal description and thus could be used as a basis for such a comparison. For TCP/IP, there is the additional problem that most real-life implementations are kernel-integrated implementations. The derivation of such kind of implementations is not the objective of our current research.

TCP/IP implementations though will not be the main application area of automated protocol implementation techniques. We see the potential application of automated implementation techniques more in the area of application protocols and especially for newly designed protocols to fast derive an implementation from the formal description which can be optimized if needed. This will help to close the "implementation gap" in the protocol development process mentioned in the introduction and allow a continuous protocol development based on formal descriptions from the design via automated code generation to testing.

References

[Abbo93] Abbott M.; Peterson, L.: *Increasing network throughput by integrating protocol layers.* IEEE/ACM Transaction On Networking 1, October 1993, pp. 600-610.

[Aho96] Aho, A. V.; Sethi, R.; Ullman, J. D.: *Compilers: principles, techniques, and tools.* Addison Wesley, 1996.

[Brau96] Braun, T.; Chrisment, I.; Diot, C.; Gagnon, F.; Gautier, L.; Hoschka, P.: *ALFred, an ALF/ILP Protocol Compiler for Distributed Application Automated Design.* Rapport de Recherche Nr. 2786, INRIA Jan. 1996.

[Clar85] Clark, D. D.: *The structuring of systems using upcalls.* Proc. 10 th ACM SIGOPS Symp. Oper. Syst. Principles, 1985, pp. 171-180.

[Clar90] Clark, D. D.; Tennenhouse, D. L.: *Architectural considerations for a new generation of protocols.* ACM SIGCOMM, 1990, pp. 200-208 .

[Fisc95] Fischer, S.; Effelsberg, W.: *Efficient Configuration of Protocol Software for Multiprocessors.* In Puigjaner, R. (ed.): High Performance Networking VI, Chapman & Hall, 1995, pp. 195-210.

[Gotz96] Gotzhein, R.; Bredereke, J.; Effelsberg, W.; Fischer, S.; Held, T.; Koenig, H.: Improving the Efficiency of Automated Protocol Implementation Using Estelle. Computer Communications 19 (1996), pp. 1226-1235.

[Held95] Held T.; König, H.: *Increasing the Efficiency of Computer-aided Protocol Implementations.* In Vuong, S., Chanson, S. (eds.): Protocol Specification, Testing and Verification XIV, Chapman & Hall, 1995, pp. 387-394.

[Henk97a] Henke, R.; König, H.; Mitschele-Thiel, A.: *Derivationof Efficient Implementations from SDL Specifications Employing*

Data Referencing, Integrated Packet Framing and Activity Threads. In Cavalli, A.; Sarma, A. (eds.): SDL'97 Time for Testing. Elsevier, 1997, pp. 397-414.

[Henk97b] Henke, R.; Mitschele-Thiel, A.; Koenig, H.: On the Influence of Semantic Constraints on the Code Generation from Estelle Specifications. In Mizuno, T. et al. (eds.): Formal Description Techniques and Protocol Specification, Testing and Verification (FORTE X/PSTV XVII'97), Chapman & Hall, 1997, 399-414.

[Hint97] Hintelmann, J.; Westerfeld, R.: *Performance Analysis of TCP's Flow Control Mechanisms Using Queueing SDL.* In Cavalli, A.; Sarma, A. (eds.): SDL'97 Time for Testing. Elsevier, 1997, pp. 69-84.

[ITU93] ITU-T. Z.100: *Specification and Description Language (SDL).* ITU, 1993.

[Lang99] Langendörfer, P.; König, H.: *COCOS - A Configurable SDL Compiler for Generating Efficient Protocol Implementations.* In Proceedings 9th SDL Forum, Elsevier, 1999.

[Leue96] Leue, S.; Oechslin, P.: *On Parallelizing and Optimizing the Implementation of Communication Protocols.* IEEE/ACM Trans. on Networking, 4 (1), 1996.

[Mans97] Mansurov, N.; Chernov, A.; Ragozin A.: *Industrial Strength Code Generation from SDL.* In Cavalli, A.; Sarma, A. (eds.): SDL'97 Time for Testing. Elsevier, 1997, pp. 415-430.

[Svob89] Svobodova, L.: *Implementing OSI Systems.* IEEE Journal on Selected Areas in Communications, 7(7),1989, pp. 1115-1130.

[Tele98] Telelogic Malmö AB: *SDT 3.2 User's Guide.* SDT 3.2 Reference Manual. 1998.

A GRAPHICAL PARALLEL COMPOSITION OPERATOR FOR PROCESS ALGEBRAS

Hubert Garavel and Mihaela Sighireanu
INRIA Rhône-Alpes / VASY team
655, avenue de l'Europe
38330 Montbonnot St Martin
France
hubert.garavel@inria.fr, mihaela.sighireanu@inria.fr, http://www.inrialpes.fr/vasy

Abstract Process algebras are suitable for describing networks of communicating processes. In most process algebras, the description of such networks is achieved using parallel composition operators. Noticing that the parallel composition operators commonly found in process algebras are often limited in expressiveness and/or difficult for novice users, we propose a new parallel operator that allows networks of communicating processes to be described easily, in a simple and well-structured manner. We illustrate on various examples (token-ring network and client-server protocol) the theoretical and practical merits of this operator.

Keywords: Concurrency, E-LOTOS, Formal Description Technique, LOTOS, Process Algebra, Protocol.

1. INTRODUCTION

Process algebras have been designed as a theoretical framework for the study of concurrency. Classical examples of process algebras are: ACP [BK84], CCS [Mil80; Mil89], CSP [Hoa85], MEIJE [dS85], etc. There also exist specification languages combining process algebraic concepts with features borrowed from functional or imperative programming languages, e.g., the OCCAM [Cam89] language based on CSP, the μCRL [GP91] language based on ACP, and the LOTOS [ISO88] language which combines the best features of CSP and CCS.

Process algebras have undeniable advantages: expressiveness, compositionality, formal semantics given in terms of Labelled Transition Systems (LTS) [Par81] using structural operational semantics [Plo81; GV92], verification algorithms based on behavioural equivalences and preorders, refinement methods, etc. Process algebras have been used successfully many times to model the behaviour of real systems. In addition, simulators, model-checkers, and theorem-provers are available for analyzing process algebraic descriptions, e.g., [DG95; CMS95; FGK+96].

In spite of these advantages, the usual process algebras suffer from limitations in terms of usability (because of their steep learning curve, they often require a substantial training effort), readability (process algebraic descriptions are sometimes difficult to understand), and coverage (important aspects of system description, such as timing, probabilistic aspects, and priorities, are not addressed, although various extensions have been discussed in the literature).

Fortunately, work is going on to extend and improve the mainstream process algebras. In particular, the International Standardization Organization (ISO) has been working since 1992 on the definition of a revised version of the LOTOS language. This revised version, named E-LOTOS and currently at the stage of Final Committee Draft [Que98], includes new features suitable for increasing both the expressiveness and user-friendliness of the language. The work on E-LOTOS has generated many proposals for enhancing both the data type part and the behaviour part of LOTOS (see, e.g., [GS98] for an overview and a discussion on these issues).

In this paper, we focus our attention on the improvement of the parallel composition operators of LOTOS. Although we assume some basic knowledge of LOTOS, our proposals could certainly be applied to other process algebras.

The paper is structured as follows. Section 2. introduces basic definitions and notations. Section 3. suggests to replace the binary parallel composition operators found in most process algebras with a new n-ary operator, better suitable for an easy description of networks of communicating processes. Section 4. proposes further enhancements to this operator, by relaxing the *maximal cooperation* paradigm used in process algebras such as CSP or LOTOS. Section 5. illustrates the usefulness of this parallel operator on a concrete application, the ODP[1] trading function [ISO95a]. Finally, Section 6. gives some concluding remarks.

2. BASIC DEFINITIONS AND NOTATIONS

In the sequel, we use the following notations borrowed from the value-passing process algebras (especially, LOTOS) terminology.

We denote with $B_1, B_2, \ldots$ the algebraic terms constructed using the standard behavioural operators (inaction, action prefix, choice, etc.); these terms are called *behaviours* or *processes*[2]. For our purpose, an exact syntactic definition of behaviours is not required. We denote with "$B_1 = B_2$" the syntactic identity of terms B_1 and B_2.

We denote with $G_1, G_2, \ldots$ the identifiers corresponding to communication points; these identifiers are called *gates*. We define two particular gates: τ, which denotes a non-observable event, and δ, which is used to express the synchronized termination of concurrent behaviours. We denote with $\widehat{G_1}, \widehat{G_2}, \ldots$ the sets of gates.

We denote with $L_1, L_2, \ldots$ the tuples of the form $\langle G, v_1, \ldots, v_n \rangle$, where G is a gate and $v_1, \ldots, v_n$ a (possibly empty) list of typed values; these tuples are called *actions* or *labels*. We denote with $gate(L)$ the gate corresponding to the first element of the tuple L.

Structural operational semantics defines how a behaviour is translated into a (possibly infinite) *labelled transition system* [Par81], which represents all the possible evolutions of the behaviour. The labelled transition system is defined by a transition relation noted "$B_1 \xrightarrow{L} B_2$", which expresses that B_1 can perform an action L and mute to B_2 afterwards.

3. FROM BINARY TO N-ARY PARALLEL COMPOSITION OPERATORS

In LOTOS and most process algebras, parallel composition operators play a central role in the description of concurrent systems. Basically, there are two main uses of parallel composition:

- Parallel composition is the natural mean to describe a set of distributed components that execute concurrently and communicate with each other by message passing: in such an approach, the operands of a parallel composition operator correspond to physically distributed entities. In the taxonomy proposed by [VSS88], this use of parallel composition is called the *resource-oriented specification style.*

- Parallel composition can also be used for refinement purpose. Typically, a given (possibly sequential) component can be divided into a set of sub-components, each of which expresses temporal constraints on the occurrences of certain events. These sub-components are combined together using parallel composition, which acts as the logical conjunction of the corresponding temporal constraints, thus leading to a constrained behaviour. In such case, parallel composition expresses neither physical distribution nor concurrency, but rather a logical modularization of a complex component. In the taxonomy of [VSS88], this use of parallel composition is called the *constraint-oriented specification style.*

Because of this double use of parallel composition, we believe that a suitable parallel composition operator must support *multiway synchronization*, i.e., rendezvous synchronization involving more than two behaviours:

- As far as resource-oriented style is concerned, multiway synchronization is not really necessary: two-way synchronization is sufficient to describe the communication between an emitter and a receiver. Most process algebras allow such *handshaking* synchronization, with some differences with respect to the form of value exchanges that take place during the synchronization.

- But, as far as constraint-oriented style is concerned, multiway synchronization is mandatory. For instance, a controller for a robot with n degrees of freedom can be expressed as the parallel composition of n sub-processes, each sub-process controlling the motion of the robot with respect to a given degree of freedom; to perform a given mission (e.g., moving the robot from one location to another one), all sub-processes have to synchronize.

With the notable exception of CCS, most process algebras (ACP, CSP, MEIJE, LOTOS...) support multiway synchronization, which is clearly a desirable feature. Yet, many process algebras rely on (associative) binary parallel composition operators to express multiway synchronization between n processes. This can be explained by the fact that the original motivation behind the design of process algebras was the search for a "minimal" model of concurrency.

For instance, LOTOS has three parallel operators, noted "B_1 |[$\widehat{G}$]| B_2", "B_1 ||| B_2", and "B_1 || B_2", respectively. The first operator is the most general: it expresses that B_1 and B_2 execute concurrently and synchronize only on the gates of $\widehat{G} \cup \{\delta\}$. The second and third operators are particular cases of the former: "|||" corresponds to the case where $\widehat{G}$ is empty (fully asynchronous execution) and "||" to the case where $\widehat{G}$ is the set of all visible gates (fully synchronous execution). From our experience in describing complex, industrial systems using LOTOS, we believe that expressing parallel composition using binary operators has major drawbacks:

- For a given network of concurrent processes, there are usually several different algebraic terms representing this network. For instance, the network shown on Figure 1.1 can be described using two (equivalent) LOTOS terms, e.g.,

$$((B_1 \;|[G_1]|\; B_2) \;|[G_1, G_3, G_4]|\; B_3) \;|[G_2, G_3, G_4]|\; (B_4 \;|[G_2]|\; B_5)$$

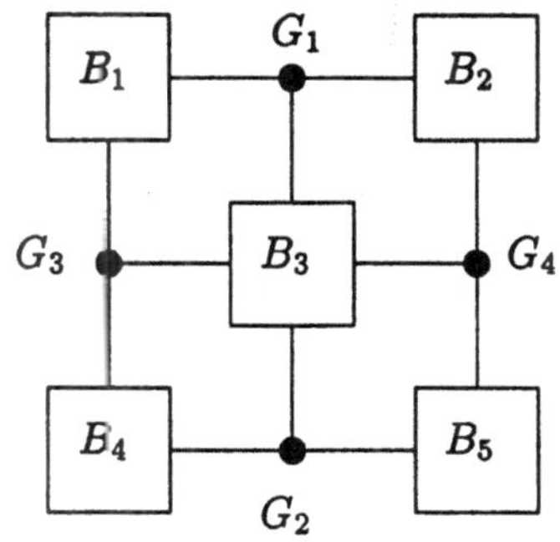

Figure 1.1

or

$$((B_1 \;|[G_1]|\; B_2) \;|||\; (B_4 \;|[G_2]|\; B_5)) \;|[G_1, G_2, G_3, G_4]|\; B_3$$

The absence of a canonical form is practically unfortunate, as an algebraic description will strongly depend on the style adopted by its author, thus leading to a lack of uniformity. Moreover, the problem of determining whether two terms are equivalent (for all possible sub-terms B_1, B_2, etc.) is decidable, but not immediate in the general case, as it implies to solve a system of boolean equations [Kar94; Kar97].

- There are some process networks that can not be expressed as algebraic terms. For instance, the network on Figure 1.2 can not be expressed using LOTOS parallel composition, because it involves two-by-two synchronization on the same gate G, whereas LOTOS would force all three processes to synchronize on G using a three-way rendezvous (this is called the *maximal cooperation* paradigm). Sufficient conditions for a process network to be translated into a LOTOS behaviour expression are studied in [Bol90]. Although the example of Figure 1.2 may seem somehow artificial, Section 5. will show that networks involving two-by-two synchronization are useful for describing client-server architectures.

Besides these theoretical issues, there are also pragmatic considerations against binary parallel composition operators. The main argument is that binary operators create a discrepancy between the graphical representation of process networks (always present in the designer's mind)

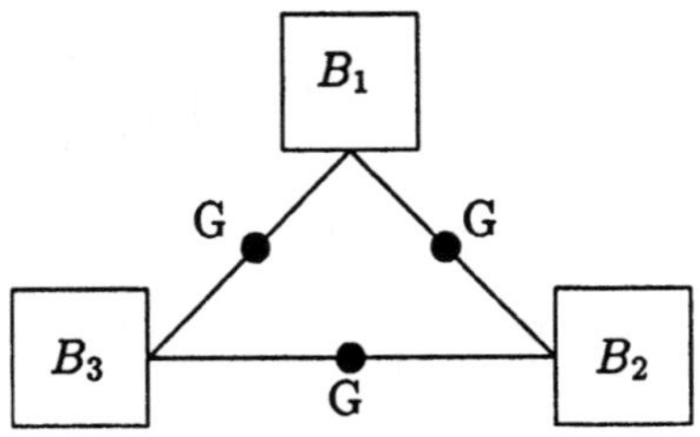

Figure 1.2

and the textual representation as an algebraic term. On the one hand, it is not easy for novice users to write an algebraic term corresponding to a given network of concurrent processes. On the other hand, given an algebraic term, it is not always immediate to infer the corresponding network.

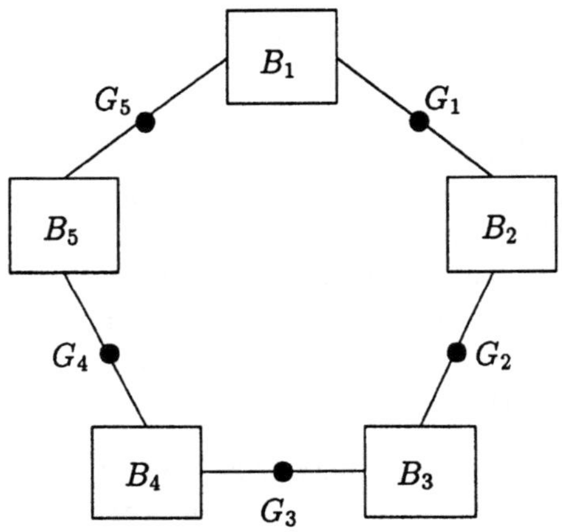

Figure 1.3

Some network topologies are particularly tricky to express using binary operators. For instance, the simplest algebraic term for representing the token-ring network shown on Figure 1.3 is:

$$(B_1 \ \texttt{|[}G_1\texttt{]|} \ B_2 \ \texttt{|[}G_2\texttt{]|} \ B_3 \ \texttt{|[}G_3\texttt{]|} \ B_4) \ \texttt{|[}G_4, G_5\texttt{]|} \ B_5$$

which is particularly non-intuitive because the circular symmetry of the network cannot be preserved during the translation to an algebraic term. In this respect, the difficulties inherent to the process algebraic approach have to be compared with graphical formalisms such as SDL [IT92] or

Statecharts [Har87], in which the user simply has to draw the desired network.

For these reasons, we suggested to introduce in E-LOTOS a new n-ary parallel composition operator that would replace the binary operators of LOTOS. Based on an early suggestion by [Bri88], we made several iterative proposals [Gar95; SG96], before our proposal was accepted for being included in E-LOTOS. The basic syntax of the n-ary parallel composition operator is:

```
par
      Ĝ1 → B1
   || Ĝ2 → B2
   || ...
   || Ĝn → Bn
endpar
```

This operator describes a network of $n \geq 1$ concurrent behaviours $B_1, ..., B_n$. We define I to be the set $\{1, ..., n\}$. To each behaviour B_i is associated an *interface* consisting of a set of gates $\widehat{G_i}$ on which B_i must synchronize. Each $\widehat{G_i}$ can be empty; in such case the arrow before B_i can be omitted.

We later discovered that such an extended parallel composition operator had been already proposed in, at least, two occasions: [Bol90] suggests that process networks could be described using an n-ary operator (noted *MaxCoop*) and gives rules for translating process networks into strongly equivalent LOTOS behaviour expressions ; [DS92; DS95] include in CSP a similar operator, the semantics of which is expressed in terms of traces.

Instead, we define the n-ary parallel composition operator by means of two rules of structured operational semantics. The first rule expresses that any behaviour B_i can execute asynchronously any action L whose gate G does not belong to the interface $\widehat{G_i}$ and is different from the termination gate δ (this encompasses the case where $L = \tau$), while the other behaviours B_j with $j \neq i$ do not evolve:

$$\frac{(\exists L)\ (\exists i \in I)\ B_i \xrightarrow{L} B_i' \ \wedge\ gate(L) \notin \widehat{G_i} \cup \{\delta\} \ \wedge\ (\forall j \in I \backslash \{i\})\ B_j' = B_j}{\textbf{par}\ \widehat{G_1} \to B_1 ... \widehat{G_n} \to B_n\ \textbf{endpar} \xrightarrow{L} \textbf{par}\ \widehat{G_1} \to B_1' ... \widehat{G_n} \to B_n'\ \textbf{endpar}}$$

The second rule expresses that a behaviour B_i wanting to execute an action L labelled by a gate $G \in \widehat{G_i} \cup \{\delta\}$ must synchronize with all the other behaviours B_j such that $G \in \widehat{G_j} \cup \{\delta\}$:

$$\frac{(\exists L)\ (\forall i \in I)\ (\mathit{if}\ gate(L) \in \widehat{G_i} \cup \{\delta\}\ \mathit{then}\ B_i \xrightarrow{L} B_i'\ \mathit{else}\ B_i' = B_i)}{\textbf{par}\ \widehat{G_1} \to B_1 ... \widehat{G_n} \to B_n\ \textbf{endpar} \xrightarrow{L} \textbf{par}\ \widehat{G_1} \to B_1' ... \widehat{G_n} \to B_n'\ \textbf{endpar}}$$

This operator solves the aforementioned problems of binary operators by establishing a direct mapping between process networks and their textual representation, thus paving the way for tools that automatically perform the translation from graphical networks to algebraic terms and vice versa. For instance, the networks of Figures 1.1 and 1.3 can be expressed as follows:

$$\begin{array}{l}\textbf{par}\\ \quad\ \ G_1, G_3 \rightarrow B_1\\ \quad || \ G_1, G_4 \rightarrow B_2\\ \quad || \ G_1, G_2, G_3, G_4 \rightarrow B_3\\ \quad || \ G_2, G_3 \rightarrow B_4\\ \quad || \ G_2, G_4 \rightarrow B_5\\ \textbf{endpar}\end{array}$$

and:

$$\begin{array}{l}\textbf{par}\\ \quad\ \ G_1, G_5 \rightarrow B_1\\ \quad || \ G_1, G_2 \rightarrow B_2\\ \quad || \ G_2, G_3 \rightarrow B_3\\ \quad || \ G_3, G_4 \rightarrow B_4\\ \quad || \ G_4, G_5 \rightarrow B_5\\ \textbf{endpar}\end{array}$$

respectively.

As far as expressiveness is concerned, it is obvious that the general parallel operator "$|[\widehat{G}]|$" of LOTOS can be obtained as a particular case of the n-ary operator:

$$B_1 \ |[\widehat{G}]| \ B_2 \ = \ \textbf{par} \ \widehat{G} \rightarrow B_1 \ || \ \widehat{G} \rightarrow B_2 \ \textbf{endpar}$$

Reciprocally, the n-ary operator is strictly more expressive than the "$|[\widehat{G}]|$" operator of LOTOS. We prove this proposition using the process network shown on Figure 1.4 (which does not satisfy the sufficient conditions of [Bol90] because, for instance, the three processes can access gate G_1 but do not synchronize altogether on this gate). This process can easily be described using the n-ary operator:

$$\begin{array}{l}\textbf{par}\\ \quad\ \ G_2, G_3 \rightarrow B_1\\ \quad || \ G_1, G_3 \rightarrow B_2\\ \quad || \ G_1, G_2 \rightarrow B_3\\ \textbf{endpar}\end{array}$$

but cannot be expressed using the LOTOS binary parallel operators: to describe this network, one must first synchronize two processes together,

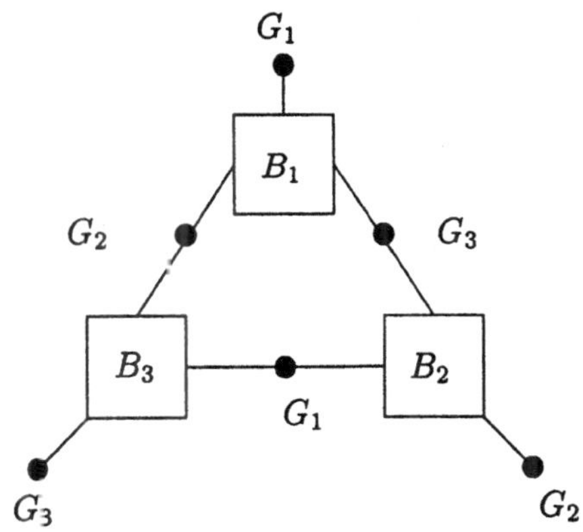

Figure 1.4

then synchronize the result with the third process; assuming that B_1 and B_2 are to be synchronized first (which can be done without loss of generality because the network is symmetric), it is mandatory to synchronize them on gate G_3; then, the resulting term "B_1 |[G_3]| B_2" has to be synchronized with B_3 on gates G_1 and G_2. But the term obtained does not correspond to the network of Figure 1.4 where process B_1 can perform actions on gate G_1 independently from process B_3.

As a side remark, it is worth noticing that the network of Figure 1.4 can only be expressed by combining the "|[$\widehat{G}$]|" operator of LOTOS together with process instantiation. Technically, this can be done by defining an auxiliary process P with auxiliary gates G'_1, G'_2, G'_3, and by a clever instantiation of this process so as to rename the auxiliary gates into G_1, G_2, G_3 respectively:

> $P[G_1, G_2, G_3, G_1, G_2, G_3]$
> **where**
> **process** $P[G_1, G_2, G_3, G'_1, G'_2, G'_3] :=$
> ($B_1[G'_1, G_2, G_3]$ |[G_3]| $B_2[G_1, G'_2, G_3]$) |[G_1, G_2]| $B_3[G_1, G_2, G'_3]$
> **endproc**

The same result could be achieved using the relabelling operator existing in other process algebras, such as ACP or CSP (in LOTOS, the process instantiation performs relabelling implicitly). However, this solution is probably too tricky for most users; in any case, the n-ary parallel operator is simpler and more intuitive.

Finally, we slightly extend the n-ary operator by allowing to specify a set $\widehat{G_0}$ of synchronization gates common to all processes B_i (assuming that $\widehat{G_0} \cap \widehat{G_i} = \emptyset$ for each $i \in I$). This extension is practically helpful for avoiding redundant lists of gates; it is simply defined as a syntactic shorthand (where "⊎" denotes disjoint union of sets):

$$\left(\begin{array}{l}\textbf{par } \widehat{G_0} \textbf{ in} \\ \quad \widehat{G_1} \rightarrow B_1 \\ \,||\; \widehat{G_2} \rightarrow B_2 \\ \,||\; \ldots \\ \,||\; \widehat{G_n} \rightarrow B_n \\ \textbf{endpar}\end{array}\right) =_{def} \left(\begin{array}{l}\textbf{par} \\ \quad \widehat{G_0} \uplus \widehat{G_1} \rightarrow B_1 \\ \,||\; \widehat{G_0} \uplus \widehat{G_2} \rightarrow B_2 \\ \,||\; \ldots \\ \,||\; \widehat{G_0} \uplus \widehat{G_n} \rightarrow B_n \\ \textbf{endpar}\end{array}\right)$$

4. FROM MAXIMAL TO "*M* AMONG *N*" COOPERATION

Although superior to the LOTOS parallel composition operator, the n-ary operator described in Section 3. is not expressive enough to represent certain process networks, such as the one of Figure 1.2. This limitation is unfortunate, because it precludes several networks of practical interest from being modelled, especially the case where a pool of n processes synchronize two by two on the same gate. Although CCS permits such "2 among n" synchronization, other process algebras, such as CSP or LOTOS, do not allow it, because they rely on the maximal cooperation paradigm.

Based on our practical experience, we suggest to extend the n-ary operator in order to allow "m among n" synchronization, i.e., when a set of n processes synchronize m by m on the same gate (with $m \leq n$). Our extended operator is based on our previous proposals [Gar95; SG96] submitted to the E-LOTOS standardization Committee. This operator has the following syntax:

$$\begin{array}{l}\textbf{par } g_1\#m_1, g_2\#m_2, \ldots, g_p\#m_p \textbf{ in} \\ \quad \widehat{G_1} \rightarrow B_1 \\ \,||\; \widehat{G_2} \rightarrow B_2 \\ \,||\; \ldots \\ \,||\; \widehat{G_n} \rightarrow B_n \\ \textbf{endpar}\end{array}$$

where $g_1, \ldots, g_p$ is a (possibly empty) list of gates and where $m_1, \ldots, m_p$ are natural numbers in the range *1, ..., n* associated to these gates. Each clause "$\#m_j$" is optional: if omitted, m_j has the default value n. We define $\widehat{G_0}$ to be the set of gates $\{g_1, \ldots, g_p\}$ and we require that $\widehat{G_0} \cap \widehat{G_i} = \emptyset$ for $i \in I$. Notice that we do not require the gates $g_1, \ldots, g_p$ to be pairwise distinct.

Informally, the semantics of this operator is the following. As regards the gates of $\widehat{G_1} \cup ... \cup \widehat{G_n} \cup \{\delta\}$, this operator behaves exactly as the one described in Section 3.. As regards the gates of $\widehat{G_0}$, this operator specifies that processes $B_1, ..., B_n$ can perform m_j among n synchronization on each gate g_j. Two special cases are of interest: if $m_j = 1$, each process B_i can execute asynchronously an action on gate g_j; if $m_j = n$, all processes B_i have to synchronize on gate g_j.

To provide a formal semantics, we introduce a predicate noted "$G \triangleright J$", where $J \subseteq I$, that is true iff the processes in $\{B_i \mid i \in J\}$ can synchronize together on gate G. Obviously, for a given gate G, there may be several subsets J such that $G \triangleright J$. This predicate is defined as follows:

- $\delta \triangleright I$, meaning that all concurrent processes must synchronize on the termination gate δ, as in LOTOS;
- $(\forall j \in \{1, ..., p\})\ (\forall J \subseteq I \mid card(J) = m_j)\ \ g_j \triangleright J$, meaning that each gate g_j achieves m_j among n synchronization;
- $(\forall G \in \widehat{G_1} \cup ... \cup \widehat{G_n})\ \ G \triangleright \{i \in I \mid G \in \widehat{G_i}\}$, meaning that all processes having G in their interfaces must synchronize on G;
- $(\forall i \in I)\ \ (\forall G \notin \widehat{G_0} \cup \widehat{G_i} \cup \{\delta\})\ \ G \triangleright \{i\}$, meaning that each process B_i can perform asynchronously any gate neither mentioned in $\widehat{G_0}$ nor in its interface $\widehat{G_i}$ (δ excepted and τ included).

Using this predicate, the operational semantics of our parallel operator can be defined with a single semantic rule:

$$\frac{(\exists L)\ \ (\exists J \subseteq I)\ \ (gate(L) \triangleright J)\ \wedge\ ((\forall i \in J)\ \ B_i \xrightarrow{L} B_i')\ \wedge\ ((\forall i \in I \backslash J)\ \ B_i' = Bi)}{\textbf{par}\ g_j\#m_j ...\ \textbf{in}\ \widehat{G_i} \rightarrow B\ ...\ \textbf{endpar} \xrightarrow{L} \textbf{par}\ g_j\#m_j ...\ \textbf{in}\ \widehat{G_i} \rightarrow B_i' ...\ \textbf{endpar}}$$

Using this operator, the process network of Figure 1.2 can be specified using 2 among 3 synchronization:

```
par G#2 in
  B1 || B2 || B3
endpar
```

More complex process networks, such as the one on Figure 1.5, in which the same gate G has various degrees of synchronization, can also be described:

```
par G#2, G#3 in
  B1 || B2 || B3
endpar
```

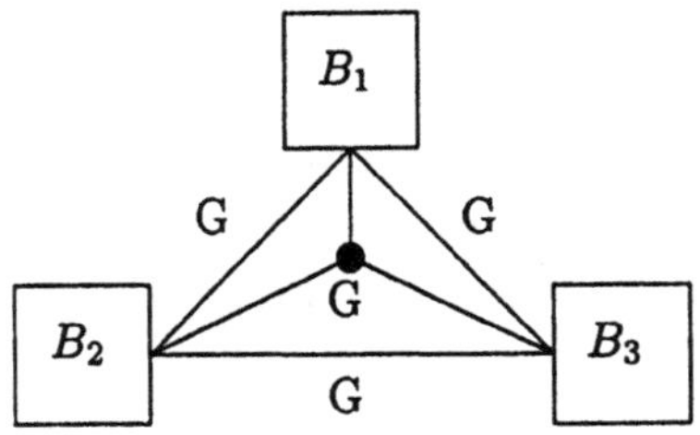

Figure 1.5

5. APPLICATION

In this Section, we illustrate the application of our parallel operator to the description of the ODP trading function. ODP [ISO95b] is a standard framework for distributed applications. Within ISO, E-LOTOS has been developed in the same working group as ODP, with the intent of being the formal description technique for distributed applications. This explains that ODP-related problems have been a constant source of inspiration for E-LOTOS designers.

Our proposals for introducing "2 among n" synchronization in E-LOTOS [Gar95; SG96] was motivated by the highly dynamic nature of ODP systems: processes can be created and destroyed dynamically, and binary communications between processes can be established dynamically. Although it has been argued that such behaviours could only be described by means of mobile process calculi, such as the π-calculus [MPW92a; MPW92b], we believe that the most salient aspects of ODP systems can be captured in the framework of an usual process algebra, such as LOTOS, extended with our new parallel operator. A comparative study of both approaches can be found in [FNLL96].

The ODP trading function is a typical example of ODP systems: this function is defined informally in an ISO standard [ISO95a]. A formal description in E-LOTOS of the essential features of the trading function can be found as an appendix of the E-LOTOS definition document [Que98, Annex A.3]. In this paper, we focus on the architectural description of the trading function, so as to explain how our parallel operator can be used to describe dynamic communication patterns.

The ODP *trader* is a computer process that establishes a relationship between a pool of *objects* within an open and dynamically changing distributed system. For simplicity, we assume that there is an upper bound n on the number of objects in the system. Each object can act either as a *service provider* (or *server*), as a *client*, or even as both.

On the one hand, a server must inform the trader of the services it is ready to offer. Advertising a service offer is called *export*. The trader keeps in a database all the export requests sent by the servers. On the other hand, a client may ask the trader about available services. Requesting knowledge about a particular service is called *import*. The trader matches the clients' service requests with its database of service offers and, if possible, selects an appropriate server. The identification of this server is sent back to the client, which can then contact directly the server without further interaction with the trader.

The interesting issue in this architecture is that a client can eventually communicate with a server the identity of which was unknown to him before asking the trader. In mobile process calculi, this situation can be described using a dynamic creation of mobile gate(s) and/or agent(s). However, alternative approaches are possible, which avoid the complexity of dynamic gate/agent creation. We can model the behaviour of the whole system by the following parallel composition. Let E (export), I (import), and W (work) be three gates used for server-trader, client-trader, and client-server communication respectively. Let "$O[E, I, W](i)$" and "$T[E, I]$" be two processes representing the ith object and the trader, respectively. The whole architecture can be described by the following term:

```
par E,I in
  T[E,I] || par W#2 in
                O[E,I,W](1) || ... || O[E,I,W](n)
            endpar
endpar
```

It is worth noticing that the pool of n objects could be expressed in a more concise way using an extended parallel operator that iterates over a finite set of values (such an operator was proposed in [Gar95] and introduced in E-LOTOS).

The behaviour of the trader can be described with the following LOTOS process, where *request* and *reply* are two enumerated values indicating the direction of the messages exchanged on gates I and W (according to the syntax of LOTOS, "[]" denotes the choice operator, "? $x : s$" denotes the receipt of a value of type s to be stored in variable x, and "! v" denotes either the transmission of value v or the receipt of a value that has to be equal to v):

```
process T[E, I](d : DataBase) : noexit :=
  E ?j: Object ?s: Service;
    T[E, I](add_to_database(d, j, s))
  []
  I ?i: Object !request ?s: Service;
    I !i !reply !search_server_in_database(d, s)
      T[E, I](d)
endproc
```

Assuming that the ith object is a client asking the trader for some service s provided by the jth server, and then requesting this server directly, its behaviour can be described as follows (we assume that there is always a server available for the requested service):

```
process O[E, I, W](i : Object) : noexit :=
  I !i !request !s;
    I !i !reply ?j: Object;
      W !j !i !request !s...;
        W !j !i !reply !s...;
          ...
endproc
```

Similarly, assuming that the jth object is a server advertising a given service s to the trader, then answering a client request, its behaviour can be described as follows:

```
process O[E, I, W](j : Object) : noexit :=
  E !j !s;
    W !j ?i: Object !request !s...;
      W !j !i !reply !s...;
        ...
endproc
```

6. CONCLUSION

Taking into account that the binary parallel composition operators found in usual process algebras (such as ACP, CCS, μCRL, LOTOS, and the early versions of CSP) are not fully appropriate for describing complex synchronization architectures, we suggest to extend the LOTOS parallel composition operator "$|[\widehat{G}]|$" in two directions:

- First, we propose to replace the binary operator with an n-ary operator that directly reflects the graphical structure of process networks. From the examples given, it is clear that the n-ary operator is simpler to use by novice users, easier to read (because the structure of process networks is preserved), strictly more expressive, and appropriate for an automatic translation from graphical networks to algebraic terms and vice-versa.

Although similar proposals can be found elsewhere [Bol90; DS92; DS95], which is a sign of soundness, our approach is slightly different because our n-ary operator can describe process networks not tackled in [Bol90] (for instance, the one of Figure 1.4) and because we adopted structured operational semantics (as in [ISO88]), rather than traces (as in [DS92; DS95]).

- Second, we increased the expressiveness of this new operator by relaxing the maximal cooperation requirement of CSP and LOTOS, in order to support "m among n" synchronization. Taking the ODP trading function as an example, we show that the new operator is user-friendly, intuitive, and practically useful for the description of networks with mobility and dynamic reconfiguration capabilities.

Our research benefited from discussions with our colleagues in the framework of the E-LOTOS standardization Committee. The parallel operator presented in this paper is a refined version of a previous proposal, which we submitted to ISO [Gar95; SG96] and which has been integrated in the current version of E-LOTOS [Que97].

Notes

1. Open Distributed Processing
2. Here, we slightly deviate from the meaning of *process* in LOTOS

References

J. A. Bergstra and J. W. Klop. Process Algebra for Synchronous Communication. *Information and Computation*, 60:109–137, 1984.

T. Bolognesi. A Graphical Composition Theorem for Networks of Lotos Processes. In IEEE Computer Society, editor, *Proceedings of the 10th International Conference on Distributed Computing Systems, Washington, USA*, pages 88–95. IEEE, May 1990.

Ed Brinksma. *On the Design of Extended LOTOS, a Specification Language for Open Distributed Systems.* PhD thesis, University of Twente, November 1988.

J. Camillieri. An Operational Semantics for OCCAM. *International Journal of Parallel Programming*, 18(5):149–167, October 1989.

Rance Cleaveland, Eric Madelaine, and Steve Sims. A Front-End Generator for Verification Tools. In Uffe H. Engberg, Kim G. Larsen, and Arne Skou, editors, *Proceedings of TACAS'95 Tools and Algorithms for the Construction and Analysis of Systems (Aarhus, Denmark)*, May 1995. Also available as INRIA Research Report RR-2612.

D. Dams and J. F. Groote. Specification and Implementation of Components of a μCRL Toolbox. Technical Report Logic Group Preprint Series 152, Utrecht University, December 1995.

Robert de Simone. Higher-level synchronising devices in MEIJE–SCCS. *Theoretical Computer Science*, 37:245–267, 1985.

J. W. Davies and S. A. Schneider. A Brief History of Timed CSP. Technical Monography PRG-96, Oxford University, 1992.

J. W. Davies and S. A. Schneider. A Brief History of Timed CS. *Theoretical Computer Science*, 138(2):243–271, February 1995.

Jean-Claude Fernandez, Hubert Garavel, Alain Kerbrat, Radu Mateescu, Laurent Mounier, and Mihaela Sighireanu. CADP (CÆSAR/ALDEBARAN Development Package): A Protocol Validation and Verification Toolbox. In Rajeev Alur and Thomas A. Henzinger, editors, *Proceedings of the 8th Conference on Computer-Aided Verification (New Brunswick, New Jersey, USA)*, volume 1102 of *Lecture Notes in Computer Science*, pages 437–440. Springer Verlag, August 1996.

A. Février, E. Najm, G. Leduc, and L. Léonard. Compositional Specification of ODP Binding Objects. In *Proceedings of the 6th IFIP/ICCC Conference on Information Network and Data Communication, INDC'96, Trondheim, Norway*, June 1996.

Hubert Garavel. A Wish List for the Behaviour Part of E-LOTOS. Rapport SPECTRE 95-21, VERIMAG, Grenoble, December 1995. Input document [LG5] to the ISO/IEC JTC1/SC21/WG7 Meeting on Enhancements to LOTOS (1.21.20.2.3), Liège (Belgium), December, 18–21, 1995.

J. F. Groote and A. Ponse. Proof theory for μ-CRL. Technical Report CS-9138, CWI Amsterdam, 1991.

Hubert Garavel and Mihaela Sighireanu. Towards a Second Generation of Formal Description Techniques – Rationale for the Design of E-LOTOS. In Jan-Friso Groote, Bas Luttik, and Jos van Wamel, editors, *Proceedings of the 3rd International Workshop on Formal Methods for Industrial Critical Systems FMICS'98 (Amsterdam, The Netherlands)*, pages 187–230, Amsterdam, May 1998. CWI. Invited lecture.

J. F. Groote and F. W. Vaandrager. Structured Operational Semantics and Bisimulation as a Congruence. *Information and Computation*, 100(2):202–260, October 1992.

D. Harel. StateCharts: A Visual Formalism for Complex Systems. *Science of Computer Programming*, 8(3):231–274, 1987.

C. A. R. Hoare. *Communicating Sequential Processes*. Prentice-Hall, 1985.

ISO/IEC. LOTOS — A Formal Description Technique Based on the Temporal Ordering of Observational Behaviour. International Standard 8807, International Organization for Standardization — Information Processing Systems — Open Systems Interconnection, Genève, September 1988.

ISO/IEC. ODP Trading Function. Draft International Standard 13235, ISO — Information Processing Systems, Genève, June 1995.

ISO/IEC. Open Distributed Processing — Reference Model. International Standard 10746, ISO — Information Processing Systems, Genève, 1995.

ITU-T. Specification and Description Language (SDL). ITU-T Recommendation Z.100, International Telecommunication Union, Genève, 1992.

Pim Kars. Representation of Process-Gate Nets in LOTOS and Verification of LOTOS Laws: the Boolean Algebra Approach. In Dieter Hogrefe and Stefan Leue, editors, *Proceedings of the 7th International Conference on Formal Description Techniques for Distributed Systems and Communication Protocols FORTE'94 (Bern, Switzerland)*, October 1994.

P. Kars. *Process-Algebraic Transformations in Context*. PhD thesis, University of Twente, June 1997.

Robin Milner. *A Calculus of Communicating Systems*, volume 92 of *Lecture Notes in Computer Science*. Springer Verlag, 1980.

Robin Milner. *Communication and Concurrency*. Prentice-Hall, 1989.

R. Milner, J. Parrow, and D. Walker. A calculus of mobile processes I. *Information and Computation*, 100(1):1–40, September 1992.

R. Milner, J. Parrow, and D. Walker. A calculus of mobile processes II. *Information and Computation*, 100(1):41–77, September 1992.

David Park. Concurrency and Automata on Infinite Sequences. In Peter Deussen, editor, *Theoretical Computer Science*, volume 104 of *Lecture*

Notes in Computer Science, pages 167–183. Springer Verlag, March 1981.

G. D. Plotkin. A structural approach to operational semantics. DAIMI FN-19 FN-19, Computer Science Department, Aarhus University, 1981.

Juan Quemada, editor. Committee Draft on Enhancements to LOTOS (E-LOTOS). ISO/IEC JTC1/SC21/WG7 Project 1.21.20.2.3, January 1997.

Juan Quemada, editor. Committee Draft on Enhancements to LOTOS (E-LOTOS). ISO/IEC FCD 15437, April 1998.

Mihaela Sighireanu and Hubert Garavel. E-LOTOS User Language. Rapport SPECTRE 96-06, VERIMAG, Grenoble, October 1996. In ISO/IEC JTC1/SC21 Third Working Draft on Enhancements to LOTOS (1.21.20.2.3). Output document of the edition meeting, Kansas City, Missouri, USA, May, 12–21, 1996.

C. Vissers, G. Scollo, and M. van Sinderen. Architecture and Specification Style in Formal Descriptions of Distributed Systems. In S. Aggarwal and K. Sabnani, editors, *Proceedings of the 8th International Workshop on Protocol Specification, Testing and Verification (Atlantic City, NJ, USA)*, pages 189–204. IFIP, North-Holland, 1988.

GUARANTEEING LIVENESS IN AN OBJECT CALCULUS THROUGH BEHAVIORAL TYPING

Elie Najm [a,1], Abdelkrim Nimour [b], Jean-Bernard Stefani [c,1]

[a]Ecole Nationale Superieure des Telecomunications
[b]Schlumberger - Test & Transactions - Smart Cards Products
[c]France Telecom-CNET

[a]Elie.Najm@enst.fr, [b]knimour@slb.com, [c]jeanbernard.stefani@cnet.francetelecom.fr

Abstract: A typing discipline for concurrent objects is defined. A decidable type system is constructed which features a liveness property of well-typed object configurations: all pending request messages are treated within a finite number of steps (modulo fairness). The type system is exhibited on *OL2*, a variant of the asynchronous π-calculus. The present paper builds on previous work. It extends the safety results of [11]. It also provides for the incremental extension of well-typed (and well-behaved) configrations: when the code of an object is type checked in the context of an interface repository, it can be added a existing configuration while preserving the liveness property.

INTRODUCTION

We are concerned with guaranteeing liveness properties in configurations of objects. We advocate a now well-established approach based on behavioral typing. The aim is to define a typing discipline and an associated type system such that well-typed programs yield well-behaving programs. We base our analysis on an object calculus *OL2*, akin to the asynchronous π-calculus [16, 10] or the actors model [1]. In our setting, the problem we tackle can be stated as follows: considering an *OL2* well-typed closed object configuration C, then C satisfies a liveness property whereby for any configuration C' derived from C, and for any pending message M contained in C', there is configuration C'' derived from C' where M has been absorbed by its target object. Furthermore, we handle configuration extensions. The type system is such that

[1]This work has been partially supported by RNRT project MARVEL

well-typed objects are safely added to running configurations and liveness is preserved in the extended configuration. The paper is structured as follows: section 2 is an informal introduction to *OL2*, section 3 presents the operational semantics of *OL2*, section 4 discusses the syntax and semantics of the interface type system, section 5 presents is devoted to a small example, section 6 introduces the static semantics, the rules of which are given in the appendix, section 7 presents the main results of the paper and section 8 is the conclusion.

THE *OL2* CALCULUS

OL2 is an object-based calculus, i.e., it provides for the description of objects with features such as interfaces, clients, servers, service offers, method invocation, etc. It is now a well-established result that object calculi are based on more foundational theories. Indeed, *OL2* is an avatar of the asynchronous π-calculus with the usual, implementation oriented [16, 4] restrictions: (i) restriction of choice to input-prefix processes; (ii) specialization of gates to input or output; (iii) tagging of messages (and actions) with constant labels. In object terminology, such message labels are called methods names; they act as input selectors. *OL2* can also be seen as extension of the Actors language whereby actors may have multiple addresses.

We start with an informal presentation of the calculus, gradually introducing the different constructs, then we will turn to the formal syntax and semantics. *OL2* describes configurations of objects running in parallel and exchanging messages. Each object offers services on a set of interfaces. Each object may invoke the services of other objects by sending messages to the interfaces providing these services. The syntax of a service offer on an interface of an object is as follows:

$$u[m() = B_1] \qquad \text{(Ex-1)}$$

The behavior of this object is to wait for the message m on the interface u and then proceed as specified in B_1. The invocation of this service m by an object is written:

$$u.m() > B_2 \qquad \text{(Ex-2)}$$

We will so distinguish between two roles that can be possessed by an object for a given interface:

- client role: an object that has the client role of an interface can invoke the services offered by this interface.
- server role: an object that has the server role of an interface can accept messages (and thus it offers services) on this interface.

For instance, the behavior expression of the example Ex-1 describes an object which has the server role on u whereas the one of example Ex-2 has the client role. Note that an object may have both roles for the same interface: a typical case for newly created interfaces. Interface roles may be exchanged between objects. For instance:

$$u.m(!v) > B \qquad \text{(Ex-3)}$$

is the behavior of sending the client role of v to the object holding the server role of u. In a second example:

$$u.m(?v) > B \qquad \text{(Ex-4)}$$

we describe the behavior of sending the server role of v. An object can also send both roles:

$$u.m(!?v) > B \qquad \text{(Ex-5)}$$

The static semantics of *OL2* enforces certain restrictions on passing roles between objects. A general restriction, which is independent of the type of the interfaces, is that server roles can be held by at most one object at a time. Thus, when a server role is sent to another object, the sending object loses this role. In Ex-4, object B is not allowed to offer services on interface v (or else it is not well-typed). There are additionnal restrictions on interfaces that are imposed by their types. An interface type, T, is a tuple (μ, k, ρ, x) where:

- μ is the public/private mode (or mode for short). An interface can be declared public or private. For a private interface, the client role can, at a given instant, be possessed by at most one object, i.e., a sender of a client role of a private interface loses that role (in Ex-3, if v has been declared private then the static semantics ensures that there can be no invocation of v in B). For public interfaces there is no such restriction.

- k is the mobility kind of the interface (or kind for short). An interface can be declared mobile or stabile. For stabile interfaces, the server role cannot be sent. There is no such restriction for mobile interfaces. Ex-4 is not well-typed if v is stabile, and is well-typed otherwise.

- ρ is the role of the interface: client (noted !), server (noted ?) or both (noted !?). The role type of a newly created interface is !?.

- x is the behavior automaton of the interface. It prescribes the allowable succession of messages that the interface can handle. Interface behavior types are discussed in a subsequent section.

An interface creation is given by the syntax:

$$\textbf{new}\ u : T\ \textbf{in}\ B$$

the behavior of this expression is to create a new interface u of type T and then to behave like B. An example of interface creation:

$$\textbf{new}\ u : \textbf{public mobile}\ !?x\ \textbf{in}\ B$$

When offering a service on an interface, an object must declare the type of the interfaces that it expects to receive. An example of service offer:

$$u[m(w : T) = B]$$

The behavior of this expression is to wait, on interface u, for the message m containing an interface of type T. The received interface is bound to w.

In *OL2* the behavior of objects can be named. This is achieved as in the following example (where we consider a behavior parametrized with two interfaces):

$$A[u:T_1, v:T_2] = B_A$$

An object can then be instantiated by providing the actual parameters. Name instantiation is similar to the *become* operation in Actors. Object creation is another feature of *OL2*. It is different from object instantiation in that the creating object has its proper continuation that runs in parallel with the created object. For instance, the expression:

$$\textbf{create } A[?w_1, !w_2] > B$$

evolves to a new one where A and B run in parallel.
In all the examples we presented until now only one message was handled by reception actions. In the general case, more than one service may be available on an interface and a reception action is written:

$$u[m_1(\widetilde{v}_1) = B_1, \ldots, m_n(\widetilde{v}_n) = B_n]$$

It is possible for an object to offer services on more than one interface at the same time. This is written :

$$u_1[\ldots] + \ldots + u_n[\ldots]$$

At any given instant, for any given object, all messages that have been sent by other objects and that are waiting to be processed are stored in a collection of buffers. This is represented as follows:

$$\left\langle B \,, Q \right\rangle$$

where Q is a set of buffers indexed by the set of server interfaces of object B:

$$\left\langle B \,, \{u_1 = q_1 \,; \cdots ; \, u_k = q_n\} \right\rangle$$

each of the buffers q_i is a list of messages of the form: $m_1(\widetilde{v}_1), \cdots, m_n(\widetilde{v}_n)$. Depending on the mode of the interface, the associated list of messages is either unordered (when the interface is public) or ordered (when the interface is private). This reflects the semantics of message transfer between a client and a server interface: the transfer is order preserving (FIFO) in the case of private interfaces and non order preserving (Bag) in the case of public interfaces. Finally, a running configuration of objects, R, is expressed using the commutative and associative operator $|$:

$$\left\langle B_1 \,, Q_1 \right\rangle | \left\langle B_2 \,, Q_2 \right\rangle | \ldots | \left\langle B_n \,, Q_n \right\rangle$$

and interfaces that are internal to a configuration, R, are expressed by the new operator:

$$\textbf{new } u_1 \textbf{ in } \cdots \textbf{new } u_n \textbf{ in } R$$

$$
\begin{array}{rcl}
T & ::= & \mu\ k\ \rho\ x \\
I & ::= & u{:}T \\
Dcl & ::= & A[\widetilde{I}] = B \\
B & ::= & a > B \;\Big|\; \sum_{i=1}^{n} u_i[Abs_i] \;\Big|\; A[\widetilde{\rho u}] \;\Big|\; \textbf{new}\ I\ \textbf{in}\ B \;\Big|\; \mathbf{0} \\
a & ::= & u.m(\widetilde{\rho v}) \;\Big|\; \textbf{create}\ A[\widetilde{\rho u}] \\
Abs & ::= & m_1(\widetilde{I_1}) = B_1, \ldots, m_n(\widetilde{I_n}) = B_n
\end{array}
$$

Terminals	Meaning
A	behavior name
u, v, w	interface names
T	interface type
$m, m_1, \ldots, m_n$	method names
μ	interface mode: **public** or **private**
k	interface kind: **mobile** or **stabile**
x	interface behavior type
ρ	interface role: client (!), server (?) or both (!?)

Summary of the Syntax of *OL2* objects

$$
\begin{array}{rcl}
R & ::= & \left\langle B\,,\,Q \right\rangle \\
 & \mid & R|R \\
 & \mid & \textbf{new}\ u\ \textbf{in}\ R \\
Q & ::= & \{u_1 = q_1\ ;\ \cdots\ ;\ u_n = q_n\}
\end{array}
$$

Summary of the Syntax of *OL2* running configurations

OPERATIONAL SEMANTICS

We need first to define how names are bound in *OL2*. Variables u, $u_1, ..., u_n$ occurring in an object or a running configuration are bound if they occur in a sub-expression having one of the forms: **new** $u : T$ **in** B, **new** u **in** R, $m(u_1, ..., u_n) = B$, $A(u_1 : T_1, ..., u_n : T_n) = B$; otherwise the occurences are free. The sets $fn(B)$ and $fn(R)$ are defined accordingly, and so is *alpha-conversion*, denoted $\equiv_\alpha$.

We present the operational semantics of object configurations in two steps. We first define a structural congruence relation and then we give a reduction relation that specifies how configurations evolve. *OL2* is statically typed so will omit all the information about types in the dynamic semantics.

Structural Congruence

The first structural rules state that the choice operator between receptions is commutative and associative:

$$\Big\langle u[Abs_1] + v[Abs_2]\ ,\ Q \Big\rangle \equiv \Big\langle v[Abs_2] + u[Abs_1]\ ,\ Q \Big\rangle$$

$$\Big\langle (u[Abs_1] + v[Abs_2]) + w[Abs_3]\ ,\ Q \Big\rangle \equiv \Big\langle u[Abs_1] + (v[Abs_2] + w[Abs_3])\ ,\ Q \Big\rangle$$

The parallel operator is also commutative and associative and 0 is its neutral element:

$$R_1 \mid R_2 \equiv R_2 \mid R_1\ ,\ (R_1 \mid R_2) \mid R_3 \equiv R_1 \mid (R_2 \mid R_3)\ ,\ R \mid 0 \equiv R$$

The order of the introduction of the interfaces is meaningless:

$$\textbf{new}\ u_1\ \textbf{in}\ (\textbf{new}\ u_2\ \textbf{in}\ R) \equiv \textbf{new}\ u_2\ \textbf{in}\ (\textbf{new}\ u_1\ \textbf{in}\ R)$$

The two last rules are the π-calculus scope extrusion and the alpha-conversion:

$$(\textbf{new}\ u\ \textbf{in}\ R_1) \mid R_2 \equiv \textbf{new}\ u\ \textbf{in}\ (R_1 \mid R_2) \quad \text{if } u \text{ is not free in } R_2$$

$$R_1 \equiv R_2 \text{ if } R_1 \equiv_\alpha R_2$$

Reduction Rules

We now define the reduction rules that specify how a configuration can evolve by making a single and atomic step. In our reduction rules we apply, on the buffers of pending messages, the operations *push*, *pop* and *top*. The semantics of these operations depend on the type of the buffer (queue or bag).

The reduction relation is defined by the following rules:

$$\frac{\begin{array}{l} R_1 = \Big\langle B_1\ ,\ \{u = q\} \cup Q_1 \Big\rangle \\ R_2 = \Big\langle u.m(\rho_1 v_1, \cdots, \rho_k v_k) > B_2\ ,\ Q_2 \Big\rangle \\ J = \{j \mid ? \in \rho_j\} \\ Q_2 = \bigcup_{j \in J} \{v_j = q_j\} \cup Q'_2 \\ q' = push(m(\rho_1 v_1, \cdots, \rho_k v_k), q) \end{array}}{R_1 \mid R_2 \longrightarrow \Big\langle B_1\ ,\ \{u = q'\} \cup Q_1 \bigcup_{j \in J} \{v_j = q_j\} \Big\rangle \mid \Big\langle B_2\ ,\ Q'_2 \Big\rangle}$$

The invocation of the service m of the interface u located in the object R_1 puts this message (m) in the buffer corresponding to this interface in R_1. The buffers associated to the servers that are migrating (q_i) move also from R_2 (the invoker) to R_1 (the invoked). Note that these buffers are transfered to the invoked object R_1 before the message m is absorbed by this invoked object R_1.

$$\frac{\begin{array}{c} Q = \{u = q\} \cup Q' \\ top(q) = m(\widetilde{pv}) \\ q' = pop(q) \end{array}}{\left\langle u[m(\tilde{u}) = B + \Sigma m_i(\tilde{u}_i) = B_i] \text{ , } Q \right\rangle \longrightarrow \left\langle B[\tilde{v}/\tilde{u}] \text{ , } \{u = q'\} \cup Q' \right\rangle}$$

A reception action consists in consuming the first message of the corresponding queue.

$$\frac{A[\tilde{u}] \stackrel{Dcl}{=} B}{\left\langle A[\tilde{v}] \text{ , } Q \right\rangle \longrightarrow \left\langle B[\tilde{v}/\tilde{u}] \text{ , } Q \right\rangle}$$

We simply replace the instantiation of the object by the corresponding behavior.

$$\frac{A[\tilde{u}] \stackrel{Dcl}{=} B}{\left\langle \textbf{create } A[\tilde{v}] > B' \text{ , } Q \right\rangle \longrightarrow \left\langle B[\tilde{v}/\tilde{u}] \mid B' \text{ , } Q \right\rangle}$$

The created object runs is parallel with the continuation of the creating behavior.

$$\frac{R_1 \longrightarrow R_1'}{R \mid R_1 \longrightarrow R \mid R_1'}$$

This rule states that if a subconfiguration can evolve to a new one then the whole configuration can evolve too.

$$\frac{R \longrightarrow R'}{\textbf{new } u \textbf{ in } R \longrightarrow \textbf{new } u \textbf{ in } R'}$$

The reduction rule concerning the interface creation operator (equivalent to the restriction operator of the π-calculus) is straightforward. In a labeled-transition semantics it would have been more complicated.

$$\frac{R_1 \equiv R_1' \longrightarrow R_2' \equiv R_2}{R_1 \longrightarrow R_2}$$

This rule states that configurations that are equivalent (according to $\equiv$) behave equally.

INTERFACE TYPES

Syntax of Interface Behavior Types

A behavior type is a triple (E, x, r), noted $E \triangleright rx$, where E is a finite state automaton describing the behavior of the interface, x is the initial state of the automaton and r is the set of capabilities of the interface ($r \subset \{?, !\}$). Following the definition introduced by Nierstrasz ([14]) we define E by a set of equations of the form $x_i = e_i$ where each x_i is a type variable that appears once and only once in the left-hand side of a equation and each e_i is an expression defined by the following syntax:

$$e ::= \sum_{i=1}^{n} m_i(\tilde{r}_i \tilde{x}_i); y_i \text{ where } i \neq j \Rightarrow m_i \neq m_j \text{ for all } i, j \in [1..n]$$

Note that interface behavior types are deterministic. We denote ε the empty summation ($n = 0$). The variables $X, Y, \ldots$ range over the set IBTypes of interface behavior types.

Note. In the following we will consider, without loss of generality[1], that all our behavior types are defined in the same environment, E. We will sometimes omit this environment when writing a behavior type, i.e., $E \triangleright rx$ will be written rx. We also sometimes refer to this environment as a type repository.

Semantics of Interface Behavior Types

As in [11], we define a transition relation that specifies the behavior of a behavior type:

$$\frac{(x = m(r_1x_1, \cdots, r_nx_n); y \,+\, e) \in E}{rx \xrightarrow[E]{\rho\, m(r_1x_1,\cdots,r_nx_n)} ry} \text{ for some } \rho \in r$$

Example. Let us consider the memory cell case. We have an environment E defined as follows:

$$\begin{array}{lcl} \text{Free\_cell} & = & write(!x); \text{Cell} \\ \text{Cell} & = & read(!y); \text{Cell} \\ & + & release(); \text{Free\_cell} \end{array}$$

An free cell (server) can perform an action "write" and then behave like a cell: $?\text{Free\_cell} \xrightarrow[E]{?write(x)} ?\text{Cell}$. The Cell can then perform either a "read" action - without changing its state or a "release" action - which puts it in the initial state: $?\text{Cell} \xrightarrow[E]{?release()} ?\text{Empty\_cell}$

We define $Succ(X)$ as the set of all the types that X can evolve to in one step:

Definition 1 $Succ(X) = \{Y \in \text{IBTypes},\ \exists\, \rho, m, \widetilde{X}$ *such that* $X \xrightarrow[E]{\rho\, m(\widetilde{X})} Y\}$

Interface Type Equivalence

Two behavior types are equivalent iff they behave equally, i.e., they are bisimilar:

Definition 2 *(Interface type equivalence)*
Two interface types X_1 and X_2 are equivalent, noted $X_1 \sim X_2$ iff:

i) $X_1 \xrightarrow[E]{\rho\, m(\widetilde{Y_1})} X_1'$ *implies* $X_2 \xrightarrow[E]{\rho\, m(\widetilde{Y_2})} X_2'$ *and* $X_1' \sim X_2'$, $\widetilde{Y_1} \sim \widetilde{Y_2}$

ii) $X_2 \xrightarrow[E]{\rho\, m(\widetilde{Y_2})} X_2'$ *implies* $X_1 \xrightarrow[E]{\rho\, m(\widetilde{Y_1})} X_1'$ *and* $X_2' \sim X_1'$, $\widetilde{Y_2} \sim \widetilde{Y_1}$

[1] Two behavior types defined in two distinct environments can be simply redefined in the union of these environments with an appropriate renaming of type variables.

Although this definition is recursive, it is well-formed.
A behavior type is uniform if it has always the same behavior.

Definition 3 *X is uniform iff:*

$$Y \in Succ(X) \Rightarrow X \sim Y \textit{ and } Y \textit{ is uniform.}$$

In *OL2* all the interfaces declared public are checked by the static semantics to have this property which ensures that all pending messages are (or will be) processable, although they may come from different clients and in an arbitrary order.

Interface behavior subtyping

Informally, an interface type X_1 is a subtype of a type X_2 iff:

client case: the processable messages of X_1 are a subset of the processable messages of X_2.

server case: the processable messages of X_1 are a superset of the processable messages of X_2.

This is to guaranty the safe substitution of a supertype by its subtype.

Definition 4 *(Interface subtyping)*
An interface type X_1 is a subtype of an interface type X_2, noted $X_1 \preceq X_2$ iff:

$$i)\ X_1 \xrightarrow[E]{!m(\widetilde{Y_1})} X_1' \textit{ implies } X_2 \xrightarrow[E]{!m(\widetilde{Y_2})} X_2' \textit{ and } X_1' \preceq X_2',\ \widetilde{Y_1} \preceq \widetilde{Y_2}$$

$$ii)\ X_2 \xrightarrow[E]{?m(\widetilde{Y_2})} X_2' \textit{ implies } X_1 \xrightarrow[E]{?m(\widetilde{Y_1})} X_1' \textit{ and } X_1' \preceq X_2',\ \widetilde{Y_1} \preceq \widetilde{Y_2}$$

Note that if X_1 and X_2 have both client and server roles then they are equivalent.

EXAMPLE

We present a small buffer example intended to show how interface types can be defined and used in a configuration of objects. We demonstrate also how coordination can be modeled using private interfaces.

Our initial step is to declare the interface type for a one place buffer that stores elements (interfaces) of type `elem`:

```
x\_buff = put(!x\_elem)      ; x\_full
x\_full = get(!x\_r\_elem)   ; x\_buff
```

Interfaces of type `x_r_elem` return elements of type `elem`:

```
x_r_elem = ret(!x_elem) ; 0
```

A buffer and a client

An *OL2* object that manages a one place buffer can be now written:

```
Buffer[buf: private stabile ?x_buff] =

  buf [ put(e: private stabile !x_elem) =
                                   Full_Buffer[?buf, !e]
      ]
```

This object starts with an initial server interface buf which is ready to accept a put method. Method put takes an argument e which is the client role of an interface of type x_elem. After accepting method put the object becomes a Full_Buffer which has the continuation of the server role for buf and a client role for interface e. Full_Buffer is ready for method get which takes an interface r that is used as a target for returning the value e:

```
Full_Buffer[buf: private stabile ?x_full,
             e: private stabile !x_elem
           ] =
             buf [ get(r: private stabile !r_x_elem) =
                                  r.ret(!e) > Buffer[?buf]
                 ]
```

Note how, at the initial state of Full_Buffer, the behavior type of interface buf is ?x_full. We can now define a client that uses our buffer:

```
Buffer_Client [ buf: private stabile !x_buff,
                   e: private stabile !x_elem
              ]=
               buf.put(!e) >
               new r: private stabile !? r_x_elem in
               buf.get(!r) >
               r [ ret(e: private stabile !x_elem) =
                   Buffer_Client[!buf, !e]
                 ]
```

Object Buffer_Client is parametrized with the (client) interface of the buffer and the (client) interface of the element that is to be stored in the buffer. It starts by invoking method put and then it creates a new interface that is used in the argument of the invocation of method get.

A buffer and two clients

We turn now to modelling two coordinated clients for our buffer: a producer and a consumer. We will take advantage of the private interface buf to synchronize these two objects: the client role for buf is passed between the two objects - the producer after storing an element in the buffer, and the consumer after getting this element. The Producer object is as follows:

```
Producer [ buf: private stabile !x_buff,
             e: private stabile !x_elem,
          prod: private stabile ?x_empty,
          cons: private stabile !x_full
         ]=
           buf.put(!buf) >
           prod [ token(buf: private stabile !x_buff) =
                  Producer[!buf,!e,?prod,!cons]
                ]
```

and the Consumer object is given hereafter:

```
Consumer [ cons: private stabile ?x_full,
           prod: private stabile !x_empty
         ]=
           cons [ token(buf:private stabile !x_full) =
                     new r: private stabile !? r_x_elem in
                     buf.get(!r) >
                     r [ ret(e: private stabile !x_elem) =
                         prod.token(!buf) >
                         Consumer[?cons, !prod]
                       ]
                ]
```

The previous objects make use of interface behavior types x_full and x_empty which are simply defined by the following equations:

```
x_full = token(!x_full) ; x_empty
x_empty = token(!x_buff) ; x_full
```

The running configuration for our three objects can now be given:

$$\textbf{new}\ \texttt{prod}\ \textbf{in new}\ \texttt{cons}\ \textbf{in new}\ \texttt{buf}\ \textbf{in new}\ \texttt{e}\ \textbf{in}$$

$$\left\langle \texttt{Buffer[?buf]}\ ,\ \emptyset \right\rangle$$

$$|$$

$$\left\langle \texttt{Producer[!buf,!e,!prod,?cons]}\ ,\ \emptyset \right\rangle \mid \left\langle \texttt{Consumer[?cons, !prod]}\ ,\ \emptyset \right\rangle$$

STATIC SEMANTICS

The type system that we propose enforces the following properties:

- interfaces that are declared to be private are privately used. That means that when a role (either client or server) is sent, it can no more be used by the sender.
- the service offered by public interfaces is uniform.
- the server role of stabile interfaces do not migrate.
- server roles of mobile interfaces migrate only towards stabile interfaces.
- objects behave according to the behavior types of their interfaces.
- when at a reception state objects are ready to receive on all of their interfaces.

It is the conjunction of these properties that ensures the liveness result stated in a subsequent. Let us consider, for instance, the tricky deadlock where a server interface is stuck in a buffer because the target server interface is stuck also in a buffer. This can occur in the configuration, R:

$$\left\langle u.m(?u') > B \text{ , } \emptyset \right\rangle \mid \left\langle u'.m'(?u) > B' \text{ , } \emptyset \right\rangle$$

which will evolve to the deadlocked configuration, R':

$$\left\langle B \text{ , } \{u' = m'(?u)\} \right\rangle \mid \left\langle B' \text{ , } \{u = m(?u')\} \right\rangle$$

Thanks to the typing rules, configuration R is not well typed, and thus the deadlock is avoided. The static semantics rules are given in the annex.

RUN-TIME LIVENESS

In the present section we only consider those configurations where all objects possess the so called input-well-guarded recursion, i.e., where recursive definitions of objects must go through at least one receiving action. In fact, this property could be dealt with in the type system. Furthermore, we consider closed configurations, i.e., configurations with no free interfaces.

Theorem 1 *(Subject reduction)*
Let R and R' be object configurations. If $\Gamma \vdash R$ and $R \longrightarrow R'$ then there exists an environment Γ' such that $\Gamma' \vdash R'$

Outline of the proof: It is an extension of the subject reduction property introduced in [11]. This result is essentially due to the following invariants that are ensured by the typing rules:

- The services available at the public servers are uniform.
- There is no duplication of server roles (public and private) and no duplication of private clients. So a private client interface type and its server interface type will evolve concomitantly.

Theorem 2 *(Liveness)*
Let R_1 be a well-typed closed configuration ($\emptyset \vdash R_1$). For all configurations, R_2, such that $R_1 \xrightarrow{*} R_2$, if R_2 contains a non-empty buffer for a server interface u ($u = q$) then there exits a configuration R_3 such that $R_2 \xrightarrow{*} R_3$ where the message top(q) has been consumed.

Outline of the proof: By induction on the length of the derivation $R_1 \xrightarrow{*} R_2$. This result is due to the conjunction of the static semantic rules:

- rule 1 enforces progress as prescribed by the interface behavior type,
- rule 5 enforces that all server interfaces are opened when the object is in a ready to receive state,
- input-well-guardedness guarantees that all objects either come to a stop or come to a ready to receive state within a finite number of steps,
- rule 2 avoids deadlocks due to cross sending of stabile server interfaces: it avoids that servers become indefinitely unavailable when they migrate in buffers towards other migrating server interfaces,

Liveness preserving configuration extension. The liveness theorem may seem restrictive to closed configurations. In fact it is also valid when dealing with configuration extension. Indeed, the theorem implies the nice feature that any well-typed running configuration can be extended at any time with the addition of well-typed objects (which may hold the client role of some public interfaces of the running configuration) and whereby the resulting new configuration possesses the liveness property stated above. This feature follows from a corollary to theorems 1 and 2.

Corollary 1 *(Object introduction)*
Let R_1 be a well-typed closed running configuration possessing the public interface u:

$$\emptyset \vdash \left\langle \textbf{\textit{new}}\ u\!:\!(\textbf{\textit{public}}, k, !?, x)\ \textbf{\textit{in}}\ B_1\ ,\ Q \right\rangle$$

Let R_2 be a well-typed initial configuration possessing a client role of the public interface u:

$$u\!:\!(\textbf{\textit{public}}, k, !, z) \vdash \left\langle B_2\ ,\ \emptyset \right\rangle \textit{ and } !z \preceq !x$$

Then the extension of the configuration R_1 with the configuration R_2 is also well-typed and so enjoys the liveness property:

$$\emptyset \vdash \left\langle \textbf{\textit{new}}\ u\!:\!(\textbf{\textit{public}}, k, !?, x)\ \textbf{\textit{in}}\ (B_1 \mid B_2)\ ,\ Q \right\rangle$$

Liveness preserving extension is a valuable feature which allows the smooth extension of running configurations. Note that the operation of *introducing an object* in a *running configuration* assumes the existence of two repositories, namely, a type repository and an interface repository. The type repository is a public database defining all behavior type identifiers. The interface repository associates to each public interface of the running configuration its (uniform) type. The interface repository is a database that can be consulted by new objects that are candidates for extending the running configuration (a broker can be used as an intermediate object for accessing the interface type repository, but this is beyond the scope of the present paper). Note that it is sufficient to check that the newly introduced object is well-typed using contextual information from the type and interface repositories. There is no need to re-inspect the code of the objects in the running configuration and to check the well-typedness of the parallel composition of the new enlarged configuration.

CONCLUSION

We defined a calculus endowed with a typing system that guarantees a progress property in well-typed object configurations. The eminent feature of *OL2* resides in the way its interfaces are declared and consistently managed. Thus, interfaces can be either public or private, mobile or stabile, client or server. The discipline enforced on the use of the interfaces is such that, modulo fairness, in all well-typed configurations, all pending messages are eventually processed by their target interfaces. The typing system of *OL2* can be adapted to a wide variety of languages that have the same communication paradigm. In another paper [13], we extended the safety result of a previous work [11] to infinite automata type systems. Besides the liveness guarantees brought by the present type system, another result is achieved which allows for the incremental, liveness preserving, introduction of new well-typed objects to existing configurations. This provides for the extension of configurations the code of which is no more available. Only the interface types are considered. This last property has important implications in open software design.

Type systems for concurrent object oriented languages is an active research topic. Many authors have tackled this issue in the realm of the π-calculus [10] and the actors [1] paradigms. Concerning the former, a wide variety of typing systems have been proposed that deal with the problem of channel typing. The simplest one [9] just checks the arity of the channels. This type system has been extended such that it can handle polymorphism and type inference [5, 23, 21] and subtyping [15, 16]. None of these typing systems handle dynamic service behavior.

The importance of distinguishing public from private interfaces has been identified by [14], but, without giving it a formal treatment. [14] has also introduced the concept of non-uniform service availability and has used traces to specify the constraints on the ordering of the messages that can be handled by a channel (an interface). Our work extends [14] in two ways: by formally introducing the concept of privacy of interfaces and by allowing message types to include parameter types.

Takeuchi et al. [20] define a typed process calculus called $\mathcal{L}$ with the notion of *session*: "a semantically atomic chain of communication actions" between two processes. In *OL2* a session is represented by a communication between two objects using a private interface. But unlike *OL2*, the session channels are static and the roles of the partners of a session cannot be passed. In *OL2*, an object can pass its client or server role of a private interface and so delegate to another object the continuation of a "session".

In [17] the author provides a typing system which handles the dynamic behavior of actors. [17] features the concept of type spliting allowing the introduction of a form of parallelism in the types. The type language is based on traces, but the possibility of having recursion and non terminating behavior are not explicitly treated.

The work that is the closest to ours is perhaps [8]. It is based on the asynchronous π-calculus and has a typing system that ensures a certain form of deadlock freeness. The main difference between [8] and our approach is that only a restricted form of behaviors, without loops, is allowed in the liveness fragment of the calculus proposed in [8].

TYCO [22, 18] is another interesting endeavor which is worth mentionning. TYCO is a calculus that is built for the purpose of experimenting behavioral type issues. In [19] a type discipline is exhibited on TYCO which is less restrictive than the one presented here but which guarantees weaker properties.

A recent interesting deadlock freeness result was achieved in [3] on a process calculus which is more expressive than *OL2* and which unifies the π and λ calculi. This result, in fact, is more about verifying complex configurations using typing techniques. Our aim is different and we are more concerned with open configurations and issues of extending applications at run time.

The technical treatment of the contexts in the static semantic of *OL2* has been inspired from [7]. In this version of the π-calculus the authors use the linear capabilities of some special channel to ensure that they are used (at most) once. We use a similar mechanism to ensure that there is no duplication of the roles of private interfaces.

References

[1] G. A. Agha, I. A. Mason, S. F. Smith and C. L. Talcott, *A Foundation for Actor Computation*, J. Functional Programming 1 (1), 1993.

[2] R. Amadio and L Cardelli, *Subtyping recursive types*, ACM-TOPLAS, 15(4):575--631, 1993.

[3] Gerard Boudol. *Typing the use of resources in a concurrent calculus*. ASIAN'97, the Asian Computing Science Conference, Kathmandu, Nepal, LNCS 1345 (1997) 239-253.

[4] Cedric Fournet, Georges Gonthier. *The reflexive chemical abstract machine and the join-calculus*. POPL'96.

[5] Simon J. Gay. *A sort inference algorithm for the polyadic π-calculus*. Twentieth ACM Symposium on Principles of Programming Languages, January 1993.

[6] Kohei Honda. *Types for Dyadic Interaction*. CONCUR'93, LNCS 612, Springer-Verlag.

[7] Naoki Kobayashi, Benjamin C. Pierce, David N. Turner. *Linearity and the Pi-Calculus*. Technical report, Department of Information Science, University of Tokyo and Computer Laboratory, University of Cambridge, 1995.

[8] Naoki Kobayashi *A Partially Deadlock-free Typed Process Calculus* Twelfth IEEE Symposium on Logic in Computer Science (LICS'97).

[9] Robin Milner. *The polyadic π-calculus: a tutorial.* Technical Report ECS-LFCS-91-180, Laboratory for Foundations of Computer Science, Department of Computer Science, University of Edinburgh, UK, October 1991.

[10] Robin Milner, Joachim Parrow, David Walker. *A calculus of mobile processes (Part I and Part II).* Information and Computation, 100:1-77, 1992.

[11] E. Najm, A. Nimour *A Calculus of Object Bindings.* in proceedings of Second conference on Formal Methods for Object-based Open Systems, Chapman and Hall, 1997.

[12] Elie Najm, Jean-Bernard Stefani. *A formal semantics for the ODP computational model.* Computer Networks and ISDN Systems , Vol 27, 1995.

[13] Elie Najm, Abdelkrim Nimour, Jean-Bernard Stefani. *Infinite Types for Distributed Objects Interfaces.* Proceedings of third IFIP conference on Formal Methods for Open Object-based Distributed Systems - FMOODS'99 - Firenze, Italy, February 99. Kluwer.

[14] Oscar Nierstrasz. *Regular Types for Active Objects.* Object-Oriented Software Composition. O. Oscar Nierstrasz, D.Tsichitzis (Ed.), Prentice Hall, 1995

[15] Benjamin C. Pierce, David Sangiorgi. *Typing and subtyping for mobile process.* Mathematical Structures in Computer Science, 1995.

[16] Benjamin C. Pierce, David N. Turner. *PICT Language Definition.* Available electronically, 1995.

[17] Franz Puntigam. *Types for active objects based on Trace Semantics.* FMOODS'96, Chapmann and Hall.

[18] Antonio Ravara, Vasco T. Vasconcelos. *Behavioral Types for a Calculus of Concurrent Objects.* Euro-Par'97, LNCS. Springer-Verlag, 1997.

[19] Antonio Ravara, Pedro Resende, and Vasco T. Vasconcelos. *Towards an Algebra of Dynamic Object Types.* In Workshop on Semantics of Objects as Processes, BRICS Notes Series, NS-98-5. 1998.

[20] Kaku Takeuchi, Kohei Honda, Makoto Kubo. *An Interaction-based Language and its Typing System.* PARLE'94, LNCS 818, Springer-Verlag.

[21] David N. Turner. *The π-calculus: Types, polymorphism and implementation.* Ph.D. Thesis, LFCS, University of Edinburgh, 1995.

[22] Vasco T. Vasconcelos. *Typed Concurrent Objects* ECOOP'94, LNCS 821, Springer Verlag.

[23] Vasco T. Vasconcelos, Kohei Honda. *Principal typing schemes in a polyadic π-calculus.* CONCUR'93, July 1993.

Appendix

STATIC SEMANTICS RULES

Preliminary definitions and notations

In the treatment of the static semantics we need to associate a type to the interfaces of the objects being considered. A context Γ is a set of bindings of the form: $u : T$ or $A : (T_1, \ldots, T_n)$ where each interface types is defined in a "Type Repository", $\mathbb{TR}$, which is the global set of behavior type equations. The static semantics is given using the judgments given in table 1.

judgment	meaning
$\Gamma \vdash u : T$	in the context Γ, the interface u has type T
$\Gamma \vdash A : (T_1, \ldots, T_n)$	in the context Γ, the object A has type $(T_1, \ldots, T_n)$
$\Gamma \vdash \left\langle B \, , \, Q \right\rangle$	in the context Γ, the behavior $\left\langle B \, , \, Q \right\rangle$ is well-typed
$\Gamma \vdash R$	in the context Γ,the configuration R is well-typed
$\Gamma \vdash Dcl \Rightarrow \Gamma'$	in the context Γ, the declaration Dcl yields context Γ'

Table 1 Judgments of the typing rules

The context extension, noted "$\Gamma, u{:}T$", is defined such that $\Gamma, u{:}T \vdash u{:}T$.

In order to avoid interface server role duplication and private interface client role duplication we define a partial function, $\oplus$, over interface types. The following equations gives the summation ($\oplus$) of two interface types. The function $\oplus$ is undefined in all the other cases.

$$\begin{aligned}
(\textbf{public}, k, x, r_1) \oplus (\textbf{public}, k, x, r_2) &= (\textbf{public}, k, x, r_1 \cup r_2) \text{ if } ? \notin (r_1 \cap r_2) \\
(\textbf{private}, k, x, r_1) \oplus (\textbf{private}, k, x, r_2) &= (\textbf{private}, k, x, r_1 \cup r_2) \text{ if } (r_1 \cap r_2) = \emptyset
\end{aligned}$$

We extend $\oplus$ to contexts as follows: $(u_1 : T_1, \ldots, u_n : T_n) \oplus (u_1 : T_1', \ldots, u_n : T_n')$ denotes the context $(u_1 : T_1 \oplus T_1', \ldots, u_n : T_n \oplus T_n')$ We will write $(\Gamma, u : (\mu, k, r, x)) + r'u$ as a shorthand for $\Gamma, u : ((u, k, r, x) \oplus (\mu, k, r', x))$

Typing Rules

The basic idea underlying our typing rules is to guaranty that each object uses the interfaces in a way compatible with their declared behavior type. Our rules ensure also that there is no duplication of the roles of interfaces, except for the client role of the public interfaces.

$$\frac{\{u \mid \Gamma \vdash u : (\mu, k, r, x) \text{ with } ? \in r \text{ and } x \neq \varepsilon\} = \emptyset}{\Gamma \vdash \left\langle \mathbf{0} \, , \, \emptyset \right\rangle} \qquad (1.A.1)$$

Rule 1 introduces the notion of receiving obligation. If the behavior type of a server can perform an action (is different from ε) then this server cannot stop.

$$\frac{\begin{array}{l}\Gamma,\ u:(\mu,k,r,x_2)\ \vdash\ \left\langle B\ ,\ Q\right\rangle \\ \Gamma\ \vdash\ \widetilde{v}:(\widetilde{\mu},\widetilde{k},\widetilde{r},\widetilde{y}) \\ rx_1 \xrightarrow[\mathbb{TR}]{!m(r_1y_1,\cdots,r_ny_n)} rx_2 \\ \Gamma+\widetilde{\rho v}\ \text{defined} \\ ?\in r_j \Rightarrow k=\text{stabile, for } j\in[1..n]\end{array}}{\Gamma,\ u:(\mu,k,rx_1)+\widetilde{\rho v}\ \vdash\ \left\langle u.m(\widetilde{\rho v})>B\ ,\ Q\right\rangle} \tag{1.A.2}$$

Rule 2 could be read like this: if $u.m(\widetilde{\rho v}) > B$ is well-typed in a context where u has the client role of an interface of type X and where the interfaces v_i have the roles $\{\rho_i\}$ then B is well-typed in a context where the type of u has evolved to Y and where the roles of the interfaces v_i have been updated.

$$\frac{\begin{array}{l}\Gamma,u:(\mu,k,X_1),\widetilde{v}_1:(\widetilde{\mu}_1,\widetilde{k}_1,\widetilde{Y}_1)\ \vdash\ \left\langle B_1\ ,\ Q\right\rangle \\ \qquad\vdots \\ \Gamma,u:(\mu,k,X_n),\widetilde{v}_n:(\widetilde{\mu}_n,\widetilde{k_n},\widetilde{Y}_n)\ \vdash\ \left\langle B_n\ ,\ Q\right\rangle \\ X \xrightarrow[\mathbb{TR}]{\text{messages}(Q,u)^*} X' \\ X' \xrightarrow[\mathbb{TR}]{?m_1(\widetilde{Y}_1)} X_1 \quad \cdots \quad X' \xrightarrow[\mathbb{TR}]{?m_n(\widetilde{Y}_n)} X_n \\ Succ(X)=\{X_1,\ldots,X_n\}\end{array}}{\Gamma,u:(\mu,k,X')\ \vdash\ \left\langle u?[m_1(\widetilde{v}_1:(\widetilde{\mu}_1,\widetilde{k}_1,\widetilde{Y}_1))=B_1\ldots m_n(\widetilde{v}_n:(\widetilde{\mu}_n,\widetilde{k}_n,\widetilde{Y}_n))=B_n]\ ,\ Q\right\rangle} \tag{1.A.3}$$

In rule 3, the condition $X \xrightarrow[\mathbb{TR}]{\text{messages}(Q,u)^*} X'$ where messages(Q,u) is the ordered list of messages for the interface u in the the buffer Q is here to take into account the delay of the server interface comparing to the client interface. This delay is induced by the message buffering at the server. In this rule it is important that $Succ(X) = \{Y_1,\ldots,Y_n\}$. This ensures that all the messages that can be processed by an interface of type X are handled by the reception action. Having the status of all the v_i(s) to *false* ensures that the servers are not immediately sent after their reception.

$$\frac{\Gamma,u:(\mu,k,!?,x)\ \vdash\ R}{\Gamma\ \vdash\ \textbf{new}\ u:(\mu,k,!?,x)\ \textbf{in}\ R} \tag{1.A.4}$$

Rule 4 states that a newly created interface has both roles: client and server.

$$\frac{\begin{array}{l}\Gamma_1 \vdash u_1[Abs_1] \ldots \Gamma_n \vdash u_n[Abs_n] \\ J = \{j \mid \Gamma \vdash u_j : (\mu, k, r, x) \text{ and } ? \in r\}\end{array}}{\Gamma_1 \oplus \cdots \oplus \Gamma_n \vdash \sum_{j \in J} u_j[Abs_j]} \quad (1.A.5)$$

Rule 5 ensures that the object opens up for all interfaces for which it possesses the server role.

$$\frac{\begin{array}{l}\Gamma \vdash (A[\widetilde{u}:(\widetilde{\mu},\widetilde{k},\widetilde{X})] = B) \Rightarrow \Gamma \\ \Gamma \vdash B\end{array}}{\Gamma \vdash (A[\widetilde{u}:(\widetilde{\mu},\widetilde{k},\widetilde{X})] = B) \Rightarrow \Gamma, \widetilde{u}:(\widetilde{\mu},\widetilde{k},\widetilde{X}), A:((\widetilde{\mu},\widetilde{k},\widetilde{X}))} \quad (1.A.6)$$

In rule 6, the first condition is here to allow recursive definition. It ensures that the behavior B is typed in a context that already contains the bindings needed for the instantiation of A.

$$\frac{\begin{array}{l}\Gamma \vdash \widetilde{u}:(\widetilde{\mu},\widetilde{k},\widetilde{X}) \\ \Gamma \vdash A:((\widetilde{\mu},\widetilde{k},\widetilde{Y})) \\ \widetilde{X} \preceq \widetilde{Y} \\ \Gamma + \widetilde{pu} \text{ defined}\end{array}}{\Gamma \vdash A[\widetilde{pu}]} \quad (1.A.7)$$

Rule 7 states states that an interface having a subtype of another can replace it in an object instantiation.

$$\frac{\begin{array}{l}\Gamma \vdash B \\ \Gamma \vdash \widetilde{u}:(\widetilde{\mu},\widetilde{k},\widetilde{X}) \\ \Gamma \vdash A:((\widetilde{\mu},\widetilde{k},\widetilde{Y})) \\ \widetilde{X} \preceq \widetilde{Y} \\ \Gamma + \widetilde{pu} \text{ defined}\end{array}}{\Gamma + \widetilde{pu} \vdash A[\widetilde{pu}] > B} \quad (1.A.8)$$

Rule 8 is similar to the precedent except that here we must be careful about how B is going to use the interfaces $\widetilde{u}$.

$$\frac{\begin{array}{l}\Gamma_1 \vdash R_1 \\ \Gamma_2 \vdash R_2 \\ \Gamma_1 \oplus \Gamma_2 \text{ defined}\end{array}}{\Gamma_1 \oplus \Gamma_2 \vdash R_1 | R_2} \quad (1.A.9)$$

Finally, rules 9 states that there is no duplication of the roles of a client interface when composing configurations.

PART FIVE

Model Checking

BLACK BOX CHECKING

DORON PELED
Bell Laboratories
600 Mountain Ave.
Murray Hill, NJ 07974, USA

MOSHE Y. VARDI[1]
Rice University
Department of Computer Science
Houston, TX 77005, USA

MIHALIS YANNAKAKIS
Bell Laboratories
600 Mountain Ave.
Murray Hill, NJ 07974, USA

Abstract Two main approaches are used for increasing the quality of systems: in *model checking*, one checks properties of a known design of a system; in *testing*, one usually checks whether a given implementation, whose internal structure is often unknown, conforms with an abstract design. We are interested in the combination of these techniques Namely, we would like to be able to test whether an implementation with unknown structure satisfies some given properties. We propose and formalize this problem of *black box checking* and suggest several algorithms. Since the input to black box checking is not given initially, as is the case in the classical model of computation, but is learned through experiments, we propose a computational model based on games with incomplete information. We use this model to analyze the complexity of the problem. We also address the more practical question of finding an approach that can detect errors in the implementation before completing an exhaustive search.

Keywords: Formal methods, Model checking, Specification, Testing, Verification.

1 INTRODUCTION

Model checking [5] and *testing* [16] are two complementary approaches for enhancing the reliability of systems. Model checking usually deals with checking whether the *design* of a finite state system satisfies some *properties* (e.g., mutual exclusion or responsiveness). On the other hand, testing is usually applied to the *actual system*, often

[1]Supported in part by NSF grants CCR-9628400 and CCR-9700061, and by a grant from the Intel Corporation. Part of this work was done when this author was a Varon Visiting Professor at the Weizmann Institute of Science.

without having access to, or knowledge of its internal structure. It checks whether a *system* (or an *implementation*) *conforms* with some design (i.e., informally, has the same behaviors). Even if access to the internal structure of the tested system is possible, it is not always a good idea to use it when performing tests, as this may lead to a bias in the testing process. Furthermore, the whole system may be very large (e.g., millions of lines of code), while we are interested only in specific aspects of it; think for example of a typical telephony switch and suppose we want to check that the implementation of a particular feature, such as call waiting, meets certain correctness properties. Extracting the part of the code that is relevant from the whole system, especially in the case of large legacy systems, is most probably infeasible (and is itself subject to errors). Suppose one is interested in checking specific properties of some system such as a communication switch or protocol system. Model checking would be appropriate for checking properties of a model of the system, but not checking the system itself. On the other hand, testing methods can compare the system with some abstract design, but usually are not used for checking specific properties.

One motivation for the current work is the case where acceptance tests need to be performed by a user who does not have access to the design, nor to the internal structure of the checked system. Our aim is thus to combine the two approaches, hence checking automatically *properties* of finite state systems whose *structure is unknown*. Of course, a completely hidden structure cannot be effectively checked. Thus, the following properties are assumed:

- A bound n on the number of states of the checked system is known.
- The tester can always **reset** the system to its (unique) initial state.
- The input alphabet Σ of the checked system is known.
- An experiment consists of repeatedly applying an input from Σ or a **reset** to the current state. An indication of whether the input was possible (enabled) from the current state is available.
- If an input α was possible from the current state, the system makes the move. Otherwise, it stays in the current state. No backtracking is available (but the tester can simulate backtracking by resetting and repeating the successful prefix of the experiment).
- The checked system is *deterministic* in the sense that from each state it can move with any given input to at most one successor state.

We do not assume that the size of the system is known precisely: n is only an upper bound. In particular, we would like to study the effect of the possibility that the bound on the number of states n may be much bigger than the actual number of states. This is the case when the number of states is only estimated. In practice, the system that is being checked may be large and have multiple functions, while the property may concern a specific aspect of the system. Although the system as a whole may be quite big, large parts of it may be irrelevant to the property, and the system may be equivalent to a much smaller finite state machine as far as checking the property is concerned. In this case, n should be taken to be an estimate on the logical complexity (the control structure) of the system with respect to the property at hand. Our methods can be used also if no bound n is available, by running the algorithms to the extent that the available time and space resources allow; the guarantees in this case depend on the time spent.

Following the automata-theoretic approach to model-checking [12, 23], the (negation of the) checked property is directly given as, or translated into, a finite automaton on infinite words, usually a Büchi automaton [3]. Then, both the system and (the complement of) the checked property are represented using automata. An example of a system that is based on such principles is Spin [8, 10], where the specification is given by an automaton called a *never claim* that recognizes the *bad* (or *disallowed*) computations. In order to check whether the system under consideration satisfies the checked property, we intersect the automaton representing the system with the automaton representing the disallowed computations. Any sequence in the intersection is a counterexample for the checked property, while the absence of any counterexample means that the property is satisfied.

The problem we study here is a variant of the above model checking problem. We are given the automaton that represents the computations not allowed by the checked property. But the internal structure of the checked system is not revealed, and only some experiments, as described above, are allowed on it. Still, we want to check whether the system satisfies the given property. We call this problem *black box checking*. To simplify the discussion, we will not deal here with machines with output. Their treatment and the results are similar to the ones presented here. We will present on-the-fly algorithms that are aimed at quickly detecting errors in a checked system.

The choice of an appropriate computational model is central to the issue of black box checking. Unlike standard decision problems, the input is not given here at the beginning of the computation, but is learned through a sequence of experiments. We propose a computational model based on games with incomplete information, and use this model to analyze the complexity of the problem.

Our methods combine techniques from model checking, conformance testing and learning theory. All three areas have been actively pursued for a number of years and there is an extensive body of literature. Model checking has been a vibrant area of research for more than 15 years with the development of the theory and a number of software tools. Most tools check properties of finite state models expressed in some formal notation. One tool that is directed at the checking of software systems without a model is VeriSoft [7]: it is aimed at checking state invariants (assertions) of communicating processes, using partial order reduction methods for space exploration. For a recent book on model checking see [5].

The study of testing black box automata was initiated in Moore's classical paper from 1956 [15], where he defined and studied several problems including the machine identification problem (infer the state transition diagram of an unknown black box automaton). He also posed the fault detection or conformance testing problem (checking that the black box conforms to a specified design automaton). This problem has been studied in the subsequent years by many researchers, obtaining good bounds on the lengths of the tests needed, as well as efficient algorithms that check for conformance for different types of automata (machines with a distinguishing sequence or with reset, or in general without) [4, 9, 24, 25]. In the last 15 years, there has been a lot of work on conformance testing in the protocols community, with a large number of papers, many of them based on the black box automaton testing models and methods. Early surveys of the work in the 50's and 60's can be found in e.g., [11, 22], and surveys of the more recent results and related work on protocol testing can be found in [13, 20]. Finally, there is substantial work in the learning community on the problem of learning finite automata (i.e. machine identification) with the help of a teacher. Efficient algorithms for learning different types of automata in this framework have been developed in [2, 14, 19].

2 PRELIMINARIES

AUTOMATA THEORETIC MODEL-CHECKING

A Büchi automaton is a quintuple $(S, S_0, \Sigma, \delta, F)$, where S is a finite set of states, $S_0 \subseteq S$ are the initial states, Σ is the finite alphabet, $\delta \subseteq S \times \Sigma \times S$ is the transition relation, and $F \subseteq S$ are the accepting states. A run over a word $a_1 a_2 \ldots \in \Sigma^\omega$ is an infinite sequence of states $s_1 s_2 s_3 \ldots$, with $s_1 \in S_0$, such that for each $i > 0$, $(s_i, \alpha_i, s_{i+1}) \in \delta$. A run is *accepting* if at least one accepting state occurs in it infinitely many times. A word is accepted by a Büchi automaton exactly when there exists a run accepting it. The *language* $\mathcal{L}(A)$ of a Büchi automaton A is the set of words that it accepts. Two automata are equivalent when they accept the same language.

An *implementation* automaton $B = (S^B, S_0^B, \Sigma, \delta^B, S^B)$ has several restrictions. We assume that the number of states $|S^B|$ is bounded by some value n, that S_0^B is a singleton $\{\iota\}$, and that $F^B = S^B$, namely, all the states are accepting.

We can view an implementation machine in our model as a Mealy machine: at each state v and for each input a, the machine outputs 0 if the transition is not enabled, and then remains in the same state, and 1 if it is enabled. Furthermore, we assume that the implementation automaton is deterministic, i.e., if $(s, a, t) \in \delta^B$ and $(s, a, t') \in \delta^B$, then $t = t'$.

For a *specification* automaton $P = (S^P, S_0^P, \Sigma, \delta^P, F^P)$, we will denote the number of states $|S^P|$ by m. Let the size of the alphabet Σ, common to the implementation and the specification, be p. As we mentioned in the introduction, we can easily extend the framework of this paper, and the results to implementation machines with arbitrary output (i.e., Mealy machines), and specification machines that describe the legal input-output behaviors.

The *intersection* (or *product*) $B \times P$ is $(S^B \times S^P, S_0^B \times S_0^P, \Sigma, \delta', S^B \times F^P)$, where

$$\delta' = \{((s, s'), \alpha, (t, t')) | (s, \alpha, t) \in \delta^B \wedge (s', \alpha, t') \in \delta^P\}.$$

Thus, the intersection contains (initial) states that are pairs of (initial, respectively) states of the individual automata. The transition relation relates such pairs following the two transition relations. The accepting states are pairs whose second component is an accepting state of P. We have that $\mathcal{L}(B \times P) = \mathcal{L}(B) \cap \mathcal{L}(P)$.

A **reset** is an additional symbol of S^B, not included in Σ, allowing a move from any state to the initial state. An *experiment* is a finite sequence $\alpha_1 \alpha_2 \ldots \alpha_{k-1} \in$

$(\Sigma \cup \{\textbf{reset}\})^*$, such that there exists a sequence of states $s_1 s_2 \ldots s_k$ of S^B, with $s_1 \in S_0^B$, and for each $1 \leq j < k$, either

1. $\alpha_j = \textbf{reset}$ and $s_{j+1} = \iota$ (a **reset** move), or
2. $(s_j, \alpha_j, s_{j+1}) \in \delta^B$ (an automaton move), or
3. there is no $t \in S^B$ such that $(s_j, \alpha_j, t) \in \delta^B$ and $s_{j+1} = s_j$ (a disabled move).

GAMES OF INCOMPLETE INFORMATION

The computation model for experiments on black box automata is not the standard one, in which the input is known from the beginning of the computation. Here, part of the input is hidden, and its structure is studied through experiments.

The relevant computational model is related to games of incomplete information [1, 18], where an $\exists$-*player* plays against a *deterministic environment* (representing a degenerate version of a $\forall$-*player*). Each such game consists of a nondeterministic machine with finitely many configurations[2] C, containing the following disjoint subsets: C_i are the *initial configurations*, W^+ and W^- are the positive and negative winning configurations for the $\exists$-player, respectively. Intuitively, since we want to check properties of systems, W^+ corresponds to finding an error, and W^- corresponds to concluding that there is no error.

Let $L_\exists$ and $L_\forall$ be sets of *labels* for the $\exists$-player and the environment, respectively. Then the sets of moves are $M_\exists \subseteq C \times L_\exists \times C$, and $M_\forall \in C \mapsto L_\forall \times C$, respectively. The $\exists$-player can have a choice of moves, thus $M_\exists$ is a relation, connecting the current configuration with all possible pairs of move-labels and resulted successor configurations. The moves of the environment $M_\forall$ are deterministic, and thus are defined as a function from the current configuration into the unique transition label and successor configuration. No move can originate in a winning configuration. Moreover, any two different moves from the same configuration must have different labels. The two players make moves in alternation, starting with the $\exists$-player, who makes the first move from an initial configuration. A *play* is a sequence from $(CL_\exists CL_\forall)^*C$, where each adjacent triple over $C(L_\exists \cup L_\forall)C$ conforms with a move of one of the players. A play is *winning* if it ends with a winning configuration in $W^+ \cup W^-$. There is no initial configurations starting both a play that ends with a configurations in W^+ and a play that ends with a configurations in W^-.

The incomplete information is stated by the partition of the configurations C into equivalence classes called *information sets*. The $\exists$-player cannot distinguish between configurations c_1 and c_2 that are in the same information set, denoted $c_1 \approx c_2$. Therefore, the move function $M_\exists$ must allow moves with the same labels for all the configurations that are in the same equivalence class. Furthermore, if $c_1 \approx c_2$ then $c_1 \in W^+$ ($c_1 \in W^-$, respectively) if and only if $c_2 \in W^+$ ($c_2 \in W^-$, respectively).

A *deterministic strategy* for the $\exists$-player is a function $st_\exists : C \times (L_\forall \cup \{\texttt{init}\}) \mapsto M_\exists$, such that

[2] Since the games that will be described later involve choosing an automaton and performing experiments on it, we choose to distinguish between a *configuration* of a game and a *state* of an automaton.

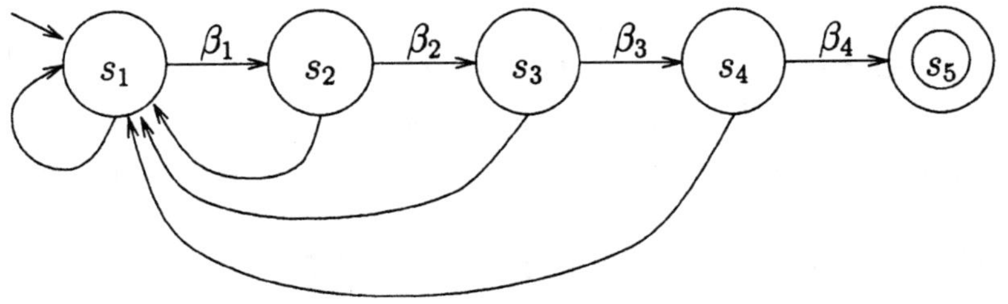

Figure 1: A combination lock automaton

1. If the $\exists$-player will keep playing $st_\exists(c, l)$ when it is his turn from configuration c and after the environment has played a move labeled with l, the sequence will end with a configuration in $W^+ \cup W^-$.
2. If $c \approx c'$, then the labels on $st_\exists(c, l)$ and $st_\exists(c', l)$ are the same.

The additional value `init` is paired with initial configurations from C_i (since there is no previous label for this configuration). A *path* played according to a strategy is an alternating sequence of configurations and labels, starting with an initial configuration. A *winning path* ends with a winning configuration. We will define the deterministic *time complexity* of a strategy as the length of the longest winning path in a strategy that ends with a configuration in $W^+ \cup W^-$.

We also define a *nondeterministic strategy* $nst_\exists : C \times (L_\forall \cup \{\texttt{init}\}) \mapsto M_\exists$ for the $\exists$-player. Let $c \in C_i$ be an arbitrary configuration such that there exists a play from c that ends with a configuration in W^+. Every play starting with c in which the $\exists$-player keeps playing his turn according to the $nst_\exists$ strategy will end with a configuration in W^+. The second constraint that was imposed on the deterministic strategy does not have a counterpart in the definition of the nondeterministic strategy. The intuition is that in the nondeterministic case, an $\exists$-player that is playing according to a nondeterministic strategy can make guesses that can distinguish between configurations that are in the same information set.

COMBINATION LOCK AUTOMATA

The following set of automata [15, 24] plays a major role in proving lower bounds on experiments with black box automata. A *combination lock* automaton [15] is a finite automaton such that there exists some complete order of the states $s_1, s_2, \ldots s_n$ with s_1 the initial state, and where the state s_n has no enabled transition. For each state s_i, $i < n$, there is a transition labeled with some $\beta_i \in \Sigma$ to s_{i+1}. For all other letters $\gamma \in \Sigma \setminus \{\beta_i\}$, there is a transition labeled with γ from s_i back to the initial state. It is called the combination automaton for $\beta_1\beta_2 \ldots \beta_{n-1}$. Figure 1 depicts a combination lock automaton for $n = 5$.

A sequence leading to a state without a successor (or even to a state where not all the letters are enabled) in a combination automaton must have a suffix of length $n - 1$

that is $\beta_1\beta_2\ldots\beta_{n-1}$. (This is a necessary, but not a sufficient condition. For example, the automaton in Figure 1 does not reach a deadlock state as a result of the sequence $\beta_1\beta_2\beta_1\beta_2\beta_3\beta_4$ when $\beta_1 \neq \beta_3$, since the second β_1 only causes it to return to its initial state.) Every path that does not contain the consecutive sequence $\beta_1\beta_2\ldots\beta_{n-1}$ is allowed (enabled) by the automaton.

3 BLACK BOX DEADLOCK DETECTION

In this section we describe a simpler problem related to black box checking. Given a deterministic finite state system B, with no more than n states, we want to check whether this machine deadlocks, namely reaches a state from which no input is possible.

In this problem, part of the model is unknown and is learned via experiments, which motivates modeling the problem as a game with incomplete information. We will also demonstrate that the deterministic and nondeterministic complexity do not have the same connections as in the standard model of computation that is based on Turing Machines.

For each implementation automaton with n or less states, there exists a single initial configuration. Each configuration in C in a play contains the same automaton as in the initial configuration, the current state of this automaton, as controlled by the moves of the $\exists$-player, and some information about the sequence of moves played so far. The current state of the automaton in an initial configuration is its initial state. The moves $M_\forall$ of the environment are labeled by `success` or `fail`. The label indicates whether the environment was successful or not in changing the state of the implementation automaton using the transition chosen according to the label of the last move of the $\exists$-player. The moves of the $\exists$-player are the possible input symbols, or a **reset** followed by a symbol (A **reset** is always successful.)

Projecting the labels of the moves of the $\exists$-player from a play ξ, we obtain an experiment over the implementation automaton in the initial configuration of ξ. If the configurations c_1 and c_2 are reachable using the prefixes of two plays ξ_1 and ξ_2 that correspond to the same experiment, then $c_1 \approx c_2$. The winning set W^+ contains only configurations that include an automaton that has a deadlock. Similarly, the winning set W^- contains only configurations that include an automaton without a deadlock.

A NONDETERMINISTIC STRATEGY

The $\exists$-player guesses in each move a label, forming a sequence of length smaller than n that brings the state of the selected machine from its initial state to some deadlock state. He then checks that this state has no enabled transitions.

Complexity: Nondeterministic time $O(n+p)$. The only information that is needed to be kept in each configuration is a counter from 0 to $n-1$ and a counter on the number of labels checked from the final state.

A DETERMINISTIC STRATEGY

The ∃-player checks systematically the possible sequences, up to length $n-1$, starting from the initial state. (Of course, there is no need to check sequences that include prefixes that have led to a failure.)

Complexity: Deterministic time $O(p^n)$.

Theorem 1 *The deterministic time complexity for black box deadlock detection is* $\Omega(p^{n-1})$.

Proof. Suppose that the initial configuration includes an implementation automaton B with n states that allows any input from any state. Consider a play ξ, played using a deterministic strategy for the ∃-player. Assume that ξ has less than p^{n-1} moves of the ∃-player, and terminates with a winning configuration c_w in W^-. Then at least one sequence $\beta_1\beta_2\ldots\beta_{n-1}$ does not appear consecutively in the experiment associated with ξ. If instead of the above automaton B, the environment would have chosen a combination lock automaton for $\beta_1\beta_2\ldots\beta_{n-1}$, the deterministic strategy would have resulted in a prefix of a play that has the same labels as ξ. Now we would have reached a configuration c' such that $c_w \approx c'$. Further, c_w is associated with an automaton without deadlocks, while c' is associated with an automaton with deadlocks. This contradicts the assumption that $c_w \in W^-$. ∎

In the standard complexity model, it is not known whether one can obtain a polynomial deterministic algorithm from a nondeterministic polynomial algorithm. Here, the (tight) lower deterministic bound is exponentially larger than the nondeterministic complexity. This justifies the use of games with incomplete information as an alternative for the common computational model of Turing machines.[3]

4 CHECKING PROPERTIES OF BLACK BOX FINITE STATE MACHINES

We address now the problem of black box checking. Namely, given a specification Büchi automaton P with m states, and a black box implementation automaton B with no more than n states, over a mutual alphabet Σ with p letters, we want to check if there is a sequence accepted by both P and B. The automaton P accepts the *bad* computations, i.e., those that are *not allowed*. Thus, if the property is given originally e.g., using a linear temporal logic (LTL) [17] property φ, then P is the automaton corresponding to $\neg\varphi$. For an efficient translation from LTL to automata, see e.g., [6]. The following simple theorem demonstrates that the current problem is at least exponential in time in the size of the automaton B.

Theorem 2 *The deterministic time complexity of black box checking is* $\Omega(p^{n-1})$.

[3] Another observation, connected with the *space complexity* of this model, which was not defined formally in this paper, is that the common space efficient strategy of binary search cannot be used here.

Proof. Similar to the proof of Theorem 1, we construct variants of a combination lock automata. The deadlock state is replaced now with a self loop labeled by γ. The symbol γ is disabled in the initial state. This removes at most half of the possible combinations (in the case where $p = 2$), so the complexity changes only by a constant factor. The property automaton P consists of two states: t_0 which is an initial state, and t_1 which is an accepting state. There is a self loop from t_0 to itself on each label from Σ, and from t_1 to itself on γ. There is also an edge labeled by γ from t_0 to t_1. Thus, the intersection is nonempty exactly if a state can be reached in the black box automaton, where γ^ω can be executed. The only such state of a combination lock black box is the state at the end of the path prescribed by the combination. ∎

4.1 AN OFF-LINE STRATEGY

A straightforward way to perform black box checking is to infer first the structure of the black box system, and then to apply model checking techniques to its newly revealed structure. The machine identification problem is a well studied problem. Typically, it is applied to automata that produce output, either at the states (Moore machines) or at the transitions (Mealy machines). As we mentioned, an implementation machine in our model can be viewed as a Mealy machine, where output 0 on a transition means that it is not enabled and output 1 means that it is enabled.[4]

It is well known that if two machines with n states are not equivalent, then there is an input of length at most $2n - 1$ that distinguishes them. This implies that any machine with at most n states is completely characterized by its output on all input strings of length $2n - 1$. That is, a black box is uniquely determined by applying all such p^{2n-1} input strings. A p-ary tree of depth $2n - 1$ can be constructed from the responses of the black box, and it can be minimized to produce the minimal machine M consistent with these outputs [22]. Then we can use model checking to check whether M satisfies the given property P. The length of the test sequence (or in terms of games with incomplete information, the length of the corresponding play), which gives us the time complexity, is $O(np^{2n-1})$. If implemented in the straightforward way, the space complexity is also exponential (to record the tree all the input strings and their output), but the minimization can be done incrementally in polynomial space. The time for the model checking is comparatively small, $O(pmn)$ where m is the size of the property automaton P (which is typically very small).

The complexity of this method is not that far off the lower bound, and in the worst case one may indeed need to identify in effect the black box machine in order to check a property. However, intuitively it is clear that in many cases this method can be wasteful in that it does not take advantage of the property to avoid doing a complete identification. For example, suppose that the property is that some error indication label γ never occurs. The property automaton P, representing the bad computations is

[4] By this convention, the output provides some partial information also on the next state, namely if the output is 0 then we know that the state does not change. This is not important for what follows (i.e. all the methods apply to any Mealy machine) but it can be used to do obvious optimizations on the tests. For example, if we apply input a and it is not enabled, then it is pointless to try again a until an enabled transition has been performed.

in this case a simple 2 state automaton. Obviously in this example, there is no reason to wait until we reconstruct the full black box automaton before we check the property. The sensible thing to do would be to check the assertion (i.e., try to see if γ is enabled in the current state) as we go along during the test, and if it gets violated at some point, then an error has been found and the check is complete.

In general, it would be obviously beneficial to use the property automaton on the fly to detect errors as early as possible and prune the test. Notably, if the estimate n on the number of states is much higher than the actual number of different states, or if it is accurate, but there is still a 'small' counterexample, i.e. there is a small set of states that exhibits the faulty behavior, we would like to be able to find the error without searching the whole space, if possible. This is not always as easy, especially in the case of properties that depend on the infinite behavior of the system, that is, in cases where the property automaton is a genuine Büchi automaton. We will investigate such methods in the following sections.

4.2 AN ON-THE-FLY STRATEGY

We will present now a strategy that can terminate quickly when the actual size of the automaton is much smaller than n, and an error is present (i.e., the intersection of B and P is nonempty).

A NONDETERMINISTIC STRATEGY

As before, we start with a nondeterministic version, in order to demonstrate the principle behind the on-the-fly black box checking.

According to the strategy, the $\exists$-player guesses a path σ of the automaton P, starting from an initial state, that can be partitioned into two subpaths σ_1 and σ_2, Each of which is of length smaller or equal to mn. Both subpaths end with the same accepting state t of P. Furthermore, the blackbox automaton must allow executing the transitions of $\sigma_1\,\sigma_2^{n+1}$ after a **reset**.

Correctness: Consider the unknown end state of the intersection automaton for each iteration of σ_2 in the experiment $\sigma_1\,\sigma_2^{n+1}$. Then at least one component state s of B must occur twice with the same accepting component state t of P (as there are no more than n states in the intersection that have the same component t). Thus, the path σ must include a cycle through an accepting state, which guarantees that an infinite accepting run exists in the intersection. Conversely, it is easy to see that if the intersection of B and P is nonempty, such a guess exists.

Complexity: Nondeterministic time $O(n^2m)$.

A DETERMINISTIC STRATEGY

The strategy finds a path as in the nondeterministic case, but in a systematic way. Now, a search for two paths of length bounded by mn is performed. The first path σ_1, when input to the automaton P needs to terminate with an accepting state t. The second path σ_2, when starting in state t of P, needs to terminate with t as well. For each such pair, we apply the second path n more times. That is, we try to execute the path $\sigma_1\,(\sigma_2)^{n+1}$.

If we succeed, this means that there is a cycle in the intersection through a state with a t (which is accepting) as the P component, since there are at most n ways to pair up t with a state of B. In this case, there is an infinite accepting path in the intersection.

Complexity: Deterministic time $O(n^2\, p^{2mn}\, m)$. This is because there are p^{2mn} choices of such paths. Each is of length bounded by mn, and we repeat it $n + 1$ times.

The following comments should be noted:

- Unlike the off-line strategy, the complexity of this strategy depends on the number of states m of P. Typically however m is small, or even fixed, when talking about a fixed property.
- For properties that can be specified by automata on finite strings (i.e., depend essentially on finite computations), we need to search only for the first string σ_1 and the complexity is $O(n\, p^{mn}\, m)$.
- When searching for the strings σ_1, σ_2, we need only consider strings that can be extended to accepting strings of the property automaton. Furthermore, we can start by limiting the length of the subpaths that we explore and gradually increase that length as we proceed in the search. In this way, if the actual size of the the automaton B is much smaller than n, and an error occurs, it can be found much earlier than in the exhaustive strategy, as required above.

4.3 A STRATEGY BASED ON LEARNING AND TESTING

We show now that the factor m in the exponent can actually be removed. We provide below a strategy with complexity whose exponential term is $O(p^n)$. Furthermore, if the black box has an error, the time complexity will be exponential only in the actual size of the minimized version of the black box automaton.

In this algorithm, we will use an algorithm for conformance testing of a known finite automaton with a black box automaton by Vasilevskii and Chow [4, 24]. We will not repeat this algorithm here, but will use the result that its time complexity is $O(l^2\, n\, p^{n-l+1})$, where n is the assumed size of the black box automaton, $l \leq n$ is the actual size, and p is the alphabet size. Intuitively, the algorithm has to check the states and transition relation of the black box automaton, and that no error that follows a 'combination lock' occurs from any one of its node.

Another procedure that we use in our strategy is an algorithm for learning automata with **reset** using membership tests and questions to an oracle (a teacher) by Angluin [2]. In the learning algorithm, the teacher answers equivalence tests to a proposed machine and provides a counterexample in case of inequivalence. We will replace the teacher with experiments on the black box automaton. Starting from a trivial automaton, Angluin's algorithm generates successively larger candidate automata M_i, for $i = 1, 2, \ldots$ (the number of states in each conjectured automaton is monotonically increasing). It asks the teacher for equivalence. If equivalence does not hold, it uses a counterexample provided by the teacher, queries some more strings, and then generates the next conjectured automaton with more states, until it reaches the correct number of states. At this point the conjectured automaton is the correct one.

We modify Angluin's algorithm as follows. Our modification can use two kinds of counterexamples, provided by the teacher:

1. A *simple* counterexample of the form $\sigma \in \Sigma^*$, meaning that σ belongs to one of the checked automata, but not the other.
2. A pair of words $\sigma_1, \sigma_2 \in \Sigma^*$ such that $\sigma_1 \sigma_2^n$ belongs to one of the checked automata, but not the other.

We construct a sequence of automata $M_1, M_2, \ldots$ that attempt to converge into the black box automaton B. Membership queries are just experiments on the black box B. For equivalence queries, suppose we have a conjectured automaton M_i for the black box. First, we check if M_i generates a word accepted also by the specification automaton P, namely if $\mathcal{L}(M_i) \cap \mathcal{L}(P) \neq \emptyset$. If the intersection is not empty, it must contain an ultimately periodic word of the form $\sigma_1\sigma_2^\omega$ [21]. We input **reset** $\sigma_1\sigma_2^{n+1}$ to the black box B. If this experiment succeeds, then there is an error as $\mathcal{L}(B) \cap \mathcal{L}(P)$ contains $\sigma_1\sigma_2^\omega$ and thus is not empty. If it fails, then this gives a counterexample for the equivalence of B with M_i. We use this in Angluin's algorithm to generate the next candidate automaton with more states.

If M_i does not generate any word accepted by P, we check whether M_i conforms with B. Let k be the number of states of M_i. We start the conformance test between M_i and B assuming B has k states and apply the Vasilevskii-Chow algorithm. If the conformance test fails, we use the counterexample in Angluin's learning algorithm to generate M_{i+1}. If the conformance test succeeds, we repeat it with $k+1$, $k+2$, $\ldots$, n. If n is reached, we declare that the black box satisfies the checked property.

This strategy is described in Figure 2. The procedure call $VC(M_i, k)$ calls the Vasilevskii and Chow algorithm for conformance testing M_i with the black box automaton B, assuming that B has no more than k states. VC returns (**true**,–) if the conformance test succeeds. If it fails, it returns (**fail**,σ), where σ is a word that is in one of the automata B or M_i but not the other. The procedures $ANGLUIN$ accepts the previous attempted automaton, and a counterexample, and returns a new attempted automaton. In the first call to $ANGLUIN$ in the strategy, it is executed with an empty automaton, and the second parameter (the counterexample) is ignored.

Complexity: Suppose that the minimum equivalent automaton M_i of the black box has l states. If the black box has an error, then the strategy will produce the automaton M_i in time $O(l^3\,p^l)$, and declare the error. This is because some earlier conjectured automaton may have an empty intersection with P, in which case the strategy will need to do conformance testing until a string that distinguishes the conjectured automaton from B is found.

If the black box satisfies the property, then the strategy will generate M_i in time $O(l^3\,p^l)$, but then it will have to spend $O(l^3\,p^{n-l+1})$ more time to verify that the black box is indeed equivalent to M_i (by conformance checking M_i with B). It should be noted that the complexity of doing the exhaustive check for trying to distinguish the conjectured automaton from B is dominating over the complexity of the other tests prescribed by the above strategy. Moreover, since $O(p^n + p^{n-1}) = O(p^n)$, we need to consider only the last exhaustive check. Thus, we have the following theorem.

Theorem 3 *Black box checking, for a black box automaton B with l states, where l is unknown but is smaller than some bound n, and a property automaton P with m states can be done in time*

```
         M_1 := ANGLUIN(empty, -);
         i := 1;
learn:   X := M_i × P;
         if L(X) = ∅ then
              k := number of states of M_i
              loop
                   (conforms, σ) := VC(M_i, k);
                   k := k + 1
              until k > n or ¬conforms;
              if conforms then WIN(+);
         else
              let σ1, σ2 be s.t. σ1 σ2^ω ∈ L(X);
              if B allows reset σ1 σ2^n then WIN(-)
                   else σ := prefix of σ1 σ2^n not allowed by B;
         M_{i+1} := ANGLUIN(M_i, σ);
         i := i + 1;
         goto learn;
```

Figure 2: A strategy using learning

- $O(l^3 p^l + l^2 mn)$, *when there is an error (i.e., the intersection of B and P is nonempty), and*
- $O(l^3 p^l + l^3 p^{n-l+1} + l^2 mn)$, *when there is no error.*

If we do not have a bound n on the number of states of the automaton B, we can run the algorithm as long as time permits. Consider first a property characterized by an automaton on finite strings (such as deadlock freedom and other safety properties). If we ever encounter an error, i.e. find a string σ accepted by both the black box and the property automaton P then it is a true error and we can stop the test. If there is an error and B has size l (or the smallest counterexample has length l), we are sure to find the error within time $O(l^3 p^l + l^2 m)$. Conversely, if after the allocated time no error has been found, then this means that either the black box is correct, or else the smallest possible counterexample and the size of the black box must exceed a certain bound, which depends on the time spent on the test.

Suppose that we have a genuine Büchi automaton that depends on infinite behaviors. Suppose that at some point the conjectured black box automaton M_i has a nonempty intersection with the property automaton P, and let $\sigma_1\sigma_2^\omega$ be a string in the intersection. If the conjectured automaton M has l states at this point and P has m states, then the strings σ_1, σ_2 have length at most lm. We can input **reset**σ_1 to the black box followed by repeated applications of σ_2 until either the black box does not accept it or we run out of time. In the first case, we have found at least one new state and we continue the algorithm as before. In the second case, that is, if we run out of time after executing r repetitions of σ_2, then we can conclude that either there is an error, or the size of the implementation machine exceeds r.

Finally, if the conjectured automaton has an empty intersection with the property automaton then we perform conformance testing for increasingly larger values of the bound n on the black box. At the end we can place again a lower bound on the size of the black box or conclude that it is otherwise correct.

Let us comment finally on the exponential lower bound derived from the combination lock automata. Obviously these are rather pathological, worst case examples. The 'average' automata are much better behaved and do not exhibit this nasty performance bottleneck. This can be formalized by considering a probabilistic model for machines with output. Formally, there has been extensive work studying the properties of random machines [22]. The usual model of a random Mealy machine on l states is defined as follows. For each state and input symbol choose the next state and output uniformly at random. For the average machine, polynomial time will suffice to find an error. In the following statement, 'almost all machines' means that the probability tends to 1 as the size goes to infinity.

Theorem 4 *For almost all black box machines with l states, if an error is present, it will be found after a test of length $O(lp \log^2 l + l^2 nm)$.*

This can be shown using the following two nice properties of almost all random machines: (1) if a state q can reach another state q' then it can reach it in $O(\log_p l)$ steps; (2) any two states can be distinguished by input strings of length $\log_p \log_q l$ [22]. Of course, if there is no error and we want to make sure that we do not have any other automaton at hand with at most n states, then we still would need to do the conformance testing (at a cost exponential in the difference $n - l$) in order to be certain of the correctness.

5 CONCLUSIONS

We defined the problem of black box checking, showed lower bounds and provided three strategies for solving the problem. The lower bound in Theorem 2 implies that the complexity of black box checking is exponential in the estimated size of the unknown automaton. For comparison, checking the emptiness of the intersection of the same automata (now both structures are given) is in NLOGSPACE-complete. In conformance testing, one checks whether a given known automaton P of length l is equivalent to a black box automaton B of length bounded by some $n \geq l$. Vasilevskii and Chow [24, 4] showed a lower and upper bound of $O(l^2 np^{n-l+1})$ for conformance testing with reliable **resets**. When $n = l$, namely the *actual* size of the black box automaton is known, this is a much more tractable complexity than that of black box checking. Thus, if a model (abstract design) is available or feasible to construct, then a good strategy for the developer of a system is to separately do a conformance test of an abstract design against the system, and then model check the design with respect to various properties. However, when a model is not available or n can be considerably bigger than l this approach does not help.

It is quite clear that the off-line strategy is suboptimal as it does not take advantage of the property at hand. On the other hand, the on-the-fly strategies, while still exponential, may work in practice in some important cases. One case is when an error exists and the estimate n is much higher than the actual size of the checked system or the size of a portion of the system that provides a counterexample. Another case is when the specification automaton P limits the possible bad executions considerably. An example for a "helpful" specification will be that P specifies sequences of the form $\alpha^*(\beta+\gamma)$. An example of an "unhelpful specification" is $X^*\alpha$, where X allows any letter of Σ except α.

The last strategy, based on learning, uses the property P while trying to learn the structure of B. Thus, an error may be found before completing the construction of a minimized automaton equivalent to B. It is also possible that no explicit bound is given on the size of the black box automaton. In this case, we can use the strategy as long as we are willing to spend time.

There is a number of issues in black box checking that deserve further investigation. Some open problems are finding strategies for partially specified automata, or known automata where the actual implementation deviates from the known design in no more than k changes ('implementation errors'). Another problem is to develop an algorithm for black box checking when reliable **reset** moves are not available. It is possible that similar techniques can be used by combining the learning algorithm of [19] with the conformance testing algorithm of [25] for machines without reset.

References

[1] R. Alur, C. Courcoubetis, M. Yannakakis, Distinguishing tests for nondeterministic and probablistic machines, Symposium on Theory of Computer Science, 1995, ACM, 363–372.

[2] D. Angluin, Learning Regular Sets from Queries and Counterexamples, Information and Computation, 75, 87–106 (1978).

[3] J. R. Büchi, On a decision method in restricted second order arithmetic, Proceedings of International Congress on Logic, Methodology and Philosophy of Science, Palo Alto, CA, USA, 1960, 1–11.

[4] T. S. Chow, Testing software design modeled by finite-state machines, IEEE transactions on software engineering, SE-4, 3, 1978, 178–187.

[5] E.M. Clarke, O. Grumberg, D. Peled, Model Checking, MIT Press, to appear 1999.

[6] R. Gerth, D. Peled, M. Vardi, P. Wolper, Simple on-the-fly automatic verification of linear temporal logic, Protocol Specification Testing and Verification, 1995, Chapman & Hall, 3–18, Warsaw, Poland.

[7] P. Godefroid, Model checking for programming languages using VeriSoft, Proc. 24th ACM Symp. on Progr. Lang. and Sys., 174-186, 1996.

[8] G. J. Holzmann, The model checker SPIN, IEEE transactions on Software Engineering, 23(5):279–295.

[9] F. C. Hennie, Fault detecting experiments for sequential circuits, Proc. 5th Ann. Symp. Switching Circuit Theory and Logical Design, 95-110, 1964.

[10] G. J. Holzmann, D. Peled, The State of Spin, 8th International Conference on Computer Aided Verification, Springer Verlag, LNCS, 1102, 385-389, 1996, New Brunswick, NJ, USA.

[11] Z. Kohavi, Switching and Finite Automata Theory, 1978, McGraw Hill.

[12] R. P. Kurshan, Computer-Aided Verification of Coordinating Processes : The Automata-Theoretic Approach, Princeton University Press, 1995.

[13] D. Lee, M. Yannakakis, Principles and methods of testing finite state machines - a survey, Proceedings of the IEEE, 84(8), 1090–1126, 1996.

[14] O. Maler, A. Pnueli, On the learnability of infinitary regular sets, Information and Computation 118 (1995), 316-326.

[15] E. F. Moore, Gedanken-experiments on sequential machines, Automata Studies, Princeton University Press, 1956, 129–153.

[16] G. J. Myers, The Art of Software Testing, Wiley International, 1979.

[17] A. Pnueli, The temporal logic of programs, $18th$ IEEE symposium on Foundation of Computer Science, 1977, 46–57.

[18] J.H. Reif, The complexity of two-player games of incomplete information, Journal of computer and system sciences, 29, 1984, 274–301.

[19] R. L. Rivest, R. E. Schapire, Inference of finite automata using homing sequences, Information and Computation 103, 299-347, 1993.

[20] D. P. Sidhu, T. K. Leung, Formal methods for protocol testing: a detailed study, IEEE Trans. Sw Eng., 15, 413-426, 1989.

[21] W. Thomas, Automata on infinite objects, Handbook of Theoretical Computer Science, MIT Press, J. van Leeuwen (Ed.), 135–192.

[22] B. A. Trakhtenbrot, Y. M. Barzdin, Finite Automata: Behavior and Synthesis, North Holland, 1973.

[23] M. Y. Vardi, P. Wolper, An automata-theoretic approach to automatic program verification, Proceedings of the First Symposium on Logic in Computer Science, Cambridge, UK, 322-331.

[24] M. P. Vasilevskii, Failure diagnosis of automata, Kibertetika, no 4, 1973, 98–108.

[25] M. Yannakakis, D. Lee, Testing finite state machines: fault detection, J. Computer and Syst. Sci., 50, 209-227, 1995.

AN OPTIMIZING COMPILER FOR EFFICIENT MODEL CHECKING

Yifei Dong, C.R. Ramakrishnan
Department of Computer Science
SUNY at Stony Brook
Stony Brook, NY 11794-4400, U.S.A.
ydong@cs.sunysb.edu, cram@cs.sunysb.edu

Abstract We present an optimizing compiler that translates high-level specifications in a language based on value-passing CCS into *rules* representing the global transition relation of the underlying automata. The transitions can be computed from these rules in unit time (modulo indexing) during verification. The compiler takes time that is, in the worst case, quadratic in the number of parallel components of the system. The compiler incorporates several optimizations to reduce the state space of the generated automata. We present experimental results which show that our technique reduces the memory and time needed for verification significantly: by more than an order of magnitude in some cases.

Keywords: Model Checking, Specification Languages, Optimization.

1. INTRODUCTION

Specification languages supported by most verification tools have high-level features such as data structures, procedures, and parameterized processes. Model checking techniques [5, 18, 6] typically view the specification in terms of the underlying automaton: labeled transition system or Kripke structure [16]. Verification tools first *compile* the input specifications into a compact automata representation. For example, SPIN [11] compiles PROMELA specifications into local automata, one for each process; CÆSAR [10] translates LOTOS specifications [2] into Petri nets. At verification time, these representations are used by the explicit-state model checkers to construct the global automaton.

In contrast to the above tools, Murφ [8] provides a guarded command language for specifying a system directly as rules defining its global transition relation. This permits the user to optimize the global automata by using several encoding tricks (such as grouping many individual computations and actions together into a single transition). A recent study of the performance of various verification tools shows that the low-level specification enables Murφ to be considerably faster, and consume much less memory, than other tools [9].

It would hence be ideal to combine the verification efficiency afforded by low-level specifications with the ease of correct encoding enabled by high-level specifications. Many tools support user annotations to reduce the state space of global automata. For instance, using *atomic* and *d_step* constructs of PROMELA, a SPIN user can group many computation steps into one transition. However, the user must ensure that the optimizations are correct. Moreover, lower-level optimizations (as can be done in a Murφ specification) can reduce the state space even further.

An Optimizing Compiler for Model Checking. In this paper we describe an optimizing compiler that efficiently translates high-level specifications (based on value-passing CCS [17]) into a compact representation of the global automaton. We use several optimizations to reduce the state space of the global automaton *without relying on user annotations.* The salient features of the compilation technique are:

- The compiler translates value-passing CCS expressions into (Horn-clause like) rules that represent the global transition relation. It precomputes possible synchronizations at compile-time while still maintaining polynomial (quadratic) bounds for compile time and space. Transitions of the global automaton can be computed from these rules in unit time (modulo indexing) during verification.

- The compiler is derived directly from the structural operational semantics of value-passing CCS expressions. In contrast, the translation for LOTOS programs described in [10] is not directly based on LOTOS operational semantics [2].

- Consecutive computation steps are grouped into single atomic transitions *even across basic block boundaries.* In contrast *atomic* annotations in SPIN can group together only computations that lie within a single block. Our aggressive optimization reduces the state space by more than an order of magnitude in some examples, thereby improving verification performance.

Our Approach and its Effectiveness. We describe the compilation technique using the XMC system [19], a model checker for modal mu-calculus [13] and XL, a process description language. XL, described formally in Section 2, extends the value-passing CCS with parameterized processes, sequential composition and procedure calls.

The XMC system is built using the XSB tabled logic programming system [22]. The core of the system consists of two predicates, `trans`:*State* × *Action* × *State* which computes the transition relation by interpreting XL process terms, and `models`:*State* × *Formula* which verifies whether a state represented by a process term models a given modal mu-calculus formula. The two predicates directly encode the structural operational semantics of XL and modal mu-calculus respectively. This encoding reduces model checking to logic program query evaluation. Furthermore, the goal-directed query evaluation mechanism of the XSB logic programming system yields a local model checker. Using XMC, we can verify systems with more than a hundred thousand reachable states in times that are comparable to SPIN and Murφ. A detailed description of the XMC system can be found in [7].

The compiler for XL is described in Section 3. We implemented the compiler in the XMC system and evaluated its effectiveness using benchmark programs with different characteristics. The results of the experiments are presented in Section 4. The compiler speeds up the XMC model checker by factors of up to 15, while reducing space requirements by factors of 6 or more. Using the compiler we can verify high-level specifications of several examples in the XMC system as fast as verifying equivalent low-level specifications in Murφ. In Section 5, we describe the applicability of our technique to other tools.

2. THE SPECIFICATION LANGUAGE XL

XL is based on the value passing CCS [17]. Values and computations are represented as Prolog terms and predicates respectively. Thus the specifications can use recursive data structures and computations.

The syntax of XL specifications are described by the grammar shown in Figure 1.1. In the figure, *Proc* is a (parameterized) process name, represented as a term (e.g., `channel(N, Buf)`). *Comp* is a term (e.g., `X is Y+1`) representing a computation. A terminating null process is represented by `true`, the empty computation. A process `if`(C,S_1,S_2), behaves like S_1 if computation C succeeds, and like S_2 if C fails. The computation C in an `if` expression is assumed to leave the bindings of variables unchanged. A process `in(chan(`t`))` inputs a value that matches term t over port `chan`; `out(chan(`t`))` outputs the value repre-

E	$\longrightarrow$	*Comp*	(computation)
	|	`in(`*Term*`)`	(input communication)
	|	`out(`*Term*`)`	(output communication)
	|	`zero`	(unit deadlocked process)
	|	E `o` E	(sequential composition)
	|	E `#` E	(choice)
	|	`if(`*Comp*`,` E`,` E`)`	(conditional)
	|	E | E	(parallel composition)
	|	E `\` *PortSet*	(restriction)
	|	E `@` *PortMap*	(relabeling)
	|	*Proc*	(process invocation)
Def	$\longrightarrow$	(*Proc* `::=` E)*	(process definitions)

Figure 1.1 Syntax of XL

sented by term t over port chan. Process invocations may be recursive; in fact, since the language provides no iterative constructs, recursion is the only way to specify loops in processes. As in CCS, relabeling and restriction are used to derive instances of a generic process. *PortSet* is a set of terms that represent the restricted ports, and *PortMap* is a list of pairs of terms (each pair of the form s/t) that denotes the mapping defined by relabeling. The language features are illustrated by the complete specification of the Alternating Bit Protocol [21] in Figure 1.2.

Operational Semantics of XL. In the following, we assume familiarity with the notions of terms and substitutions [15]. A substitution θ is a *unifier* of two terms t_1 and t_2 if $t_1\theta = t_2\theta$. The most general unifier of t_1 and t_2 is denoted by $mgu(t_1, t_2)$.

Following CCS, the structural operational semantics of XL can be specified in terms of labeled transition systems. In Figure 1.3 the semantics of an XL process is specified in terms of transitions in a labeled transition system where each state is represented by a process/substitution pair E, θ. We use $E, \theta \xrightarrow{\alpha} E', \theta'$ to denote that a process E under substitution θ can make a transition with label α to become process E' under substitution θ'. For computation C, $[\![C]\!]_L\theta$ is defined as the substitution θ' that is the result of evaluating the Prolog query $C\theta$. We say that $[\![C]\!]_L\theta = \{\}$ if the query $C\theta$ fails.

The semantics of communication primitives is specified by rules 1 and 2, sequential composition by rules 3 and 4, choice by rules 5 and 6, and the conditional by rules 7 and 8. Rules 9 and 10 denote autonomous transitions in a parallel composition. Rule 11 captures synchronization by matching the transition labels of the two components. Values

```
chan ::= in(get(Data)) o ( out(put(Data)) # out(drop) ) o chan.

send(Seq) ::=
% Seq is the sequence number of the next frame to be sent
    out(dataOut(Seq)) o
    (   ( in(ackIn(AckSeq)) o
          if( AckSeq == Seq
              , NSeq is 1-Seq o send(NSeq)      % successful ack, next message
              , send(Seq))                      % unexpected ack, resend message
        ) # send(Seq)).                         % upon timeout, resend message

recv(Seq) ::=
% Seq is the expected sequence number of the next frame to be received
    in(dataIn(RecSeq)) o
    if( RecSeq == Seq
        , ( NSeq is 1-Seq o out(ackOut(RecSeq)) o recv(NSeq) )
        , out(ackOut(RecSeq)) o recv(Seq)).   % unexpected seq, resend ack

abp ::=
    ( send(0) @ [s2r_in(X) / dataOut(X), r2s_out(X) / ackIn(X)]
    | chan @ [s2r_in(X) / get(X), s2r_out(X) / put(X)]    % sender -> receiver
    | chan @ [r2s_in(X) / get(X), r2s_out(X) / put(X)]    % receiver -> sender
    | recv(0) @ [s2r_out(X) / dataIn(X), r2s_in(X) / ackOut(X)]
    ) \ {s2r_in(_), s2r_out(_), r2s_in(_), r2s_out(_)}.
```

Figure 1.2 Specification of the Alternating Bit Protocol in XL

transmitted by synchronizing in(t) with out(t') are represented by the matching substitution σ (i.e., σ such that $t' = t\sigma$).

Process invocation is represented by rule 12: when a process P' is invoked under substitution θ, a definition P: :=E can be chosen such that P and P' unify, with the most general unifier composed with the current substitution θ. As is normal in such cases, we assume that P: :=E and P' are standardized apart before the unifier is computed: i.e., variables in P and E are suitably renamed to avoid capture.

Rule 13 specifies the semantics of restriction; the rule is made identical to that of basic CCS by overloading the meaning of membership '$\in$' as follows: given a list L of terms, $s \in L$ if there is a term t in L such that s =in(t) or s =out(t). Similarly, rule 14 uses a relabeling function as follows: given a list F of pairs of terms, if s/t is a member of F then F(in(t)) =in(s) and F(out(t)) =out(s); if there is no s such that s/t is a member of F then $F(t) = t$.

3. COMPILING XL

Given an XL program, the compiler derives a set of rules that concisely describes the corresponding transition relation. The conciseness

$$(1)\quad \frac{}{\texttt{in}(t),\theta \xrightarrow{\texttt{in}(t\theta)} \texttt{true},\theta} \qquad (2)\quad \frac{}{\texttt{out}(t),\theta \xrightarrow{\texttt{out}(t\theta)} \texttt{true},\theta}$$

$$(3)\quad \frac{E,\theta \xrightarrow{\alpha} E',\theta'}{E \circ F,\theta \xrightarrow{\alpha} E' \circ F,\theta'} \qquad (4)\quad \frac{[\![Comp]\!]_L\theta = \theta' \ \wedge \ E,\theta' \xrightarrow{\alpha} E',\theta''}{Comp \circ E,\theta \xrightarrow{\alpha} E',\theta''}$$

$$(5)\quad \frac{E_1,\theta \xrightarrow{\alpha} E_1',\theta'}{E_1 \,\#\, E_2,\theta \xrightarrow{\alpha} E_1',\theta'} \qquad (6)\quad \frac{E_2,\theta \xrightarrow{\alpha} E_2',\theta'}{E_1 \,\#\, E_2,\theta \xrightarrow{\alpha} E_2',\theta'}$$

$$(7)\quad \frac{[\![C]\!]_L\theta \neq \{\} \ \wedge \ E_1,\theta \xrightarrow{\alpha} E_1',\theta'}{\texttt{if}(C,\ E_1,\ E_2),\theta \xrightarrow{\alpha} E_1',\theta'} \qquad (8)\quad \frac{[\![\texttt{not}(C)]\!]_L\theta \neq \{\} \ \wedge \ E_2,\theta \xrightarrow{\alpha} E_2',\theta'}{\texttt{if}(C,\ E_1,\ E_2),\theta \xrightarrow{\alpha} E_2',\theta'}$$

$$(9)\quad \frac{E_1,\theta \xrightarrow{\alpha} E_1',\theta'}{E_1 \mid E_2,\theta \xrightarrow{\alpha} E_1' \mid E_2,\theta'} \qquad (10)\quad \frac{E_2,\theta \xrightarrow{\alpha} E_2',\theta'}{E_1 \mid E_2,\theta \xrightarrow{\alpha} E_1 \mid E_2',\theta'}$$

$$(11)\quad \frac{E_1,\theta_1 \xrightarrow{\alpha} E_1',\theta_1' \ \wedge \ E_2,\theta_2 \xrightarrow{\beta} E_2',\theta_2' \ \wedge \ \{\alpha,\beta\} \equiv \{\texttt{in}(t),\texttt{out}(t\sigma)\}}{E_1 \mid E_2,\ \theta_1\theta_2\sigma \xrightarrow{\texttt{tau}} E_1' \mid E_2',\ \theta_1'\theta_2'\sigma}$$

$$(12)\quad \frac{E,\theta \xrightarrow{\alpha} E',\theta'}{P',\sigma\theta \xrightarrow{\alpha} E',\sigma\theta'} \quad (P\texttt{::=}\ E, \sigma = mgu(P,P'))$$

$$(13)\quad \frac{E,\theta \xrightarrow{\alpha} E',\theta'}{E \setminus L,\theta \xrightarrow{\alpha} E' \setminus L,\theta'} \quad (\alpha \notin L) \qquad (14)\quad \frac{E,\theta \xrightarrow{\alpha} E',\theta'}{E \bullet F,\theta \xrightarrow{F(\alpha)} E' \bullet F,\theta'}$$

Figure 1.3 Operational Semantics of XL

requirement is especially important since the transition relation may be exponential in the size of the input program.

We follow the traditional approach for concisely representing large (even infinite) automata: separating *control* from *data*. A (possibly infinite) set S of states of an automaton is characterized by a single *control state* π, which is analogous to a program counter value. Associated with each control state π are a set of variables, denoted by $vars(\pi)$. For convenience, we use a term that contains $vars(\pi)$ to represent the control state π itself. Each state in the labeled transition system can then be represented by $\pi\sigma$, i.e. a control state π under a substitution σ. Transitions between states $\pi_s\sigma_s$ and $\pi_d\sigma_d$ can be captured by a transition rule of the form $\texttt{trans}(\pi_s,\ \alpha,\ \pi_d,\ c)$, where α is the label and c relates σ_s and σ_d. Formally, given a set of transition rules $\mathcal{T}$, the underlying transition relation is computed using the inference rule:

$$\frac{\texttt{trans}(\pi_s,\ \alpha,\ \pi_d,\ c) \in \mathcal{T} \ \wedge \ [\![c]\!]_L\theta_s = \theta_d}{\pi_s\theta_s \xrightarrow{\alpha} \pi_d\theta_d} \tag{1.1}$$

We compile each process expression into an automaton with two distinguished control states: *entry* and *exit*. The compiler is described in

$$[\![Comp]\!]\ entry\ exit = \{\ \mathtt{trans}(entry,\ \mathtt{i},\ exit,\ Comp)\ \} \qquad \cdots (1)$$

$$[\![\mathtt{in}(t)]\!]\ entry\ exit = \{\ \mathtt{trans}(entry,\ \mathtt{in}(t),\ exit,\ \mathtt{true})\ \} \qquad \cdots (2)$$

$$[\![\mathtt{out}(t)]\!]\ entry\ exit = \{\ \mathtt{trans}(entry,\ \mathtt{out}(t),\ exit,\ \mathtt{true})\ \} \qquad \cdots (3)$$

$$[\![E_1 \circ E_2]\!]\ entry\ exit = [\![E_1]\!]\ entry\ mid\ \cup [\![E_2]\!]\ mid\ exit \qquad \cdots (4)$$

where *mid* is a new control state such that
$vars(mid) = (vars(entry) \cup vars(E_1)) \cap (vars(exit) \cup vars(E_2))$.

$$[\![E_1 \# E_2]\!]\ entry\ exit = [\![E_1]\!]\ entry\ exit\ \cup [\![E_2]\!]\ entry\ exit \qquad \cdots (5)$$

$$[\![\mathtt{if}(C, E_1, E_2)]\!]\ entry\ exit \qquad \cdots (6)$$

$$= \{\ \mathtt{trans}(entry,\ \mathtt{i},\ iftrue,\ C),\ \mathtt{trans}(entry,\ \mathtt{i},\ iffalse,\ \mathtt{not}(C))\ \} \cup [\![E_1]\!]\ iftrue\ exit \cup [\![E_2]\!]\ iffalse\ exit$$

where *iftrue* and *iffalse* are new control states such that
$vars(iftrue) = (vars(entry) \cup vars(C)) \cap (vars(exit) \cup vars(E_1))$
$vars(iffalse) = (vars(entry) \cup vars(C)) \cap (vars(exit) \cup vars(E_2))$

$$[\![E_1 | E_2]\!]\ entry\ exit = \{\ \mathtt{trans}(s_1 \times V_2,\ \alpha_1,\ d_1 \times V_2,\ c_1) : \mathtt{trans}(s_1,\alpha_1,d_1,c_1) \in [\![E_1]\!]\ entry\ exit \text{ and } V_2 \text{ is a fresh variable}\} \qquad \cdots (7a)$$

$$\cup \{\ \mathtt{trans}(V_1 \times s_2,\ \alpha_2,\ V_1 \times d_2,\ c_2) : \mathtt{trans}(s_2,\alpha_2,d_2,c_2) \in [\![E_2]\!]\ entry\ exit \text{ and } V_1 \text{ is a fresh variable}\} \qquad \cdots (7b)$$

$$\cup \{\ \mathtt{trans}((s_1 \times s_2)\theta,\ \mathtt{tau},\ (d_1 \times d_2)\theta,\ (c_1,c_2)\theta) : \mathtt{trans}(s_1,\alpha_1,d_1,c_1) \in [\![E_1]\!]\ entry\ exit \wedge \mathtt{trans}(s_2,\alpha_2,d_2,c_2) \in [\![E_2]\!]\ entry\ exit \qquad \cdots (7c)$$

and θ is the most general substitution such that
$\{\alpha_1, \alpha_2\} \equiv \{\mathtt{in}(t),\ \mathtt{out}(t\theta)\}$ for some term t $\}$.

$$[\![E \backslash L]\!]\ entry\ exit = \{\ \mathtt{trans}(\rho_{E,L}(s),\ \alpha,\ \rho_{E,L}(d),\ c) : \mathtt{trans}(s,\alpha,d,c) \in [\![E]\!]\ entry\ exit \wedge \alpha \notin L\ \} \qquad \cdots (8)$$

where $\rho_{E,L}$ maps every control state except *entry* and *exit* to a new control state.

$$[\![E@F]\!]\ entry\ exit = \{\ \mathtt{trans}(\rho_{E,F}(s),\ F(\alpha),\ \rho_{E,F}(d),\ c) : \mathtt{trans}(s,\alpha,d,c) \in [\![E]\!]\ entry\ exit\ \} \qquad \cdots (9)$$

where $\rho_{E,F}$ maps every control state except *entry* and *exit* to a new control state.

$$[\![Proc]\!]\ entry\ exit = \{\ \mathtt{trans}(entry,\ \mathtt{i},\ ep(Proc)(\langle exit \rangle \oplus subterms(Proc)),\ \mathtt{true})\ \} \qquad \cdots (10a)$$

$$\cup\ [\![definitions(Proc)]\!] \qquad \cdots (10b)$$

where *ep*(*Proc*) is the name of the start control state of the automaton corresponding to *Proc*, *subterms*(*t*) is the sequence of immediate subterms of term *t*, and *definitions*(*P*) is the set of all definitions $P' ::= E'$ in the input specification such that P unifies with P'.

$$[\![Proc ::= E]\!] = [\![E]\!]\ ep(Proc)(\langle Cont \rangle \oplus subterms(Proc))\ \ Cont \qquad \cdots (11)$$

where *entry* and *exit* are new control states, *ep* and *subterms* are as defined in the previous rule and *Cont* is a fresh variable.

Figure 1.4 Compilation Rules for XL

Figure 1.4 in terms a semantic function $[\![E]\!]$ *entry exit* which maps process expression E to a set of the transition rules. In the figure, we use $\langle \ldots \rangle$ to denote sequences and $\oplus$ to denote concatenation of sequences.

The compiler views computation and process invocation in XL as state-changing steps that are not observable. Such steps are represented by transitions with a special label, '`i`', denoting internal transitions (see equations 1, 6, and 10). Since `i`-transitions are unobservable, we can define a form of equivalence called `i`-equivalence. Informally, two processes P and Q are `i`-equivalent if P and Q have identical branching behaviors when `i`-transitions are ignored. Under `i`-equivalence the correctness of the compiler w.r.t. XL semantics can be readily shown.

The first six equations in Figure 1.4 directly encode the semantic rules 1–8 in Figure 1.3. The remaining equations are explained below.

Compiling Process Invocations. The control and data flow associated with calls and returns are uniformly handled by passing the current continuation as the first argument to the called process (the term *exit* in equation 10a and corresponding parameter *Cont* in equation 11) and "jumping" to the continuation at the end of a process (equation 11). The continuation-passing style [1] naturally limits the number of `i`-transitions introduced by the compiler. Parameter passing and variable renaming are delegated to the underlying Logic Programming engine as follows. We encode the caller state with the arguments, i.e., subterms in the call, (equation 10a). We then access the parameters in the callee using the subterms in the pattern defining the callee (equation 11). Note that for each process expression E, the compilation rules associate the set of *all* `trans` rules representing the transitions of E. If E contains a process invocation, say P', then $[\![E]\!]$ contains the `trans` rules corresponding to the process P' as well (due to equation 10b). This convention considerably simplifies the compilation of expressions with parallel, restriction and relabeling operations.

An expression with restriction $E\backslash L$ is compiled by discarding any transitions with labels in L from a fresh copy of the automaton for E. Similarly, an expression with relabeling $E@F$ is compiled by suitably renaming all transitions in a fresh copy of the automaton for E.

Compiling Parallel Composition. If E_1 and E_2 have automata with control states drawn from S_1 and S_2 respectively, the automaton corresponding to $E_1 \,|\, E_2$ has control states drawn from $S_1 \times S_2$. Equations 7a and 7b correspond to E_1 and E_2 making autonomous moves respectively. Equation 7c precomputes synchronizations *without* incurring an exponential blow up of transition rules, as explained below. In the compilation of $E_1 \,|\, E_2$, let the automata for E_1 and E_2 have n_1 and n_2 non-`tau` transitions and m_1 and m_2 `tau` transitions respectively. The automaton constructed for $E_1 \,|\, E_2$ using equation 7 has $n_1 + n_2$ non-tau transitions (from equations 7a and 7b); $m_1 + m_2$ (from 7a and 7b) $+n_1n_2$ (from 7c) tau transitions. Note that the product operation in 7c takes only non-tau transitions and produces only tau transitions. Hence, the results of the product cannot be fed into a product operation again, thereby averting the exponential blow-up. In fact, a rough analysis of the product operation yields an upper bound of n^2N^2 transitions in a parallel composition of N automata with at most n transitions each. However, the number of transition rules generated is usually much smaller than the above bound. For instance, the number of transition rules for the leader election protocol as well as the sieve of Eratosthenes grow linearly with the number of processes.

The compiler is implemented in the XSB tabled logic programming system [22] by simply encoding the set equations in Figure 1.4 as a Horn clause program and evaluating it using tabled resolution [3]. Termination of the compiler can be readily shown by induction on the structure of process expressions, whenever there is no parallel composition within a recursive process definition. An XL specification that overlaps parallel composition with recursion, (e.g., `p(s(N)) ::= q | p(N)`) cannot be handled by the compilation scheme. The scheme can be extended to compile parameterized processes (such as `p(N)` above) whenever the value of the parameter is known at compile time. However, extending the scheme to *dynamically* compile processes that are created at verification time remains an interesting open problem.

Reducing Internal (i-) Transitions. Since i-transitions are not observable, we can reduce the system state space modulo i-equivalence, i.e., replacing sequences of transitions of the form $\mathtt{i}^*\alpha\mathtt{i}^*$ with α whenever the system's branching behavior is not changed. We call such an optimization as i-elimination. Note that i-transitions are only generated by sequential statements, namely computation, conditional, and process invocation. Therefore it is sufficient to describe i-elimination only for sequential components. It should be noted that state space reduction in sequential components translates into corresponding (multiplicative) reduction in the state space of a parallel processes.

In our formulation of i-elimination, we need to distinguish between i-transition rules that arise from computations (equations 1 and 6) and those that arise from process invocation (equation 10). We use labels $\mathtt{i}_c$ and $\mathtt{i}_p$ to denote these two kinds of transitions respectively whenever such a distinction is necessary. The i-elimination optimization can be defined as a sequence of rewrites on the set of transition rules $\mathcal{T}$ using the following rules:

Forward elimination: Let $I = \texttt{trans}(s,\texttt{i}_c,t,c) \in \mathcal{T}$. Then, $\mathcal{T}{:=}(\mathcal{T} - \{I\}) \cup \{\ \texttt{trans}(s,\alpha,t'\theta,(c \wedge c'\theta)) \mid \texttt{trans}(s',\alpha,t',c') \in \mathcal{T}$ and there exists a substitution θ such that $t = s'\theta\}$.

Backward elimination: Let $I = \texttt{trans}(s,\ \texttt{i},\ t,\ c) \in \mathcal{T}$ such that $\forall\ \texttt{trans}(s,\ \alpha'',\ t'',\ c'') \in \mathcal{T} - \{I\}$, the computations c and c'' are mutually exclusive, i.e., $\forall$ substitutions σ $[\![c]\!]_L\sigma \cap [\![c'']\!]_L\sigma = \{\}$. Let $R = \texttt{trans}(s',\ \alpha,\ t',\ c') \in \mathcal{T}$ such that $\alpha \neq \texttt{i}_p$, and θ be a substitution such that $t'\theta = s$. Then,
$\mathcal{T}{:=}(\mathcal{T}-\{I,R\}) \cup \{\texttt{trans}(s',\alpha,t,(c' \wedge c\theta)), \texttt{trans}(s',\alpha,t',(c' \wedge \texttt{not}(c)\theta))\}$

Eliminating $\texttt{i}_p$-transitions by forward elimination, which is currently not permitted, will be analogous to inlining. This is explicitly avoided since additional conditions are needed to ensure that the inlining terminates even in the presence of recursion in processes. Also note that forward elimination may exponentially increase the size of rule sets since shared computations may be repeated on independent transitions; however, the benefits of the optimization appear to outweigh the potential (although rare) blow ups. For backward elimination, the mutual exclusion condition on I is needed to preserve the system's branching behavior.

Each application of an i-elimination optimization can be shown correct with respect to i-equivalence. We can also show that any sequence of i-eliminations will terminate. Although presented as an off-line optimization, both elimination steps can be performed while applying the compilation rules by suitably merging the rule sets whenever new control states are generated.

Live Variable Optimization. Since each control state is associated with a set of variables, it is imperative that we keep only the "needed" variables in each state. We compute an upper approximation of the needed set as follows. Consider the creation of a new control state *mid* as an intermediate state in a sequential composition (equation 4). The variables in the expression E_1 as well as the variables in the control state *entry* are used before reaching *mid*. Similarly, the variables in E_2 and *exit* are potentially used after leaving *mid*. The definition of *vars*(*mid*) is the intersection of these two sets, and forms an upper approximation of the needed variable set.

Removing dead variables, i.e., those that are not needed, from state vectors not only lowers memory requirements for the state space search, but may reduce the size of the state space itself. Values of dead variables may increase the number of states in the system by discriminating between two instances of what is essentially one state. This increase may be avoided by setting the dead variables to a default value manually,

```
trans(abp(A,chan_1(B,C),D,E), out(drop), abp(A,chan_0(C),D,E), true).
trans(abp(A,B,chan_1(C,D),E), out(drop), abp(A,B,chan_0(D),E), true).
trans(abp(sender_1(A,B),C,D,E), i, abp(sender_0(A,B),C,D,E), true).
trans(abp(send_0(A,B),chan_0(C),D,E), tau,
      abp(send_1(A,B),chan_1(A,C),D,E), true).
trans(abp(send_1(A,B),C,chan_1(D,E),F), tau,
      abp(send_0(G,B),C,chan_0(E),F), (D == A,  G is 1 - A)).
trans(abp(send_1(A,B),C,chan_1(D,E),F), tau,
      abp(send_0(A,B),C,chan_0(E),F), (not(D == A))).
trans(abp(A,chan_1(B,C),D,receiver_0(E,F)), tau,
      abp(A,chan_0(C),D,receiver_4(B,G,F)), (B == E,  G is 1 - E)).
trans(abp(A,chan_1(B,C),D,receiver_0(E,F)), tau,
      abp(A,chan_0(C),D,receiver_3(B,E,F)), (not(B == E))).
trans(abp(A,B,chan_0(C),receiver_4(D,E,F)), tau,
      abp(A,B,chan_1(D,C),receiver_0(E,F)), true).
trans(abp(A,B,chan_0(C),receiver_3(D,E,F)), tau,
      abp(A,B,chan_1(D,C),receiver_0(E,F)), true).
```

Figure 1.5 Transition Rules for the Alternating Bit Protocol

and is a fairly common encoding trick adopted by experienced users. It is encouraging that such workarounds can be effectively replaced by a simple program analysis (see Section 4 for details).

Finally, note that the control states of the global transition relation are maintained as a tree with '×' as internal nodes and the control states of the component sequential processes as the leaves. When the structure of the concurrent process is known at compile time, the tree structure can be collapsed into a tuple, yielding the *state vector* representation. Currently the compiler generates rules in terms of state vectors whenever possible, but without compressing components of the vector itself.

In Figure 1.5, the transition rules generated by the compiler for the alternating bit protocol (Figure 1.2) are shown as an illustration.

4. EXPERIMENTAL RESULTS

Consistent with the spirit of the XMC system, the XL compiler was implemented starting with an encoding of the equations in Figure 1.4 as a tabled logic program. The program was subsequently modified to implement the optimizations described in the previous section. The transition rules generated by the compiler are used to compute the global transitions as and when required by the XMC model checker.

In the following, we evaluate the effectiveness of the XL compiler, and compare the performance of the XMC system, with and without compilation, and the performance of Murφ [8]. Performance measurements for Murφ were taken using version 2.70, since Murφ versions 3.0 and above do not support verification of liveness formulas.

The Benchmarks. The following examples were used as benchmarks:

`i-Protocol` (*i-p* in the tables) is a sliding window protocol in the GNU UUCP stack. The sender and receiver processes have complex internal (sequential) structure. Presence of a livelock was checked for sliding window sizes (WS) of 1 and 2, each with and without the livelock.

`Rether` [4] is an Ethernet protocol that supports real-time traffic. Compared to `i-Protocol`, each node of `rether` is a relatively simple process, while the communication patterns are more complex (each node can communicate with any other). The protocol was tested with 5 nodes and up to 3 real-time slots per Ethernet frame. Two properties, deadlock-freedom (*dlf*) and starvation-freedom (*ns*), were verified.

`Leader` and `sieve` are taken from the SPIN example suite. For `leader`, the property verified was that exactly one leader is elected in any run. For `sieve`, we verified that the sequence produced by the final filter has a specific value at a given position. The example specifications, runtime parameters used, and experimental data are available from `http://www.cs.sunysb.edu/~lmc/compiler`.

Performance Measurement and Evaluation Criteria. All measurements were made on a Sun Enterprise 4000 with 2 GB main memory running Solaris 5.2.6. The times measured were CPU times reported by the different systems. For Murφ, verification time is the CPU time used by the executable obtained by compiling the generated C++ code using g++ (v2.8.1) (with `-O4` for max. optimization). The times for Murφ were obtained *without* using symmetry reduction. Verification times for XMC is the CPU time needed to answer the corresponding `models` query. Space usage for XMC was measured as the sum of the maximum space used in each of the memory areas of XSB: the table space, the four Prolog stacks, as well as permanent space (which includes space for dynamic code). Space usage for Murφ was taken as the sizes of the state hash table reported by its statistics.

Analysis of Experimental Data. Table 1.1 lists the time and space performance of XMC (with and without compilation) and Murφ on the benchmark examples. Observe from the table that compilation speeds up XMC by up to a factor of 15 for `sieve`, factors of 5 or better for `i-Protocol` and around a factor of 4 for `leader`. On the other hand, the compiler slows the verification of `rether`. This is due to the severe indexing overheads added by the presence of "dead" rules: those that specify transitions from states that are unreachable at verification time.

Note that all times for XMC with compilation are comparable to verification using Murφ. XMC's local model checker explores only a subset

Benchmark	*Time (sec)*			*Space (MB)*		
	XMC	*XMC/c*	*Murφ*	*XMC*	*XMC/c*	*Murφ*
i-p(1, bug)	0.98	0.05	1.76	6.18	0.52	0.30
i-p(1, fixed)	99.72	12.82	7.33	388.85	18.31	1.61
i-p(2, bug)	1.30	0.31	35.61	13.01	0.78	5.58
i-p(2, fixed)	out of mem	214.36	139.83	out of mem	198.06	26.71
rether(dlf)	0.19	0.22	0.20	0.78	0.58	0.03
rether(ns)	0.38	0.47	0.49	0.87	0.64	0.10
sieve(3)	1.54	0.19	0.14	6.07	1.07	0.02
sieve(5)	15.57	1.20	0.92	55.07	7.42	0.17
sieve(7)	130.12	8.71	8.36	437.88	61.32	1.03
leader(3)	0.13	0.04	0.03	0.85	0.47	0.01
leader(5)	2.82	0.59	0.39	13.27	2.44	0.06
leader(7)	68.66	12.90	12.63	294.47	44.56	1.87

Table 1.1 Performance of XMC (with and without compilation) and Murφ

of reachable states for the buggy versions of `i-Protocol`, but explores all reachable states for other examples. In contrast, Murφ performs global checking, and hence explores all reachable states for all examples.

The compilation time for XMC (not shown in the table) ranges from 0.1s to 0.2s, and is typically much smaller than the verification time. This includes the time to preprocess and load XL specifications, translate them to rules, and load the rules. For Murφ, the time needed to generate the executable ranges from 7s for `leader` to 11s for `i-Protocol`.

Observe that compilation also reduces memory requirements, by factors ranging from 2 to 15. In contrast to the original XMC system, the transition relation is precomputed by the compiler into a set of Prolog facts and need not be tabled (cached). Again, the savings in `rether` is minimal, since XSB's permanent space, where the transition rules are stored, dominates memory usage.

The performance of XMC relative to SPIN is not shown in the table. For both buggy versions of `i-Protocol`, XMC with compilation and SPIN (v3.2.3) show very similar performance in terms of time (0.05s and 0.31s for XMC vs. 0.04s and 0.4s for SPIN, for WS=1 and 2 respectively) as well as space (0.52M and 0.78M vs. 1.0M and 2.3M respectively). For fixed version with window size 1, SPIN takes 495s and consumes 1.1GB of memory; for window size 2, we have been unable to obtain SPIN numbers without using bit-state hashing or hash compaction, both of which compute only approximate answers. Although partial order reduction [12] does not change the verification performance for `i-Protocol`,

Benchamrk	*Time (sec)*		*Space (MB)*		*# of states*		*# of trans.*	
i-p(1, Bug)	1.72	(34x)	1.98	(4x)	9K	(40x)	18K	(57x)
i-p(1, Fixed)	300	(23x)	224	(12x)	188K	(13x)	664K	(13x)
i-p(2, Bug)	2.41	(8x)	2.57	(3x)	12K	(8x)	23K	(9x)
i-p(2, Fixed)				out of memory				
rether(dlf)	0.21	(1x)	0.59	(1x)	341	(1x)	371	(1x)
rether(ns)	0.49	(1x)	0.66	(1x)	341	(1x)	371	(1x)
sieve(3)	2.11	(11x)	10.22	(10x)	8K	(14x)	28K	(20x)
sieve(5)	47.8	(40x)	45.15	(6x)	123K	(31x)	574K	(47x)
sieve(7)				out of memory				
leader(3)	0.03	(1x)	0.53	(1x)	88	(1.3x)	170	(1.4x)
leader(5)	0.72	(1.2x)	3.42	(1.4x)	1456	(1.7x)	5K	(1.7x)
leader(7)	21.5	(1.7x)	81.3	(1.8x)	26K	(2.1x)	116K	(2.2x)

Table 1.2 Effect of *not* eliminating i-transitions across block boundaries

it substantially reduces the search space for `leader`, making SPIN significantly outperform XMC, even with compilation. Currently, XMC does not perform partial order reduction; integrating such search-space reduction techniques into XMC is a topic of future work.

Finally, note that the space usage for Murφ is significantly lower than XMC with or without compilation. State vectors are compressed into bit-vectors in Murφ, thereby reducing space usage; for instance, a state in `i-Protocol` (WS=2) is stored using only 56 bits (without hash compaction). In contrast, the XL compiler generates uncompressed state vectors, leading to higher memory requirement.

Effectiveness of the Optimizations. Elimination of i-transitions leads to considerable reductions in the state space. The effect of performing this optimization across block boundaries is of special interest, since this cannot be done by user annotations (such as *atomic* and *d_step* in SPIN) alone. We measured this effect by turning off i-elimination whenever the candidate transitions spanned basic blocks, but performed elimination in all other cases. Table 1.2 shows the degradation in time and space performance, as well as relative state space sizes when the optimization is turned off. The savings due to this optimization are substantial in all examples except `rether`, again due to the simplicity of the sequential components of the protocol.

As noted in Section 3, eliminating dead variables from state representation can reduce the state space size, in addition to lowering the memory requirements. With dead variable elimination, we observed a 7% drop in memory requirements across all examples. For `i-Protocol`, the elimination reduced the state space by 5%, but there was no change in state space for the other examples.

5. DISCUSSION

We have shown that compiling high-level specifications into global transition rules improves model checking performance. We also showed that such compilation can be performed very efficiently. Although presented as a compilation technique for XL, the technique can be readily adapted to translate high-level specifications written in other formalisms also. We showed that combining computations across block boundaries, an optimization that cannot be done based on user annotations alone, reduces state space significantly. This optimization can be introduced to improve the performance of any explicit-state model checker. In fact, this optimization has been recently added to SPIN 3.3.0 and reportedly has led to significant performance gains (see [20]).

While we have offered preliminary evidence of the importance of optimizing compilation, its full power remains to be exploited. For instance, can powerful state-space reduction techniques such as partial order reduction [12, 14] be used at the transition rule level to eliminate entire families of transitions in one step, at compile time? What techniques and optimizations are most useful for reducing the space requirements for verification? We believe that answers to these questions will considerably improve the performance of current model checkers.

Acknowledgements We thank Gerard Holzmann for his help in collecting accurate figures on SPIN's performance on the i-Protocol, and Xiaoqun Du for encoding and evaluating the i-Protocol on SPIN and original XMC. We also thank Rance Cleaveland, Owen Kaser, I.V. Ramakrishnan, Scott Smolka, and David Warren for fruitful discussions.

This research was supported in part by grants CDA–9303181, EIA–9705998, and CCR–9711386 from the National Science Foundation.

References

[1] A.W. Appel. *Compiling with Continuations*. Cambridge University Press, 1992.

[2] T. Bolognesi and E. Brinksma. Introduction to the ISO specification language LOTOS. In P.H.J. van Eijk, C.A. Vissers, and M. Dias, editors, *The Formal Description Technique LOTOS*, pages 23–76. North-Holland, 1989.

[3] W. Chen and D. S. Warren. Tabled evaluation with delaying for general logic programs. *Journal of the ACM*, 43(1), January 1996.

[4] T. Chiueh and C. Venkatramani. The design, implementation and evaluation of a software-based real-time ethernet protocol. In *Proceedings of ACM SIGCPMM'95*, 1995.

[5] E. M. Clarke and E. A. Emerson. Design and synthesis of synchronization skeletons using branching-time temporal logic. In D. Kozen, editor, *Workshop on Logic of Programs, LNCS 131*, pages 52–71, Yorktown Heights, 1981.

[6] E. M. Clarke, E. A. Emerson, and A. P. Sistla. Automatic verification of finite-state concurrent systems using temporal logic specifications. *ACM TOPLAS*, 8(2), 1986.

[7] B. Cui, Y. Dong, X. Du, K. Narayan Kumar, C.R. Ramakrishnan, I.V. Ramakrishnan, A. Roychoudhury, S.A. Smolka, and D.S. Warren. Logic programming and model checking. In *Programming Languages, Implementations, Logics and Programming (PLILP'98)*, 1998.

[8] D. L. Dill. The Murφ verification system. In *Computer Aided Verification (CAV'96)*, 1996.

[9] Y. Dong, X. Du, Y.S. Ramakrishna, C.R. Ramakrishnan, I.V. Ramakrishnan, S.A. Smolka, O. Sokolsky, E.W. Stark, and D.S. Warren. Fighting livelock in the i-protocol: A comparative study of verification tools. In *Tools and Algorithms for the Construction and Analysis of Systems, (TACAS '99)*, 1999.

[10] H. Garavel and J. Sifakis. Compilation and verification of LOTOS specifications. In *Protocol Specification, Testing and Verification (PSTV '90)*, 1990.

[11] G. J. Holzmann. The model checker SPIN. *IEEE Transactions on Software Engineering*, 23(5), May 1997.

[12] G. J. Holzmann and D. Peled. An improvement in formal verification. In *Formal Description Techniques (FORTE '94)*, pages 177–194. Chapman and Hall, 1995.

[13] D. Kozen. Results on the propositional μ-calculus. *Theoretical Computer Science*, 27:333–354, 1983.

[14] R. Kurshan, V. Levin, M. Minea, D. Peled, and H. Yenigün. Static partial order reduction. In *Tools and Algorithms for the Construction and Analysis of Systems (TACAS '98)*, 1998.

[15] J. W. Lloyd. *Foundations of Logic Programming*. Springer Verlag, 1984.

[16] Z. Manna and A. Pnueli. *Temporal Verification of Reactive Systems: Safety*. Springer Verlag, 1995.

[17] R. Milner. *Communication and Concurrency*. Prentice Hall, 1989.

[18] J. P. Queille and J. Sifakis. Specification and verification of concurrent systems in Cesar. In *International Symposium on Programming, LNCS 137*, Berlin, 1982.

[19] Y. S. Ramakrishna, C. R. Ramakrishnan, I. V. Ramakrishnan, S. A. Smolka, T. Swift, and D. S. Warren. Efficient model checking using tabled resolution. In *Computer-Aided Verification (CAV '97)*, 1997.

[20] SPIN Newsletter, Number 25, May 1999. Available from `http://netlib.bell-labs.com/netlib/spin/news/news25.html`.

[21] A. S. Tanenbaum. *Computer Networks*. Prentice Hall, 1996.

[22] XSB. The XSB logic programming system, 1998. Available from `http://www.cs.sunysb.edu/~sbprolog/XSB/`.

EXPLOITING SYMMETRY WHEN MODEL-CHECKING SOFTWARE

(EXTENDED ABSTRACT)

PATRICE GODEFROID
Bell Laboratories, Lucent Technologies
263 Shuman Boulevard
Naperville, IL 60566, USA
email: god@bell-labs.com

Abstract We study how to exploit symmetry induced by identical processes or data structures when systematically exploring the state spaces of concurrent software applications such as implementations of communication protocols. Existing model-checking symmetry reduction methods are based on equivalence classes of states, and assume that every system state can easily be encoded by a unique string of bits. When dealing with processes described by software programs written in full-fledged programming languages such as C, C++ or Java, this assumption is not valid anymore. Indeed, the state of each process is determined by the values of all the memory locations that can be accessed by the process and influence its behavior (including activation records associated to procedure calls). This amount of information is typically far too large and complex to be efficiently computed at each step of a state-space exploration.

We develop in this paper a simple theory based on equivalence classes of sequences of transitions for representing symmetries in a system. We then present a state-space exploration algorithm for exploiting symmetries on transitions which does not rely on explicit encodings of system states. This algorithm can be used to dynamically prune the state spaces of implementations of concurrent reactive software systems in a reliable way.

Keywords: software testing, model checking, state-explosion problem, symmetry, partial-order methods.

1 INTRODUCTION

Systematic state-space exploration, also often referred to as "model checking", is an effective method for checking the correctness of concurrent reactive systems. State-space exploration tools for software systems have traditionally been restricted to the exploration of the state space of an abstract description of the system, specified in a modeling language (e.g., [15, 5, 22]). Once a model of a new software application has been thoroughly analyzed, it can also be used as the core of the implementation of the application, as can be done with software development environments for languages such as SDL [16] and VFSM [9].

VeriSoft [11] is a recent tool which extends the scope of systematic state-space exploration to concurrent systems in which processes execute arbitrary code written in general-purpose programming languages such as C or C++. VeriSoft systematically explores the state space of a concurrent software application by controlling and observing the execution of the actual code of all its components. By broadening the scope of systematic state-space exploration from modeling languages to programming languages, VeriSoft eliminates one major obstacle to a wider use of these techniques, namely the need to build a model of the software application to be analyzed. This new tool and approach have been applied successfully for analyzing several software products developed in Lucent Technologies, such as telephone-switching applications and implementations of network protocols (e.g., see [12]). Since VeriSoft can typically generate, execute and evaluate thousands of tests per minute, it can quickly reveal behaviors that are virtually impossible to detect using conventional testing techniques.

A key originality of VeriSoft is that it systematically explores state spaces without storing any explicit encoding of intermediate states in memory. Indeed, when dealing with processes described by software programs written in full-fledged programming languages such as C, C++ or Java, the fundamental assumption used in traditional model-checking that every system state can easily be encoded by a unique string of bits is not valid anymore. Since the state of each process is determined by the values of all the memory locations that can be accessed by the process and influence its behavior (including activation records associated to procedure calls), computing unambiguous encodings of system states and then saving these in memory is typically far too complex and expensive to be performed at each step of a state-space exploration, even for simple software applications.

We investigate in this paper how to exploit symmetry induced by identical processes or data structures for verifying properties of a software application without completely exploring its state space. Existing model-checking symmetry reduction methods [2, 8, 17] define equivalence classes of system states from symmetries in a (model of a) system. During state-space exploration, the successors of a state need not be explored if the successors of an equivalent "symmetric" state have already been explored. Unfortunately, detecting symmetric states requires nontrivial manipulations of state encodings which would not be tractable in the context considered here.

We develop in this work an alternative approach that makes it possible to exploit symmetries in a system without manipulating and comparing encodings of its reachable states. This paper is organized as follows. In the next two sections, we briefly recall the main principles of the framework introduced in [11] and of existing symmetry reduction methods. In Section 4, we describe how to transpose equivalence relations on states to equivalence relations on sequences of transitions for representing symmetries in a system. We then study properties of the transposition. In Section 5, we discuss how to exploit symmetries on transitions in practice. We then present in Section 6 a state-space exploration algorithm that can avoid exploring symmetric parts of a state space without using any explicit encoding of system states. This algorithm uses an original combination of partial-order and symmetry methods. We illustrate our ideas with a simple example in Section 7, and present our conclusions in Section 8.

2 BACKGROUND

We briefly recall in this section the main ideas of the framework introduced in [11]. We consider a concurrent system S composed of a finite set of *processes* and a finite set of *communication objects*. Each process executes a sequence of *operations* described in a sequential program written in a full-fledged programming language such as C or C++. Such sequential programs are deterministic: every execution of the program on the same input data performs the same sequence of operations. We assume that processes communicate with each other by performing operations on communication objects, such as shared variables, semaphores, and FIFO buffers. We assume that operations on communication objects are executed atomically: namely, no process can execute an operation on a given communication object while another process is currently doing so. Operations on communication objects are called *visible operations*, while other operations are by default called *invisible*. The execution of an operation is said to be *blocking* if it cannot currently be completed; for example, waiting for the reception of a message blocks until a message is received. We assume that only visible operations may be blocking.

A concurrent system is said to be in a *global state* when the next operation to be executed by every process in the system is a visible operation. Every process in the system is always expected to eventually attempt execution of a visible operation. (If a process does not attempt to perform a visible operation within a given amount of time, an error can be reported at run time.) This assumption implies that initially, after the creation of all the processes of the system, the system can reach a first and unique global state s_0, called the *initial global state* of the system. A *process transition* is one visible operation followed by a finite sequence of invisible operations performed by a single process and ending just before a visible operation. The *state space* of the concurrent system is then defined as the global states that are reachable from the initial global state s_0, and of the transitions that are possible between these.

It has been proved [11] that *deadlocks* and *assertion violations* of a concurrent system that satisfies the above assumptions can be detected by exploring global states only. Deadlocks are states where the execution of the next operation of every process in the system is blocking. Assertions can be specified by the user with the special operation "VS_assert". This operation can be used in the code of any process, and is considered visible. It takes as its argument a boolean expression that can test and compare the value of variables local to the process.

When "VS_assert(expression)" is executed, the expression is evaluated. If the expression evaluates to false, the assertion is said to be *violated.*

VeriSoft is a tool for systematically exploring the state space of a concurrent system as defined above. Systematic state-space exploration is performed by controlling and observing the execution of all the visible operations of all the concurrent processes of the system. The execution of the system processes is controlled by an external process, called the *scheduler.* This process observes the visible operations performed by processes inside the system, and can suspend their execution. By resuming the execution of (the next visible operation of) one selected system process in a global state, the scheduler can explore one transition between two global states in the state space of the concurrent system. By reinitializing the system, the scheduler can explore alternative paths in the state space. There are exactly two sources of nondeterminism in the concurrent systems we consider here: concurrency and "VS_toss" operations. VS_toss is a special operation which simulates nondeterminism and is convenient for modeling the environment in which an open system operates and for specifying test drivers. Since the VeriSoft scheduler has complete control over nondeterminism, it can always reproduce any scenario leading to an error found during a state-space search.

Since the states of programs can be very complex (because of pointers, dynamic memory allocation, large data structures of various shapes, recursion, etc.), the VeriSoft scheduler does not attempt to compute any representation for the reachable states of the system being analyzed, and hence performs a systematic state-space exploration without storing any intermediate states in memory. It is shown in [11] that the key to make this approach tractable is to use a new search algorithm built upon existing state-space pruning techniques known as partial-order methods [10]. For finite acyclic state spaces, this search algorithm is guaranteed to terminate and can be used for detecting deadlocks and assertion violations without incurring the risk of any incompleteness in the verification results. In practice, VeriSoft can be used for systematically and efficiently testing the correctness of any concurrent system, whether or not its state space is acyclic. Indeed, it can always guarantee, from a given initial state, complete coverage of the state space up to some depth.

3 SYMMETRY ON STATES

The state space of a concurrent system as defined in the previous section can be represented by a *transition system.* Given a nonempty finite set P of *atomic propositions*, we define a transition system as a tuple $M = (S, L, R, s_0)$ where

- S is a set of global states,
- $L : S \rightarrow 2^P$ is an *interpretation* function that associates with each state the set of atomic propositions that are true in the state,
- $R \subseteq S \times S$ is a set of transitions,
- s_0 is the initial global state.

A transition $t = (s, s')$ is said to be *enabled* in state s. The set of enabled transitions in a state s is denoted by *enabled*(s). We write $s \xrightarrow{t} s'$ to mean that the execution of the transition t leads from the global state s to the global state s', while $s \stackrel{w}{\Rightarrow} s'$ means that the execution of the finite sequence w of transitions leads from s to s'. If $s \stackrel{w}{\Rightarrow} s'$, s' is said to be *reachable* from s.

We now recall the main principles of symmetry-based reduction methods [2, 8, 17]. The basic idea is that symmetries in the system induce an equivalence relation on states of the state space of the system. While performing model-checking, one can discard a state s if one has already explored an equivalent state s'. Precisely, we have the following.

A *permutation* σ on a set S is a bijection of S into itself. Let $Perm(S)$ be the group of all permutations of S [13].

Definition 1 Given a transition system $M = (S, L, R, s_0)$, a subgroup G of $Perm(S)$ is called a *symmetry group* of M if for all permutations σ in G,

- $(s_1, s_2) \in R$ iff $(\sigma(s_1), \sigma(s_2)) \in R$, and
- $L(s) = L(\sigma(s))$.

■

The symmetry group G defines an equivalence relation $\equiv$ on S as follows: $s_1 \equiv s_2$ if $\exists \sigma \in G : s_2 = \sigma(s_1)$. The equivalence class $[s] = \{\sigma(s) | \sigma \in G\}$ of s under $\equiv$ is called the *orbit* of s under G.

Definition 2 Given a transition system $M = (S, L, R, s_0)$ and a symmetry group G of M, a *quotient transition system* for M modulo G is a transition system $M_{[s]} = (S', L', R', s'_0)$ where

- $S' = \{[s] | s \in S\}$,
- $L' : S' \to 2^P$ is such that $\forall s \in S : L'([s]) = L(s)$,
- $R' = \{([s], [s']) | (s, s') \in R\}$,
- $s'_0 = [s_0]$.

■

Since $s \equiv s'$ implies $L(s) = L(s')$, atomic propositions do not distinguish equivalent states. It is then easy to prove that s_0 and $[s_0]$ are bisimilar.

Definition 3 Let $M_1 = (S_1, L_1, R_1, s_0^1)$ and $M_2 = (S_2, L_2, R_2, s_0^2)$ be transitions systems. A binary relation $\mathcal{B} \subseteq S_1 \times S_2$ is a *bisimulation relation* if $(s_1, s_2) \in \mathcal{B}$ implies:

- $L_1(s_1) = L_2(s_2)$,
- if $(s_1, s'_1) \in R_1$, then there is some $s'_2 \in S_2$ such that $(s_2, s'_2) \in R_2$ and $(s'_1, s'_2) \in \mathcal{B}$, and
- if $(s_2, s'_2) \in R_2$, then there is some $s'_1 \in S_1$ such that $(s_1, s'_1) \in R_1$ and $(s'_1, s'_2) \in \mathcal{B}$.

Two states s_1 and s_2 are *bisimilar*, denoted $s_1 \sim s_2$, if they are related by some bisimulation relation. ■

Theorem 1 *Let $M = (S, L, R, s_0)$ be a transition system and $M_{[s]} = (S', L', R', s'_0)$ be a quotient transition system for M modulo a symmetry group G of M. Then, we have $s_0 \sim s'_0$.*

It is well-known [14] that bisimilar states cannot be distinguished by formulas of propositional modal fixpoint logic [24], also known as the propositional μ-calculus [19]. Therefore, the quotient transition system $M_{[s]}$ can be used instead of the full transition system M for model-checking formulas of this logic or of any of its fragments, such as linear-time temporal logic (LTL) [20] and computation-tree logic (CTL) [3].

In practice, the quotient transition system of a system is usually generated on-the-fly using a *canonicalization function* ζ. This function maps each state s into a unique representative $\zeta(s)$ of the equivalence class $[s]$. Whenever a state s is visited during state-space exploration, the value $\zeta(s)$ is saved in memory, usually in a hash-table [17]. This type of search can also be done "symbolically", for instance using Binary Decision Diagrams [1] for representing sets of states [2]. When a state s' equivalent to s is visited later during the search, the search does not explore successors of s'. Various schemes have been proposed for efficiently implementing canonicalization functions [17]. The complexity of this problem is discussed in [4].

4 SYMMETRY ON TRANSITIONS

Storing states in memory in one form (hash-table) or another (BDDs) and using canonicalization functions assumes that every system state can easily be encoded by a unique string of bits. When dealing with processes described by arbitrary programs written in full-fledged programming languages, this assumption is clearly not valid anymore.

Of course, nothing prevents us from systematically searching the state space of a concurrent system without storing any intermediate states in memory, as is done in VeriSoft. This is called a *state-less search* in [11]. In the context of a state-less search, a state is identified only by the sequence of transitions that were executed from the initial state s_0 in order to reach that state. In other words, the sequence of transitions that leads to a state from s_0 can be viewed as a "representation" of that state. Obviously, this representation is not unique since many different paths from s_0 may lead to a same state.

We now discuss how symmetry can be exploited for pruning state spaces in conjunction with a state-less search. We start by transposing the equivalence relation on states defined in the previous section to an equivalence relation on transitions.

Definition 4 Let $t = (s, s')$ denote a transition of a transition system M, and let σ denote a permutation of a symmetry group G of M. We write $\sigma(t)$ to denote the transition $(\sigma(s), \sigma(s'))$. We then define the relation $\equiv$ on transitions as follows: $t \equiv t'$ if $\exists \sigma \in G : t' = \sigma(t)$. ■

Lemma 1 *The relation $\equiv$ on transitions is an equivalence relation.*

We can also extend the definition of $\equiv$ on transitions to sequences of transitions as follows.

Definition 5 Let $w = t_1 t_2 \ldots t_n$ denote a nonempty sequence of transitions of a transition system M, and let σ denote a permutation of a symmetry group G of M. We write $\sigma(w)$ to denote the sequence of transitions $\sigma(t_1)\sigma(t_2)\ldots\sigma(t_n)$. We then define the relation $\equiv$ on nonempty sequences of transitions as follows: $w \equiv w'$ if $\exists \sigma \in G : w' = \sigma(w)$. ■

Lemma 2 *The relation $\equiv$ on sequences of transitions is an equivalence relation.*

As before, we denote by $[t]$ the equivalence class $[t] = \{\sigma(t) | \sigma \in G\}$ of t under $\equiv$, while $[w]$ denotes the equivalence class $[w] = \{\sigma(w) | \sigma \in G\}$ of w under $\equiv$.

We define a quotient transition system based on an equivalence relation on sequences of transitions as follows.

Definition 6 Given a transition system $M = (S, L, R, s_0)$ and a symmetry group G of M, a *quotient transition system* for M modulo G defined with an equivalence relation $\equiv$ on sequences of transitions is a transition system $M_{[w]} = (S', L', R', s'_0)$ where

- $S' = \{[w] | s_0 \overset{w}{\Rightarrow} s \text{ in } M\}$,
- $L' : S' \rightarrow 2^P$ is such that $L'([w]) = L(s)$ with $s_0 \overset{w}{\Rightarrow} s$ in M,
- $R' = \{([w], [wt]) | s_0 \overset{w}{\Rightarrow} s \text{ in } M \text{ and } t = (s, s') \in R\}$,
- $s'_0 = \epsilon$ (the empty word).

■

We now show that the initial states of M and $M_{[w]}$ are bisimilar.

Theorem 2 *Let $M = (S, L, R, s_0)$ be a transition system and $M_{[w]} = (S', L', R', s'_0)$ be a quotient transition system for M modulo a symmetry group G of M as defined in Definition 6. Then, $s_0 \sim s'_0$.*

Proof: Proofs of lemmas and theorems are omitted in this extended abstract. ■

Since the initial states of M and $M_{[w]}$ are bisimilar, $M_{[w]}$ can be used for model-checking temporal properties such as those discussed in the previous section.

From the previous theorem and Theorem 1, we immediately obtain by transitivity that the initial states of $M_{[w]}$ and $M_{[s]}$ are also bisimilar. We now compare the sizes of both types of quotient transition systems.

It is easy to prove that equivalent sequences of transitions from the initial state s_0 always lead to equivalent states.

Lemma 3 *Let $M = (S, L, R, s_0)$ be a transition system and let σ be a permutation of a symmetry group G of M. For any sequences w and w' of transitions such that $s_0 \stackrel{w}{\Rightarrow} s$ and $s_0 \stackrel{w'}{\Rightarrow} s'$, $w' = \sigma(w)$ implies $s' = \sigma(s)$.*

Since each equivalence class $[w]$ of sequences of transitions from s_0 can be mapped to the equivalence class $[s]$ of states such that $s_0 \stackrel{w}{\Rightarrow} s$, it is easy to see from Definitions 6 and 2 that $M_{[w]}$ will always contain at least as many states and transitions as $M_{[s]}$. Hence $M_{[w]}$ cannot be smaller than $M_{[s]}$.

In contrast with Lemma 3, it is possible to prove that equivalent states s and $\sigma(s)$ can be reached from s_0 by equivalent sequences of transitions only if $s_0 = \sigma(s_0)$, i.e., s_0 is symmetric with respect to permutation σ in G.

Lemma 4 *Let $M = (S, L, R, s_0)$ be a transition system and let G be a symmetry group of M. If $s_0 = \sigma(s_0)$ with $\sigma \in G$, then for any states s and s' such that $s' = \sigma(s)$, $s_0 \stackrel{w}{\Rightarrow} s$ implies $s_0 \stackrel{\sigma(w)}{\Rightarrow} s'$.*

Putting it all together, we now state the exact condition required for $M_{[w]}$ and $M_{[s]}$ to be identical. ($\Rightarrow$ denotes logical implication.)

Theorem 3 *Let $M = (S, L, R, s_0)$ be a transition system. Let $M_{[w]}$ and $M_{[s]}$ be quotient transition systems for M modulo a symmetry group G of M as defined in Definition 6 and Definition 2 respectively. Then $M_{[w]}$ and $M_{[s]}$ are identical if and only if the two following conditions are satisfied in M:*

1. $\forall s \in S : (s_0 \stackrel{w}{\Rightarrow} s \wedge s_0 \stackrel{w'}{\Rightarrow} s) \Rightarrow (\exists \sigma \in G : w' = \sigma(w))$.
2. $\forall \sigma \in G : (s_0 \stackrel{w}{\Rightarrow} s \wedge s_0 \stackrel{w'}{\Rightarrow} \sigma(s)) \Rightarrow (\exists \sigma' \in G : s_0 = \sigma'(s_0) \wedge \sigma(s) = \sigma'(s))$.

$M_{[w]}$ and $M_{[s]}$ are identical when there is a one-to-one correspondence between equivalence classes on states and equivalence classes on sequences of transitions. This is the case for instance when the state space is isomorphic to a tree and $|[s_0]| = 1$ (i.e., the initial state is symmetric with respect to every permutation in G).

In summary, exploring exactly one sequence w of transitions per equivalence class $[w]$ is sufficient to visit at least once every equivalence class of states $[s]$ and transitions $[t]$ in a transition system. If exactly one sequence w per equivalence class $[w]$ is explored, redundant visits of a same equivalence class of states or transitions are due only to the exploration of multiple paths that are not equivalent with respect to $\equiv$. In practice, such redundant visits are usually unavoidable with a state-less search since multiple unequivalent paths may very well lead to the same state or to equivalent states.

5 EXPLOITING SYMMETRY ON TRANSITIONS

In order to exploit the ideas presented in the previous section, we need a practical way to determine when transitions (elements of R) are symmetric without computing and comparing explicit encodings of the states they originate from or lead to. We propose to use a labeling function for representing actions performed during the execution of transitions, in conjunction with permutation functions on actions for modeling symmetries in the system.

Precisely, we first extend the definition of transition systems to allow labels on transitions.

Definition 7 A *labeled transition system* (LTS) $M = (S, L, R, s_0, \Sigma, \ell)$ is a transition system (S, L, R, s_0) extended with a set Σ of *actions*, and a labeling function $\ell : R \rightarrow \Sigma$. ■

The action of a transition t, denoted $\ell(t)$, identifies the visible operation executed during the transition. This identifier includes the type of visible operation (system call) executed, values of parameters if any, and the identity of the process executing the operation. By construction, we guarantee that, for any action a in Σ and any state s in S, there is at most one transition labeled with a from s. Therefore, any transition from a given state can be uniquely identified by its label. Note that the following results also hold for transitions (pairs of states) labeled with more than one label, as long as these labels satisfy the previous assumption.

Symmetries in the system can then be modeled by permutation functions π_σ on actions (elements of Σ). Specifically, a function π_σ on actions is associated with each permutation in G as follows.

Definition 8 Let $M = (S, L, R, s_0, \Sigma, \ell)$ be a LTS and let σ be a permutation of a symmetry group G of M. A partial function $\pi_\sigma : \Sigma \rightarrow \Sigma$ on actions is a *valid permutation function on actions* for permutation σ in G iff the following property holds for all actions a in Σ:

- if $\pi_\sigma(a)$ is defined, then $\forall t = (s, s') \in R : \ell(t) = a \Rightarrow \ell(\sigma(t)) = \pi_\sigma(a)$.

■

By definition, $\pi_\sigma(a)$ is unique when it is defined.

In practice, the user can declare symmetries in a system by providing values of permutation functions π_σ on actions. Standard classes of permutations can be pre-packaged in order to facilitate this step. For instance, various classes of process symmetries can easily be specified using simple manipulation functions on process indices (such as swapping, circular rotation, etc.) which can be applied uniformly to any action name, independently of the type of operation executed. An example of function π_σ on actions is given in Section 7.

Of course, declaring symmetries to the verification system has to be done with care since the validity of the verification results relies on the validity of these assumptions. Conservative approximations that are sufficient for identifying symmetric transitions but may miss other symmetries can be used.

The following rule exploits symmetries on actions in order to prune transitions during state-space exploration.

Definition 9 (Pruning Rule) In any state s reached during state-space exploration, for any pair of transitions t and t' enabled in s such that

$$\exists \sigma \in G : \ell(t') = \pi_\sigma(\ell(t)) \wedge s = \sigma(s),$$

explore only one of these two transitions from state s. ■

The correctness of this rule is based on the following theorem.

Theorem 4 *Let s be a state of a LTS $M = (S, L, \Sigma, R, s_0)$ and let G be a symmetry group on M. If t and t' are transitions in enabled(s) such that $\exists \sigma \in G : \ell(t') = \pi_\sigma(\ell(t)) \wedge s = \sigma(s)$, then $t' = \sigma(t)$.*

From Theorem 2 of the previous section, we already know that exploring (sequences of) transitions that are equivalent with respect to $\equiv$ is unnecessary in order to verify properties as those considered in Section 3. When the condition stated in Definition 9 is satisfied, we have $t \equiv t'$ by Theorem 4, and it is thus sufficient to explore only one of these transitions.

In order to implement the rule above, we need a practical way to check whether a reachable state s satisfies the condition $s = \sigma(s)$ without using any explicit encoding of s. We now discuss how this can be done assuming we know that $s_0 = \sigma(s_0)$.

Let s be a state reachable from s_0 by a nonempty sequence $w = t_1 t_2 \dots t_n$ of transitions. Assuming $s_0 = \sigma(s_0)$, $s = \sigma(s)$ can be proved by showing that $s_0 \stackrel{\sigma(w)}{\Rightarrow} s$. If for all $1 \leq i \leq n$, $\ell(t_i) = \pi_\sigma(\ell(t_i))$, then $t_i = \sigma(t_i)$ for all t_i in w, and $s = \sigma(s)$. Otherwise, let $t_i = (s_{i-1}, s_i)$ denote the first transition along w such that $\ell(t_i) \neq \pi_\sigma(\ell(t_i))$, and let $t'_i = \sigma(t_i)$. Since t_i is the first such transition in w and since $s_0 = \sigma(s_0)$, we know that $s_{i-1} = \sigma(s_{i-1})$. Therefore, t'_i is enabled in s_{i-1} and labeled by $\pi_\sigma(\ell(t_i))$, but was not explored because of the pruning rule defined above. In other words, the exploration of $\sigma(w)$ stopped in s_{i-1}. Checking that $\sigma(w)$ leads from s_0 to s may thus seem problematic.

Fortunately, there is a way to check whether $\sigma(w)$ leads to s without using any explicit encoding for s or without even executing $\sigma(w)$ in the first place: if w and $\sigma(w)$ are different interleavings of a same partial ordering of transitions, then they are guaranteed to lead to a same state, and hence $s = \sigma(s)$. In the next section, we use partial-order methods to solve this problem and complete our algorithm.

6 COMBINING SYMMETRY AND PARTIAL-ORDER REDUCTIONS

Partial-order methods [10] denote another family of algorithms for tackling the state explosion problem which limits the efficiency and applicability of verification by state-space exploration. The basic idea behind partial-order methods is that many paths in the state space of a concurrent system correspond simply to different execution orders of the same process transitions (defined in Section 2). If process transitions are "independent", for instance because they are executed by noninteracting processes, then changing their execution order will not modify their combined effect. This notion of independency between process transitions and its complementary notion, the notion of dependency, can be formalized by the following definition (adapted from [18]).

Definition 10 Let $\mathcal{T}$ be the set of process transitions of the system and $D \subseteq \mathcal{T} \times \mathcal{T}$ be a binary, reflexive, and symmetric relation. The relation D is a *valid dependency relation* for the system iff for all $t_1, t_2 \in \mathcal{T}$, $(t_1, t_2) \notin D$ (t_1 and t_2 are *independent*) implies that the two following properties hold for all global states s in the state space of the system:

1. if t_1 is enabled in s and $s \xrightarrow{t_1} s'$, then t_2 is enabled in s iff t_2 is enabled in s' (independent process transitions can neither disable nor enable each other); and

2. if t_1 and t_2 are enabled in s, then there is a unique state s' such that $s \stackrel{t_1 t_2}{\Rightarrow} s'$ and $s \stackrel{t_2 t_1}{\Rightarrow} s'$ (commutativity of enabled independent process transitions).

■

This definition characterizes the properties of possible "valid" dependency relations for the process transitions of a given system. In practice, it is possible to give easily checkable syntactic conditions that are sufficient for process transitions to be independent. In a concurrent system as defined in Section 2, dependency can arise between transitions of different processes that access the same communication objects. We refer the reader to [10] for a detailed presentation of this topic.

Following the work of Mazurkiewicz [21], one can use the notion of independent process transitions to define an equivalence relation on sequences of process transitions: two sequences of process transitions are equivalent if they can be obtained from each other by successively permuting adjacent independent process transitions. Thus, given a valid dependency relation, sequences of process transitions can be grouped into equivalence classes which Mazurkiewicz calls traces. By construction, all the sequences of process transitions of a Mazurkiewicz trace executed from a given state all lead to the same final state (e.g., see Theorem 3.10 in [10]).

Thanks to this property and assuming we know $s_0 = \sigma(s_0)$, we can check whether $s_0 \stackrel{\sigma(w)}{\Rightarrow} s$ by checking whether w and $\sigma(w)$ (computed by applying π_σ successively to the labels of the transitions in w) can be obtained from each other by successively permuting adjacent independent process transitions. In the worst-case, this can be done in time $O(|P||w|)$ where $|P|$ is the number of processes in the system and $|w|$ is the length of w. Since the purpose of the approach developed in this work is to make possible the systematic analysis of all executions of a concurrent system up to some depth, w is usually short (e.g.,

```
1   Initialize: Stack is empty;
2   Search() {
3     DFS(∅);
4     }
5   DFS(set: Sleep) {
6     T = Persistent_Set()\Sleep;
7     T = Sym(T, Stack);
8     while T ≠ ∅ do {
9       take t out of T;
10      push (t) onto Stack;
11      Execute(t);
12      DFS({t' ∈ Sleep | t' and t are independent});
13      pop t from Stack;
14      Undo(t);
15      Sleep = Sleep ∪ {t};
16      };
17    }
```

Figure 1: State-less search using symmetry and partial-order reductions

less than a few hundred process transitions), and the computation is fast. If the result of this check shows that w and $\sigma(w)$ belong to the same Mazurkiewicz trace, we can safely conclude that $s = \sigma(s)$. Otherwise, we have to assume by default that $s \neq \sigma(s)$. This prevents the pruning of any transition in s using permutation σ.

Putting it all together, we can now present in Figure 1 a complete algorithm for performing a state-less search using symmetry and partial-order reductions. This algorithm is an extension of the algorithm of [11]. The only difference is the addition of line 7. This algorithm performs a selective depth-first search (DFS) in the state space of a concurrent system and uses two partial-order reduction techniques: ***persistent sets*** and ***sleep sets***. A detailed presentation of these algorithmic techniques can be found in [10, 11]. The data structure *Stack* contains the sequence of process transitions that leads from the initial global state s_0 to the current global state being explored. A set *Sleep* is associated with each global state reached during the search, i.e., with each call to the procedure DFS. The sleep set associated with a global state s is a set of process transitions that are enabled in s but will not be explored from s. The sleep set associated with the initial global state s_0 is the empty set. Each time a new global state s is encountered during the search, a call to DFS is executed. The sleep set that has to be associated with s is passed as argument. The rules for computing the sleep set of a new state are given in lines 12 and 15.

In lines 6 and 7, a new set of process transitions is selected to be explored from s in two steps. Persistent_Set() returns a persistent set in the current global state s that is nonempty if there exist process transitions enabled in s.

In line 7, symmetric process transitions are eliminated using the new function Sym described in details below. In line 11, a process transition t is executed from s. The procedure Execute(t) returns after a new global state has been reached by the concurrent system. Then all the process transitions of *Sleep* that are independent with t are passed into the sleep set associated to that new global state (line 12). Once the search from that new state (and hence the corresponding call to DFS) is completed, the exploration of the other process transitions selected to be explored from s may proceed. The concurrent system is then brought back to the global state s in line 14. (This can be done by reinitializing the system and reexecuting the sequence of process transitions in *Stack*, for instance.) Next, process transition t, i.e., the last process transition explored from s, is added to *Sleep* in line 15. Note that this algorithm does not compute, store or manipulate any explicit state encoding.

The originality of the algorithm of Figure 1 is thus the use of the function Sym in line 7. The function Sym checks whether any process transition of its first argument T can be pruned using the rule of Definition 9. It takes as arguments a subset T of the set of enabled process transitions in s and the sequence of process transitions formed by the current depth-first search *Stack*. When the function Sym finds a permutation σ such that $\ell(t') = \pi_\sigma(\ell(t))$ with t and t' in T, it then checks whether $s = \sigma(s)$ by examining the current path *Stack* of process transitions being explored and its symmetric image $\sigma(Stack)$ following the procedure described earlier in this section. The function Sym finally returns a subset (not necessarily proper) of its argument T which determines the process transitions to be explored from the current state s.

When both symmetry and partial-order reduction methods are applied, the properties preserved by their combination is the intersection of the properties preserved by each type of reduction alone. Since the algorithm of [11] preserves deadlocks and assertion violations, let us now discuss how the use of symmetry reductions affects this specific class of properties. By definition of a symmetry group, if a global state s is a deadlock, so are all the states s' in $[s]$. Thus, exploring at least one state per equivalence class $[s]$ using symmetry reductions is sufficient to detect a deadlock in the system. In contrast, the detection of an assertion violation using symmetry methods is guaranteed only for assertions that are violated in all the states of an equivalence class $[s]$. Let us call such assertions "symmetry-preserving assertions". For instance, any assertion that does not distinguish a process from any other is symmetry-preserving under any process symmetry.

Formally, the correctness of the algorithm of Figure 1 is established by the following theorem.

Theorem 5 *Consider a concurrent system as defined in Section 2, and assume its state space is finite and acyclic. Then, if there exist deadlocks in the state space of the system, the algorithm of Figure 1 will visit at least one of them. Moreover, if there exists a global state in the state space of the system where a symmetry-preserving assertion is violated, then there exists a global state visited by the algorithm of Figure 1 where the same assertion is violated.*

```
/* phil.c : dining philosophers (version without loops) */

#include <stdio.h>
#include <sys/types.h>
#include <sys/ipc.h>
#include <sys/sem.h>

#define N 2

philosopher(i)
     int i;
{
  printf("philosopher %d thinks\n",i);
  semwait(i);              /* take left fork */
  semwait((i+1)%N);        /* take right fork */
  printf("philosopher %d eats\n",i);
  semsignal(i);            /* release left fork */
  semsignal((i+1)%N);      /* release right fork */
  exit(0);
}

main()
{
  int semid, i, pid;

  semid = semget(IPC_PRIVATE,N,0600);

  for(i=0;i<N;i++)
    semsetval(i,1);
  for(i=0;i<(N-1);i++) {
    if((pid=fork()) == 0)
      philosopher(i);
    };
  philosopher(i);
}
```

Figure 2: A simple example of concurrent C program

7 EXAMPLE

In this section we illustrate our ideas with a simple example. Consider the concurrent C program shown in Figure 2. This program represents a concurrent system composed of two processes. It describes the behavior of these processes as well as the initialization of the system. This example is inspired by the well-known dining-philosophers problem, with two philosophers. The two processes communicate by executing the visible operations *semwait* and *semsignal* on two semaphores that are identified by the integers 0 and 1 respectively. The

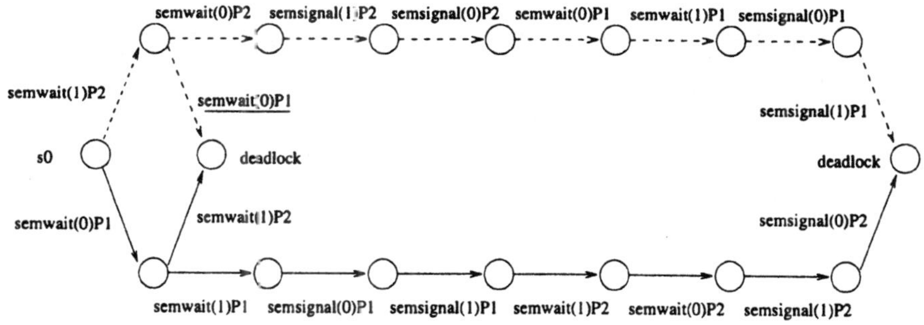

Figure 3: State space of the two-dining-philosophers system

value of both semaphores is initialized to 1 (with the operation *semsetval*). By implementing these operations using actual UNIX semaphores, the program above can be compiled and run on any UNIX machine. The operation *exit* is a visible operation whose execution is always blocking.

The state space of this system is shown in Figure 3, where global transitions are labeled with the visible operation of the corresponding process transition. Since all the processes are deterministic, nondeterminism (i.e., branching) in A_G is caused only by concurrency.

The dining-philosophers system contains symmetry: swapping the states of the two processes and semaphores yields an equivalent state. We can model this symmetry by a permutation function π_σ on actions as defined in Definition 8[1]:

- $\pi_\sigma(semwait(x)_{P1}) = semwait((x+1)\%2)_{P2}$,
- $\pi_\sigma(semwait(x)_{P2}) = semwait((x+1)\%2)_{P1}$,
- $\pi_\sigma(semsignal(x)_{P1}) = semsignal((x+1)\%2)_{P2}$,
- $\pi_\sigma(semsignal(x)_{P2}) = semsignal((x+1)\%2)_{P1}$,
- $\pi_\sigma(exit(0)_{P1}) = exit(0)_{P2}$, and
- $\pi_\sigma(exit(0)_{P2}) = exit(0)_{P1}$.

When using this permutation function with the state-space exploration algorithm of Figure 1, the dotted transitions of Figure 3 are not explored thanks to symmetry reductions. For this example, half of the transitions need not be explored with the new algorithm, which is the maximum reduction one can obtain with this type of process symmetry and two processes. Note that, for this example, using partial-order reduction techniques (persistent sets and sleep sets) alone would avoid exploring only one transition in the state space (the transition whose label is underlined is pruned thanks to sleep sets).

Table 1 compares the number of transitions explored by a state-less search, a state-less search using partial-order reductions (SL+PO) as described in [11],

[1]The modulus binary operator % returns the remainder from the division of its first argument by its second argument.

N	State Space	State Less	SL+PO	SL+PO+SYM
2	(18)	18	17	9
3	(123)	1,680	75	37
4	(708)	386,816	275	133
5	(3,765)	?	959	461

Table 1: Number or transitions explored

and a state-less search using partial-order reductions and symmetry reductions (SL+PO+SYM) as described in Figure 1. Results are given for the previous program with various numbers N of processes. The numbers between parentheses in the column entitled "State Space" (column 2) give the total number of transitions in the state space. The run time of the three algorithms considered is proportional to the number of explored transitions.

One clearly sees from the numbers of transitions explored by a simple state-less search (column 3) that the run-time cost of not storing visited states in memory is very high. However, this cost can be dramatically reduced by using partial-order reduction techniques (column 4), as extensively discussed in [11]. Moreover, the use of symmetry reductions yields additional substantial reductions (column 5). For this example, the number of explored transitions is reduced roughly by a factor of 2 when exploiting symmetry.

Note that the maximum theoretical symmetry reduction with this type of process symmetry is by a factor of N, where N is the number of processes in the system, since the size of each equivalence class of states and transitions is at most N. In contrast, partial-order reductions reduce the number of explored transitions of a state-less search by a factor much greater than N for $N > 2$ in this example. Therefore, a state-less search using symmetry reductions alone (i.e., without partial-order reductions) could not yield such a huge reduction.

8 CONCLUSIONS

We have developed a simple framework for exploiting symmetry in a software application in order to safely avoid exploring parts of its state space when searching for deadlocks and assertion violations. Our framework is based on equivalence classes of sequences of transitions instead of the traditional equivalence classes of states. We then presented a state-space exploration algorithm for exploiting symmetries on transitions which does not rely on explicit encodings of system states. This algorithm combines both symmetry and partial-order reduction methods in an original way. It can be used to systematically and efficiently explore the state spaces of implementations of concurrent reactive software systems.

Note that it has been known for some time that symmetry and partial-order reduction methods are essentially orthogonal, and hence compatible and complementary. The combination of state-based symmetry reductions with partial-order reductions was already suggested and studied in [23, 6, 7].

ACKNOWLEGMENTS

I wish to thank Ramin Hojati, Norris Ip, Lalita Jagadeesan, Denis Leroy and Prasad Sistla for interesting discussions on combining partial-order and symmetry methods. I am also thankful to Glenn Bruns and the anonymous referees for helpful comments on this paper.

References

[1] R.E. Bryant. Symbolic boolean manipulation with ordered binary-decision diagrams. *ACM Computing Surveys*, 24(3):293–318, 1992.

[2] E. Clarke, T. Filkorn, and S. Jha. Exploiting symmetry in temporal logic model checking. In *Proc. 5th Conference on Computer Aided Verification*, volume 697 of *Lecture Notes in Computer Science*, pages 450–462, Elounda, June 1993. Springer-Verlag.

[3] E. M. Clarke and E. A. Emerson. Design and Synthesis of Synchronization Skeletons using Branching-Time Temporal Logic. In D. Kozen, editor, *Proceedings of the Workshop on Logic of Programs,* Yorktown Heights, volume 131 of *Lecture Notes in Computer Science*, pages 52–71. Springer-Verlag, 1981.

[4] E. M. Clarke, E. A. Emerson, S. Jha, and A. P. Sistla. Symmetry Reductions in Model Checking. In *Proc. 10th Conference on Computer Aided Verification*, volume 1427 of *Lecture Notes in Computer Science*, pages 147–158, Vancouver, June/July 1998. Springer-Verlag.

[5] R. Cleaveland, J. Parrow, and B. Steffen. The concurrency workbench: A semantics based tool for the verification of concurrent systems. *ACM Transactions on Programming Languages and Systems*, 1(15):36–72, 1993.

[6] S. Duri, U. Buy, R. Devarapalli, and S. M. Shatz. Application and Experimental Evaluation of State Space Reduction Methods for Deadlock Analysis in Ada. *ACM Transactions on Software Engineering and Methodology*, 3(4):340–380, October 1994.

[7] E. A. Emerson, S. Jha, and D. Peled. Combining Partial Order and Symmetry Reductions. In *Proceedings of the Third International Workshop on Tools and Algorithms for the Construction and Analysis of Systems (TACAS'97)*, volume 1217 of *Lecture Notes in Computer Science*, pages 19–34, Twente, April 1997. Springer-Verlag.

[8] E. A. Emerson and A. P. Sistla. Symmetry and model checking. In *Proc. 5th Conference on Computer Aided Verification*, volume 697 of *Lecture Notes in Computer Science*, pages 463–478, Elounda, June 1993. Springer-Verlag.

[9] A. R. Flora-Holmquist and M. Staskauskas. Formal validation of virtual finite state machines. In *Proc. Workshop on Industrial-Strength Formal Specification Techniques (WIFT'95)*, pages 122–129, Boca Raton, April 1995.

[10] P. Godefroid. *Partial-Order Methods for the Verification of Concurrent Systems - An Approach to the State-Explosion Problem*, volume 1032 of *Lecture Notes in Computer Science*. Springer-Verlag, January 1996.

[11] P. Godefroid. Model Checking for Programming Languages using VeriSoft. In *Proceedings of the 24th ACM Symposium on Principles of Programming Languages*, pages 174–186, Paris, January 1997.

[12] P. Godefroid, R. S. Hanmer, and L. J. Jagadeesan. Model Checking Without a Model: An Analysis of the Heart-Beat Monitor of a Telephone Switch using VeriSoft. In *Proceedings of ACM SIGSOFT ISSTA'98 (International Symposium on Software Testing and Analysis)*, pages 124–133, Clearwater Beach, March 1998.

[13] J. A. Green. *Sets and Groups - A First Course in Algebra*. Routledge and Kegan Paul, 1965.

[14] M. Hennessy and R. Milner. Algebraic laws for nondeterminism and concurrency. *Journal of the ACM*, 32(1):137–161, 1985.

[15] G. J. Holzmann. *Design and Validation of Computer Protocols*. Prentice Hall, 1991.

[16] G. J. Holzmann and J. Patti. Validating sdl specifications: An experiment. In *Proc. 9th IFIP WG 6.1 International Symposium on Protocol Specification, Testing, and Verification*. North-Holland, 1989.

[17] C. N. Ip and D. L. Dill. Better verification through symmetry. In D. Agnew, L. Claesen, and R. Camposano, editors, *Proceedings of the 1993 Conference on Computer Hardware Description Languages and their Applications*, April 1993.

[18] S. Katz and D. Peled. Defining conditional independence using collapses. *Theoretical Computer Science*, 101:337–359, 1992.

[19] D. Kozen. Results on the Propositional Mu-Calculus. *Theoretical Computer Science*, 27:333–354, 1983.

[20] Z. Manna and A. Pnueli. *The Temporal Logic of Reactive and Concurrent Systems: Specification*. Springer-Verlag, 1992.

[21] A. Mazurkiewicz. Trace theory. In *Petri Nets: Applications and Relationships to Other Models of Concurrency, Advances in Petri Nets 1986, Part II; Proceedings of an Advanced Course*, volume 255 of *Lecture Notes in Computer Science*, pages 279–324. Springer-Verlag, 1986.

[22] K. L. McMillan. *Symbolic Model Checking.* Kluwer Academic Publishers, 1993.

[23] A. Valmari. Stubborn sets for of colored petri nets. In *Proceedings of the 12th International Conference on Apllications and Theory of Petri Nets*, pages 102–121, Gjern, 1991.

[24] M.Y. Vardi. Why is modal logic so robustly decidable? In *Proceedings of DIMACS Workshop on Descriptive Complexity and Finite Models.* AMS, 1997.

PART SIX

Verification

User-Friendly Verification

PAO-ANN HSIUNG and FARN WANG
Institute of Information Science, Academia Sinica, Taipei, TAIWAN.
E-mail: {eric,farn}@iis.sinica.edu.tw
URL: http://www.iis.sinica.edu.tw/~eric/sgm/

Key words: timed automata, TCTL, model-checking, state-space explosion, verification tool, concurrent real-time systems

Abstract: Model checking often faces the problem of reducing the large exponential sizes of state-space representations. Several reduction techniques such as bisimulation equivalence, partial-order semantics, and symmetry-based reduction have been proposed, but existing tools do not completely allow a user the flexibility in manipulating state spaces. We propose a new *user-friendly* verification environment where a user has full control on what techniques to apply and in what sequences to apply them. We have implemented the environment in a tool called *State-Graph Manipulators* (SGM). SGM packages verification techniques into efficient, reusable, modular *manipulators*, that act on high-level state-space representations called *state-graphs*. Further, we are also proposing four new state-space reduction techniques, namely *variable shielding*, *read-write analysis*, *internal transitions bypassing*, and *sibling transition multiplicity reduction*. They are implemented into SGM and experiments have been conducted to show their usefulness.

1. INTRODUCTION

In the recent few years due to several breakthroughs, *model-checking* [2], a formal approach to the verification of concurrent systems, has become more and more popular. Model-checking has been used to verify real-time systems. *Timed automata* [4] has been widely used as the system model for verifying real-time properties specified in *Timed Computation Tree Logic* (TCTL) [11]. Concurrency is generally modeled as the interleaving of computation sequences [15]. This causes state-space explosions and the large sizes of state-space representations become unmanageable, thus hindering verification. Both the degree of concurrency and the complexity of systems verifiable are limited. Several techniques have been proposed in the literature for reducing the state-space representation size, including symmetry-based reductions [10], partial-order reductions [15], bisimulation equivalences [16], and minimization techniques [3].

Such reduction techniques usually require years of studying to master and are typically implemented with sophisticated data-structures for state-graphs. Moreover, since different verification tasks may need different ways of attacking the state-explosion problem, for each specific verification task the optimal combination of reduction techniques that reduces the state-space representations most and in the least time might be *different*. Even for an engineer properly knowledgeable of verification theory, to experiment with different combinations of the various reduction techniques may painfully take too much time and resources a project can sustain. This need for experimentation necessitates an environment where a designer can easily change the combination of reduction techniques applied to a given verification task. For this purpose, we have developed a new tool called the *State-Graph Manipulators* (SGM).

SGM adopts and presents a high-level view of all verification intricacies through a graphical user-interface (GUI). Various different sophisticated verification technologies are packed into efficient *manipulators* that act on high-level data-objects representing state-spaces. These data-objects are called *state-graphs* as defined later in Section 4. With a friendly GUI, users can choose manipulators, from the menu bar, during state-graph constructions and compare the effects of applying different *sequences of manipulators* to state-graphs.

Section 2 describes verification-related tools and their techniques. Section 3 describes the verification framework, the verification procedure language, and the three interfaces of SGM. Section 4 describes the seven manipulators implemented in SGM. Section 5 gives the experimental results for several application examples. Section 6 gives the final conclusion.

2. PREVIOUS WORK

This section gives a brief account of some mature tools that are widely known and of the verification techniques they have used. It is worth noticing that although such tools are numerous, yet very few tools have been, are, or will be providing high-level perceptions. NuSMV [6], a new version of the well-known *Symbolic Model Verifier* (SMV), is one tool that will be adopting an approach similar to ours, but they do not yet have any published results on the benefits of using high-level techniques, nor is there any description of concrete implementations in [6].

Murϕ Verification System [9] is a language-based verifier tool for finite-state verification of concurrent systems with symmetry-based reduction. **SPIN** [12] is an automated protocol validation system using PROMELA with refining dependencies for efficient partial-order verification. **KRONOS** [7] is a well-known verification tool with minimization algorithms, inactive clock reduction [8], and clock equality [8] reduction techniques implemented. **UPPAAL** [5] is a widely-used verification tool for real-time systems with a graphical interface, simulation, quotient construction, minimizations, trivial equation elimination, and equivalence reduction implemented. **SMV** [14] is a well-known tool for model-checking finite-state systems against CTL (*Computation Tree Logic*) specifications. A new version called NuSMV has implemented automatic variable ordering and cache configuration [6].

From the above, we notice that although different techniques have been implemented in the various well-known tools, yet if a user needs to apply two or more different techniques (implemented in different tools) to a verification task, then the user will have to spend a great effort trying to either translate the output results of one tool to the input format of another tool or make some strong assumptions of technique application that may be invalid. Pain staking efforts could be avoided if various compatible techniques could be collected, implemented, and integrated into a single environment in which users can flexibly fine tune their verification problems. SGM is proposed with this motivation in mind.

3. SGM: A USER-FRIENDLY VERIFICATION FRAMEWORK

SGM provides a user-friendly verification environment through a *flexible* input language and three user interfaces: *graphical*, *interactive*, and *batch*.

We adopt the framework of model-checking in which a system is described by a set of concurrent timed automata [2] with dense-time semantics and a timing property is specified in TCTL (*Timed Computation Tree Logic*) [2,11] extended from CTL. To compute the compact global state-space representations (i.e. *state-graphs*), first, the set of timed automata is merged, two at a time. Second, during each merge iteration different sequences of reduction techniques may be applied to the intermediate state-graphs. Finally, the system is verified by model-checking the global state-graph against the given TCTL specification.

SGM allows a user to describe his/her system of timed automata, the property to be verified, and the actions to be performed on *state-graphs* (defined in Section 4). SGM helps the designer in experimenting, online or offline, with the application of different sequences of reduction techniques to the intermediate state-graphs during the composition of a global state-graph. For a given verification task, this helps in finding a good sequence of *manipulators* (proved and implemented verification techniques), which reduces the state-graph sizes and increases verification scalability.

We have developed seven manipulators in SGM, namely *state-graph merging, timed symmetry-reduction* [17], *clock-shielding, variable-shielding, read-write analysis, internal transition bypassing*, and *sibling transition multiplicity reduction*. Of the above, the first three were proposed by other authors and ourselves, the other four are new reduction techniques. These seven techniques will be discussed in Section 4. Other reduction techniques can be easily hooked into our tool as a new manipulator. Thus, researchers can take advantage of the framework to experiment with how one's reduction technique works in collaboration with other existing techniques.

3.1 Verification Procedure Language

The input language of SGM is called *Verification Procedure Language* (VPL), which consists of three parts for system description, specification, and state-graph manipulation. System description corresponds to the timed automata model of real-time systems. Specification corresponds to a TCTL formula. Manipulation corresponds to a list of actions on state-graphs.

We will use *Fischer's Mutual Exclusion Protocol* (FMEP) as a running example in this paper. FMEP is a benchmark example that has been used for state-graph reduction technique evaluation in several literatures [13]. A generic timed automaton for the *i*th process and the corresponding SGM input for a 11-process system obeying FMEP are shown in Fig. 1. Fischer's protocol says that given a system of *n* concurrent processes, the time taken by each process for checking (or reading) whether a *lock* variable is zero must be lesser than that for assigning (or writing) its index value to the *lock* variable. We must prove that such a system will never allow more than one process into the critical section (mode M_4). VPL allows easy change of a system's degree of concurrency by only changing the first line.

Theoretically, this part of the input corresponds to the timed automata model of a real-time system. A timed automaton (TA) is composed of various *modes* interconnected by *transitions*. Variables are distinguished into *clock* and *discrete*, where variables of the former type increment at a uniform rate and can be reset on a transition, while variables of the latter type change values only when assigned a new value on a transition. A TA may remain in a particular mode as long as the values of all its variables, including clock and discrete, satisfy a *mode predicate*, which is a conjunction of clock constraints and boolean propositions. In the following, N and R represent the sets of non-negative integers and non-negative real numbers, respectively.

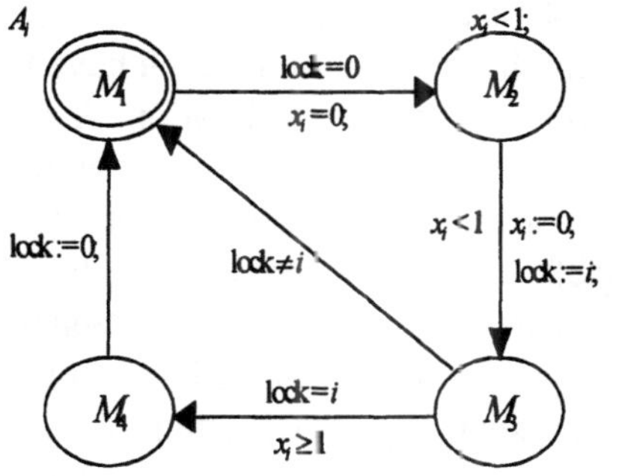


```
automata A : i = 1..11;
clock x[1..total(A)]; register lock;
automaton A[i] {initially M1 and lock=0 and x[i]=0;
     mode M1 { invariant True;
       when lock=0 may goto M2; x[i]:=0;}
     mode M2 { invariant x[i]<1;
       when x[i]<1 may goto M3; x[i]:=0; lock:=i;}
     mode M3 { invariant True;
       when x[i]>=1 and lock=i may goto M4;
       when ~(lock=i) may goto M1;}
     mode M4 { invariant True;
       when True may goto M1; lock:=0; }
```

Figure 1. Fischer's Mutual Exclusion Protocol

Definition 1: Mode Predicate. Given a set C of clock variables and a set D of discrete variables, the syntax of a *mode predicate* η over C and D is defined as follows: η := false | $x \sim c \mid x - y \sim c \mid d \sim c \mid \eta_1 \wedge \eta_2 \mid \neg\eta_1$, where x, $y \in C$, $\sim \in \{\leq, <, =, \geq, >\}$, $c \in$ N, $d \in D$, η_1, η_2 are mode predicates.

Let $B(C, D)$ be the set of all mode predicates over C and D. A TA may go from a mode to another, that is perform a transition, when the triggering condition (specified as a mode predicate) is satisfied by the current valuation of the clock and discrete variables. On a transition, some clocks may be reset to zero and some discrete variables may be assigned new integer values.

Definition 2: Timed Automaton. A *Timed Automaton* (TA) is a tuple $A_i = (M_i, m_i^0, C_i, D_i, \chi_i, E_i, \tau_i, \rho_i)$ such that: M_i is a finite set of modes, $m_i^0 \in M$ is the initial mode, C_i is a set of clock variables, D_i is a set of discrete variables, χ_i: $M_i \rightarrow B(C_i, D_i)$ is an *invariance* function that labels each mode with a condition true in that mode, $E_i \subseteq M_i \times M_i$ is a set of transitions, τ_i: $E_i \rightarrow B(C_i, D_i)$ defines the transition triggering conditions, and ρ_i: $E_i \rightarrow 2^{C_i \cup (D_i \times \mathrm{N})}$ is an *assignment* function that maps each transition to a set of assignments such as resetting some clock variables and setting some discrete variables to specific integer values.

As far as temporal property specification is concerned, SGM uses TCTL. A TCTL formula has the following syntax. $\phi ::= \eta \mid \exists\Box\phi' \mid \exists\phi' U_{\sim c}\phi'' \mid \neg\phi' \mid \phi' \vee \phi''$. Here, η is a mode predicate in $B(C, D)$, ϕ', ϕ'' are TCTL formulae, $\sim \in \{<, \leq, =, \geq, >\}$, and $c \in$ N. Due to page-limit, we do not elaborate on the semantics of a TCTL formula. Details can be found in [11].

The specification begins with the keyword verify. Users can easily specify a TCTL formula using the user-friendly keywords. For example, all_paths means *for all paths in the state-graph starting from the initial mode*. For our running example of 3-automata FMEP, the TCTL specification is as follows.

```
verify all_paths (henceforth
    not{(mode(A[1])=M4) and (mode(A[2])=M4)} and not{(mode(A[1])=M4) and
    (mode(A[3])=M4)} and not{(mode(A[2])=M4) and (mode(A[3])=M4)});
```

In the last part of SGM input, state-graphs are variables which have to be declared first. Then, a list of manipulations follows the declaration. A *simple* manipulation can be either the merging of two state-graphs such as g[3] := merge_graph(g[1], g[2]);or the application of a single reduction technique on some state-graph such as shield_clock(g[3]);, or model-checking a state-graph such as model_check(g[3]);, or printing a state-graph such as print_graph(g[3]);, where g[1], g[2], g[3] are all state-graph variables. More complex programming constructs are also provided, such as *for-loops*, and *if-then-else* statements for more dynamic selections of manipulator sequences. These are omitted due to page-limit.

State-graph manipulation for FMEP example is as follows.

```
manipulation
 graph g[1..total(A)-1];          -- new graph declarations
 for(i:=1; i<total(A); i++) {     -- for each declared graph g[i]
  if(i = 1) { g[1] := merge_graph(A[1], A[2]); }
  else { g[i] := merge_graph(g[i-1], A[i+1]); }
  shield_clock(g[i]);
  normalize_region(g[i]); }
 model_check(g[total(A)-1]);      -- model-check global graph
```

3.2 User Interfaces

Besides a human-readable input language, user-friendly verification is achieved by SGM through three user interfaces including graphical, interactive, and batch. A user has to first create a text input file using VPL.

In the graphical mode, as shown in Fig. 2 after a user loads input file, SGM displays the system as a set of boxes where each box represents a state-graph. Each box has some basic visible information including its size and component processes. A detailed information of each state-graph (box) can be obtained by opening the boxes. After selecting two state-graphs (boxes), one can merge them through a *Merge Graph* command in the SGM menu and a new state-graph (box) is created, which represents their merger. A state-graph (box) can be selected for applying any manipulator in the SGM Reduce menu. The manipulators will be described in detail in the next section.

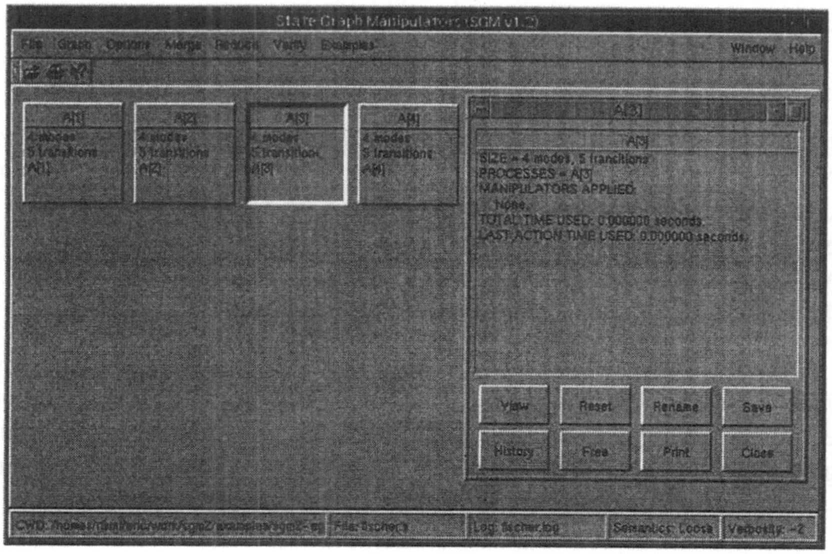

Figure 2. SGM Graphical User Interface

In the interactive mode, SGM uses a command shell to accept commands from the user one at a time. This mode is useful when windows cannot be displayed, for example, over a terminal connection. The commands can be classified as graph commands, shell commands, and miscellaneous commands. Using the commands, a user can change his/her sequence of manipulators based on initial experimental results.

In the batch mode, SGM needs an input file that contains all the three parts of VPL, including the manipulation part, so that SGM can run all the listed actions in a batch.

4. MANIPULATORS

The main factor that makes SGM user-friendly is the complete flexibility provided to a user in manipulating state-graphs through different permutations of manipulators. Any technique that acts on state-graphs can be implemented as a *manipulator*. We first define state-graphs and regions.

Definition 3: State-Graphs. Given a concurrent real-time system S of n timed automata (n>0) with process identifiers (also called indices) $\{1,2,\ldots,n\}$, $A_i = (M_i, m_i^0, C_i, D_i, \chi_i, E_i, \tau_i, \rho_i)$, a state-graph $G=(q,Q,T)$ is a graph representation of a state-space, such that $q \in V$, is a starting node of the state-graph, Q includes all the nodes in the state-graph, where each node represents a region (Definition 4) of the state-space and is semantically a collection of states satisfying some conditions in $B(\cup_{\{i=1\}}^{n} C_i, \cup_{\{i=1\}}^{n} D_i)$, and T includes all the arcs in the state-graph, where each arc represents a possible transition of the system.

Definition: Regions A *region* is a collection of states such that: (1) they have the same *mode-vector label* (μ), which gives the mode names that each process timed automata is in, (2) they satisfy the same timing constraint in the form of a *zone* (ζ), which is a *Difference-Bound Matrix* (DBM) [1], and (3) they have the same *set of literals* (L), representing the discrete variables' valuation. Thus, a region can be uniquely represented by a triple (μ, ζ, L).

A manipulator checks the characteristics of each region in a state-graph, performs some comparisons between region characteristics or regions themselves, merges identical regions/transitions, removes redundant regions/transitions, etc. Three types of manipulators are supported: merging of state-graphs, reduction techniques, and model-checking algorithms.

4.1 State-Graph Merging

This manipulator called **merge_graph()** is not a reduction technique, but a necessary construction technique for composing two state-graphs into a single state-graph that represents the concurrent behaviors of the two components. In SGM, concurrency is modeled using interleaving semantics, that is, the execution of two or more concurrent transitions of a system is modeled as a set of interleaved computations. All possible interleavings of the concurrent transitions must be considered for a complete analysis. Although, reductions on interleavings are possible such as using the partial-order reduction technique, yet **merge_graph()** is implemented as a pure composition operator so that reduction effects can later be compared.

4.2 Timed Symmetry Reduction

This is an extension of the *non-timed* symmetry-based reduction technique proposed recently by Emerson and Sistla [10], such that *clocks* and their *readings* are also considered for symmetry reductions. We have independently implemented our *timed-symmetry reduction technique* using a normalization scheme in the form of a manipulator called **normalize_region()**. This manipulator is an indispensable one when the system is a set of *concurrent symmetric processes*.

Our implementation of the timed-symmetry reduction technique adopts a *sorting procedure* on the process indices by considering all the data-structures of a region: mode-vectors, zones, and literal sets. After sorting, each region can be represented by a normalized form, that is mode-vectors, zones, and literal sets are all normalized. If an existing region is itself in the normal form, then it is used for representing all regions with the same normal form. If no such normal form region exists, then a new normal region is created. In this way, the manipulator reduces the number of regions. Transitions will be duplicated when regions with the same normal form are merged into a single region, thus further reduction is possible by considering the redundancy in transitions (see subsection 4.7). Finally, after all regions and transitions are reduced, we have a unique representation for each region.

4.3 Clock Shielding

This manipulator was individually devised and implemented in SGM by the authors, but there is already published literature on *clock activity* [8], which is semantically the same as that implemented in our manipulator. Though their semantics is the same as ours, yet we adopt a different strategy in detecting *inactive clocks* (clocks that do not affect future evolution of a system are called inactive in [8]). We call them *shielded clocks* instead, because the clocks themselves are not inactive, they progress like all other clocks, but their readings are not useful currently so we instead *shield* them from external observation. The method presented in [8] was a fix-point iteration strategy where the set of inactive clocks is obtained after a few iterations. Our method is by tracing all the paths in a state-graph starting from the region under consideration. Briefly, the manipulator works as follows: if a clock is not in the TCTL specification and if it is never read before it is reset, or will never ever be read again, then the clock can be shielded. As demonstrated by Daws and Yovine, this is a useful reduction technique. Our implementation and experimental results using SGM further show how this technique works in collaboration with other reduction techniques implemented as manipulators.

4.4 Variable Shielding

Applying the same concept of clock shielding to the discrete variables, we have constructed a new manipulator called **shield_variable()**. A discrete variable is said to be *shieldable in a mode* if it does not occur in a given TCTL specification and along all possible out-going paths starting from that mode either its value will never be read again or it will be assigned a new value before being read. This indicates the value of a shieldable discrete variable in a mode has no effect on the global behavior of a system.

By shielding some variables in the modes of a state-graph, some modes become identical in all respects and can thus be identified into a single mode.

Resulting multiple transitions can be reduced using *Sibling Transition Multiplicity Reduction* technique as described later in this section.

4.5 Read-Write Analysis

This is a new reduction technique that reduces the state-graph through an analysis of the discrete variable values as they are changed during transitions. Recall that each region is associated with a *literal set* (Definition 4). A literal in the set is basically of the form ($v \neq k$), where v is a global discrete variable and k is any integer value. By analyzing all the possible values that a process automaton writes to, or does not write to a global discrete variable, this manipulator, called **read_write()**, computes the literal set of each region. The literal set represents the values that each global discrete variable does not possibly have in a region. Through such an analysis and literal sets formulation, transitions that have triggering conditions conflicting with the literal sets will never be triggerable and can thus be eliminated from further consideration. Lemma 1 explains the reduction technique implemented in **read_write()**.

Suppose we have an intermediate state-graph composed from state-graphs for processes with indices in set H. For a given global variable y, we let $D_{\{H:y\}}$ be the set of values written to y by processes with identifiers in H but NOT by processes without identifiers in H.

LEMMA 1 Suppose we are given a variable y and a finite run segment $(\nu_h,t_h)(\nu_{\{h+1\}},t_{\{h+1\}})\ldots(\nu_k,t_k)$ such that for all i, $h \leq i < k$, ν_i goes to $\nu_{\{i+1\}}$ without making assignment to y on a transition from a process with identifier in H.

- If we enter state ν_h with an assignment $y:=a;$, then for all $h \leq i < k$, $t_I \leq t \leq t_{\{i+1\}}$, $\nu_i+t \models \wedge_{b\in(D\{H:y\}-\{a\})} y\neq b$.
- If we enter state ν_h without an assignment to y but with a triggering condition $y=a$, then for all $h\leq i<k$, $t_i\leq t\leq t_{\{i+1\}}$, $\nu_i+t \models \wedge_{b\in(D\{H:y\}-\{a\})} y\neq b$.
- If we enter state ν_h without an assignment to y but with a triggering condition $y\neq a$ with $a\in D_{\{H:y\}}$, then for all $h \leq i < k$, $t_i \leq t \leq t_{\{i+1\}}$, $\nu_i+t \models y\neq a$.

Proof: Since the values in $D_{\{H:y\}}$ are not going to be written to y by other processes with identifiers not in H, and we have the full knowledge that processes with identifiers in H will neither write values in $D_{\{H:y\}}$ to y along the segment, thus the lemma holds.

4.6 Bypass Internal Transition

During compositional verification, an intermediate state-graph represents the concurrent behavior of k processes, $1 \leq k \leq n$. These k processes are called *internal*, while the rest are called *external*. Some behavior of internal processes in a state-graph such as the occurrence of a transition might not be observable by external processes. Such behaviors do not affect the global evolution of a system. *Bypass Internal Transition* (BIT) is a reduction manipulator that detects the occurrence of a transition in an intermediate state-graph, that is not observable by external processes. BIT detects internal behavior by checking (1) if the invariance condition of a node v implies that of a successor node v', where (v, v') is a transition under consideration, (2) if $\tau(v, v')$ is implied by the invariance condition of v, (3) if (v, v') only accesses variables not accessed by external processes, (4) if (v, v') does not reset any clocks, and (5) v' is not related to a given TCTL specification. Once such a transition is detected, BIT bypasses the transition by eliminating (v, v') and by adding new transitions between v and the successor nodes of v'.

BIT can achieve reduction of state-graph sizes because if a transition (v, v') is bypassed, it may happen that v' becomes unreachable from the initial node of a state-graph. The BIT manipulator will be illustrated in the Ring Network example of Section 5.

4.7 Sibling Transition Multiplicity Reduction

Reduction techniques such as symmetry reduction, clock shielding, etc. often results in more than one transition between a pair of nodes in a state-graph. Out of the identical transitions between a pair of nodes, only one need be left in the state-graph. But, there is still some redundancy left in the transitions. *Sibling Transition Multiplicity Reduction* (STMR) is a reduction manipulator that detects such redundancy by checking if two transitions are corresponding transitions of two similar processes and their common source node is symmetric with respect to the two process indices. On detecting such transitions only one is left in the state-graph by STMR. This is because when a computation of a concurrent system reaches the source node of the corresponding transitions, the system cannot distinguish between the two processes.

Applying STMR after each of the other manipulators helps reduce state-graph sizes to a large extent because the reductions are accumulated through the sequence of manipulator applications.

5. APPLICATION EXAMPLES

Several academic as well as industrial examples were verified using SGM. The three application examples presented here include: (1) Fischer's timed mutual exclusion protocol (FMEP) [13], (2) Rules of graphical user interface for a simple calculator, and (3) ring network token passing.

5.1 Fischer's Timed Mutual Exclusion Protocol

This example was used for illustration throughout the article. The timed automaton for the *i*th process was shown in Fig. 1 and system description using VPL were described in subsection 3.1. As shown in Table 1, after experimenting with five different sequences of the manipulators in SGM we notice that for Fischer's mutual exclusion protocol [13], although both the sequences (D) and (E) have the same final effect, that is, they reduce the intermediate state-graphs to the same size, yet the *decrease rate* is not the same. The first sequence decreases the state-graph sizes more quickly than the second sequence. Further, comparing the time taken by the two sequences for state-graph reductions, we see it is also the first sequence that uses a shorter time. Thus, we conclude the first sequence is a better manipulation of the state-graphs.

Table 1. Fischer's Mutual Exclusion Protocol (11 processes)

	#Modes #Transitions Construction Time (sec)									
n	**2**	**3**	**4**	**5**	**6**	**7**	**8**	**9**	**10**	**11**
A	70	1239	28593	O/M	O/M	O/M	O/M	O/M	O/M	O/M
	160	4013	120850							
	0.06	3	-							
B	31	182	1024	5804	O/M	O/M	O/M	O/M	O/M	O/M
	70	578	4149	28451						
	0.07	1	13	-						
C	36	224	1270	O/M	O/M	O/M	O/M	O/M	O/M	O/M
	80	743	6357							
	0.08	3.3	186							
D	16	39	76	130	204	301	424	576	760	219
	35	109	247	472	810	1290	1944	2822	3937	1063
	0.07	0.6	3	10.7	35	108	276	-	-	-
E	16	39	76	130	204	301	424	576	760	219
	35	109	247	472	810	1290	1944	3400	4786	1174
	0.08	0.9	4.7	18	61	212	487	-	-	-

A: {mg, rw}, **B**: {mg, rw, sc}, **C**: {mg, rw, nr}, **D**: {mg, rw, sc, nr}, **E**: {mg, rw, nr, sc}, O/M: Out of Memory, mg = merge_graph(), rw = read_write(), sc = shield_clock(), nr = normalize_region()

5.2 Rules of graphical user interface for a calculator

This is a real project example from the Institute of Information Science, Academia Sinica, Taiwan. The project goal was to develop a generator of graphical user interfaces (GUI). The example considered here is a GUI for a simple calculator. The generator created a set of condition/action rules governing the behavior of a calculator GUI. Due to the large number of rules, it was difficult to verify if a resulting GUI behaved in the same way as a real calculator. It was also difficult to comprehend how large the state space would be and how the state space could be reduced. Thus, SGM came handy in such a situation. We collaborated with the project members to verify the GUI rules created by their generator.

The set of rules was transformed into a corresponding set of timed automata and input to SGM. Each rule was modeled by a single timed automaton with one mode and one or more looping transitions. The rule condition was mapped to a triggering condition of the transitions. The action part was mapped to a set of transition assignment statements. The set of automata obtained from the rules is shown in Fig. 3.

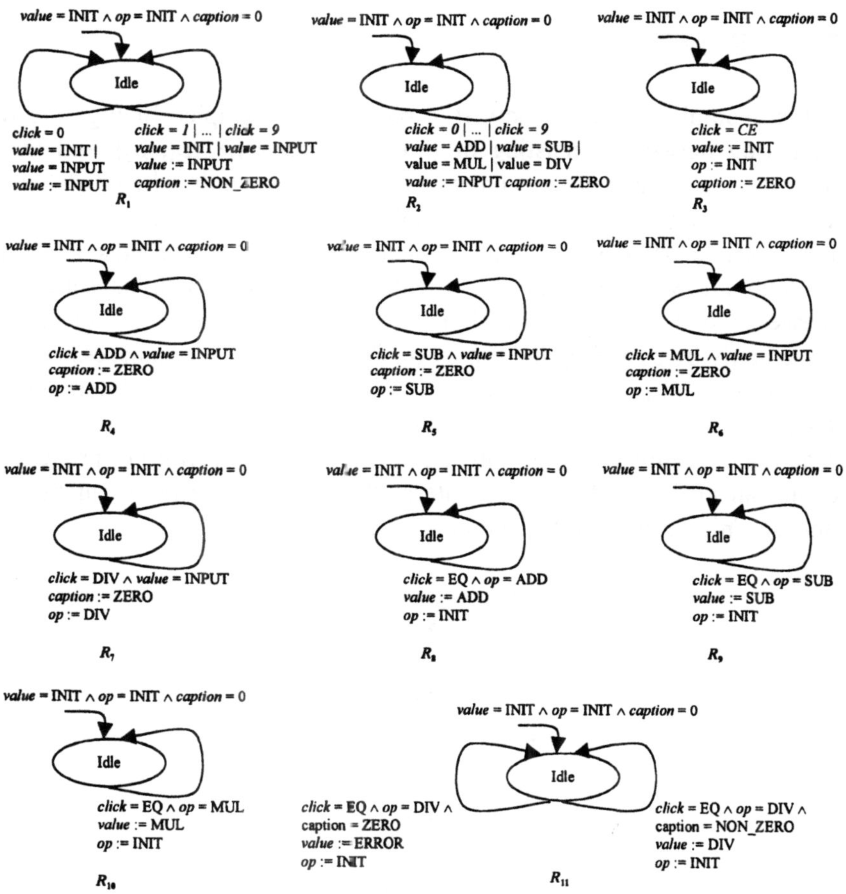

Figure 3. Calculator GUI Example

As shown in Table 2, on applying the manipulators **read_write()** and **shield_variables()** after each **merge()**, we found a significant reduction in state-graph sizes. The reduction was as much as 73.6% for transitions and 24.7% for modes.

Table 2. Rules of Calculator GUI

	#Modes #Transitions										Time (sec)
n	**2**	**3**	**4**	**5**	**6**	**7**	**8**	**9**	**10**	**11**	
A	11	11	22	33	44	55	69	83	97	90	12.9
	660	671	1364	2079	2816	3575	4554	5561	6596	1229	
B	11	11	21	31	41	51	61	71	81	90	27.5
	650	561	1091	1641	2211	2801	2711	2421	1931	1229	
C	10	10	10	19	28	37	55	64	73	83	32.6
	590	510	519	1005	1509	2031	2444	2182	1740	1133	

A:{mg},B:{mg,rw},C:{mg,rw,sv}, mg=merge_graph(),rw=read_write(),sv=shield_variables()

5.3 Ring Network

This example illustrates how BIT reduces state-graphs. In Fig. 11, we have a ring network consisting of three processes $\{p_0, p_1, p_2\}$ and three tokens $\{t_0, t_1, t_2\}$. Each process enters a critical section only when its token is true. Initially, only process p_0 has its token t_0 set as true, thus p_0 enters the critical section. After p_0 leaves the critical section, it sets t_0 to false and t_1 to true, thus allowing process p_1 to enter the critical section. Likewise, processes p_1 and p_2 behave in a similar manner. We observe that token t_0 is accessed only by p_0 and p_2 and not by p_1. Similarly, token t_1 is accessed only by p_1 and p_0, and t_2 by only p_2 and p_1. When we construct an intermediate state-space representation for processes p_0 and p_2, token t_0 becomes internal. The action of reading t_1 along transition (*Idle*, *Critical-Section*) of process p_0 becomes an internal action in the intermediate state-space representation of p_0 and p_2. This action is not observable by external process p_1.

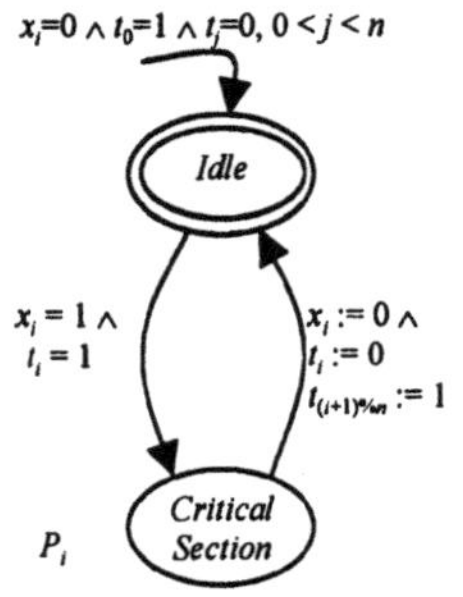

Figure 4. Ring Network

From Table 6, we can observe that the BIT manipulator in sequence (B) reduces the intermediate state-graph sizes and achieves a greater scalability such that 7 processes in a ring network can be verified as compared to that of only 6 processes without applying BIT. Sequence (C) can achieve a greater reduction but at the cost of time.

Table 3. Ring Network

	#Modes / #Transitions						Total time
n	2	3	4	5	6	7	(seconds)
A	32/64	221/663	1364/5450	7842/38460	O/M	-	N/A
B	14/28	63/202	302/1440	1508/9600	7567/61220	18/18	125
C	12/26	51/1840	244/1355	1254/9433	6574/63974	18/18	1270

A:{mg},B:{mg,rw,bit},C:{mg,rw,bit,sv}, mg=merge_graph(), rw=read_write(), bit=bypass_internal_transition(), sv=shield_variables()

6. CONCLUSION

We have successfully developed a user-friendly verification environment as a state-graph manipulation tool called *State-Graph Manipulators* (SGM) for the specification and verification of real-time systems which are modeled as timed automata and model-checked against TCTL specifications. We have also proposed four new state-graph reduction techniques: Variable-Hiding, Read-Write, BIT, and STMR. SGM allows system designers to experiment with different sequences of manipulators that best fit a particular verification task at hand. At the same time, SGM allows verification researchers to experiment with how a new reduction technique developed by him/her would collaborate with other existing techniques. We expect that SGM would be a useful tool to both the verification expert as well as the verification layman (one who just wants to see how much his/her verification task could be best tuned for efficiency and scalability).

Other existing reduction techniques will be gradually implemented into SGM. Further, new reduction techniques and their interaction with existing techniques will be investigated using SGM

Acknowledgments

We acknowledge the help provided by Mr. Ruey-Cheng Chen and Mr. Chao-Chi Chang of National Taiwan University for the implementation of the SGM Graphical User Interface. We also acknowledge Prof. Yue-Sun Kuo for providing us the Calculator GUI rules example.

References

[1] R. Alur, C. Courcoubetis, D. Dill, N. Halbwachs, and H. Wong-Toi, "An implementation of three algorithms for timing verification based on automata emptiness," In *Proc. IEEE Intl Conf Real-Time Systems Symposium*, 1992.

[2] R. Alur, C. Courcoubetis, N. Halbwachs, and D. Dill, "Modeling checking for real-time systems," In *Proc IEEE Logics in Computer Science*, 1990.

[3] R. Alur, C. Courcoubetis, N. Halbwachs, D. Dill, and H. Wong-Toi, "Minimization of timed transition systems," In *Proc Intl Conf CONCUR'92*, LNCS, volume 630, pages 340-354, August 1992.

[4] R. Alur and D. Dill, "Automata for modeling real-time systems," *Theoretical Computer Science*, 126(2):183-236, April 1994.

[5] J. Bengtsson, F. Larsen, K.Larsson, P. Petterson, Y. Wang, and C. Weise, "New generation of UPPAAL," In *Procs of the Intl Workshop on Software Tools for Technology Transfer* (STTT'98), July 1998.

[6] A. Cimatti, F. Clarke, E Giunchiglia, and M. Roveri, "NuSmv: a reimplementation of smv," In *Procs of the Intl Workshop on Software Tools for Technology Transfer* (STTT'98), July 1998.

[7] C. Daws, A. Olivers, S. Tripakis, and S. Yovine, "The tools KRONOS," In *Hybrid System III*," Lecture Notes in Computer Science, volume 1066, pages 208-219, 1996.

[8] C. Daws and S. Yovine, "Reducing the number of clock variables of timed automata," In *Proc Real-Time Systems Symposium*, pages 73-81, December 1996.

[9] A. J. Dill, D. L.Drexler, A. J. Hu, and C. H. Yang, "Protocol verification as a hardware design aid," In *Procs of the IEEE Intl Conf on Computer Design: VLSI in Computers and Processors*, 1992.

[10] E.A. Emerson and A.P. Sistla, "Utilizing symmetry when model-checking under fairness assumptions: An automata-theoretic approach," *ACM Trans on Programming Languages and Systems*, 19(4):617-638, July 1997.

[11] T.A. Henzinger, X. Nicollin, J. Sifakis, and S. Yovine, "Symbolic model checking for real-time systems," In *Proc IEEE Logics in Computer Science*, 1992.

[12] G. J. Holzmann, *Design and Validation of Computer Protocols*, Prentice Hall, 1991.

[13] L. Lamport, "A fast mutual exclusion algorithm," *ACM Trans. on Computer Systems*, 5(1):1-11, February 1987.

[14] K. L. McMillan, *Symbolic Model Checking*, Kluwer Academic Publisher, 1993.

[15] D.A. Peled, "All from one, one for all: On model checking using representatives," In *Proc of the 5th Intl Conf on Computer-Aided Verification*, Lecture Notes in Computer Science, volume 697, pages 409-423, 1993.

[16] S. Tripakis and S. Yovine, "Analysis of timed systems based on time-abstracting bisimulations," In *CAV'96*, Lecture Notes in Computer Science, volume 1102, 1996.

[17] F. Wang and P.-A. Hsiung, "Automatic verification on the large," In *Proc 3rd IEEE High-Assurance Systems Engineering Symposium* (HASE'98), November 1998.

Specification and Verification of Synchronous Hardware using LOTOS

Ji He and Kenneth J. Turner
Computing Science and Mathematics, University of Stirling, Stirling FK9 4LA, Scotland

Keywords: Digital Logic, Hardware Description, LOTOS, Verification

Abstract: This paper investigates specification and verification of synchronous circuits using DILL (Digital Logic in LOTOS). After an overview of the DILL approach, the paper focuses on the characteristics of synchronous circuits. A more constrained model is presented for specifying digital components and verifying them. Two standard benchmark circuits are specified using this new model, and analysed by the CADP toolset (Cæsar/Aldébaran Development Package).

1. INTRODUCTION

1.1 Background

DILL (Digital Logic in LOTOS [14,16,17,25]) is an approach for specifying digital circuits using LOTOS (Language Of Temporal Ordering Specification [12]). DILL offers higher-level abstractions for describing hardware using a macro library for typical components and designs. DILL is used to formally specify digital hardware, using LOTOS at various abstraction levels. DILL addresses functional and timing aspects, supported by a library of common components and circuit designs, and using standard LOTOS tools.

The new work reported here allows synchronous circuits to be specified and verified. Two hardware verification benchmarks are used as examples. The paper extends the applicability of LOTOS in hardware design, and so is of interest to the LOTOS community. Of necessity some background in

LOTOS and hardware is required. The paper demonstrates the possibility of hardware verification using LOTOS, although some limitations will be discussed.

LOTOS supports rigorous specification and analysis, unlike semi-formal HDLs (Hardware Description Languages) such as VHDL (VHSIC Hardware Description Language [10]). LOTOS is neutral with respect to whether a specification is to be realised in hardware or software, allowing hardware-software co-design [22]. LOTOS inherits a well-developed verification theory from the field of process algebra, and has a theory for testing and test derivation. There is good support from general-purpose LOTOS toolsets such as CADP (Cæsar/Aldébaran Development Package [7]), LITE (LotoSphere Integrated Tool Environment) and LOLA/TOPO (LOTOS Laboratory). Most of these tools have been used in analysing DILL specifications. DILL is actually realised through translation into LOTOS.

This paper elaborates a DILL approach for modelling and verifying synchronous circuits. Synchronous design is chosen here as it is the main approach for digital technology. Control by clock signals makes it easier to abstract away from timing information. The current standard for LOTOS does not support quantified timing, although the authors have developed Timed DILL [16] for hardware timing analysis, using ET-LOTOS [19] as a basis.

Section 2 discusses how DILL models hardware, particularly synchronous circuits. Two case studies then demonstrate that DILL can successfully specify and verify standard benchmark designs. The Single Pulser in section 3 ensures that a switch causes well-defined pulses. The Bus Arbiter in section 4 grants bus access to only one client at a time among several.

1.2 Hardware Description and Verification

Hardware description has been studied extensively. Languages such as VHDL, Verilog [11] and ELLA [2] are commonly used in industry. These languages are semi-formal because their semantics is based on simulation models. Other HDLs do have formal semantics, e.g. CIRCAL (Circuit Calculus [21]), HOL [9] and Ruby [18]. DILL most closely resembles CIRCAL in that both have a behavioural basis in process algebra. At a low level of specification, the true concurrency semantics of CIRCAL are perhaps more appropriate than the interleaving semantics of LOTOS. However, the integrated data typing in LOTOS makes it much more expressive than CIRCAL. In the authors' experience, DILL can be used successfully at a variety of abstraction levels. However, CIRCAL appears to be less effective at higher levels. For example, describing the behaviour of a synchronous circuit in CIRCAL requires the corresponding Mealy or Moore machine to be defined and then translated into CIRCAL.

Much of the early work on hardware verification used theorem-proving. Although quite general, this requires a significant amount of human guidance during verification. More recently model-checking, language containment and reachability analysis have attracted attention. Approaches using FSM (Finite State Machine) models can be automated, but they do not yet scale up to realistic hardware designs. The trend is to combine theorem-proving and model-based approaches so as to achieve generality as well as automated support. LOTOS verification approaches tend to be state-based using an LTS (Labelled Transition System). Current LOTOS tools offer model checking and reachability analysis, together with equivalence or preorder checking. DILL can thus exploit a range of verification techniques.

Various researchers have studied the use of LOTOS for hardware description. The initial work at Stirling [25] overlapped independent work in Ottawa [6]. The European project FORMAT [5] studied the translation of LOTOS to VHDL. Other hardware applications of LOTOS have included bus protocols [3,23] and hardware synthesis [27].

The new DILL model for synchronous circuits has been evaluated on two standard benchmark circuits [24] that are intended for comparing different approaches to hardware verification. The machine used by the authors for verification was a SUN (300 MHz CPU, 128 MB memory).

1.3 Verification with CADP

The authors used CADP to verify DILL hardware. CADP accepts full standard LOTOS, using Cæsar.ADT for the data part of LOTOS and Cæsar for the behavioural part. The result is an LTS that can be used for verification. Aldébaran performs verification using the LTS or a network of LTSs (i.e. a finite state machine connecting several LTSs by LOTOS parallel and hiding operators). XTL (Executable Temporal Language) is a functional-like programming language that allows compact implementation of temporal logic operators. Several temporal logics such as ACTL (Action-based Computational Temporal Logic [4]) have been embedded in XTL. To partially solve the problem of state space explosion, CADP uses advanced verification techniques such as compositional generation, on-the-fly comparison, and a BDD (Binary Decision Diagram) representation of LTSs. These techniques permit verification of relatively large specifications.

CADP supports verification through bisimulation and temporal logic property checking. For verifying DILL (LOTOS) specifications, ACTL is an obvious candidate because the semantics of LOTOS is also based on actions. ACTL is also more understandable than the μ-calculus. The modal operators of HML (Hennessy-Milner Logic) are also employed in verification for convenience. The subset of temporal operators used later in the paper is as follows. *A*, *B* and *C* are action sets, while *F* and *G* are formula sets.

- **ACTL_NOT_TO_UNLESS** (A, B, C): this can be read as 'not *A* to *B* unless *C*'. After an action satisfying *A* in the current state, all paths leading to an action satisfying *B* must also satisfy *C*.
- **AG** (F): all reachable states must satisfy *F*.
- **AU_A_B** (F, A, B, G): this is the *until* operator $\mathcal{U}$. A restricted form is used in this paper: **AU_A_B (true,** *A*, *B*, **true**). This means that for the current state, each of its paths should have the following property: the actions along the path satisfy *A until* there is an action that satisfies *B*.
- **BOX** (A, F): for the current state, all outgoing actions (if any) that satisfy *A* must result in states satisfying *F*.
- **EVAL_A** (A): yields a state set corresponding to action *A*.
- **EX_A** (A, F): from the current state, there exists an *A* that can lead to a state satisfying *F*.
- **WDIA** (A, F): from the current state there exists a path with possible preceding internal actions and *A*, leading to a state satisfying *F*.

2. MODELLING APPROACH

2.1 General Approach

The basic philosophy of DILL is that it should be easy for the hardware engineer to translate a circuit schematic into a LOTOS specification, and then to analyse and verify the properties of this specification. There is thus a need for a component library. The library is available online for research purposes [15] and is summarised in Table 1.

It is possible to describe logic designs at different levels of abstraction, and to compare a higher-level design with a more detailed one. DILL does not give refinement guidelines, since these will be motivated by normal hardware design procedures. Components in the original DILL library were specified by progressively combining simpler components. This approach is termed structural since it reflects how a component is constructed. As the philosophy of DILL is to enable circuits, including library components, to be specified at different abstraction levels, higher level specification is also needed. This is termed the behavioural style. It specifies only what the component should do, not how it is constructed. Adding a new component to the DILL library does, of course, need reasonable knowledge of LOTOS. However the existing library provides simple patterns to follow as examples.

Having abstract (behavioural) as well as design (structural) specifications in the library is helpful in both bottom-up and top-down design.

Table 1. DILL Library

Component	Variants
Adder	2/4 inputs, behavioural/structural, half/full/parallel/ripple
And, ...	2/3/4//8 inputs, 0/1-active tri-state enable
Clock	-
Comparator	1/4/8/*n* inputs, behavioural/structural
Counter	behavioural/structural
Decoder	2/3 inputs, behavioural/structural, 0/1-active outputs, BCD/Decimal/Excess-3/Gray
Demultiplexer	1/2 inputs, behavioural/structural
Delay	dynamic/general/hold/inertial/pure/setup/width/edge
Divider	2/4/8 inputs, behavioural/structural, positive/negative edge trigger
Encoder	4/8 inputs, behavioural/structural, 0/1-active outputs
Flip Flop	D/JK/MS/RS/T, behavioural/structural, positive/negative edge trigger, preset, preclear, lockout
Inverter	1/4/8 inputs, 0/1-active tri-state enable
Latch	D/RS, 1/4/8 bits, behavioural/structural, preset, preclear, clocked
Memory	behavioural/structural
Multiplexer	2/4 inputs, 1/8/*n*-bit, behavioural/structural
One, ...	source of logic 1/0, sink
Parity	8 inputs, behavioural/structural
Register	4/8/*n* bits, behavioural/structural, positive/negative edge trigger, load enable/preclear, tri-state output, bucket brigade/pass-on/shift
Repeater	/4/8 inputs, 0/1-active tri-state enable

Since component specifications are translated into LOTOS, the designer must be familiar with how to combine LOTOS behaviour expressions. Fortunately the relationship between a circuit design and its DILL representation is straightforward, and does not require detailed LOTOS knowledge. The principal method of connecting components is to compose their behaviours in parallel. The synchronisation rules of LOTOS allow components to be connected in a natural way.

LOTOS, like most specification languages, deals only with discrete events. It is therefore signal changes that are modelled in asynchronous (unclocked) design. However in synchronous circuits, changes in signal level are controlled by clock pulses (except for components such as level-triggered flip-flops). Signal levels can thus be treated as maintained during a clock cycle, and so correspond to one LOTOS event per clock cycle in the synchronous case.

Wires or tracks between components are not normally represented explicitly in DILL. A component's ports (e.g. its pins) are represented by LOTOS gates. (The term 'gate' will be qualified as it has different meanings in hardware and LOTOS.) and To 'wire up' two ports, their LOTOS gates are merely synchronised. Since LOTOS allows multi-way synchronisation, it is easy to connect one output to several inputs. In high-speed circuits, the transmission time over a wire may be modelled as a delay. Multi-bit signals or multi-wire connections (e.g. buses) and multi-component assemblies (e.g. memory arrays) are supported by DILL.

2.2 Synchronous Circuit Model

A piece of combinational logic merely combines its inputs to produce outputs; it is referred to as a stage in the following. Sequential logic incorporates feedback, so the state of an output depends on previous inputs. Synchronous circuits, as one form of sequential design, are distinguished from asynchronous circuits through control by a global clock.

The classical synchronous circuit model is shown in figure 1. In this model, the combinational logic provides the primary outputs and internal outputs according to the primary inputs and internal inputs. Internal outputs are then fed into state hold components to produce the internal inputs. Changes of the internal inputs are synchronised with the clock, in other words they are changed only at a particular moment of the clock cycle (usually its transition). The internal inputs determine the state of the whole circuit.

For a synchronous circuit, the designer must ensure that the clock cycle is slower than the slowest stage in a circuit. This can be done by analysing the timing characteristics of components used in the circuit. The untimed version of DILL cannot of course confirm if the clock constraint is met. As discussed in [16], Timed DILL *can* specify such constraints. However, sections 2.4 and 3 will show that properly modelling the storage components and environment ensures a DILL specification always meets the clock condition.

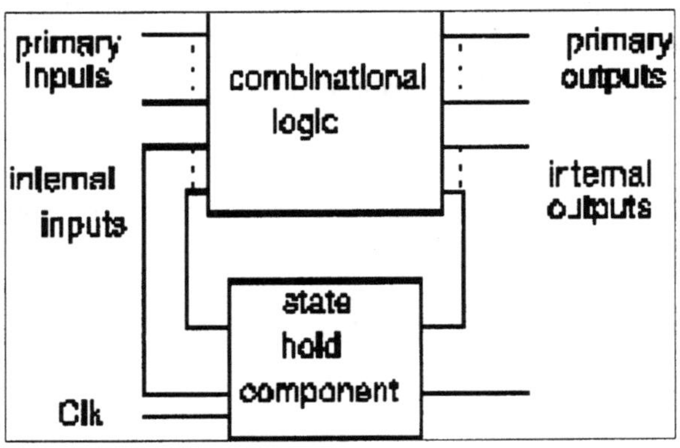

Figure 1. Synchronous Circuit Model

In synchronous design, the primary inputs are usually synchronised with the clock signal. This eases design and analysis of synchronous circuits. DILL incorporates this practice into its synchronous circuit model, assuming that the primary inputs have already been synchronised with the clock signal.

Besides the above, the DILL synchronous model has two more restrictions. It is important that there is no cyclic connection within a stage, and storage components have to be specified in the behavioural style. These restrictions are related to the way components are modelled, for otherwise a DILL specification might deadlock where a real circuit could still work. This is discussed further in section 2.4 .

2.3 Synchronous Model for Basic Logic Gates

The fundamental DILL model for basic logic gates allows an input or output port to offer an event corresponding to a signal change at any time. This model is a very generic representation of logic gates used in real world, but this may lead to non-determinism due to the lack of quantified timing [17]. The gate model therefore has to be constrained according to the environment in which the gates operate. Logic gates are presumed to be part of a synchronous design. If the clock is slow enough to let every signal settle down, it is reasonable to allow the value of each signal to change just once per clock cycle. The transient values are ignored because they do not affect circuit behaviour. The synchronous model allows basic logic gates (and thus all other components within combinational logic) to wait until all inputs occur before outputting the corresponding value.

The following example models a two-input *nand* gate. Note that inputs are interleaved, i.e. they can occur in any order. It might appear that the order of input events could be fixed since it does not influence the functionality of a component. This would result in a smaller state space when circuits are verified. Unfortunately this might cause deadlock when components are connected. Suppose that components *A* and *B* each have two inputs. Imagine that inputs are required in the order *IpA1* before *IpA2*, and *IpB1* before *IpB2*. This would lead to deadlock if the components shared inputs, with *IpA1* connected to *IpB2* and *IpA2* connected to *IpB1*. For this reason, DILL insists on fully interleaved inputs.

```
process Nand2 [Ip1, Ip2, Op] : noexit :=
  (Ip1 ?dtIp1 : Bit; exit (dtIp1, any Bit)                 (* allow one input *)
  |||
   Ip2 ?dtIp2 : Bit; exit (any Bit, dtIp2))                (* allow other input *)
 >> accept dtIp1, dtIp2 : Bit in                           (* accept both inputs *)
  (Op !(dtIP1 nand dtIp2);                                 (* output nand of inputs *)
   Nand2 [Ip1, Ip2, Op])                                   (* repeat behaviour *)
endproc (* Nand2 *)
```

2.4 Synchronous Model for State Hold Components

The gate model just discussed is not suitable for circuits with cyclic connections since these result in input-output interdependency and thus in specification deadlock. Cyclic connections are common in latches and flip-flops, so state hold components are modelled in the behavioural style. At a higher level of specification and design, problems due to cyclic connections do not arise. For synchronous circuits, two modifications are made to the fundamental DILL model. LOTOS events are considered to model signal levels rather than changes, and a constraint is added to reflect the assumption of a slow enough clock. A DFF (Delay Flip-Flop) is a simple memory element with data input *D*, clock input *Clk* and output *Q*. Its specification is as follows:

```
process DFF [D, Clk, Q] (dtD, dtClk : Bit) : noexit :=
 D ?newdtD : Bit; DFF [D, Clk, Q] (newdtD, dtClk);        (* input new data *)
 Clk ?newdtClk : Bit;                                     (* input clock pulse *)
 ([(dtClk eq 1) and (newdtClk eq 0)] →                    (* ignore negative pulse *)
   DFF [D, Clk, Q] (dtD, newdtClk)                        (* continue behaviour *)
 []
  [(dtClk eq 0) and (newdtClk eq1)] →                     (* react to positive pulse *)
   Q !dtD;                                                (* output stored data *)
   DFF [D, Clk, Q] (dtD, newdtClk)                        (* continue behaviour *)
  )
endproc (* DFF *)
```

Suppose a combinational logic circuit feeds into this flip-flop as the state hold component. If the clock signal is not constrained, it is possible that the clock moves to the next cycle before the combinational logic has settled down. The model of a synchronous circuit must exclude this possibility. After a positive-going transition of the clock signal, if the *D* input of the flip-flop has not occurred yet then the next positive-going transition of clock signal must not occur. This is ensured by the following constraint on the D flip-flop specification. The process *Cons_DFF* deals with the initial state of the flip-flop The next positive-going clock transition is handled by process *Cons_DFF_Aux*. The full specification of a D flip-flop combines *DFF* and *Cons_DFF* with the LOTOS parallel operator.

```
process Cons_DFF [D, Clk] (dtClk : Bit) : noexit :=
 D ?newdtD : Bit;                                         (* input new data *)
 Cons_DFF [D, Clk] (dtClk)                                (* continue behaviour *)
 []
 Clk ?newdtClk : Bit;                                     (* input clock pulse *)
 ([(newdtClk eq 1) and (dtClk eq 0)] →                    (* react to positive pulse *)
   Cons_DFF_Aux [D, Clk] (newdtClk)                       (* after one clock pulse *)
 []
 [(newdtClk eq 0) and (dtClk eq 1)] →                     (* ignore other pulses *)
   Cons_DFF [D, Clk] (newdtClk))                          (* continue behaviour *)
where
```

```
  process Cons_DFF_Aux [D, Clk] (dtClk : Bit) : noexit :=
   D ?newdtD : Bit; Clk !0; Clk !1;                (* input before negative pulse *)
   Cons_DFF_Aux [D, Clk] (1)                               (* continue behaviour *)
 []
   Clk !0; D ?newdtD : Bit; Clk !1;                 (* input after negative pulse *)
   Cons_DFF_Aux [D, Clk] (1)                               (* continue behaviour *)
  endproc (* Cons_DFF_Aux *)
endproc (* Cons_DFF *)
```

3. CASE STUDY: A SINGLE PULSER

The informal description of the Single Pulser appears in the standard benchmark document [24]. A Single Pulser is a clocked-sequential device with a one-bit input *I* and a one-bit output *O*. It deals with a debounced switch that is on (true) in the down position and off (false) in the up position. When the Single Pulser senses the switch being turned on, it must assert an output signal lasting one clock cycle. The circuit should not allow additional outputs until after the switch has been turned off. The benchmark also informally defines some properties that the Single Pulser must respect.

3.1 Specification

Figure 2 shows a design for the Single Pulser given in the benchmark. *P_In* is the input from the switch, and *P_Out* is the output from the circuit. It is very straightforward to represent the Single Pulser design in DILL. Because the clock is implicit in a synchronous circuit design, circuit properties may not actually refer to it. Experience shows that hiding the clock signal can make the temporal logic formulae clearer. The Single Pulser specification is as follows (omitting process gate names for brevity):

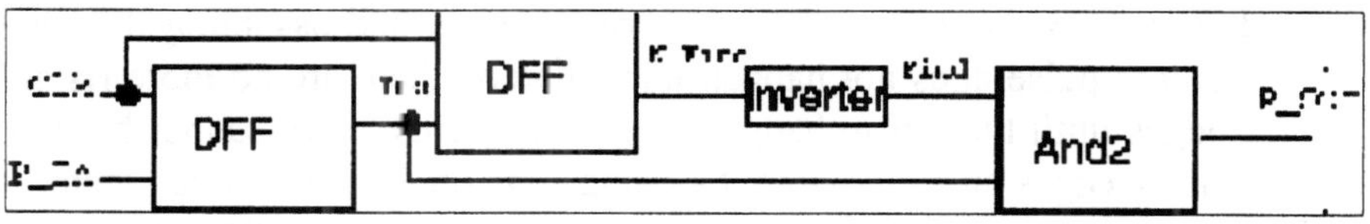

Figure 2. Single Pulser Design

```
hide Inp, N_Find, Find, Clk in                             (* hide internal gates *)
  ((Cons_DFF |[ N_Find, Inp ]| (Inverter |[ Find ]| And2))  (* flip-flop, inverter, and *)
  |[ Clk, Inp ]|                                           (* synchronised with ... *)
   Cons_DFF)                                                           (* flip-flop *)
|[ P_In, Clk, P_Out ]|                                     (* synchronised with ... *)
  Env                                                            (* the environment *)
```

The *Env* process serves as the environment constraint on the Single Pulser. It permits *P_In* to come before each positive-going clock transition, and allows the next clock cycle only after *P_Out* has occurred. Without this constraint, the properties discussed later are invalid. The constraint between *P_In* and *Clk* ensures that *P_In* is synchronised with *Clk*. The constraint between inputs and output respects the slow-clock requirement: *P_Out* must happen before the next positive-going clock transition. These assumptions are not automatically guaranteed by the circuit design, but they are required by the DILL synchronous circuit model. In outline, *Env* is specified as:

```
(P_In ? dtPIn : Bit;                                         (* pulse in *)
 Clk ! 1;                                                    (* positive-going clock *)
 (Clk ! 0; exit ||| P_Out ? dtPOUt : Bit; exit))             (* negative-going clock, pulse out *)
>>                                                           (* and then ... *)
 Env                                                         (* same environment behaviour *)
```

3.2 Verification

The formulation of properties in CADP was briefly explained in section 1.3 For brevity the properties are given only informally here; see the details in [17]. Verification of the Single Pulser was undertaken using only XTL model checking, although it is not difficult to give a higher level specification in DILL/LOTOS and then check for equivalence between the two levels. Because LOTOS events are modelled as signal levels instead of signal transitions, representing a rising edge needs two clock cycles. In the first cycle the signal should be at level 0, in the second cycle it should be at level 1. Each signal happens once and only once in a clock cycle, so the second appearance of the same signal indicates the second clock cycle.

Property 1: If *P_In* has a rising, eventually *P_Out* becomes true.

Property 2: Whenever *P_Out* is 1, it becomes 0 in the next state and remains 0 at least until the next rising edge on *P_In*.

Property 3: Whenever there is a rising edge, and assuming that the output pulse does not happen immediately, there are no more rising edges until that pulse happens. In other words, there cannot be two rising edges on *P_In* without a rising edge on *P_Out* between them.

The size of the LTS produced by Cæsar.ADT and Cæsar from the DILL specification has 295 states and 538 transitions. Aldébaran minimises the LTS to a smaller one having 97 states and 174 transitions modulo strong bisimulation. Because the resultant LTS is small, all the generation and verification steps take negligible time. Aldébaran uses the LTS to show that the DILL design is deadlock free. The XTL tool is also able to demonstrate that all the supposed properties of the circuit are valid.

4. CASE STUDY: A BUS ARBITER

In this section, the DILL approach is evaluated using another benchmark circuit. For brevity, the specifications are not given here but can be found in [17]. The purpose of the Bus Arbiter is to grant access on each clock cycle to a single client among a number of clients requesting use of a bus. The inputs to the arbiter are a set of request signals, each from a client. The outputs are a set of acknowledge signals, indicating which client is granted access during a clock cycle. The documentation also defines some properties that the Bus Arbiter must respect. Thse are given informally and also in CTL (Computational Temporal Logic). Besides listing the properties to be fulfilled, the benchmark documentation also gives an arbitration algorithm in plain English. Finally the gate level implementation of the Bus Arbiter is provided as a circuit diagram.

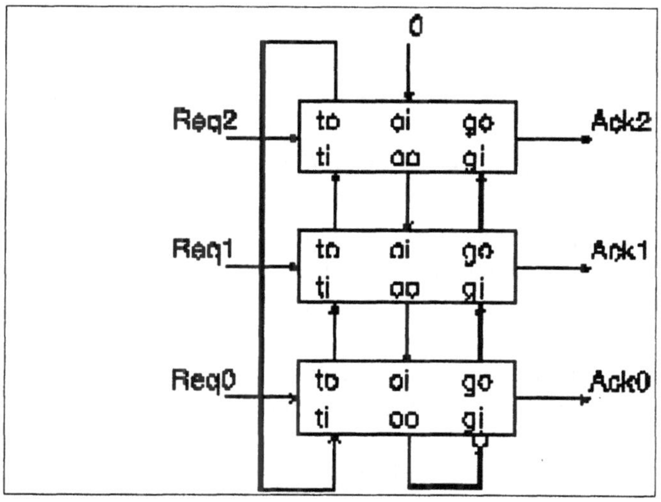

Figure 3. Bus Arbiter With Three Cells

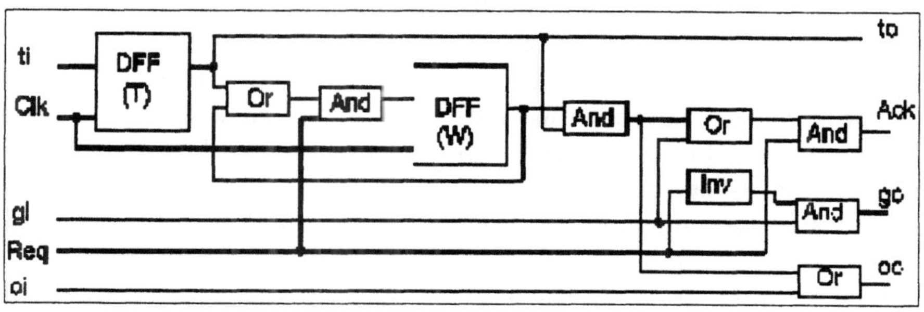

Figure 4. Design of An Arbiter Cell

4.1 Higher-Level Specification in LOTOS

LOTOS supports specification at various levels of abstraction. Although the benchmark circuits have been studied by many researchers, as far as the authors knowledge there has not been a formal specification of the arbitration algorithm used in the design. With LOTOS, it is possible to provide such a higher-level specification. There are two clear benefits of this formalisation. Firstly, better understanding of the algorithm can be gained from rigorous specification. Secondly, correctness of the algorithm itself can be ensured before the circuit is built and verified. Flaws in the algorithm will be more time-consuming to fix if they are found only after implementation.

The arbitration algorithm embodied in the design is a round-robin token scheme with priority override. Normally the arbiter grants access to the highest priority client: the one with the lowest index number among all the requesting clients. However as requests become more frequent, the arbiter is designed to fall back on a round-robin scheme, so that every requester is eventually acknowledged. This is done by circulating a token in a ring of arbiter cells, with one cell per client. The token moves once every clock cycle. If a client's request persists for the time it takes for the token to make a complete circuit, that client is granted immediate access to the bus.

Translating the algorithm to LOTOS is quite straightforward, mainly using LOTOS value expressions. For example each cell has two associated variables: *token* indicates if the token is in the cell, and *waiting* indicates if the client's request has persisted for a completed token cycle. Circulating the token, (re)setting the waiting variable and so on correspond to LOTOS value expressions. For an arbiter with three cells, the LOTOS specification has 79 lines (including comments) for the behavioural specification.

4.2 Lower-Level Specification in DILL

The design of the arbiter consists of repeated cells. Each cell is in charge of accepting request signals from a client, and sending back acknowledgements to the same client. Figure 3 shows an arbiter with three cells. Figure 4 shows the design of each cell. The first cell is slightly different because it is assumed that the token is initially in the first cell.

The principle of the circuit will not explained in detail here. Briefly, the *ti* (token in) and *to* (token out) signals are for circulation of the token. The *to* output of the last cell connects to the *ti* input of the first cell to form a ring. The *gi* (grant in) and *go* (grant out) signals are related to priority. The grant of cell i is passed to cell $i+1$, meaning no client of index $\leq i$ is requesting. Hence a cell may assert its acknowledge output if its grant input is asserted. The *oi* (override in) and *oo* (override out) signals are used to override the priority. When the token is in a persistent requesting cell, its corresponding client will get access to the bus. The *oo* signal of the cell is set to 1. This signal propagates down to the first cell and resets its grant signal through an inverter. As a consequence the *gi* signal of every cell is reset, in other words the priority has no effect during this clock cycle. Within each cell, register *T* stores 1 when the token is present; register *W* (waiting) is set to 1 when there is a persistent request. Initially the token is assumed to be in the first cell.

The components of each cell are in the DILL library, so specification of a cell is very easy. The specification of an arbiter with three cells is obtained by connecting three such processes. As for the Single Pulser, there is also an environment constraint in the structural specification of the arbiter to meet the conditions of the synchronous circuit model discussed in section 2.2.

Since the properties that the arbiter must fulfill are given in the benchmark documentation, it is obvious that the verification should consist of model checking these properties. Equivalence checking is also performed since two levels of specifications are identified.

4.3 Verification

Section 1.3 explained how to formulate properties in CADP. They are translated into action-based temporal logic (ACTL and HML). The following properties refer to client 0; the formulae for other clients have a similar form.

Property 1: No two acknowledge outputs are asserted in the same clock cycle (safety).

```
AG (                                              (* for all states ... *)
 not (                                        (* it is not the case that ... *)
  EX_A (                                          (* there exists action *)
   EVAL_A (Ack0 !1)                              (* Ack0 !1 leading to ... *)
    (WDIA (EVAL_A (Ack1 !1), true) or             (* action Ack1 !1 or *)
     WDIA (EVAL_A (Ack2 !1), true)))))              (* action Ack2 !1 *)
```

Property 2: Every persistent request is eventually acknowledged (liveness).

```
AG (                                              (* for all states ... *)
 BOX (                                  (* after all its outgoing action *)
  EVAL_A (Req0 !1),                              (* which is Req0 !1 ... *)
   AU_A_B (true, true,                                      (* until ... *)
    (EVAL_A (Ack0 !1) or                      (* eventually Ack0 !1 ... *)
     EVAL_A (Req0 !0)), true)))                      (* unless Req0 !0 *)
```

Property 3: Acknowledge is not asserted without request (safety).

```
AG (                                                  (* for all states *)
 ACTL_NOT_TO_UNLESS (      (* not Req0 !0, Ack0 !1 unless Req0!1 *)
  EVAL_A (Req0 !0),                                   (* after Req0 !0 *)
   EVAL_A (Ack0 !1),                         (* Ack0 !1 is impossible ... *)
    EVAL (Req0 !1)))                              (* unless after Req0 !1 *)
```

To verify the higher-level specification against the temporal logic formulae, the LTS of the specification was produced first. Cæsar generates an LTS with 3649 states and 7918 transitions. Aldébaran reduces this to 379 states and 828 transitions with respect to strong bisimulation. Both generation and reduction take seconds. The temporal logic formulae are then checked against the minimised LTS. Each is verified as true within 1 minute.

The real challenge comes when the lower-level DILL specification is verified. The state space is so large that direct generation of the LTS from the LOTOS specification is impractical. As mentioned before, there are several advanced techniques implemented in CADP to tackle the problem of state space explosion. Nevertheless, using on-the-fly verification of the arbiter also fails after considerable run-time. CADP does not currently support the direct generation of BDDs from a LOTOS specification.

Compositional generation was tried out to verify the arbiter. Basically the idea is that of 'divide and conquer'. A LOTOS specification is divided into several smaller specifications to make sure that it is possible for Cæsar to generate an LTS for each of them. Then Aldébaran is used to reduce these LTSs with respect to a suitable equivalence relation. The minimised LTSs are then combined using the LOTOS parallel operator (and also the hide operator if necessary) to form a network of communicating LTSs (the CADP term). At this stage, an LTS might be produced from the network, or on-the-fly verification might be performed against the network. In order to get valid verification results, special attention must be given to the equivalence relation that is used. The relation must be a congruence at least with respect to the compositional operators, here the LOTOS parallel and hide operators. The relation must also preserve the properties to be verified. This ensures that the resulting network of communicating LTSs will respect the same properties as the original LOTOS specification.

Among the benchmark properties, the first and the third concern safety while the second concerns liveness. Safety equivalence [1] preserves safety properties, while branching bisimulation equivalence [26] preserves liveness properties when there are no livelocks in specifications. Both of these equivalences are congruences with respect to the parallel and hide operators. These two equivalences are thus appropriate to compositional generation.

The arbiter design was divided into three pieces, one per cell. After about seven minutes, an LTS that is safety equivalent to the LOTOS specification of the design is generated. The two safety properties were verified to be true against this LTS, implying that the design also satisfies these safety properties. Verification of the formulae takes just seconds. However generating the LTS which is branching equivalent to the design takes almost one day, after which the liveness property is also verified to be true.

Before checking equivalence, a suitable equivalence must be chosen. For most systems, observational equivalence is an obvious choice. Informally it means that two systems have exactly same behaviour in terms of the observable actions. For hardware systems, testing equivalence (two specifications pass or fail exactly the same external tests) is also used as a criterion in some approaches such as CIRCAL. The algorithm for testing equivalence is not implemented in CADP, so the stronger notion of observational equivalence was used when checking the Bus Arbiter.

As before, compositional generation was used to generate the LTS for the design. This time each cell was reduced with respect to observational equivalence, since this is a congruence for the parallel and hide operators. Generating the LTS took about eight minutes. This LTS was expected to be observationally equivalent to that for the higher-level specification. However Aldébaran discovered that they are not! Table 2 is one of the sequences given as a counter-example. (The Aldébaran output has been rendered more readable here.) This sequence indicates that in the first three clock cycles only client 0 requests the bus; both the high-level specification and the low-level design grant access to this client. In the fourth cycle, client 0 cancels its request but client 1 begins to request access. At this point the two levels of specifications are different: the lower-level specification offers 0 for *Ack1*, whereas the higher-level specification offers 1 for *Ack1*.

Table 2. A Counter-Example generated by Aldébaran

	Cycle 1	Cycle 2	Cycle 3	Cycle 4
Req0	1	1	1	0
Req1	0	0	0	0
Req2	0	0	0	0
Ack0	1	1	1	
Ack1	0	0	0	**0** or **1**
Ack2	0	0	0	

After step-by-step simulation of the counter-example, it was soon discovered that the circuit may not properly reset the *oo* (override out) signal to *0*. Suppose a cell has been requesting access, so its *W* register is set to *1*. If the cell cancels the request in the very clock cycle that the token happens to arrive. In this situation, because the client has already cancelled its request it should be possible for another client to get the bus. However, the design sets the *oo* signal to override the priority as if this client were still requesting. This prevents any other client from accessing the bus in this clock cycle.

Fixing the problem was much easier than finding it. The correction was to connect the *Req* signal to the *And* gate that follows the *W* register. (See [17] for the revised circuit diagram.) The output of the *And* gate guarantees that the *oo* signal is always correctly set or reset according to the request signal in the current clock cycle. This modified design was verified to be observationally equivalent to the higher-level algorithmic specification.

As mentioned in section 2.2, in DILL the inputs are assumed to be synchronised with the clock signal. Suppose that the *Req* signal in figure 4 is always ready before the active clock transition, i.e. is not synchronised with the clock. In this case the problem discussed above might not happen. As the benchmark documentation does not state if inputs are synchronised with the clock or not, it is believed that the modified design is more robust.

5. CONCLUSION

With the new approach to specifying synchronous, it is possible to verify standard hardware benchmarks – here, the Single Pulser and the Bus Arbiter. In comparison with other techniques applied to the same case studies, e.g. COSPAN [8] and CIRCAL [20], DILL is much more convenient for giving a higher-level specification. This is not so surprising since LOTOS is an expressive language. CIRCAL, by way of contrast, gives an abstract view of a synchronous circuit by directly specifying its corresponding finite state machine, which is not always a natural representation of circuit behaviour.

Being based on process algebra, DILL specifications can be verified by equivalence and preorder checking. This is distinctive in that most hardware verification systems are based on theorem proving or model checking. The former needs human assistance to complete a proof. The latter needs specialised expertise since temporal logic specifications are not easy to write. In contrast, equivalence or preorder checking makes it possible to write the specification in the same formalism as the implementation, here DILL (or really, LOTOS). The correctness of a DILL specification can be easily checked by simulation tools. Another benefit of equivalence checking is seen in the Bus Arbiter case study. As a classical verification benchmark, the Bus Arbiter has been investigated using many approaches. But as far as the authors know, the defect reported in section 4.3 is a new discovery.

However, the size of the circuit that can be effectively verified is small compared to that handled by other mature hardware verification tools. COSPAN can verify an arbiter with four cells with the consumption of about 1 MB memory, due to a symbolic representation using BDDs and efficient reduction techniques [8]. CIRCAL is reported to generate the state space of an arbiter with up to 40 cells using reasonable computing resources, although the actual memory used was not reported [20]. Again this is due to the BDD representation of the CIRCAL specification. Note that CIRCAL was not in fact used to verify the arbiter formally. [20] just gives a test pattern to show that even if all clients request the bus, only one can gain access to the bus in each clock cycle. CIRCAL does not have the functionality of temporal logic model checking. Because of its limited power in specifying higher-level behaviour, equivalence checking was not used in the CIRCAL case study. CADP on the other hand consumes more than 100 MB of memory to produce the state space of a three-cell arbiter. Although the resulting state space is relatively small, the intermediate stages of generation need considerable memory.

There are two main reasons for this performance limitation. One comes from the modelling language LOTOS and the other comes from CADP. Firstly, for synchronous circuits the order in which signals occur during a clock cycle is not so important. So it is reasonable to imagine that the inputs happen together and then output occurs. But when modelling such circuits in DILL, independent (interleaved) inputs are allowed so the state space is considerably enlarged. Secondly, CADP is a tool under development and currently some of its features are mainly based on explicit state exploration. Because CADP cannot produce the minimised state space in the first place, large amounts of memory have to be consumed before a smaller LTS can be produced by minimisation. On-the-fly algorithms are of some help, but they apply only in particular situations. For example, on-the-fly observational equivalence checking is not supported by CADP. Also CADP does not offer a BDD representation of LOTOS specifications, although BDDs are used to represent intermediate data types in some algorithms. Fortunately CADP is currently being actively improved by the CADP developers.

REFERENCES

[1] A. Bouajjani, J. C. Fernandez, *et al.* Safety for branching time semantics. In *Automata, Languages and Programming*, LNCS 510, pages 76–92. Springer-Verlag, Berlin, 1991.

[2] R. Boulton, M. J. C. Gordon *et al.* The HOL verification of ELLA designs. TR 199, University of Cambridge Computer Laboratory, Aug. 1990.

[3] G. Chehaibar, H. Garavel, *et al.* Specification and verification of the PowerScale bus arbitration protocol: An industrial experiment with LOTOS. TR 2958, INRIA, Le Chesnay, Aug. 1996.

[4] R. De Nicola and F. Vaandrager. Action versus state based logics for transition systems. In *Semantics for Systems of Concurrent Processes*, LNCS 469, pages 407–419. Springer-Verlag, Berlin, 1990.

[5] C. Delgado Kloos, T. de Miguel *et al.* VHDL generation from a timed extension of the formal description technique LOTOS with the FORMAT project. *Microprocessing and Microprogramming*, 38:589–596, 1993.

[6] M. Faci and L. M. S. Logrippo. Specifying hardware in LOTOS. In *Proc. Computer Hardware Description Languages and Their Applications XI*, pages 305–312. North-Holland, Amsterdam, Apr. 1993.

[7] J.-C. Fernández, H. Garavel, *et al.* CADP (Cæsar/Aldébaran development package): A protocol validation and verification toolbox. In R. Alur and T. A. Henzinger, editors, *Proc.*

Computer-Aided Verification VIII, LNCS 1102, pages 437–440. Springer-Verlag, Berlin, Aug. 1996.

[8] K. Fisler and R. P. Kurshan. Verifying VHDL designs with COSPAN. In *Formal Hardware Verification Methods and Systems in Comparison*, LNCS 1287, pages 206–247. Springer-Verlag, Berlin, 1997.

[9] C. A. R. Hoare and M. J. C. Gordon, editors. *Mechanized Reasoning and Hardware Design*. Prentice Hall, Englewood Cliffs, 1992.

[10] IEEE. *VHSIC Hardware Design Language*. IEEE 1076. Institution of Electrical and Electronic Engineers Press, New York, 1993.

[11] IEEE. *IEEE Standard Hardware Design Language based on the Verilog Hardware Description Language*. IEEE 1364. Institution of Electrical and Electronic Engineers Press, New York, 1995.

[12] ISO/IEC. *Information Processing Systems – Open Systems Interconnection – LOTOS – A Formal Description Technique based on the Temporal Ordering of Observational Behaviour*. ISO/IEC 8807. International Organization for Standardization, Geneva, 1989.

[13] ISO/IEC. *Information Processing Systems – Open Systems Interconnection – Enhancements to LOTOS*. International Organization for Standardization, Geneva, Apr. 1998.

[14] Ji He and K. J. Turner. Extended DILL: Digital logic with LOTOS. TR CSM-142, Computing Science and Mathematics, University of Stirling, UK, Nov. 1997.

[15] Ji He and K. J. Turner. DILL (Digital Logic in LOTOS) translator. *http://www.cs.stir.ac.uk/~kjt/software/dill.html*, Jan. 1998.

[16] Ji He and K. J. Turner. Timed DILL: Digital logic with LOTOS. TR CSM-145, Computing Science and Mathematics, University of Stirling, Apr. 1998.

[17] Ji He and K. J. Turner. Modelling and verifying synchronous circuits in DILL. TR CSM-152, Computing Science and Mathematics, University of Stirling, Feb. 1999.

[18] G. Jones and M. Sheeran. Circuit design in Ruby. In J. Staunstrup, editor, *Formal Methods For VLSI* Design, pages 13–70. Elsevier Science Publishers, Amsterdam, 1990.

[19] L. Léonard and G. Leduc. An introduction to ET-LOTOS for the description of time-sensitive systems. *Computer Networks and ISDN Systems*, 28:271–292, May 1996.

[20] G. A. McCaskill and G. J. Milne. Sequential circuit analysis with a BDD based process algebra system. TR HDV-25-93, Computer Science, University of Strathclyde, Jan. 1993.

[21] G. J. Milne. *The Formal Specification and Verification of Digital Systems*. McGraw-Hill, New York, 1994.

[22] L. Sánchez Fernández, M. L. López *et al.* Co-design at work: The Ethernet bridge case study. *Current Issues in Electronic Modelling*, 8, Apr. 1996.

[23] M. Sighireanu and R. Mateescu. Validation of the link layer protocol of the IEEE-1394 serial bus ('Firewire'): An experiment with E-LOTOS. TR 3172, Institut National de Recherche en Informatique et Automatique, Le Chesnay, May 1997.

[24] J. Staunstrup and T. Kropf. IFIP WG10.5 benchmark circuits. *http://goethe.ira.uka.de/hvg/benchmarks.html*, July 1996.

[25] K. J. Turner and R. O. Sinnott. DILL: Specifying digital logic in LOTOS. In R. L. Tenney, P. D. Amer, and M. Ü. Uyar, editors, *Proc.Formal Description Techniques VI*, pages 71–86. North-Holland, Amsterdam, 1994.

[26] R. J. van Glabbeek and W. P. Weijland. Branching time and abstraction in bisimulation. TR CS R8911, Centrum voor Wiskunde en Informatica, Amsterdam, 1989.

[27] K. Yasumoto, A. Kitajima *et al.* Hardware synthesis from protocol specifications in LOTOS. In S. Budkowski, E. Najm, and A. Cavalli, editors, *Proc. Formal Description Techniques XI/Protocol Specification, Testing and Verification XVIII*. Chapman-Hall, London, 1998.

PROVABLY CORRECT HARDWARE COMPILATION USING TIMING DIAGRAMS

Michael Schenke
Universität Oldenburg, Postfach 2503, D-26111 Oldenburg
Michael.Schenke@Informatik.Uni-Oldenburg.DE

Michael Dossis
The author has worked on the OMI/HORN ESPRIT project and is now an independent consultant.

Abstract In this paper we present a framework within which hardware implementations are proven correct from specifications given in an *OCCAM-like language* called *Handel* by the use of a robust set of mathematical transformational laws. The semantic basis for Handel and its hardware implementations are simple functions of time which are called *timing diagrams*. This basis allows us to denote the abstract properties of the Handel programs and hence the implementations in a modal logic, called *Duration Calculus*. The semantic treatment by one model for all three levels including the abstract properties, Handel and the level of gates, is one of the outstanding features of our approach. The delicate mathematical model which is used is able to cope with the complex form of parallelism used in Handel and with the detailed treatment of the relation between parallelism and timing. An immediate benefit of this approach is that Handel is a language already in use by hardware designers for specification in addition to the potential for easy integration of the proposed transformations with existing hardware compilation systems.

*This work was partially funded by the EPSRC in the project "Design and Transformational Implementation of Provably Correct Real Time Systems" and by the Leibniz Programme of the German Research Council (DFG) under Ol 98/1-1. A more complete version of the paper including more examples, proofs and drawings can be found at ftp://theoretica.Informatik.Uni-Oldenburg.DE/pub/Papers/msmd97-p1.ps.gz and ...-p2.ps.gz

The introduction of so-called *quick* operators utilises the non-determinism of implementations and enables to realise implementations of sequential and parallel Handel operations, which are improved over conventional code optimisation techniques.

Keywords: Semantic foundations, Real-time aspects, Parallel systems, Hardware compilation.

1. INTRODUCTION

The current availability of specification languages with higher levels of abstraction begins to blur the difference in the engineering practices for the design of software and hardware. However, for software implementations the semantic gaps between abstract specification code and its actual implementation seem to be smaller than for hardware implementations, if the many levels of abstraction in hardware are taken into account.

The eventual hardware model generated by almost all the current automatic or semi-automatic synthesis methods is usually given in the form of gate-netlists. The process of attempting to bridge the enormous abstraction gap between the specification and its direct implementation in hardware netlists is driven in practice by design optimisations. These optimisations are based on a very diverse range of implementation constraints. The optimisations introduce a large set of additional semantics in the delivered hardware netlist, and thus they generate different semantic spaces between the specification and implementation. This does not cause any significant problems, particularly when the details of the target architecture are well-known in advance and they are fixed for a large number of applications. Experience indicates, however, that the methodology has its problems, when formal verification of the coherence between the specification and implementation, as well as of the actual transformation taking place during automatic hardware synthesis, is required. The difficulty to apply verification then, is due to the semantic inconsistencies caused by the heuristic optimisation models.

An Occam-based notation has been used by the Oxford research group for hardware specification and synthesis. This computer language is called Handel. A set of derivatives for its hardware compiler have been developed, used and currently extended by member researchers or the industrial engineers who collaborate with the hardware compilation group in Oxford.

The bulk of the research work with Handel addresses the compilation process from Occam-like programs to the produced hardware netlists, which are normally prototyped on FPGA devices. FPGAs

(Field-Programmable Gate Arrays) utilise a systolic-type array of configurable logic blocks (CLBs) and a matrix of erasable interconnect nodes determining the connections between them. This, in conjunction with the one-to-one translation of the spatial and temporal attributes of the Handel programs into the hardware architecture, allows for total control of the intended architecture by the Handel designer.

In addition, the distributed mechanism used for the synchronous hardware designs generated by the Handel compiler, combined with the systolic-type architecture of the targeted devices, improve significantly the scalability of the Handel designs. Furthermore, the compilation approach is based on a reduction of the Occam programs used for specification into their normal form. In combination with the transformational algebra CSP ?, this supports the Handel paradigm. It makes the synthesis approach fertile for further exploitation and enhancement using formal techniques, such as the one presented in this paper.

2. PREVIOUS WORK AND CURRENT APPROACH

Programs written in Handel ? specify hardware implementations. Although Handel programs resemble occam programs ? a major difference concerns the use of the parallel operator: Like occam, Handel allows the possibility of a point to point communication, but additionally in the version we are using an asynchronous communication via access to variables can take place. The only restriction on this sort of communication is that a simultaneous write access of two different processes will lead to an undetermined value.

The compiler has been constructed in such a way that certain algebraic transformation and implementation rules are fulfilled. It is the ambition of the authors' work to supply a transformational framework in which an implementation relation can be defined between Handel programs and hardware which makes these rules true, thus proving the implementation calculus used in the Handel compiler correct. Handel programs will have as much timing nondeterminism as possible. This will allow several different implementations. Step by step this nondeterminism will be reduced, until a level is reached in which no timing nondeterminism is left: the so-called *intermediate terms*. From there obvious implementation rules lead directly to hardware implementations. The validity of these rules, on which the compiler relies, is then proven in our framework. Furthermore on top of Handel we set an interval temporal logic, the duration calculus (DC) ?; ?, which allows us to prove properties of specifications and implementations. All levels are based on the same semantic model: so-called *timing diagrams*. This way is mathematically the most complex, but also the most satisfying one because all possible inconsistencies are avoided, which appear when one semantic model is changed to another one. We get the following picture of our approach:

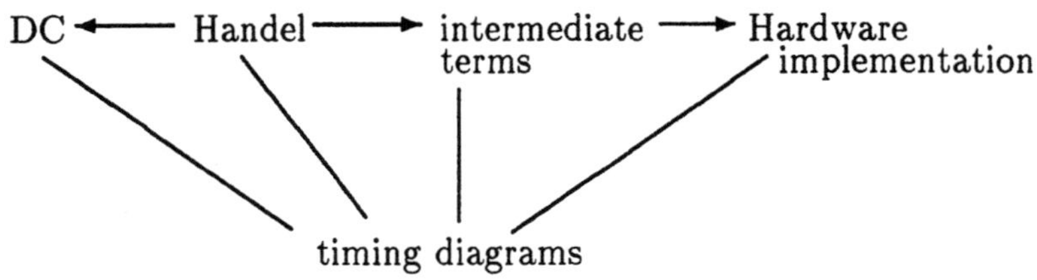

Syntactic transformation and its semantic foundations

Another novelty of our approach is the treatment of a parallel operator in a denotational way with synchronous and asynchronous communication in a many read / many write fashion. Some implementation examples, kept reasonably small for the purpose of this paperbut related to real hardware implementations, are also included.

The current approach achieves the necessary degree of delicacy in the mathematical analysis of the Handel language:
1.) It is very difficult to find a denotational semantics for a language which has a complex parallelism like the one in Handel, which combines the OCCAM-like point to point communication with communication via shared variables.
2.) If timing nondeterminism is retained and decisions for a parallel or sequential implementation of the specifications are employed at a later stage during the automatic design synthesis transformations, then as a consequence the interdependencies between the timing of the parallel and sequential composition inevitably become very complex.
3.) Any approach where the semantics of all abstraction levels is expressed in a uniform model leads inevitably to a certain degree of complexity in the description of the operators.

An approach similar to the one proposed has been taken in the ProCoS project ?, where hardware verification was only one issue among many other aspects. However, the formal analysis of Handel transformations presented in this paper was not an issue in that project. Transformation methods based on real-time refinement have attracted a considerable amount of interest. However, the present paper is among the first in which real-time refinement is used for hardware verification in a far-reaching approach from the very abstract requirement language level (DC) down to hardware implementation. Previous work using transformation algebra was pursued by He Jifeng et al., e.g. ?. The broader framework of our work is given by ?; ?. As a complementary approach one might regard self-timed circuit implementation techniques as in ?; ?; ?.

3. THE MODEL

Our models will be based on *timing diagrams*. Intuitively these are functions which attribute to each point t of time the state the system is in at t. They are used to give a semantic foundation to the DC, to Handel and to hardware. In this paper we are using a discrete version. The semantics of a Handel process P is given as a transformation on sets of timing diagrams. The meaning of this is that we describe how P affects a system which has evolved already according to some timing diagram. For reasons which will be explained later our semantics is called *Quickstep-Semantics*. We have to introduce some general notation:

A.) In the semantics presentation we assume a fixed set of channels $Chan$ and a disjoint set of program variables Var. Additionally there are three distinguished variables W, R, C which are not contained in $Chan \cup Var$. Intuitively W, R describe the sets of all variables which are written to or read from at some point in time, C provides information about the readiness of the channels for communication. By this we remain close to the tradition of CSP ?, the scientific predecessor of occam, which requires that channel communications are dealt with by means of so-called refusal or ready sets, C will contain, however, slightly more information than just a ready set.

Then we introduce the following notations and general assumptions:

• Each program variable v is equipped with a type $ty(v)$, W, R are of type $\mathcal{P}(Var)$ and C is of a special type, which contains sets of elements of the form $ch?v$ or $ch!e$.

• The elements of the set $Var \cup \{W, R, C\}$ are called observables.

• A *state*, typically called σ, is a mapping from the set of all observables where the image of each observable obs is contained in $ty(obs)$.

• For a logical term e or a logical formula b over the variables in the process, let $[\![e]\!](\sigma)$ and $[\![b]\!](\sigma)$ be their values in the state σ as usually defined.

B.) In the following we shall define functions of time. As domains of such functions we need the following sets of closed intervals:

$$\mathcal{I}ntv_{fin} = \{ \{0, \ldots, e\} \mid e \in \mathbb{N}_0 \}$$
$$\mathcal{I}ntv = \mathcal{I}ntv_{fin} \cup \{\mathbb{N}_0\}$$

Our time domain is $Time \stackrel{\text{def}}{=} \mathbb{N}_0$.

We also remind the reader of some mathematical standard notation: The functional composition $f \circ g$ is defined by $f \circ g(x) \stackrel{\text{def}}{=} f(g(x))$. The repeated application of composition f^n is defined by $f^n(x) \stackrel{\text{def}}{=} f(...f(x))$, where f has been applied exactly n times.

C.) The following set of notations are connected with timing diagrams, on which our model is built:

• For a given set Obs of observables and $I \in \mathcal{I}ntv$ a *timing diagram* over I is a mapping, typically called td, with domain $Obs \times I$ such that $\forall obs \in Obs, t \in I : td(obs, t) \in ty(obs)$.
Throughout the paper we shall consider only timing diagrams with the property that $td(W, 0)$ is equal to the set of all variables ($td(W, 0) = Var(\alpha)$ in the notation of section 5.). This means that each variable must have been set to some initial value.

• Note that for each fixed $t \in I$ the function $td(\,.\,, t)$ is a state. Here $td(\,.\,, t)$ is the function which attributes to each variable v its value $td(v, t)$ at time t.

• Let $\ell(td)$, the *length of the timing diagram td* over I, be the length of the interval I, where the arithmetic on $\mathbb{N}_0 \cup \{\infty\}$ is defined as usual, and let $I(td) = I$ and $Obs(td) = Obs$. These are projection functions which extract either component from the domain of td.

• Let $TD(Obs)$ denote the set of all timing diagrams over Obs and over any $I \in \mathcal{I}ntv$.

• Let $TD_{fin}(Obs), TD_{inf}(Obs)$ denote the set of all timing diagrams which are defined for finite and for infinite intervals, respectively.

4. DISCRETE DURATION CALCULUS

As an assertional language we are using a discrete version of the duration calculus. From the hardware design point of view this is the form we shall need, because we assume the existence of a global synchronous clock scheme. For more detailed studies of DC the reader may refer to ?. That text introduces its continuous version.

Semantically, duration formulae F are evaluated in a given interpretation $\mathcal{I}$ and a given time interval $\{b, ..., e\} \subseteq Time$. Such an interval may also be empty. For us interpretations are timing diagrams (modulo W, R, C).

State assertions P are obtained by applying propositional connectives like '$\wedge$', '$\vee$' or '$\neg$' to elementary assertions of the form $var = val$ or the expression '*true*'. For a given interpretation $\mathcal{I}$, state assertions denote functions

$$P_{\mathcal{I}} : Time \rightarrow \{0,1\}$$

For example $P_{\mathcal{I}}(t) = (v = true \wedge w = false)(t)$ is 1, iff v is true and w is false at time t in the interpretation $\mathcal{I}$.

Then the basic syntax of duration formulae apart from propositional connectives is as follows:

- **Duration:** $\int P = k$ expresses that the *duration* of the state assertion P in $\{b, ..., e\}$ is k. Semantically, duration is the measurement $\sum_b^e P_{\mathcal{I}}(t)$.
- **Chop:** The composite duration formula F_1 ; F_2 holds in $\{b, ..., e\}$ if this interval can be divided into an initial subinterval where F_1 holds and an adjacent final subinterval where F_2 holds.
- **Next:** This operator is typical for discrete time calculi. It cannot be used in a continuous setting. A duration formula $\bigcirc P$ holds at time b, iff at time $b+1$ the duration formula P holds, i.e. $\int P = 1$ in the one point interval $\{b+1\}$.

Besides this basic syntax various abbreviations can be defined like *length* (ℓ) of an interval, *sometime* ($\Diamond F$), *always* ($\Box F$). They can be used as is common in temporal logic and explained in ?. Furthermore $\lceil P \rceil$ means $\int P = \ell \wedge \ell > 0$ and $\bigcirc var = exp$ means that in the next point of time the variable *var* gets the value determined by the expression *exp*.
A duration formula F *holds* in an interpretation $\mathcal{I}$ if F evaluates to true in $\mathcal{I}$ and every interval of the form $\{0, ..., t\}$ with $t \in Time$. As usual in logic we can identify a formula F with the set of all interpretations which make F true.

5. THE SYNTAX OF HANDEL

An *alphabet* $\alpha = Chan(\alpha) \cup Var(\alpha)$ is associated with each process P. This alphabet must include all channel and variable names which occur in P with $Chan(\alpha) \cap Var(\alpha) = \emptyset$. Let $Chan(\alpha) = In(\alpha) \cup Out(\alpha)$, the sets of input and output channels and $Var(\alpha) = Wr(\alpha) \cup Re(\alpha)$, the sets of write and read variables. This does not exclude bidirectional wires or read-write variables.

The set of Handel processes P is generated by the following grammar:

$$\begin{array}{rcl} P & ::= & \texttt{SKIP} \mid \texttt{STOP} \mid \texttt{DELAY} \mid ch?v \mid ch!exp \mid x_1, ..., x_n := exp_1, ..., exp_n \\ & & \texttt{SEQ}[P, P] \mid \texttt{IF}(cond, P, P) \mid \texttt{WHILE}(cond, P) \mid \texttt{PAR}[P, P] \end{array}$$

Parallel processes:. A process $\texttt{PAR}[P_1, ..., P_n] = \texttt{PAR}[P_1, \texttt{PAR}[P_2, ..., P_n]...]$ consists of $n > 0$ processes where the interface of the subcomponents P_i is given by α_i. PAR is an associative operator in our setting. We distinguish two sorts of channels: *external channels* for which no internal communication partner exists: $\exists^1 i : ch \in Chan(\alpha_i)$ and *internal channels*, if there are exactly two i with $ch \in Chan(\alpha_i)$.

From now on we assume that all channels appearing in the interfaces of the sequential components are internal or external. A parallel process can be observed at its external channels and its variables. We shall assume the existence of external channels only in our inductive definitions; compilable Handel programs do not have them.

Intermediate terms:. In order to cope with the problems of a proper implementation of the timing, we shall need some additional syntactical constructs whose behaviour resembles the behaviour of the already known constructs but does not allow timing non-determinism. The reason for introducing intermediate terms will be made clear in the section on the transformation of Handel programs. The present section is only meant to exhibit the necessary syntax.

The set of intermediate terms is generated by the following grammar:

$$\begin{aligned} P \;::=\; & op \mid ch??v \mid ch!!e \mid x_1, ..., x_n ::= exp_1, ..., exp_n \mid \\ & \texttt{if}\,(cond, P, P) \mid \texttt{while}\,(cond, P) \end{aligned}$$

where op is an arbitrary Handel operator, and the newly introduced operators are called *slow operators.*

6. THE SEMANTICS OF HANDEL

6.1 OPERATORS ON TIMING DIAGRAMS

In order to describe the semantics of a language which has an elaborate parallel operator like Handel and includes timing, it is not surprising that we need some auxiliary operators. It is rather more surprising that we need just so few.

- By $td[n \,/\, v := exp]$ we denote the timing diagram td after a *timing diagram update* at time n of the observable v by $[\![exp]\!](td(., n-1))$, the value of the expression exp one time unit before.
- The delay operator del extends a given timing diagram by one time unit, retaining the last state. On infinite timing diagrams the delay has no effect.

$$del(td)(v,t) \overset{\text{def}}{=} \begin{cases} td(v,t) & \text{for} \quad 0 \le t \le \ell \\ td[\ell \,/\, W, R, C := \emptyset, \emptyset, td(C,\ell)](v,\ell) & \text{for} \quad t = \ell + 1 \end{cases}$$

where $\ell = \ell(td)$. At the newly created point of time $\ell + 1$ no write or read has taken place so far, so W, R are empty. The willingness for a communication has not changed, so C remains the same.

• In order to allow for a quicker update by an assignment than only at the end of a given timing diagram *td*, we give a definition for the earliest point of time where a process P writing on a variable set X and reading from a variable set Y can interfere with *td*. P must not interfere earlier than one of the variables from X or Y has been written upon and not earlier than one of the variables from X has been read for the last time:

$$time(td, X, Y) \stackrel{\text{def}}{=} max(\{1+n \mid td(W,n) \cap (X \cup Y) \neq \emptyset\} \cup \{1+n \mid td(R,n) \cap X \neq \emptyset\})$$

An earlier interference of P would have disastrous consequences.

• The following *at* operator is a little more elaborate than the pure update. It is basically the semantic equivalent of a program assignment. So in contrast to the update, a changed variable retains its value until the next explicit update. Formally, *at* is defined by a repeated use of the timing diagram update. Also the distinguished observables W, R are updated:

$$at[n \,/\, v := exp](td) \stackrel{\text{def}}{=} \begin{cases} \emptyset \quad \text{for } \ell(td) < n \vee n < time(td, vars(v), vars(exp)) \\ td\,[i \,/\, v := exp \mid i = n, ..., \ell(td)] \\ \quad [n \,/\, W := W \cup vars(v)] \\ \quad\; [n-1 \,/\, R := R \cup vars(exp)] \qquad \text{otherwise} \end{cases}$$

The first line maintains that the timing diagram cannot be updated if it is not long enough or if the update is ruled out by the *time* function. The notation in the other three lines describes an update of variable v at time n by the expression *exp*. Its effect remains noticeable at all times in $n, ..., \ell(td)$, expressed in the second line. The program variable v is written upon at time n, so W must be updated at n (third line). The program variables from *exp* are read at time $n-1$, so R must be updated at $n-1$ (fourth line).

In the semantic model we may want to use updates of the sets W, R, C. For W, R we use the same syntax, for example for W:

$$at[n \,/\, W := W \cup \{v\}](td) \stackrel{\text{def}}{=} td[i \,/\, W := W \cup \{v\} \mid i = n, ..., \ell(td)]$$

We have to say how C should be treated by *at*:

$$at[n \,/\, C := C \cup \{ch?v\}](td) \stackrel{\text{def}}{=} td[i \,/\, C := C \cup \{ch?v\} \mid i = n, ..., \ell(td)]$$

$$at[n \,/\, C := C \setminus \{ch?v\}](td) \stackrel{\text{def}}{=} td[i \,/\, C := C \setminus \{ch?v\} \mid i = n, ..., \ell(td)] \\ [n \,/\, W := W \cup \{ch?\}]$$

The first line is used when channel *ch* becomes ready for a communication at time n, the other two express a communication at n. Then readiness is withdrawn (line two) and the notation $ch?$ will indicate that the communication has taken place, similar to the way $v \in W$ means that a communication on v has taken place. By '\' we denote the set theoretic difference. Since a communication via *ch* can take place only after the last communication via *ch*, we have to make our definition more precise by

$$at[n \,/\, C := C \cup \{ch?v\}](td) \stackrel{\text{def}}{=} \emptyset$$
$$\text{if } \exists m : n < m \leq \ell(td) \wedge \{ch?, ch!\} \cap td(W, m) \neq \emptyset$$

The expression $\{ch?, ch!\} \cap td(W, m) \neq \emptyset$ means that a communication via *ch* has taken place at m. New readiness for a communication is impossible at any earlier point of time (called here n). Note that we claim strict inequality between n and m. This corresponds to the idea that a channel which has just been communicated upon is immediately ready for a new communication. The operator for output is defined in an equivalent fashion.

The following abbreviation $at[n_1 \,/\, v_1 := exp_1, ..., n_m \,/\, v_m := exp_m]$ marks a repeated use of the *at* operator at various points of time $n_1 < ... < n_m$. This format will be called the *at-format* of a mapping. The following lemma establishes a somewhat weaker property than commutativity. It allows us to consider the at-format as some sort of normal form for mappings on timing diagrams.

Lemma 6..1 (*comassi*)
For $m \leq n$

$$at[m \,/\, v_m := exp_m] \circ at[n \,/\, v_n := exp_n] \in \{\emptyset, at[m \,/\, v_m := exp_m, n \,/\, v_n := exp_n]\}$$

So either the two mappings commute or their composition amounts to the empty mapping. □

A similar law governs the interplay between the *del* and *at* operators:

Lemma 6..2 (*delassi*)
For $0 < n \in \mathbb{N}_0$ it is

$$del \circ at[n \,/\, v_n := exp_n](td) \in \{\emptyset, at[n \,/\, v_n := exp_n] \circ del(td)\}$$

So either the two mappings commute or the image of their composition is the empty set. On *td* the mappings do not commute, iff $n = \ell(td) + 1$. □

In the following we shall inductively construct the mapping which Handel programs induce on a timing diagram *td*. Basically this happens by repeated applications of the *at* and *del* operators. The *del* operator is used for extending the given diagram in the most trivial way by one time unit (if this is necessary) after the *at* operator modifications of the state take place.

6.2 SEMANTICS OF THE HANDEL OPERATORS

Now we are ready to describe the semantics of the Handel operators. They will be conceived as operators on sets of timing diagrams, which are given in the at-format, preceded by sufficiently many applications of the *del* function. The definition will be given on single diagrams and extended to sets TD of timing diagrams. The semantics will be given in a continuation style, saying how some process will act on some history. Therefore infinite timing diagrams will not be changed by the Handel operators. The slow operators are only mixed terms, whereas the quick ones, which are a part of Handel, can change the past, so it is justified to call them "quick". This has led to the name "Quickstep-Semantics".

For the slow assignment $ass \equiv$
$v ::= exp$

let the semantics be defined by

$$[\![ass]\!](td) \quad = \quad at[\ell(td) + 1 \,/\, v := exp] \circ del(td)$$

So the update can take place only at the end of the timing diagram after the extension by one unit. In contrast to the quick assignment there is hence no non-determinism when the update takes place.

For the quick assignment $ass \equiv$
$v := exp$

let the semantics be defined by

$$[\![ass]\!](td) \quad = \quad \bigcup_{i=0}^{\ell(td)} at[i \,/\, v := exp](td) \;\cup\; [\![v ::= exp]\!](td)$$

So the update can take place at any point of time in the timing diagram or, after an extension, one time unit after the end of *td*. Note that in practice most of the $at(i \,/\, v := exp)$ contribute nothing to the semantics, because an update at time i is not allowed due to the first line in the definition of *at*.

$$\begin{array}{lcl} [\![\mathtt{SKIP}]\!](td) & = & td \\ [\![\mathtt{DELAY}]\!](td) & = & del(td) \\ [\![\mathtt{SEQ}[P_1,P_2]]\!](td) & = & [\![P_2]\!]([\![P_1]\!](td)) \end{array}$$

For td let $[\![\mathtt{STOP}]\!](td)$ be the unique extension td' of td with

$$\ell(td') = \infty \ \wedge \ \forall obs, i > \ell(td) : td'(obs,i) = del^{i-\ell(td)}(td(obs,i))$$

So after $[\![\mathtt{STOP}]\!](td)$ no further change of the diagram is possible.

Semantically, each of the channel communications is, apart from the communication sets, treated almost like half an assignment, so that later their parallel composition amounts to a full assignment. For quick operators this is

$$\begin{array}{lcl} [\![ch?v]\!](td) & = & \bigcup_{m \in ty(v)} \bigcup_{i=0}^{\ell(td)} \Big(\bigcup_{j=i+1}^{\infty} at[j \,/\, C, v := C \setminus \{ch?v\}, m] \circ \\ & & \quad at[i \,/\, C := C \cup \{ch?v\}] \circ del^{j-\ell(td)} \cup \\ & & \quad [\![\mathtt{STOP}]\!] \circ at[i \,/\, C := C \cup \{ch?v\}] \Big)(td) \end{array}$$

$$\begin{array}{lcl} [\![ch!exp]\!](td) & = & \bigcup_{i=0}^{\ell(td)} \Big(\bigcup_{j=i+1}^{\infty} at[j \,/\, C, R := C \setminus \{ch!exp\}, R \cup vars(exp)] \circ \\ & & \quad at[i \,/\, C := C \cup \{ch!e\}] \circ del^{j-\ell(td)} \cup \\ & & \quad [\![\mathtt{STOP}]\!] \circ at[i \,/\, C := C \cup \{ch!e\}] \Big)(td) \end{array}$$

Here i is the point of time where the channel becomes ready and j is the point of time where the communication actually happens. Here $j - \ell(td)$ extensions of the diagram by the del operator can become necessary if the communication happens after $\ell(td)$. For $j < \ell(td)$ the del operator is applied zero times.
The infinite timing diagram introduced by $[\![\mathtt{STOP}]\!]$, stands for the case where a communication does not take place. For input additional nondeterminism is added by the various input values m.
The definition for the slow operators $ch??v$ and $ch!!exp$ coincides with the one for the quick operators, except that i is confined to $\ell(td)$.

For parallel composition we relinquish the occam view that parallelism is an intersection of the behaviours of the parallel components. Instead we construct an operator on functions in the at-format:
We assume that $[\![P_j]\!](td)$, $j = 1,2$ are each given by a set of elements of the form $at[i \,/\, v_i := exp_i \mid i \in M \subseteq \mathbb{N}_0] \circ del^k$. (So M is the set of points in time where variables are changed.) Then we construct

$$[\![\mathtt{PAR}[P_1,P_2]]\!](td) = \{m_1 \,||\, m_2 \mid m_j \in [\![P_j]\!](td)\} \ \cap \ \mathrm{SYN}$$

With SYN as below and the additional definition for a set S

$$at[n\,/\,v := S] = \bigcup_{m \in S} at[n\,/\,v := m]$$

we define

$$at[i\,/\,v_i := exp_i \mid i \in M_1 \subseteq \mathbb{N}_0] \circ del^k \,||\, at[i\,/\,w_i := exp'_i \mid i \in M_2 \subseteq \mathbb{N}_0] \circ del^l = at[i\,/\,v_i := exp_i \,||\, w_i := exp'_i \mid i \in M_1 \cup M_2] \circ del^{max}$$

Here $max = max(k, l)$ indicates that at first the shorter timing diagram is extended. Furthermore

$$v_i := exp_i \,||\, w_i := exp'_i \quad = \quad v_i := exp_i \quad \text{for } i \in M_1 \setminus M_2$$

and analogously for $i \in M_2 \setminus M_1$. For $i \in M_1 \cap M_2$ let

$$v_i := exp_i \,||\, w_i := exp'_i \quad = \quad v_i, w_i := exp_i, exp'_i$$

the normal parallel assignment for program variables for $v_i \neq w_i$. Simultaneous assignment to the same variable from both components leads to total non-determinism:

$$v_i := exp_i \,||\, v_i := exp'_i \quad = \quad v_i := ty(v_i)$$

For the distinguished observables W, R the RHS contains their join. For the distinguished observable C the changes on both sides of the $||$ take place simultaneously.

Example 6..3 ()

$$v_1, v_2, C := exp_1, exp_2, C \cup \{ch_1\} \,||\, v_1, v_3, C := exp'_1, exp'_2, C \setminus \{ch_2\} = v_1, v_2, v_3, C := ty(v_1), exp_2, exp'_2, C \cup \{ch_1\} \setminus \{ch_2\}$$

Note that a simultaneous addition of some string by one component and its removal by the other is impossible. This ensures the well definedness for C. □

The set SYN makes channel communication happen:

$$\begin{aligned} \text{SYN} = \{td \mid\ & \forall ch \forall n, v, exp:\ (ch? \in td(W, n) - ch! \in td(W, n)) \land \\ & (\{ch?v, ch!exp\} \subseteq td(C, n) \Rightarrow \\ & \{ch?, ch!\} \subseteq td(W, n+1) \land td(v, n+1) = [\![exp]\!](td(n)))\} \end{aligned}$$

The first line means that a communication takes place, only if both partners take part in that communication. The second and third lines mean that if an output and an input on a channel *ch* is possible at some time n, the communication must take place as soon as possible, namely at time $n+1$. This is indicated by the communication markers $ch?$, $ch!$ in W. Additionally the update on v must be performed.

Note that in this definition hiding is implicitly contained, because a parallel process can be observed at its external channels and its variables. If two parallel processes fitted together, their common channels are not longer external. This means that they are hidden.

The semantics for `IF` does not need a further explanation, and also `WHILE` is defined as usual by a least fix-point.

Now we can explain why an update as early as possible, by quick operators, is useful: For example for variables x, y, z we might want some implementation

$$\texttt{SEQ}[x := y, y := z] \prec \texttt{PAR}[x := y, y := z]$$

(defined below). If we had only the slow semantics, the LHS would inevitably take two time units and the RHS one unit, so an implementation would be impossible. In the Quickstep-Semantics the LHS takes one or two time units and the RHS still only one unit, which is a reduction of non-determinism.

In the version available by FTP (address on page 1), several examples are provided which show that this semantics can cope with the details of the timing behaviour and the cooperation of the various operators (sequence, parallelism, assignment).

6.3 IMPLEMENTATION RELATION

In order to compare DC formulae and Handel programs, we face three problems:

- For us DC formulae are timing diagrams, whereas Handel programs are operators on timing diagrams.
- The arbitrarily given initial values at time 0 must be somehow taken into account. Remember that assignment takes time. So each initialisation on the program level can become effective not earlier than at time 1.
- Whereas DC formulae are interpreted only over infinite timing diagrams, Handel programs are mostly dealing with finite ones.

The first two problems are overcome the same way: We apply programs to diagrams of length 0. These reflect exactly the implicit initial values at time 0. To a DC formula an additional conjunct must be added which makes the initial values explicit. A diagram *td* of length 0 over variables var_i with values val_i can be conceived as a state assertion

$$\lceil td \rceil \equiv \lceil \forall i : var_i = val_i \rceil$$

Then the DC formula $\lceil td \rceil; true$ says that the observation must start in a state in which $\lceil td \rceil$ holds.

The last problem is fixed by extending finite timing diagrams to infinite ones using the STOP operator.

So for any Handel process P and for any DC formula F we define implementation $\prec$ (and analogously its converse $\succ$)

$$F \prec P \iff_{df} \forall td : \ell(td) = 0 \Rightarrow$$
$$[\![\mathtt{STOP}]\!]([\![P]\!])(td) \subseteq \lceil td \rceil; true \wedge F \quad \text{modulo} \quad W, R, C$$

7. THE TRANSFORMATION FROM HANDEL TO HARDWARE

All hardware devices are described by DC formulae. We only give an example here.

A latch with input x, output y and latch condition *cond* conforms to the following DC formula

$$Latch(x, y, cond) \equiv \quad \Box \forall val \in ty(x) = ty(y) :$$
$$(x = val \wedge cond = true \vee y = val \wedge cond = false) \Rightarrow \bigcirc y = val$$

Similar formulae apply to other devices like gates, multiplexers, etc. The aim of our transformations is to find an architecture, expressed in DC, which implements the Handel description. The transformation method starts with some algebraic laws for intermediate terms which are used for manipulations in the CSP tradition, for example associativity of SEQ or that SKIP is the neutral element of sequential composition. We will not go into more details here.

Besides laws for semantic identity there are also laws for implementation which reduce time non-determinism, most notably the fact that the slow variant of each operator implements the quick one. This allows us to finally do away with all quick operators in the last transformation step.

7.1 DISTRIBUTED CONTROL CIRCUITS

In this section we shall describe how an intermediate term, which does not contain any quick operators, is directly implemented in hardware. For such a given process P we need a data-multiplexing part $DM(P)$, a channel arbitration part $CA(P)$ and a control part $CTRL(P)$. Then the implementation of P is given by the formula

$$IMPL(P) \;=\; DM(P) \wedge CA(P) \wedge CTRL(P)$$

Before explaining the details of these formulae, we make some assumptions:
1.) For each expression *exp* appearing in some assignment or output and for each condition *cond* appearing in some control structure there is some circuitry (gates, adders, etc.) which implements *exp*, *cond* and is described by some term *impl*(*exp*), *impl*(*cond*).
2.) For each channel *ch* there is only one variable v_{ch} in which an input can be written. So all inputs to *ch* are of the form $ch??v_{ch}$. Furthermore such a v_{ch} is not written in an assignment. These assumptions are not essential. However they make this part more readable.

By $DM(P)$ the change of data during one system tact is defined in terms of simple DC formulae for latches and multiplexers. Latches and multiplexers are triggered by expressions $ctrl(exp_i)$ (playing for example the role of *cond* in the definition of a latch) which allow or prohibit data updates. The $ctrl(exp_i)$ are defined below. The full formulae can be found in the version available by FTP.

The channel arbitration part will be given as a conjunction over the channels

$$CA(P) \;=\; \bigwedge_{ch \in Chan(P)} CA(ch) \;\wedge\; Connect_{DM-CA}(P)$$

$Connect_{DM-CA}(P)$ describes the interplay between DM and CA. Via a channel *ch* performing output by one of the expressions $ch!!exp_1, ..., ch!!exp_n$ in the program, a data transfer can take place if the input and one output are ready:

$$\begin{aligned} CA(ch) \;=\; \Box(\;& transfer_{ch} = \textstyle\bigvee_{i=1}^{n} transfer_{ch,i} \wedge \\ & \textstyle\bigwedge_{i=1}^{n} (transfer_{ch,i} = rdy_{ch!exp_i} \wedge rdy_{ch?v}) \;) \end{aligned}$$

The rdy will be determined in $CTRL(P)$. The connection between DM and CA defines some of the $ctrl(exp_i)$ allowing data update (if the data transfer is possible):

$$Connect_{DM-CA}(P) \;=\; \Box(\bigwedge_{ch,i} ctrl(exp_i) = transfer_{ch,i})$$

where the conjunction varies over all channels *ch* and all output expressions $ch!!exp_i$ appearing in the program for *ch*.

The implementation of the control part $CTRL(P)$ of a process P is done inductively by means of two DC variables s_P, f_P, which indicate the start and the final control state (one-hot encoding). The following formulae will give
1.) the connections between the start and the final control states of the subprocesses of P, given by $Connect_{SF}(P)$,
2.) the connections between the control states of the subprocesses and, respectively, the control wires of the DM part and the not yet defined wires of the CA part, given by $Connect_{CTRL-DM}(P)$ and $Connect_{CTRL-CA}(P)$.

For R = DELAY control at state s_R means next time control at final state f_R:

$$CTRL(R) \;=\; \Box\,(\bigcirc f_R = s_R) \quad (\equiv Connect_{SF}(R))$$

For $R = x ::= exp$, control changes the same way and a data update is allowed:

$$CTRL(R) = Connect_{SF}(R) \wedge Connect_{CTRL-DM}(R)$$
$$Connect_{SF}(R) = \Box\,(\bigcirc f_R = s_R) \wedge Connect_{CTRL-DM}(R) = \Box\,(s_R = ctrl(exp))$$

The last equality is another part of the delayed definitions of the $ctrl(exp_i)$ from DM. For the process R = SEQ$[Q\,1, ..., Q\,n]$ let

$$CTRL(R) \;=\; \textstyle\bigwedge_i CTRL(Q\,i) \wedge Connect_{SF}(R)$$
$$Connect_{SF}(R) \;=\; \Box\,(s_R = s_{Q\,1} \wedge f_R = f_{Q\,n} \wedge \textstyle\bigwedge_i s_{Q\,i+1} = f_{Q\,i})$$

For R = PAR$[P_1, ..., P_n]$ simultaneous termination of the P_i at f_R is achieved by consideration of $wasf_{P_i}$ and $keepf_{P_i}$ (saying that P_i has reached f_{P_i}):

$$\begin{aligned} CTRL(R) &= \textstyle\bigwedge_i CTRL(P_i) \wedge Connect_{SF}(R) \\ Connect_{SF}(R) &= \Box\,(\textstyle\bigwedge_i (s_R = s_{P_i}) \wedge (f_R = \textstyle\bigwedge_i keepf_{P_i}) \wedge \\ &\qquad (\textstyle\bigwedge_i (keepf_{P_i} = f_{P_i} \vee wasf_{P_i})) \wedge \\ &\qquad (\textstyle\bigwedge_i (\bigcirc wasf_{P_i} = keepf_{P_i} \wedge \neg f_R))) \end{aligned}$$

The formulae for `if` and `while` are dropped due to space limitations. For $R = ch??v$ (similarly for $R = ch!!e_i$) let $wasrdy_{ch?v}$ be a new internal DC variable showing readiness but no data transfer in the last step.

$$\begin{aligned} CTRL(R) = \Box\,(&(rdy_{ch?v} = wasrdy_{ch?v} \vee s_R) \wedge \\ &(\bigcirc wasrdy_{ch?v} = rdy_{ch?v} \wedge \neg transfer_{ch}) \wedge \\ &(\bigcirc f_R = transfer_{ch})) \end{aligned}$$

7.2 IMPLEMENTATION BY DISTRIBUTED CONTROL CIRCUITS

At the Handel level we only consider the data variables. The newly introduced control variables in the DC formulae describing the hardware implementation are not visible on the upper levels. So they will be hidden as usual by existential quantification. These are all variables of the form

$$mux_v, change_v, ctrl(exp), transfer_{...}, rdy_{...}, wasrdy_{...}, s_P, f_P, keepf_P, wasf_P$$

Let the set of all these variables be called $CtrVar$. As we have done for the data variables we must specify an initial state for each control variable. Since we aim at a one-hot encoding for a given process P, initially only one variable in $CtrVar$ can be true: the start variable s_P of course. Let $initctr$ be the state with

$$\lceil initctr \rceil \equiv \lceil s_P \wedge \bigwedge_{s_P \neq v \in CtrVar} \neg v \rceil$$

In the following theorem $initctr$ is used to define the initial state of those control variables which are not defined in the Handel program. Remember that the initialisation of the other variables is dealt with in the definition of "$\prec$".

Theorem 1 ()
For all Handel processes R involving only slow operators

$$R \prec \exists CtrVar : \lceil initctr \rceil ; true \wedge IMPL(R)$$

□

The proof is full of mathematical details, see the version available by FTP.

8. CONCLUSIONS AND FURTHER WORK

In this paper we have described a formal transformation framework for provably-correct hardware compilation using Handel. The outstanding features of this method are the precise mathematical treatment of a complex parallel operator and its interdependencies with sequential composition with respect to time. Our method allows us to prove abstract requirements in a logical formalism, the DC, on the basis of a uniform model. Such a uniform model is also a novelty.

The introduction of the quick operators enables a powerful mechanism for exploiting operation re-scheduling to deliver optimal implementations. Although this model is adequately general to demonstrate real applications in the near future, the clear and strict definition of the operators which specify and construct the timing diagrams, which in turn constitute the transformational basis, ensure the robustness as well as the further refinement of our model. There is ongoing work, with a number of real-world applications such as communication protocols and large coding blocks, formally designed using our framework. Results from these activities will be the subject of future publications.

References

G.M.Brown. Towards truly delay-sensitive circuit realisation of process algebra. In: G.Jones and M.Sheeran, Eds., Designing Correct Circuit, Springer-Verlag, 1991, 120–131.

M.R.Hansen, Zhao Chaochen. Lecture Notes on Logical Foundations of Duration Calculus. Formal Aspects of Computing 9, 1997, 283-330.

He Jifeng, C.A.R. Hoare, M. Fränzle, M. Müller-Olm, E.-R. Olderog, M. Schenke, M.R. Hansen, A.P. Ravn, and H. Rischel. Provably correct systems. In: H. Langmaack, W.-P. de Roever and J. Vytopil, Eds., Formal Techniques in Real-Time and Fault-Tolerant Systems, LNCS 863 (Springer-Verlag, 1994), 288–335.

He Jifeng, I.Page, J.Bowen Towards a Provably Correct Hardware Implementation of occam, CHARME93, LNCS 683 (Springer-Verlag, 1993), 214–226.

He Jifeng, Zhang Jianping. Simulation Approach to Provably Correct Hardware Compilation. In: H. Langmaack, W.-P. de Roever and J. Vytopil, Eds., Formal Techniques in Real-Time and Fault-Tolerant Systems, LNCS 863 (Springer-Verlag, 1994), 336–350.

C.A.R.Hoare. Communicating Sequential Processes. Prentice Hall, 1985.

A.J.Martin. The design of a self-timed circuit for distributed mutual exclusion. In: H.Fuchs, Ed., Proc 1985 Chapel Hill Conference VLSI, 1985, 247–260.

A.J.Martin. Compiling communicating processes into delay-insensitive circuits. Distributed Computing 1, 226-234.

I.Page, W.Luk. Compiling occam into Field Programmable Gate Arrays. In: W.Moore, W.Luk, Eds., FPGAs, Oxford Workshop on Field Programmable Logic and Applications, 1991.

M.Spivey, I.Page. How to program in Handel. Based on the Handel hardware compiler, ver. H 159.H. Revised by W.Luk, Dec.94. Oxford University Computing Laboratory, Hardware Compilation Research Group technical report, 1994.

Zhou Chaochen, C.A.R. Hoare, A.P. Ravn. A Calculus of Durations. IPL 40/5 1991, 269-276.

Invited Talk III

PART SEVEN

Test III

An enhanced model for testing asynchronous communicating systems

Myungchul Kim, Jaehwi Shin[+], Samuel T. Chanson[*], and Sungwon Kang[+]

Information and Communications University
Yousonggu, Taejon, Korea 305-348
mckim@icu.ac.kr

[+]*Korea Telecom R & D Group*
Sochogu, Seoul, Korea 137-792
{jhshin, kangsw}@sava.kotel.co.kr

[*]*Hong Kong University of Science and Technology*
Clear Water Bay, Kowloon, Hong Kong
chanson@cs.ust.hk

Abstract

This paper studies the problem of testing concurrent systems considered as blackboxes and specified using asynchronous Communicating Finite State Machines. In our previous work [Kim 96], an approach to derive test cases for concurrent systems was developed. The approach addressed the state space explosion problem by introducing a causality relation model and the concept of logical time to express true concurrency. This paper builds on and extends the previous work. The model is enhanced to improve the quality (i.e., verdict capability) of test cases by relaxing the unit-time assumption to any natural number and by presenting a computationally efficient algorithm for deriving the test cases. Finally the approach is generalized to describe timing constraints with any real numbers.

Keywords

Protocol Testing, Concurrent Systems, Logical Time, Conformance Testing, Asynchronous Communicating Finite State Machines.

1. INTRODUCTION

In order to specify precisely the concurrent behavior of communication protocols and services in a distributed system, formal description techniques such as Communicating Finite State Machines (CFSMs), Estelle, SDL and LOTOS have been developed. These techniques allow interactions among multiple machines (or modules or processes) in a concurrent system to be specified formally. To increase the confidence in protocol products implemented according to international standards, various protocol testing methodologies have been developed and are being used.

Until recently, most work on automatic test case generation has focused on single machine specifications even though the implementation may consist of multiple machines. Typically, a preprocessing phase is used to translate the multiple machine specifications into a single system specification in order to take advantage of the many existing testing methods for sequential systems. Usually reachability analysis is used in the translation process. Unfortunately, the "state space explosion" problem is often encountered during reachability analysis. In addition, it is difficult to express timing constraints using reachability analysis techniques [West 78].

To alleviate the state space explosion problem, the concept of "true concurrency" or "simultaneous execution of events" has been introduced by several researchers to reduce the number and lengths of test sequences in the test suite for concurrent systems (see for example [West 78, Itoh 83, Kajiw 85, Araka 91]). The concept works because a single event sequence using true concurrency could represent many interleaved event sequences. These methods treat concurrent systems as blackboxes. The control flow graph representing the global behavior of a given concurrent system is first constructed and conventional test case generation methods are applied to the control flow graph. To reduce the number and the lengths of test sequences, the notion of so-called reduced implementation sequences is used based on true concurrency. However, the methods for deriving the reduced implementation sequences are usually based on heuristics and are often informally described. Another aspect worth noting is that the reduced implementation sequences may reduce the verdict capability (i.e., may lead to higher probability of inconclusive verdicts).

We have presented a formal methodology of generating test sequences using true concurrency and logical time [Lampo 78, Rayna 92] for concurrent systems described as asynchronous CFSMs [Kim 96]. The resulting test sequence was called Minimal Causality Path (MCP) calculated based on the assumption that the duration of each event is exactly one unit time. In fact, the 'MCP' and 'reduced implementation sequence' turn out to be the same test sequence although the procedures for obtaining them are different. However as a result of the unit-time assumption, tests using MCPs have low verdict capabilities.

In this paper we propose to enhance the verdict capability by allowing more state space to be investigated. There is a tradeoff between the verdict capability and size of the state space explored. The first enhancement is to relax the unit-time assumption on event duration to any arbitrary integer value and present a computationally efficient method for deriving the test cases. Then we show how the approach can be generalized to describe timing constraints using a causal relation model and real time logical clock.

The rest of the paper is organized as follows: Section 2 surveys some related work in protocol testing, and Section 3 summarizes the model proposed in [Kim 96] and discusses the verdict capability of the resultant test cases. Section 4 provides an enhancement to the model by relaxing the assumption of unit time to any natural number. In addition, two fundamental properties on the event sequence of concurrent systems are presented, and a computationally efficient algorithm to deriving test cases under the relaxed logical time framework is derived based on the two properties. In Section 5, the model is generalized to express timing constraints by any real value. Finally Section 6 concludes the paper.

2. RELATED WORK

According to the ISO conformance testing methodology, an Implementation Under Test (IUT) is viewed as a blackbox for testing purposes [ISO 94]. Our work aims at testing asynchronous CFSMs which are treated as blackboxes. By comparison, most of the existing work in software testing such as [Young 88, Taylor 92, Tai 95] is based on synchronous communicating programs where the internal code is assumed to be accessible (the white-box approach). Since our work is focused on protocol testing, we present below a brief survey of some related papers on testing protocols modeled as concurrent systems.

In [Lee 93], an approach for conformance testing of protocols specified as a collection of synchronous CFSMs is presented. The approach consists of two parts: pruning and a guided random walk procedure. This allows some parts of the behavior space to be removed from consideration. Also, the model is based on interleaving in order to explore the state space.

Cavalli et al. [Caval 93] present a test sequence generation method based on Unique Input/Output sequences and signatures in a framework which allows conformance test sequences to be obtained from LOTOS specifications. The model is for synchronous communicating systems only.

Luo et al. [Luo 94] present a model of generating test sequences for concurrent systems and communication protocol modeled as communicating nondeterministic finite state machines (CNFSMs). After the CNFSMs are transformed into a single nondeterministic finite state machine (NFSM) by reachability analysis, test sequences are obtained from the resulting NFSM using the generalized Wp-method. The approach gives rise to the state space explosion problem easily because the reachability analysis in the transformation phase.

In [Ulric 95], an approach to derive a concurrent transition tour as a test suite from multi-module specifications described by a restricted class of Calculus of Communicating Systems is presented. This approach alleviates the state space explosion problem by introducing a true concurrency model. The model is for synchronous communicating systems and uses some heuristics in constructing the order of messages.

[Henni 97] proposes an approach for generating test cases in Concurrent TTCN from a system of asynchronous communicating finite state machines. An algorithm is given for generating a noninterleaving model of prime event structures (adopted from verification and concurrency theory) from a generalized model of asynchronous communicating finite state machines. The work also deals with test case generation from the prime event structure. However, computational efficiency, verdict capability and timing constraints have not been considered.

Ulric et al. [Ulric 97] presents a concurrency model, called behavior machine, and its construction algorithm. Further the paper outlines how test derivation can be based on the new concurrency model in order to derive test suites that still exhibit true concurrency between test events. The model is for synchronous communicating systems and uses the Petri net unfolding algorithm.

Our work aims at testing asynchronous CFSMs treated as blackboxes, and is a non-heuristic (i.e., exact) approach using a true concurrency model. The model is evaluated in terms of computational efficiency and verdict capability, and generalized to describe timing constraints using a causal relation model and real time logical clock to increase the verdict capability of the test cases.

3. BACKGROUND

In this Section we introduce the concept of logical clocks and summarize our previous work.

3.1 Logical clocks

Lamport showed that asynchronous distributed executions can be characterized by a partial order relation on the events produced. The relation, called the *causality* relation (or *happened before* relation) is denoted by $\rightarrow$ [Lampo 78]. Let E be a set of events produced by the execution of a distributed system. For $e, e', e'' \in E$, $e \rightarrow e'$ holds if:

1. e and e′ are events in the same process and e precedes e′,
2. e is a send event and e′ is the corresponding receive event, or
3. there exists e″ such that $e \rightarrow e''$ and $e'' \rightarrow e'$.

Two events e and e′ are said to be *causally related* if $e \rightarrow e'$ or $e' \rightarrow e$ holds. If neither $e \rightarrow e'$ nor $e' \rightarrow e$ holds, these events are *concurrent* (or *mutually independent*), denoted by $e \parallel e'$.

According to the above principles, a number of time stamping mechanisms have been devised including linear time, vector time, and matrix time [Rayna 92]. They are important in studying the behavior or distributed systems. The linear time c_i associated with process P_i, for example, can be computed as follows (all values of c_i's are initialized to zero):

- When P_i executes an internal event or a send event, the time is updated, i.e., $c_i = c_i + d$, where *d* represents the duration of the corresponding operation. In the case of a send event, the message carries the updated c_i value as the timestamp.
- When P_i executes a receive event where the message contains a timestamp *ts*, the time is computed as $c_i = max\,(c_i, ts) + d$.

3.2 Overview of our basic framework [Kim 96]

A concurrent system consists of a number of sequential systems. Each sequential system is modeled as a Finite State Machine (FSM). It is assumed that there is no global physical clock; that is, the system model is *asynchronous*. Thus, the concurrent system is represented by asynchronous CFSMs and treated as blackboxes. The environment (e.g., operators, testers, and/or subsystems which are not under test) of a concurrent system is also modeled as FSMs. The FSMs communicate with each other by messages and are assumed to contain no message loops. The underlying message passing system provides asynchronous communication and is assumed to be reliable; that is, there is no message loss, distortion or duplication. An event is the atomic execution of an operation within a system.

We assume that a concurrent system consists of subsystems which do not share variables; instead, they interact by message passing. Thus, there are no dataraces in this model. Data races occur if concurrent events have read/write or write/write access conflicts to shared variables. Since each subsystem is treated as a blackbox, only communication events such as message send and receive events are considered.

Definition 1: A Global Event (GE) is defined by the set of all (observable) events that occur at the same time in all the FSMs of a concurrent system. Namely, GE = $(e_1, \ldots, e_i, \ldots, e_n)$ where e_i is an event of the i-th FSM.

Note that since we are interested in blackbox testing, GE does not include internal events. The GE consists of the send and receive events of the subsystems and expresses true concurrency.

A path is a sequence of events performed by an FSM starting and ending in the same initial state and is denoted as $e^1 \rightarrow e^2 \rightarrow \ldots \rightarrow e^m$. We consider, for simplicity, the second visit to the initial state of an FSM as termination. This is not a limitation since subsequent behavior (if any) can be viewed as another path.

Definition 2: Paths[i] is the set of all event sequences that can be performed by the i-th FSM, i.e., Paths[i]= {Path(i,1), Path(i,2), ...}.

For the example in Figure 1 adopted from [Kajiw 85], the Paths[i] are given below:
Paths[1] = {-a → -b → +c → -d → +f, -a → -b → -c → +f}
Paths[2] = {+a → +d → -f → -g, +a → +e → -f → -g}
Paths[3] = {+b → +c → -e → +g, +b → -c → +g}

From all the possible paths of the *n* FSMs, we construct path vectors each of which represents a situation where the FSMs are communicating with one another (See Definition 3).

Definition 3: A Concurrent Path (CP) is a vector of event sequences, denoted by $CP_k = \langle Path(1,1_k), \ldots, Path(i,i_k), \ldots, Path(n,n_k) \rangle$. CP_k is an element of Paths[1] × ... × Paths[i] × ... × Paths[n] where the operator '×' is the Cartesian product such that a send event is matched with the corresponding receive event in chronological order. An element $Path(i,i_k)$ of CP_k is null if the i-th FSM is idle. If a send event does not match with the corresponding receive event during generating CP_k, event sequences already generated for the CP_k is kept.

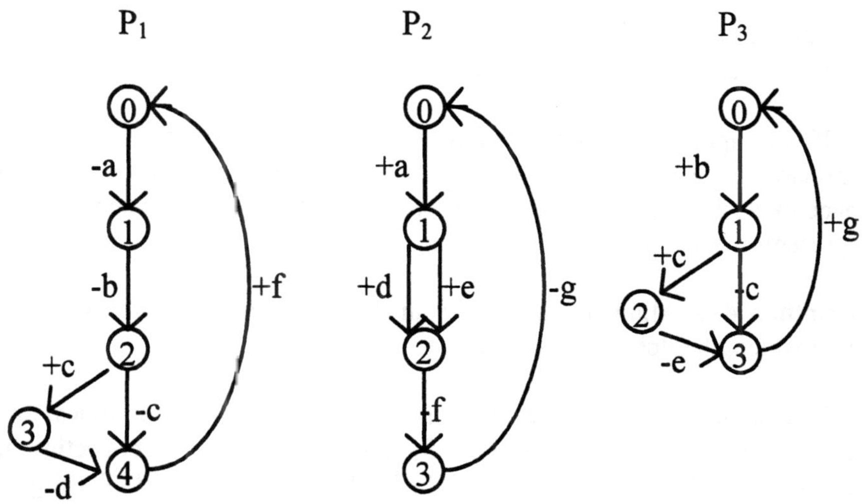

Figure 1. Example of a concurrent system

Definition 3 is illustrated using the example in Figure 1. Here, we have 6 different CP's as follows:

CP_1 = <-a → -b → +c → -d → +f , +a → +d → -f → -g, +b → -c → +g>
CP_2 = <-a → -b → -c → +f, +a → +e → -f → -g, +b → +c → -e → +g>
CP_3 = <-a → -b → +c, +a → +e, +b → +c>
CP_4 = <-a → -b → -c → +f, +a → +d, +b → -c → +g>
CP_5 = <-a → -b → +c → -d → +f, +a → +e, +b → -c → +g>
CP_6 = <-a → -b → -c → +f, +a → +d, +b → +c → -e → +g >

The CP such as CP_3, CP_4, CP_5 and CP_6 begins with the initial state but may end in a different state in case of error such as deadlock and unspecified reception.

Let us consider test sequence generation for the following Concurrent Path:
CP_2 = <-a → -b → -c→ +f, +a → +e → -f → -g, +b → +c → -e → +g>.

There are two kinds of constraints imposed by causal relationships among the events in a CP: (1) ordering imposed by each sequential FSM and (2) ordering imposed by matching send and receive events. In CP_2, we have the following constraints of the first kind:

- -a → -b → -c→ +f,
- +a → +e → -f → -g, and
- +b → +c → -e → +g.

The set of constraints of the second kind in CP_2 is:

- {-a → +a, -b → +b, -c → +c, -e → +e, -f → +f, -g → +g }.

If we apply reachability analysis to CP_2 enumerating all possible interleavings, we get 36 event sequences for testing. To alleviate this state space explosion, we construct a minimal length time-event sequence which represents the above causal relationships.

Definition 4: A Minimal Causality Path (MCP) corresponding to a CP is a minimal length Global Event sequence.

To compute Minimal Causality Paths, first calculate the linear logical time of events in a CP, then obtain GEs for each logical time by collecting all events having the same logical time. The MCP is the sequence of GEs aligned in increasing order of logical time.

If we suffix the events in MCPs with the FSM identifier, then

$$MCP_2 = -a \rightarrow (-b, +a) \rightarrow (-c_1, +b) \rightarrow +c_3 \rightarrow -e \rightarrow +e \rightarrow -f \rightarrow (+f, -g) \rightarrow +g.$$

The notation (-b, +a) means the two events occur concurrently, i.e., they are mutually independent. Therefore, the order of occurrences between the two events does not matter. The MCPs obtained in this way constitute concise and accurate test sequences for concurrent systems when the duration of events can be abstracted to unit time.

4. MODEL ENHANCEMENT

If the duration of events cannot all be set to unit time, the reduced test sequence can lead to lower verdict capability. When the trace of an event sequence conforms to a Global Event Sequence, we may conclude that the system conforms to specifications. However, when the trace does not conform, the result is inconclusive. To investigate the verdict capability quantitatively for a given GE sequence, we propose the following measure.

Definition 5 : The Verdict Capability (VC) of a test sequence or a GE sequence (GES_k) for a Concurrent Path CP_k is defined as follows:

$$VC = |EX(GESk)| \, / \, INTk,$$

where INTk is the number of paths in the reachability tree of CPk and EX(GESk) is the set of all interleaved event sequences that can be obtained from GES_k by juxtaposing the GEs in GES_k for all possible ordering. The function |E| of a set E is the cardinality of E.

Note that Definition 5 can also be used to compute the VC for MCPs because MCP is a special case of GES. For example, the VC for MCP_2 is 8/36, because the number of paths in the reachability tree of CP_2 is 36, and the number of interleaved event sequences that can be derived from CP_2 is 8. Thus, there is a high possibility of getting inconclusive verdict if we test the system in Figure 1 with respect to MCP_2. Therefore, it is necessary to devise a test sequence generation method with the desired verdict capability without incurring state space explosion.

So far, the value of d in the logical clocks for representing the duration of an operation in Section 3.1 is assumed to be unit time. For test sequence generation, if d may assume any natural number or any real number in a certain range, the verdict capability can be increased and the model will be more useful. In this section, we shall consider test sequence generation when the value of d of each event takes on a natural number between 1 and d_m. Common matrix notation is used in the following description.

Definition 6: For a concurrent system, the Concurrent Path k (CP), the Duration (D) and the Logical times (L) of the j-th event in CP_k for FSM i are denoted as $CP_k[i][j]$, $D_k[i][j]$ and $L_k[i][j]$ respectively.

If the duration of each event in a CP is known, the corresponding Global Event sequence can be constructed in the same way used for MCPs except that the duration values are not necessarily 1's. The Global Event Sequence (GES) can be viewed as a test sequence for the given concurrent system. Note that a MCP is a special type of GES in which all event durations are one unit long.

Take CP_2 in Section 3.2 as an example. According to Definition 6, D_2 (the unit time) and L_2 (corresponding to MCP_2 in Section 3.2) are represented as follows:

$$P_2 = \begin{bmatrix} -a, -b, -c, +f \\ +a, +e, -f, -g \\ +b, +c, -e, +g \end{bmatrix} \quad D_2 = \begin{bmatrix} 1, 1, 1, 1 \\ 1, 1, 1, 1 \\ 1, 1, 1, 1 \end{bmatrix} \quad L_2 = \begin{bmatrix} 1, 2, 3, 8 \\ 2, 6, 7, 8 \\ 3, 4, 5, 9 \end{bmatrix}$$

L_2 is obtained from CP_2 and D_2 by the method described in Section 3.2. Because there are three 2-tuples representing concurrent events in L_2, MCP_2 covers eight (= 2!×2!×2!) paths in the reachability tree of CP_2.

To generate the test sequence (or equivalently to generate GES for combinations of relaxed event durations), we can assign event durations D_k to the desired values. However, in the remainder of this section, for simplicity of illustration, we shall consider the case when the duration of each event in CP_2 is either 1 or 2 only.

Example 1: Given CP_2 and D_2', we calculate L_2' and the corresponding GES_2' as follows:

$$P_2 = \begin{bmatrix} -a, -b, -c, +f \\ +a, +e, -f, -g \\ +b, +c, -e, +g \end{bmatrix} \quad D_2' = \begin{bmatrix} 1, 1, 1, 2 \\ 2, 1, 1, 1 \\ 1, 1, 1, 1 \end{bmatrix} \quad L_2' = \begin{bmatrix} 1, 2, 3, 9 \\ 3, 6, 7, 8 \\ 3, 4, 5, 9 \end{bmatrix}$$

$$GES_2' = -a \rightarrow -b \rightarrow (-c, +a, +b) \rightarrow +c \rightarrow -e \rightarrow +e \rightarrow -f \rightarrow -g \rightarrow (+g, +f).$$

The values in L_2' represent the linear logical times of the corresponding events. Note the numbers '3' and '9' in L_2'. The events in CP_2 having the same value given in L_2' (e.g. -c, +a, and +b) are concurrent. Thus, the GES covers twelve (=3!×2!) paths in the reachability tree of CP_2. This example shows how the verdict capability can be improved by relaxing the event durations.

If all cases of event durations in CP_2 (which may be either 1 or 2) are considered, there are a total of 4096 (=2^{12}) different combinations of event durations for the 12 events. In general, given the maximum duration d_m and number of events M, there are d_m^M combinations to be considered. Because the number of combinations can become very large, tremendous amount of computation is needed if we calculate GESs for all the combinations. Thus, it is necessary to reduce the amount of computation by introducing some constraints.

First we define *concurrent event* and *non-concurrent event* as follows: An event e in a CP is said to be a *concurrent event* if there exists at least one other event e′ in CP such that e || e′. Otherwise e is a *non-concurrent event.*

Property 1: The position (i.e. sequential order) of a non-concurrent event in a GE sequence obtained from a CP is independent of the event duration of the other events.
Proof: The proof is straightforward from the fact that there is a causality relation between a non-concurrent event and every other event.

As a result of Property 1, we do not need to consider the event duration of non-concurrent events because they do not change the order of a GE sequence in any way.

Example 2: The events '-a', '+e', and '-f' in CP2 are non-concurrent and the other events are concurrent. Therefore, the number of combinations of event duration we need to consider is reduced to 512 (= 29) for the 9 concurrent events in CP2 instead of 4096 (= 212).

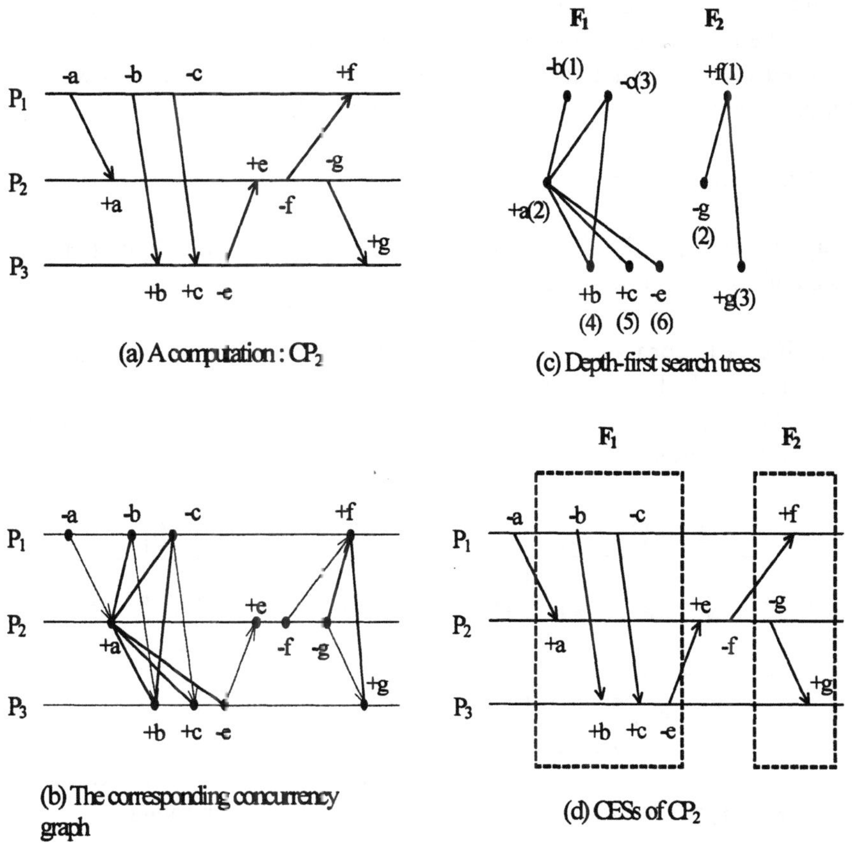

Figure 2. Example of CES construction

In order to reduce the number of combinations of event duration further, we introduce the concept of Connected Event Set (CES):

Definition 7: *Connected Event Set* (CES)
Given a CP, a concurrency graph can be constructed consisting of vertices corresponding to events in the CP and edges joining all pairs of vertices whose corresponding events are concurrent. Generally a concurrency graph has one or more disjointed subgraphs. A *Connected Event Set* is the set of all events in a maximal connected subgraph of a concurrency graph.

We shall illustrate Definition 7 with the examples in Figure 2. Given CP_2 shown in Figure 2(a), we obtain the concurrency graph in Figure 2(b) using the method described in Section 3.1. A solid dark line between two events means the two events are concurrent. Note that the concurrent relation is symmetric and non-transitive. From Figure 2(b), we can obtain the CESs shown in Figure 2(d). A CES may be constructed using an efficient search technique instead of enumerating all combinations of events. In the example, we used the depth-first search technique as shown in Figure 2(c) where the numbers in parentheses denote the order of searching.

A property of CES is given below:

Property 2: For a CP with CESs $F_1, F_2, \ldots, F_n$, given F_i and F_j such that $1 \le i, j \le n$, and $i \neq j$, the following hold:

2.1: If $e \rightarrow f$ (or $f \rightarrow e$) for $f \in F_i$, $e \in CP$ and $e \notin F_i$, then $e \rightarrow f'$ (or $f' \rightarrow e$) for any $f' \in F_i$. In other words, $e \rightarrow F_i$ (or $F_i \rightarrow e$).

2.2: If $f_i \rightarrow f_j$ for $f_i \in F_i$, $f_j \in F_j$, then $f_i' \rightarrow f_j'$ for any $f_i' \in F_i$ and $f_j' \in F_j$. In other words, $F_i \rightarrow F_j$.

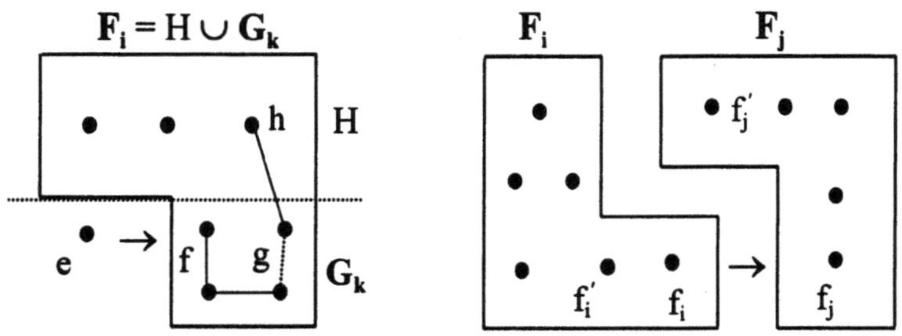

(a) Diagram illustrating Property 2.1

(b) Diagram illustrating Property 2.2

Figure 3. Diagrams showing properties of CES

Proof: **2.1** With reference to Figure 3(a), if the cardinality of F_i is N and G_k is a subset of F_i with k elements, then G_N is equal to F_i. Assume $e \rightarrow f$ holds for $e \notin F_i$ and $f \in F_i$. First, we set G_1 to $\{f\}$, then $e \rightarrow G_1$. Assume we have a set G_k satisfying $e \rightarrow G_k$ for $1 \le k \le N-1$. It is sufficient to prove that we can find a set G_{k+1} satisfying $e \rightarrow G_{k+1}$ from the known sets G_k and F_i.

Let set H be the set $F_i - G_k$. Since G_k and H together make up the set F_i, there exists at least one concurrent relation, say $g \parallel h$, between the two events $g \in G_k$ and $h \in H$. Let us construct G_{k+1} by including h in

G_k, i.e. $G_k \cup \{h\}$. Since $e \notin F_i$, either $e \rightarrow h$ or $h \rightarrow e$ hold. If we assume $h \rightarrow e$, since $e \rightarrow g$ then $h \rightarrow g$. Since this contradicts with $g \parallel h$, $e \rightarrow h$ must hold. Thus $e \rightarrow G_{k+1}$.

The proof of the converse can be obtained in a similar manner.

2.2 Refer to Figure 3(b). Assume $f_i \rightarrow f_j$ holds for $f_i \in F_i$ and $f_j \in F_j$. Then by Property 2.1, $f_i \rightarrow F_j$ holds. This in turn implies $F_i \rightarrow f_j'$ for any $f_j' \in F_j$. Thus $F_i \rightarrow F_j$.

Example 3: Consider CP_2 in Section 3.2. According to Property 2 and by applying the method given in Section 3.1, we can obtain the constraint $F_1 \rightarrow F_2$, where $F_1 = \{+a, -b, +b, -c, +c, -e\}$ and $F_2 = \{+f, -g, +g\}$ as shown in Figure 2. Hence, the order in any GE sequence for CP_2 can be represented as follows:

$$GES_2 = -a \rightarrow (+a, -b, +b, -c, +c, -e) \rightarrow +e \rightarrow -f \rightarrow (+f, -g, +g).$$

Note that the above GES consists of a sequential event and a CES in sequential order. In other words, concurrent behavior may exhibit some sequential characteristics in the high-level abstraction.

Definition 8: A linear logical base time, T_B, for a given event is defined as $T_B \equiv \max(c_i, ts)$ for a receive event, and $T_B \equiv c_i$ for a send event, where c_i is the logical time of the preceding event associated with FSM P_i and *ts* is the timestamp of the receive event as defined in Section 3.1.

Three cases of a leading event (i.e., the first event of a sequence in chronological order) in each FSM in a CES F are of interest:

- case 1: a receive event in F corresponding to a send event which occurred before F,
- case 2: a send event, and
- case 3: a receive event corresponding to a send event in F.

The following property states that the order of an event sequence and the difference in logical time between events are dependent on the event duration of the events in F but independent of the event duration of the events in the other CESs and any non-concurrent events.

Property 3: The base times (T_B) for the leading events in cases 1 and 2 in a CES F are the same and equal to the maximum linear logical time, T_{max}, among all events that occurred before F.

Proof: Let *e* be a leading event in case 1 or 2 in F, and let T_e, T_B and δ be the logical time, base time, and duration of the event *e*, respectively. Then T_e can be determined as follows:

$$T_e = T_B + \delta.$$

Since the base time of *e* is determined prior to F, the relation $T_B \leq T_{max}$ must hold. If we assume $T_B \neq T_{max}$, T_e can be represented as follows:

$$T_e = T_{max} - \varepsilon + \delta, \quad \varepsilon = T_{max} - T_B.$$

As $\varepsilon > 0$, if we set δ such that $0 < \delta \leq \varepsilon$, then $T_{max} - \varepsilon < T_e \leq T_{max}$ holds. This contradicts Property 2 which indicates $T_e > T_{max}$. Thus $T_B = T_{max}$.

According to Property 3, given a CES F, the base times of events belonging to cases 1 and 2 are equivalent to the corresponding T_{max}. Furthermore, the base times of events belonging to case 3 are determined by the logical time of the matching send events in F. Therefore, the events before F do not influence the order of event sequence or the difference in logical time between events in F. This means the event sequence and the time difference between events with $T_B = 0$ are equivalent to those with $T_B = T_{max}$, thus enabling independent derivation of GE sequences in each CES.

Example 4: In F_2 of Example 3, '-g' is a case 2 event, '+f' is a case 1 event and '+g' is a case 3 event. The base times of events '-g' and '+f' are equal to the logical time ($=T_{max}$) of the event '-f', and the base time of '+g' is equal to the logical time of the event '-g'.

Definition 9: Given a CES, a partial-GES (denoted as pGES) is defined as a sequence of GEs containing all events in the CES.

Note that there may be one or more pGESs for a given CES. Based on the results of Properties 1, 2 and 3, the following algorithm is developed to generate GE sequences for a concurrent system.

Algorithm 1: (Global Event Sequence generation for a concurrent system)
Step 1. Generate CESs for a given CP. Let these CESs be $F_1, F_2, \ldots, F_n$.
Step 2. For i = 1, 2, ..., n, do:
Generate pGESs from F_i with the initial value of T_B set to 0 for each combination of event duration. Let the output be $B_i = \{pGES_i^1, pGES_i^2, \ldots, pGES_i^{p(i)}\}$, where $pGES_i^k$ is the k-th pGES of F_i, and p(i) is the total number of pGESs.
Step 3. Obtain GESs by applying the Cartesian product $B_1 \times B_2 \times \ldots \times B_n$ and inserting non-concurrent events into the predetermined positions.

Example 5: Let us apply Algorithm 1 to GES_2 in Example 3. Since $|F_1| = 6$ and $|F_2| = 3$, 72 ($=2^6 + 2^3$) cases of event durations are considered in Step 2. As a result, 10 and 4 partial-GESs are obtained for CES F_1 and F_2, respectively. The GE sequence generated in Step 3 is given by:

$\{-a\} \times B1 \times \{+e\} \times \{-f\} \times B2$

$= \{-a\}\times\{pGES_1^{\,1},\ pGES_1^{\,2},\ \ldots,\ pGES_1^{\,10}\}\times\{+e\}\times\{-f\}\times\{pGES_2^{\,1},\ pGES_2^{\,2},\ \ldots,\ pGES_2^{\,4}\}$

$$=\{-a\}\times \underbrace{\left\{\begin{matrix} (-b,+a)\rightarrow(-c,+b)\rightarrow+c\rightarrow-e, \\ (-b,+a)\rightarrow+b\rightarrow-c\rightarrow+c\rightarrow-e, \\ (-b,+a)\rightarrow-c\rightarrow+b\rightarrow+c\rightarrow-e, \\ -b\rightarrow(-c,+a,+b)\rightarrow+c\rightarrow-e, \\ -b\rightarrow(-c,+a)\rightarrow+b\rightarrow+c\rightarrow-e, \\ -b\rightarrow(+a,+b)\rightarrow-c\rightarrow+c\rightarrow-e, \\ -b\rightarrow+a\rightarrow(-c,+b)\rightarrow+c\rightarrow-e, \\ +a\rightarrow-b\rightarrow(-c,+b)\rightarrow+c\rightarrow-e, \\ +a\rightarrow-b\rightarrow+b\rightarrow-c\rightarrow+c\rightarrow-e, \\ +a\rightarrow-b\rightarrow-c\rightarrow+b\rightarrow+c\rightarrow-e \end{matrix}\right\}}_{B_1} \times\{+e\}\times\{-f\}\times \underbrace{\left\{\begin{matrix} (+f,-g)\rightarrow+g, \\ +f\rightarrow-g\rightarrow+g, \\ -g\rightarrow+f\rightarrow+g, \\ -g\rightarrow(+f,+g) \end{matrix}\right\}}_{B_2}$$

Step 2 of Algorithm 1 provides a saving in computation complexity from $2^{\{|F1|\ +\ldots+\ |Fn|\}}$ to $2^{|F1|}+\ \ldots\ +2^{|Fn|}$, where 2 is the number of different event durations and $|F_i|$ is the number of events in CES F_i. The savings are partially offset by Steps 1 and 3 of Algorithm 1 if a CES exists. However, the computation overhead is small compared to the saving in Step 2. If there are k different event durations, the computation saving is $k^{\{|F1|+\ldots+|Fn|\}}$ minus $k^{|F1|}+\ \ldots\ +k^{|Fn|}$.

In Example 5, 72 ($=2^6+2^3$) cases of event durations are considered in Step 2 in order to generate GE sequences for CP_2. If all the combinations of event durations in CP_2 which are either 1 or 2 are considered, there will be 4096 cases. In the same fashion, we need only 72 out of 4096 cases for CP_1, 72 out of 4096 for CP_2, 8 out of 16 for CP_3, 32 out of 64 for CP_4, 64 out of 128 for CP_5, and 64 out of 128 cases for CP_6. In total, 312 cases of event durations are sufficient to generate GE sequences for all CPs where the total number of combinations of event durations in all the CPs is 8528.

It can be easily shown that the VC of test sequences obtained for CP_2 in Example 5 is 24/36 whereas the VC of MCP_2 is 2/9. This represents an improvement of nearly three times. Recall the maximum value of VC is 1. To attain VC values closer to 1, a larger maximum duration value d_m is needed. Therefore, there is a tradeoff between the amount of computation required for test sequence generation and the resultant verdict capability.

5. GENERALIZATION OF THE MODEL

In this section, we discuss how the models proposed in Section 4 can be further generalized by :

i) a technique to use real value of 'd' by the logical clocks for representing the duration of an operation, and

ii) application of the model to timed automata (used for describing the timing constraints in real-time systems).

Consider CP_2 in Section 3.2. Note that events '-b', '-c', and '-g' happened before events '+b', '+c', and '+g' respectively. The notation (x, y, z) means that events x, y and z occurred concurrently. As a result of Property 2 in Section 4, the generalized GE Sequence (GGES) corresponding to CP_2 is as follows:

$$GGES_2 = -a \rightarrow (-b \rightarrow -c, +a, +b \rightarrow +c \rightarrow -e) \rightarrow +e \rightarrow -f \rightarrow (+f, -g, +g)$$

In this formula, all the causal relations between events are represented and the duration of each event can be assigned any arbitrary value. If each event takes one unit time, $GGES_2$ becomes MCP_2 in Section 3.2 and if each event takes one or two unit time, $GGES_2$ becomes GES_2 in Section 4. In this section, we shall derive GGESs corresponding to CPs as follows:

$$GGES_1 = -a \rightarrow (-b \rightarrow +c \rightarrow -d, +a, +b \rightarrow -c) \rightarrow +d \rightarrow -f \rightarrow (+f, -g, +g)$$

$$GGES_3 = -a \rightarrow (-b, +a, +b)$$

$$GGES_4 = -a \rightarrow (-b \rightarrow -c, +a, +b \rightarrow -c)$$

$$GGES_5 = -a \rightarrow (-b \rightarrow +c \rightarrow -d, +a, +b \rightarrow -c)$$

$$GGES_6 = -a \rightarrow (-b \rightarrow -c, +a, +b \rightarrow +c \rightarrow -e,)$$

By merging GGES's while preserving the causal relation among events, the Combined GGES in Figure 4 is obtained (the algorithm is omitted here, see [Kim 98] for details)[1]. An event is labeled at an arrow head. A circular arc

1. $GGES_4$, $GGES_5$ and $GGES_6$ are not shown in the figure for simplicity.

crossing a number of arrows (e.g., arc A just after event '-a') is used to indicate that the events after event '-a' and before event '+e' (i.e., -b, -c, +a, +b, +c, -e) are in a Connected Event Set (CES) (refer to Figure 2). Both $GGES_3$ and $GGES_1$ are branched off from events in the CES of $GGES_2$. Also, there may be causal relationship among the events of different FSMs in a CES (for example, events '-b' and '+b' just after arc A).

The merged (or combined) GGES analysis shown above has advantages over the conventional reachability tree analysis for the following reasons:

- timing constraints can be described because the duration of each event and the causal relation among the events are provided, and
- concurrent behavior can be represented.

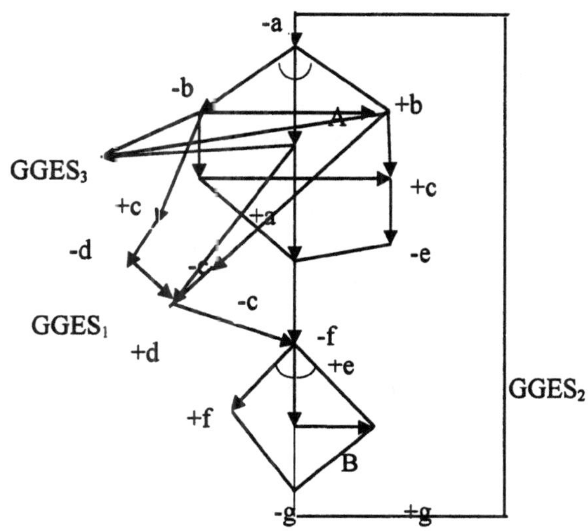

Figure 4. Combined GGES of the concurrent systems in Figure 3

The absence of these properties is a major contributor to the state space explosion problem in conventional reachability analysis as mentioned in Section 1.

Given a sequential sequence of events, $e^1 \rightarrow \ldots \rightarrow e^i \rightarrow \ldots \rightarrow e^n$ where n is an integer, let us define 'T_i' to be the time when the i-th event, e^i, occurs. Furthermore, let t_i be the time distance between e^i and the previous event, e^{i-1}. t_i may take on a range of time values, say $low_i \leq t_i \leq high_i$, where low_i and $high_i$ are any real numbers. In other words, e^i occurs after e^{i-1} within a time range between low_i and $high_i$. Now if T_i is measured from the initial event until the occurrence of the i-th event (including t_i), the timing constraints on an event are:

1. For an event, e^n, having a single directed path from the initial event, time T_n is in the range: $\Sigma_{i=1}^{n} low_i \leq T_n \leq \Sigma_{i=1}^{n} high_i$.
2. For an event, e^n with more than one (say K) directed paths from the initial event, time T_n is: MAX $(\Sigma_{i=1}^{m} low_{1,i}, \ldots, \Sigma_{j=1}^{n} low_{K,j}) \leq T_n \leq$ MIN $(\Sigma_{i=1}^{m} high_{1,i}, \ldots, \Sigma_{j=1}^{n} high_{K,j})$ where $\Sigma_{i=1}^{m} low_{1,i} \leq T_n^1 \leq \Sigma_{i=1}^{m} high_{1,i}, \ldots, \Sigma_{j=1}^{n} low_{K,j} \leq T_n^K \leq \Sigma_{j=1}^{n} high_{K,j}$ and i, j, m, n, and K are integers. T_n^k is the logical time for the k-th path to complete execution.

For example, for an event sequence: -a → -b → -c, the time T_{-c} for event '-c' is $low_{-a} + low_{-b} + low_{-c} \leq T_{-c} \leq high_{-a} + high_{-b} + high_{-c}$ because there is a directed path. Also T_{+c} is: MAX $((low_{-a} + low_{-b} + low_{-c} + low_{+c}), (low_{-a} + low_{-b} + low_{+b} + low_{+c})) \leq T_{+c} \leq$ MIN $((high_{-a} + high_{-b} + high_{-c} + high_{+c}), (high_{-a} + high_{-b} + high_{+b} + high_{+c}))$ because there are two paths from -a to +c, i.e., -a → -b → -c → +c and -a → -b → +b → +c. Because the timing constraints are specified by using the above rule after the generation of GGES, it may not be possible for some events to happen because of conflicts in the timing constraints.

By using the Combined GGES, tests can be generated dynamically. This means a test can be performed without a predetermined scenario specified with time schedules. Instead, when an event is observed, we can execute an appropriate operation to generate events satisfying the causality and timing constraints in the Combined GGES. If an observed event satisfies the constraints, the verdict is 'pass'. If not, the verdict is 'fail'. Note that sometimes it may not be possible to generate an event due to errors such as deadlocks or conflicts in the timing constraints.

6. CONCLUSIONS

In order to test and verify the concurrent behaviors of protocols running in a concurrent system, traditional approaches map multiple modules into a single module and analyze the sequential behavior of the resultant monolithic module. Other approaches apply interleaving methods for events in a concurrent system. Recently some attempts have been made to adopt a causality based method or a true concurrency model in protocol verification and testing of concurrent systems (e.g., [Probs 90, Ulric 95, Kim 96, Henni 97, Ulric 97].

In this paper, we have enhanced our previous model [Kim 96] by relaxing the assumption on logical time from unit time to any natural or real numbers in describing timing constraints. An efficient algorithm for deriving test cases for the relaxed event duration has been proposed and evaluated in terms of

Verdict Capability and computational efficiency. We have showed that the saving in testing effort can be quite substantial. We believe our model forms a sound foundation for further investigation in this area.

REFERENCES

[Araka 91] Arakawa, N. and Sonecoka, T., "A Test Case Generation Method for Concurrent Programs," *Proceedings of the 4th Int'l Workshop on Protocol Test Systems*, pp. III 35 - 47, Leidschendam, The Netherlands.

[Caval 93] Cavalli, A., Kim, S. U., and Maigron, P., "Improving Conformance Testing for LOTOS," *Proceedings of the Sixth Int'l Conference on Formal Description Techniques*, pp. 371 - 386, Boston, USA, Oct.

[Henni 97] Henniger O., "On Test Case Generation from Asynchronously Communicating State Machines," *Testing of Communicating Systems Vol 10* (also this is 10th Int't Workshop on Testing of Communicating Systems, Cheju Island, Korea) edited by M. Kim, S. Kang and et al., Chapman & Hall, pp. 255 - 271, Sept.

[ISO 94] ISO IS 9646 *OSI Conformance Testing Methodology and Framework.*

[Itoh 83] Itoh, M. and Ichikawa, H., "Protocol verification using reduced reachability analysis," *The Trans. of the IECE of Japan*, Vol. 66, No. 2, pp. 88 - 93, Feb.

[Kajiw 85] Kajiwara, M., Ichikawa, H., Itoh, M., and Yoshida, Y., "Specification and Verification of Switching Software," *IEEE Tr. on Communications*, Vol. COM-33, No. 3, pp. 193 - 198, March.

[Kim 96] Kim, M. C., Chanson, S. T., Kang S. and Shin J.W. "An Approach for testing asynchronous communicating systems," Testing of Communicating Systems (also this is 9th Int't Workshop on Testing of Communicating Systems, Darmstadt, Germany) edited by B. Baumgarten and et al. pp. 141 - 154, Sept.

[Kim 98] Kim, M. C., Shin J.W., Chanson, S. T., and Kang S. "An Approach for testing asynchronous communicating systems," Technical Report, School of Information and Computer Engineering, Information and Communications Univ., Daejon, Korea

[Lampo 78] Lamport, L., "Time, Clocks, and the Ordering of Events in a Distributed System," *CACM*, Vol.21, No. 7, pp. 558 - 565, July.

[Lee 93] Lee, D., Sabnani, K. K., Kristol, D. M. and Paul, S., "Conformance Testing of Protocols Specified as Communicating FSMs," *IEEE INFOCOM*, pp. 115 - 127.

[Luo 94] Luo, G., Bochmann, G. V., and Petrenko, A., "Test Selection Based on Communicating Nondeterministic Finite-State Machines Using a Generalized Wp-Method," *IEEE Tr. on Software Engineering*, Vol. 20, No. 2, pp. 149 - 161.

[Probs 90] Probst, D. K. and Li, H. F., "Using Partial-order Semantics to Avoid the State Explosion Problem in Asynchronous Systems," *2nd Int'l Conference on Computer-aided Verification*, pp. 146 - 155.

[Rayna 92] Raynal, M., "About Logical Clocks for Distributed Systems," *ACM Operating Systems Review*, Vol. 26, No. 1, pp. 41 - 48, Jan.

[Taylo 92] Taylor, R. N., Levine, D. L., and et al., "Structural Testing of Concurrent Programs," *IEEE Tr. on Software Engineering*, Vol. 18, Iss.3, pp. 206 - 215, March.

[Tai 95] Tai, K. C. and Carver, R. H., "Testing of Distributed Programs," *Handbook of Parallel and Distributed Computing*, ed. A. Zomaya, McGraw Hill.

[Ulric 95] Ulrich, A. and Chanson, S. T., "An Approach to Testing Distributed Software Systems," *15th Int'l Symposium on Protocol Specifications, Testing, and Verification*, pp. 121 - 136.

[Ulric 97] Ulrich, A. and Koenig H., "Specification-based testing of concurrent systems," *Int'l Con. On Formal Description Techniques and Protocol Specifications, Testing, and Verification*, pp. 7 - 22, Osaka, Japan, Nov.

[West 78] West, C. H., "General Technique for Communications Protocol Validation," *IBM J. RES. DEVELOP*, Vol. 22, No. 4, pp. 393 - 404, July.

[Young 88] Young, M., and Taylor, R. N., "Combining Static Concurrency Analysis with Symbolic Execution," *IEEE Tr. on Software Engineering*, Vol. 14, No. 10, pp. 1499 - 1511, Oct.

Interoperability Test Suite Derivation for the TCP

Soonuk Seol[†], Myungchul Kim[†], Sungwon Kang[††], Yongbum Park[†,*], and Younghan Choe[*]
suseol@icu.ac.kr, mckim@icu.ac.kr, kangsw@sava.kotel.co.kr, ybpark@pec.etri.re.kr, and yhchoe@pec.etri.re.kr
[†]*Information and Communications University:* [††]*Korea Telecom Research & Development Group:* [*]*Electronics and Telecommunications Research Institute*

Abstract

A test derivation method suitable for testing interoperability for the class of communication protocols like the ATM/B-ISDN signaling protocol was proposed in [KanK 97]. Up to now, there is no work implementing automatic derivation of the interoperability test suite in the method and analyzing its coverage. In this paper, we implement the algorithm proposed in [KanK 97], apply it to the ATM/B-ISDN signaling protocol and the TCP protocol, and analyze the application results. In this process, we (1) improve the algorithm so that it can handle features like the TCP's three-way handshaking, (2) implement the improved algorithm as a program and derive an interoperability test suite automatically as a result of applying it to a pair of communicating TCP FSMs, and (3) analyze the generated interoperability test suite in terms of transition coverage[1].

Keywords

protocol testing, interoperability testing, conformance testing, test architecture, test suite generation, transition coverage, ATM/B-ISDN, TCP, SDL

1. INTRODUCTION

To increase the confidence in protocol products implemented according to international standards, various protocol testing methodologies have been developed and applied. As an activity, conformance testing that checks whether an implementation is correct with respect to the relevant standards has been standardized by ISO [ISO 94] and applied to protocols such as N-ISDN, ATM/B-ISDN, and so on. Nevertheless, it is widely known that conformance testing has limitation in ensuring interoperability. Thus, even a pair of two conforming implementations can fail to interoperate [RafC 90, APRS 93]. Two main causes of non-interoperation of conforming

---

[1] Transition coverage is the proportion of executed transitions over the whole transitions after validating or simulating a system.

implementations are ambiguity of protocol standards and incompleteness of conformance testing [ShiK 98]. There is some research work on interoperability testing done in the past. [RafC 90] deals with interoperability test suite generation based on reachability analysis. However, it is not based on a rigorous definition of interoperability and considers only the case where lower testers exist between two IUTs. [VerB 94] describes experiences with interoperability testing of the FTAM protocol that uses only a single tester between two IUTs and thus has limited capability. [APRS 93] derives conformance test suite and interoperability test suite separately and later manually combines them to reduce the number of conformance test cases. [KanK 97] develops a method for systematically dealing with symmetric communication protocols[2] and uses an interoperability test architecture that does not observe the interface behavior between two IUTs. [ShiK 98] proposes and applies a test derivation method suitable for testing interoperability for the class of communication protocols like the ATM/B-ISDN signaling protocol. Comparing the work listed above, our work made contributions to the automatic derivation of interoperability test suites and test coverage analysis of its applications results. The two papers [KanK 97, ShiK 98], which are closely related to our work, derive interoperability test suite manually and stop with applying their methods to the ATM/B-ISDN protocol. As a step to improve the method, we enhance and implement the algorithm. Then we apply the implemented program not only to the ATM/B-ISDN signaling protocol but also to TCP protocol, and analyze the application results. In addition, we compare and analyze transition coverage of two cases: one case carried out by applying the validation in the commercial SDL tool [SDT 97] to the IUT and the other case carried out by applying the interoperability test suite utilizing the simulation facility of the commercial tool.

This paper is organized as follows: In Section 2, we state backgrounds related to this work. Section 3 introduces our enhanced algorithm by analyzing the problems in the existing algorithm. In Section 4, we apply the algorithm to the ATM/B-ISDN signaling protocol and the TCP protocol. During this process, we generate the interoperability test suites both for the ATM/B-ISDN protocol and the TCP. In Section 5, we compare transition coverages of two cases: one obtained from applying the interoperability test suite derived in Section 4 to the simulator of the commercial SDL tool and

[2] When implementations based on the same protocol specification interact each other, we define the protocol as symmetric protocol (e.g., ATM/B-ISDN signaling protocol and TCP). If protocol specifications which have implemented as IUTs are different (e.g., server/client), we define the protocol as asymmetric protocol.

the other from validation process using the validator of the tool. Finally in Section 6, we discuss the conclusion and the future work.

2. BACKGROUND

It is difficult to stimulate/observe the interactions between processes or systems because of the asynchronicity in concurrent programs or distributed systems. To avoid these difficulties, [AraS 92] imposes two principles on the black-box testing for concurrent programs. First, *scheduler-independent observation principle* makes events arise from each process without influence of scheduler algorithm. In many cases, this principle implies that the observation cannot distinguish the order of events occurring in different processes. Second, *single stimulus principle* means that a single stimulus must be given at each stable state. A stable state for concurrent programs is defined in [AraS 92] as the system state (1) that is reachable from the initial state adhering to the *single stimulus principle*, and (2) from which no change can occur without another stimulus. The system adhering to the single stimulus is considered as the system running in a slow environment [LuBP 94]. In a slow environment, inputs can be sent from the environment to the system only in situations where all the queues and all the channels are empty. According to the *single stimulus principle*, multiple stimuli and stimuli during transitions are not considered. The interoperability test suite derivation algorithm proposed by [KanK 97, ShiK 98] is based on *single stimulus principle* that is the viewpoint of regarding the response of implementations for a single input as constituting a test case.

[ShiK 98] presents three test architectures for the interoperability testing. In this paper, only two of them are considered. They are shown in Figure 1. In Test Architecture I, Tester A sends a message to IUT A and then IUT A can sends messages to IUT B and/or Tester A. If IUT B receives the message, it can also sends messages to IUT A and/or Tester B. Tester C located between IUT A and IUT B can observe/modify messages between IUT A and IUT B. In Test Architecture II, however, it is impossible to read these messages owing to absence of Tester C.

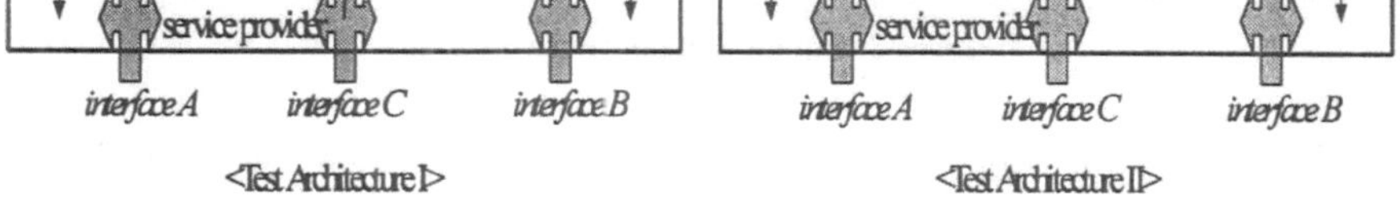

Figure 1. Interoperability test architectures.

To achieve interoperability testing, a composite Finite State Machine (FSM) need be derived. Let us represent a protocol specification with an FSM in order to formally describe the derivation process of composite FSM. Let M be the FSM for the protocol specification of an IUT. Then M can be represented as follows:

Definition 1. A FSM is a 5-tuple ($\mathbf{S}$, S_0, $\mathbf{L_{in}}$, $\mathbf{L_{out}}$, $\mathbf{Tr}$) where:

(1) $\mathbf{S} = \{ S_0, S_1, \ldots, S_{n-1} \}$ is a set of states of M,
(2) $S_0 \in \mathbf{S}$ is the initial state,
(3) $\mathbf{L_{in}} = \{ v_1, v_2, \ldots, v_m \}$ is a set of input symbols,
(4) $\mathbf{L_{out}} = \{ u_1, u_2, \ldots, u_k \}$ is a set of output symbols, and
(5) $\mathbf{Tr} \subseteq \{ S—v/U\rightarrow S' \mid S, S' \in \mathbf{S} \wedge v \in \mathbf{L_{in}} \wedge U \subseteq \mathbf{L_{out}} \}$ is a set of transitions.

Bold letters in Definition 1 represent sets. " $S—v/U\rightarrow S'$ " stands for a transition where S, S', v, and U are, respectively, the starting state, the destination state, an input symbol and a set of output symbols.
Let $\prod$ be the composition of the two FSMs, one, M_A, for the specification A and the other, M_B, for the specification B. We assume that U in Definition 1, for both M_A and M_B, consists of at most two output symbols where each symbol is for different interface. Then $\prod$ can be defined as follows:

Definition 2. The composite FSM $\prod$ is a 5-tuple ($\mathbf{S_\prod}$, $S_{\prod,0}$, $\mathbf{L_{\prod,in}}$, $\mathbf{L_{\prod,out}}$, $\mathbf{Tr_\prod}$) where:

(1) $\mathbf{S_\prod} = \{ S_{\prod,0}, S_{\prod,1}, \ldots, S_{\prod,n-1} \}$ is a set of global states, and $S_{\prod,i} = (S_A, S_B)$, where S_A and S_B are states of M_A and M_B respectively.
(2) $S_{\prod,0} \in \mathbf{S_\prod}$ is the initial state. Namely, $S_{\prod,0} = (S_{A,0}, S_{B,0})$.
(3) $\mathbf{L_{\prod,in}} = \{ v_{\prod 1}, v_{\prod 2}, \ldots, v_{\prod m} \}$ is a set of input symbols. It consists of input messages for the interface A and B. These messages are external inputs from testers. Namely, $\mathbf{L_{\prod,in}} = (\mathbf{L_{in,A,E}} \cup \mathbf{L_{in,B,E}})$.

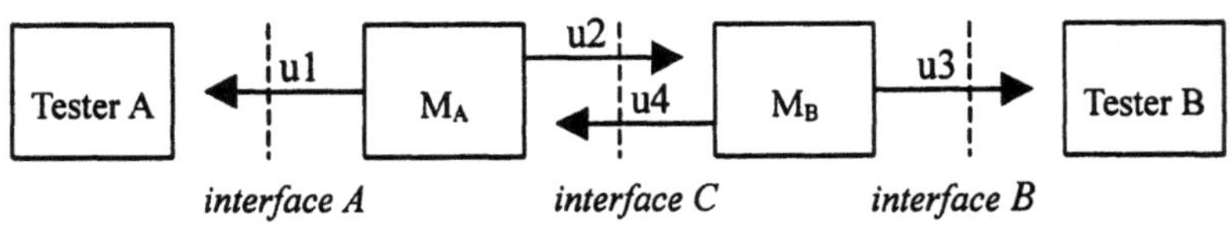

Figure 2. Output Symbols of two FSMs.

(4) $\mathbf{L_{\prod,out}} = \{ U_{\prod 1}, U_{\prod 2}, \ldots, U_{\prod k} \}$ is a set of output symbols. As shown in Figure 2, $U_{\prod i}$, an element of $\mathbf{L_{\prod,out}}$, is represented as vector <u1, u2, u3, u4>, where u1, u2, u3, and u4 represent messages transmitted by M_A via

interface A, M_A via interface C, M_B via interface B, and by M_B via interface C respectively.

(5) $\mathbf{Tr}_{\Pi} \subseteq \{ S_{\Pi} \text{—} v_{\Pi}/U_{\Pi} \rightarrow S_{\Pi}' \mid S_{\Pi}, S_{\Pi}' \in \mathbf{S}_{\Pi} \wedge v_{\Pi} \in \mathbf{L}_{\Pi,in} \wedge U_{\Pi} \in \mathbf{L}_{\Pi,out} \}$ is a set of transitions.

Note that set of transitions $\mathbf{Tr}_{\Pi}$ derived by the interoperability test suite derivation algorithm [KanK 97, ShiK 98] includes not only interoperability test cases but also conformance test cases.

3. ENHANCEMENT OF THE INTEROPERABILITY TEST SUITE DERIVATION ALGORITHM

In this section, we propose our enhanced algorithm by analyzing problems in the algorithm shown in [ShiK 98].

3.1 Problems in the Existing Algorithm

The existing algorithm proposed by [ShiK 98] has a problem that the algorithm is not applicable to a case where there are more than two messages between two IUTs for an external input. Namely, the number of internal messages caused by an external input are limited to two. These internal messages such as u2 and u4 in Figure 2 do not interact with the environment.

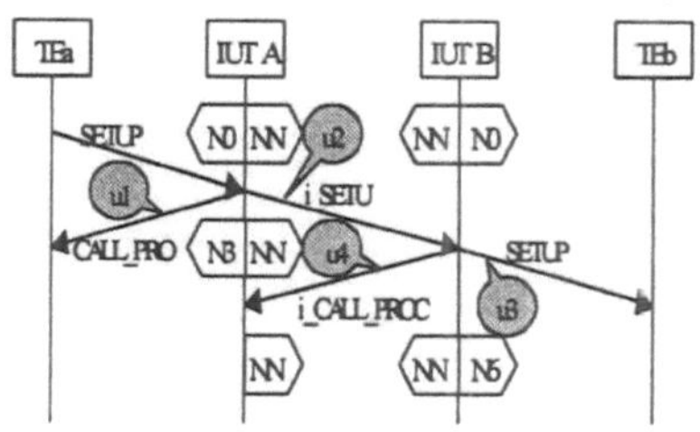

Figure 3. An example of message interaction

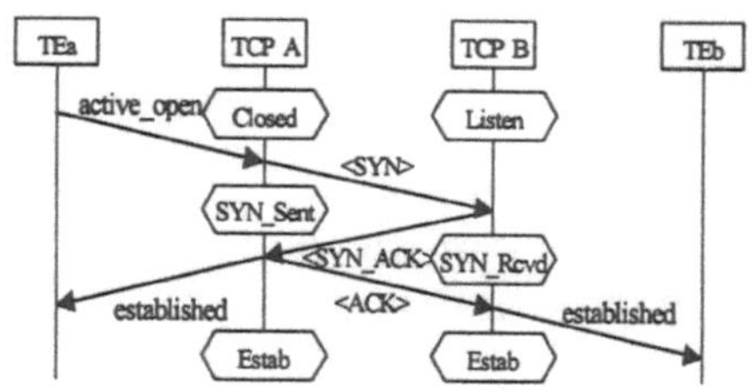

Figure 4. An example of message interaction for the ATM/B-ISDN signaling protocol for the TCP.

In the case of ATM/B-ISDN signaling protocol as shown in Figure 3, internal messages between two IUTs can be represented with two elements (i.e., u2 and u4) in a vector <u1, u2, u3, u4> defined in Definition 2 - (4).
In general protocols, however, internal messages can not be represented with this data structure when the number of messages exchanged between two IUTs is more than two. For instance, TCP connection is made with a three-way handshake. So the three messages, SYN, SYN_ACK, and ACK occur at the interface C located between TCP A and TCP B as shown in Figure 4. These messages can not be represented only with u2 and u4.

3.2 The Enhanced Algorithm

We solve the problem pointed out in the previous section by redefining the Definition 2 – (4) as follows:

Definition 2′.

(1) $W_{\Pi} = [U_{\Pi,1} \oplus \ldots \oplus U_{\Pi,i} \oplus \ldots \oplus U_{\Pi,l}]$ is the concatenation of output symbols sequence for an external input. The output symbols $U_{\Pi,i}$ is represented by the vector <u1, u2, u3, u4> where u1, u2, u3, and u4 represent messages transmitted by M_A via interface A, M_A via interface C, M_B via interface B, and by M_B via interface C respectively.

(2) $\mathbf{L}_{\Pi,out} = \{ W_{\Pi 1}, W_{\Pi 2}, \ldots, W_{\Pi k} \}$ is a set of output symbol sequences.

The following is an enhanced algorithm adhering to the Definition 2′. The statements in the dashed box are the part of original algorithm [ShiK 98], and the statements appended or modified from the algorithm are represented in shadow.

```
L_Π,in := L_in,A,E ∪ L_in,B,E
S_Π,0 := ( S_A,0, S_B,0 )
S_Π := { S_Π,0 }
Tr_Π := {}
IOPTS := {}
NEW := S_Π
while NEW≠∅ do begin
    gs_i := choose-any(NEW)
    NEW := NEW – {gs_i}
        Input := L_Π,in
        while Input≠∅ do begin
            v := choose-any(Input)
            Input := Input-{v}              [ U_out := [ , , , ] ]
            U_Π,1 :=< , , , >
            W_Π := [ U_Π,1 ]
            gs_f := state( gs_i | v )
            W_Π := output'( gs_i | v )      [ U_out := ouput( gs_i | v )
            L_Π,out := L_Π,out ∪ W_Π          L_Π,out := L_Π,out ∪ {U_out}
            Tr_Π := Tr_Π ∪ {gs_i—v/W_Π→gs_f}  Tr_Π := Tr_Π ∪ {gs_i—v/U_out→gs_f} ]
            if ( gs_f ∉ S_Π)
            then begin
                S_Π := S_Π ∪ {gs_f}
                NEW := NEW ∪ {gs_f}
            end
            if ( U_Π,1[2]≠NULL ∨ U_Π,1[4]≠NULL )   [ ( U_out[2]≠NULL ∨ U_out[3]≠NULL ) ]
```

```
25:         then
26:             IOPTS := IOPTS ∪ { gs_i—v/W_Π→gs_f }
27:     end while
28:end while
                                        { gs_i—v/U_out→gs_f }
```

Algorithm 1. Enhanced algorithm for interoperability test suite derivation.

Now we confirm that the enhanced algorithm is applicable to the case in Figure 4. In line 1 of the Algorithm 1, a set of input symbols $\mathbf{L_{\Pi,in}}$ is initialized by external inputs such as *passive_open*, *close*, *send_data*, and *active_open* which were shown in TCP FSM. By initialization steps from line 1 to 6, an initial value of the set *NEW* becomes the initial state (Closed, Closed). To explain an example in Figure 4, we assume that the start stable state (Closed, Listen) has been already added to the set *NEW* as a result of progress of the algorithm. Also, we assume that start stable state *gs_i* gets state (Closed, Listen) from the set *NEW* in line 8 and the input symbol *v* gets the input symbol *active_open* from a set of input symbols Input in line 12. Then, W_Π becomes [< , SYN, ,SYN_ACK >, < established, ACK, established, >] by the function output′() in line 16 after it is initialized in line 14. These output symbols can not be represented with the existing algorithm's data structure U_Π (i.e., <u1, u2, u3, u4>). Figure 5 summarizes the proceeding of the algorithm.

- start stable state *gs_i* = (Closed, Listen)
- input symbol *v* = *active_open*
- final stable state *gs_f* = (Estab, Estab)
- output symbol W_Π = [$U_{\Pi,1}$, $U_{\Pi,2}$], where
  $U_{\Pi,1}$ = < , SYN, , SYN_ACK> and
  $U_{\Pi,2}$ = <established, ACK, established, >
- transition Tr_Π = (Closed, Listen) — active_open / [< , SYN, , SYN_ACK>, <established, ACK, established, >] → (Estab, Estab)

Figure 5. Application of Algorithm 1 to the example in Figure 4.

Our enhanced algorithm is the generalized one which is applicable to protocols in general since the algorithm have no limitation on the number of internal messages.

The program implementing Algorithm 1 gets FSMs corresponding to two IUTs as a file, executes the data in the file, and writes its result (namely, interoperability test suite) to a file like the format in Appendixes 1 and 2. The time taken for generating each test suites is negligible.

4. INTEROPERABILITY TEST SUITE DERIVATION

In this section, we demonstrate the feasibility of Algorithm 1 in Section 3 by applying the implementation of the algorithm to ATM/B-ISDN signaling protocol used in [ShiK 98]. Then we show the generality of this algorithm through derivation of interoperability test suite for connection management (establishment/release) part of TCP.

4.1 Application to ATM/B-ISDN Signaling Protocol

The result of applying the implementation of the enhanced algorithm to ATM/B-ISDN signaling protocol is described in Appendix 1. Out of the thirty-two test cases derived, twenty-six test cases are for interoperability testing and six test cases are for conformance testing. The result of this application is the same as that of the interoperability test suite derived manually in [ShiK 98]. The test cases for conformance testing in Appendix 1 were not given in [ShiK 98].

Let us explain an example of interoperability test case corresponding to the scenario in Figure 3. The test case is equivalent to item (1) of Appendix 1 and is as follows:

(0:0,0:0) -- SETUPa/[<CALL_PROC,i_SETUP,SETUP,i_CALL_PROC>] → (3:9,6:3)

An input symbol with postfix 'a' (or 'b') represents a message transmitted by tester TEa (or TEb) via interface A (or interface B). 'i_' is used to indicate internal messages as opposed to external messages. Each element of the starting stable state (0:0, 0:0) denotes state N0:NN0 of IUTa and state N0:NN0 of IUTb, respectively. SETUPa before '/' is an input symbol to IUTa and [<CALL_PROC, i_SETUP, SETUP, i_CALL_PROC>] after '/' is output symbol sequence.

4.2 Application to TCP

In this section we demonstrate the generality of this algorithm by applying it to TCP. In this paper, we consider only the connection management part of TCP.

4.2.1 Modeling TCP interoperability testing architecture

Testers A and B of test architectures in Figure 1 correspond to Lower Testers for IUT A and IUT B, respectively. To test interoperability of TCP we need to modify Test Architecture II. The modified test architecture is in Figure 6. It has two Upper Testers that are able to access user interfaces of TCP A and TCP B, respectively. In spite of this modification, the same algorithm can be used to derive interoperability test suite. In case that we apply the algorithm to derive interoperability test suites from the point of view of the users of the interacting protocols or systems, the method can be utilized for service testing which tests service interoperability of protocols or systems under test. Upper Tester controls and observes TCP through Point of Control and Observation (PCO). For example, in the case of socket communication, Upper Tester makes use of Abstract Service Primitives (ASPs) such as *socket*, *bind*, *listen*, *accept*, and *connect* to control and observe TCP. Two testers may be merged in a local test environment whereas testers may be designed to communicate with each other through a separate communication channel in a distributed test environment.

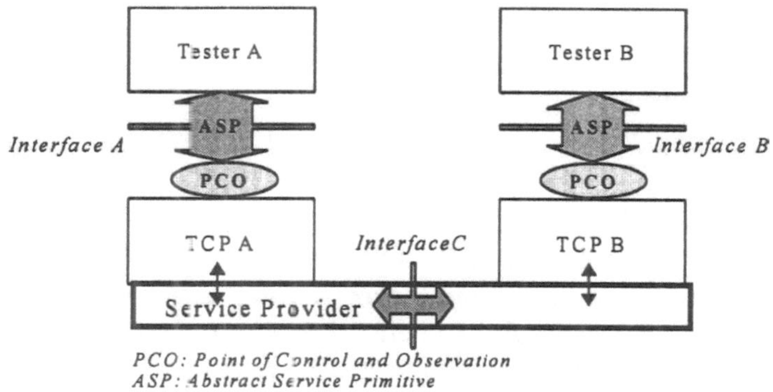

Figure 6. Architecture for TCP interoperability.

4.2.2 FSM representation of TCP

Figure 7 is an FSM for connection management part of TCP [GarR 95]. The FSM has eleven states and describes two procedures: one for connection establishment from state CLOSED to state ESTABLISHED and the other for connection release. Data transfer procedure at state ESTABLISHED is not dealt with here. Note that the FSM has simultaneous open and simultaneous close procedures. When both systems actively ask for a connection simultaneously, the procedure is called simultaneous open or simultaneous connection establishment. A simultaneous close occurs when both applications initiate a close at the same time. Input is given by the upper application program running on TCP and output is presented by one or more bit selection among URG, ACK, PSH, RST, SYN, and FIN in the 6-bit control field of TCP packet. TCP controls connection management by selection of these bits.

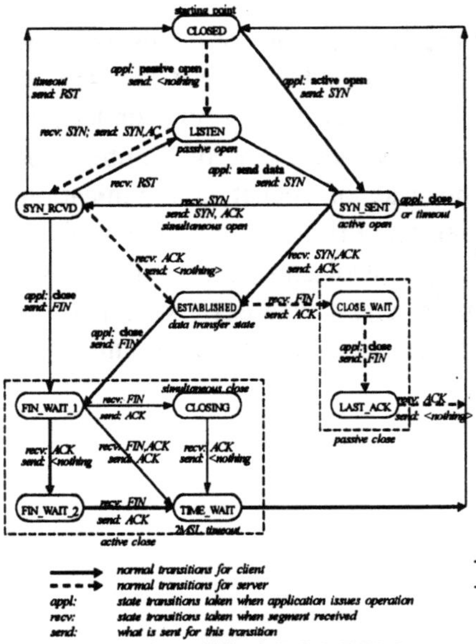

Figure 7. Simplified TCP FSM.

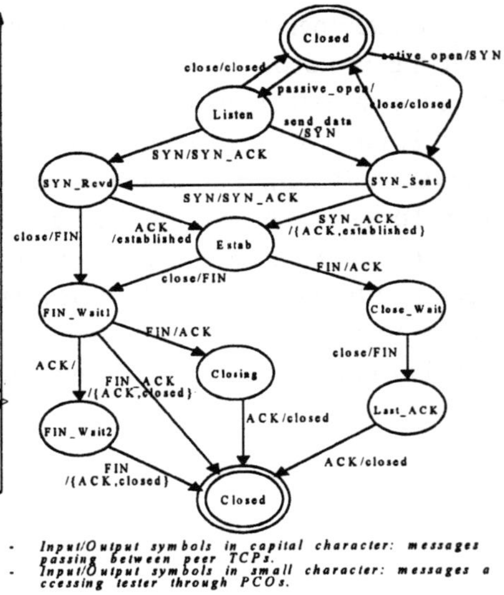

Figure 8. TCP FSM

.In this paper, we make use of the FSM in Figure 8 obtained by omitting the behavior related to timer from Figure 7. The reason for the omission is that the tester cannot control the internal behavior of IUT relating to time. This omission makes it impossible to test RST message handling the part that invokes connection re-establishment. In the FSM, there are three types of messages: messages from tester to TCP, messages between two TCPs represented in bold characters, and messages (which are responses of socket communications) from TCP to the tester through PCO. The last type messages are extracted from the basic socket system calls [Rich 91].

4.2.3 Application result

The application result of the implementation of the enhanced algorithm to the connection management part of TCP is described in Appendix 2. Out of the thirty-two test cases derived, sixteen test cases are for interoperability testing and sixteen test cases for conformance testing. The ratio of the number of the conformance test cases over that of the interoperability ones for TCP is higher than that of ATM/B-ISDN signaling protocol because of the fact that there are several paths for connection establishment in TCP. Namely TCP has a negotiation procedure returning to the initial state when there is some disagreement between peer TCPs. For example, TCP FSM in Figure 8 returns to state Closed if IUT does not receive a proper message after it transits to state SYN_Sent by receiving an active_open message from state Closed. TCP establishes its connection via passing through states Listen or SYN_Sent from the initial state whereas ATM/B-ISDN signaling protocol establishes its connection by a single SETUP message. Since TCP has local behavior without any message exchange with peer TCP as the above, more conformance test cases are needed.

Let us show an example of test case for the message interaction in Figure 4. This test case is equivalent to item (13) of Appendix 2 and as follows:

(Closed,Listen) -- active_open_a/[<,i_SYN, ,i_SYN_ACK>, <established, i_ACK, established,>] → (Estab,Estab)

5. TRANSITION COVERAGE

In this section we investigate transition coverage of the TCP interoperability test suite derived in Section 4. For this, we view the TCP FSM simulator as the IUT and carry out two case studies of transition coverage: one for that of the derived interoperability test suite and the other for that of validation by a commercial SDL tool. In the latter case, if there are no errors in the specification and its SDL model, we should get 100% coverage provided that the validation method of choice is complete. We first explain the two cases and then we compare their respective transition coverage. Also we analyze the reasons why certain transitions are not covered.

5.1 Two Case Studies of Transition Coverage Calculation

This section conducts two case studies of calculating transition coverage. The first one (Case I) is carried out by applying the commercial SDL tool to the IUT. Let us denote by TC1 the transition coverage of this case. The second case (Case II) is carried out by applying the interoperability test suite utilizing the simulation facility of the commercial tool. Let us denote by TC2 the transition coverage of this case. Since Case I is for validation and Case II is for interoperability testing, their purposes are different. However, it is meaningful to compare TC2 with TC1 because the TC1 is obtained by exercising mechanically and completely all the transitions in a step by step manner.

Case I is carried out in Bit-State mode which examines all possible transitions for each state. The Bit-state mode is a complete validation method. Case I is carried out under the validation architecture of Figure 9. This structure is designed such that input message to the two IUTs are applies from the external environment. When the Bit-State mode is used, this external input should be selected automatically by the commercial tool.

In Case I, the validation tool controls the external environment whereas in Case II the tester controls it. Thus as shown in Figure 10 the test architecture for Case II is designed such that the tester communicates with the external environment of the validation architecture. Moreover, since Case II is for interoperability testing, in order to apply one or more interoperability test cases, each test case should start in the initial state. For this, IUT′ used in the test architecture of Case II is augmented with transitions that take us to the initial state upon receiving reset message in each state.

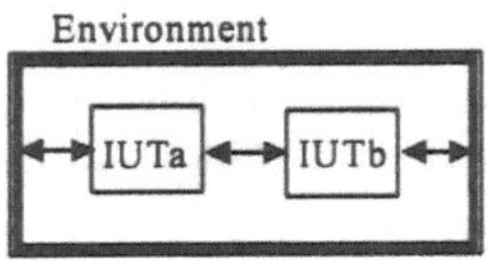

Figure 9. Validation Architecture.

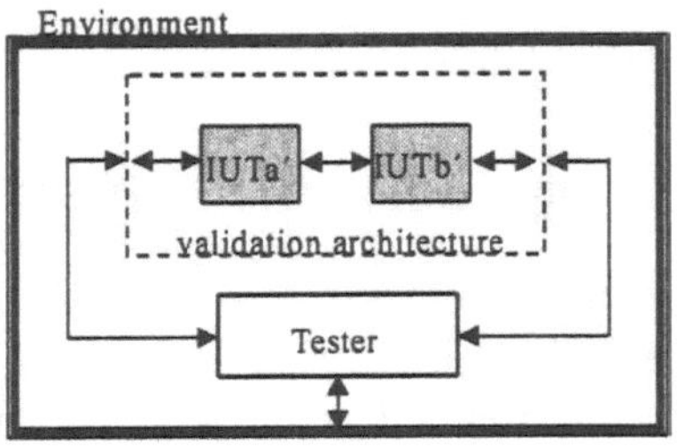

Figure 10. Test Architecture.

Since the test architecture of Case II follows the test architecture suggested in Figure 6, it requires two testers. But we assume that one tester performs the role of two testers and the underlying layers are reliable and correct and consider only Protocol Data Unit (PDU) for TCP.
Since Case II is for interoperability testing, Case II comparing with Case I has the following properties:

P1. There are transitions from any state to the initial state when the IUT receives reset signal from the tester. Uncovered transitions out of those transition executed by reset signal should be excluded in calculating transition coverage because those ones are added for testing purpose.

P2. The test cases are for interoperability testing. Thus, uncovered transition out of those transitions executed for conformance testing should be excluded in calculating the transition coverage of the interoperability testing.

5.2 The Comparison Result

Figure 11 depicts the SDL model of TCP FSM shown in Figure 8. The two target IUTs of interoperability testing works based on the same FSM. Therefore, we examine transition coverage with respect to one IUT's FSM only, that for the other one being the same. The total transition coverage for the whole system of IUTs is twice that of one IUT.

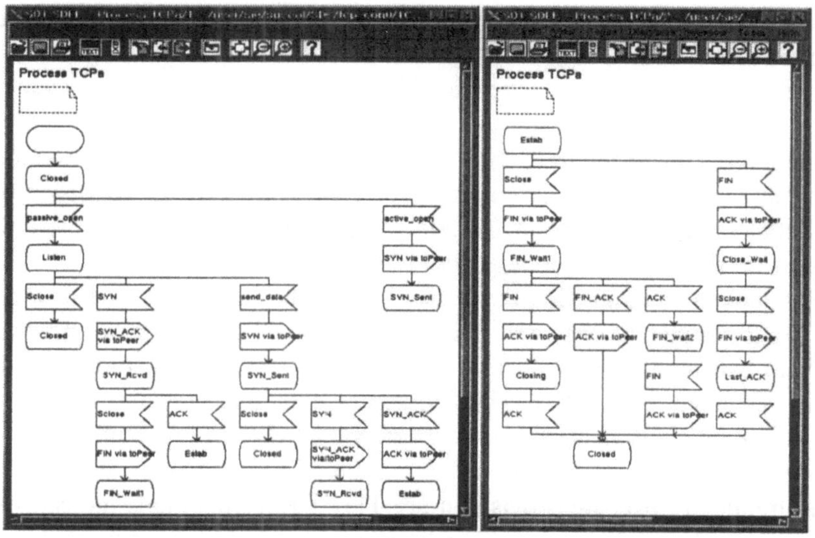

Figure 11. SDL Description[3].

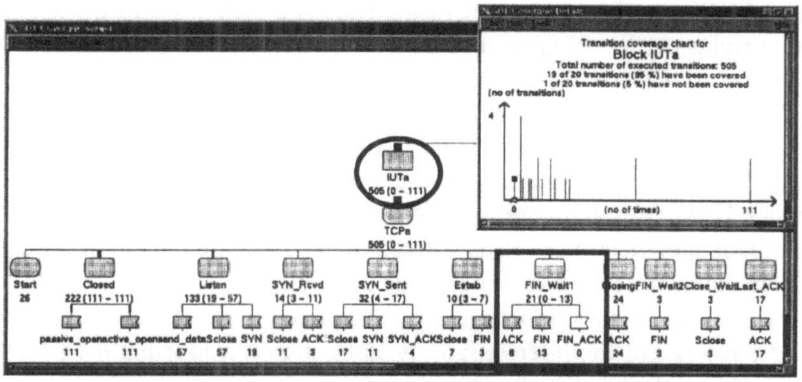

Figure 12. Transition coverage tree and chart for Case I.

The transition coverage for Case I obtained by validation under the architecture of Figure 9 is shown in Figure 12. The left-hand side figure of Figure 12 depicts the part for the IUTa among the whole transition coverage tree. The size of shaded parts indicates the ratio of the executed transitions. In order to illustrate the numbers, let us take the example of "505(0-111)" in the circle, in which:

- 505 transition have been executed,

[3] The word 'close' is reserved in the commercial SDL tool. Thus we prefix it with 'S'.

- the least frequently executed transition in block IUTa has been executed zero times, and
- the most frequently executed transition in block IUTa has been executed 111 times.

In the lower part of the tree, there is one transition that was never executed. It is represented by 0 in the rectangle of Figure 12. It is the transition that was not executed upon receiving FIN_ACK in FIN_Wait1 state. Note that there is no shade for the input symbol FIN_ACK.

Let us look at the transition coverage chart on the right-hand side of Figure 12. Total number of executed transitions is 505, the number of transitions that are executed at least once is 19 (95%) and that of transitions that are never executed is 1 (5%). The horizontal axis indicates the number of executions of the transition under consideration. When this number is 0, it means that the transition was never executed. The execution times of the most frequently executed transition are 111. The vertical axis indicates the number of relevant transitions. The transition that was never executed is the transition from FIN_Wait1 to Closed state among those marked with thick arrows in IUT FSM of Figure 14. sIn the TCP FSM[4] of Figure 7 shown in a book [GarR 95], there are three paths from FIN_Wait1 state to Closed state: (1) upon receiving FIN, move to Closing state and then move to Closed state upon receiving ACK, (2) upon receiving ACK, move to FIN_Wait2 state and then move to Closed state upon receiving FIN and (3) upon receiving FIN_ACK, move to Closed state. In the last case, FIN_ACK means that among the control bits of the packet both FIN bit and ACK bit are set. Even though there are three possibilities, there is no state sending FIN_ACK corresponding to (3) and therefore this transition cannot be executed. That is, the protocol specification itself is not complete. Thus the transition coverage would be 100% if we do not consider the transition.

Let us examine the transition coverage for Case II. Here we want to calculate the test coverage of the interoperability test suite derived using the program realizing Algorithm 1. It is shown in Figure 13.

The left-hand side of Figure 13 depicts only the part with 0% coverage for IUTa among the whole transition coverage tree. As shown, there are 13 transitions in IUTa that are never executed. The chart in the right-hand side of Figure 13 represents that total number of executed transitions is 41, and 17 (57%) transitions are covered and 13 (43%) ones are not covered out of the total 30 transitions. Note that 10 transitions triggering by reset input out of the total 30 ones were added for testing purpose (See property P1 in Section 5.1). The 10 transitions consist of 7 ones out of the 13 uncovered

---

[4] In the TCP FSM of RFC [RFC 81], on which the FSM of [GarR 95] is based, there is no transition to take from FIN_Wait1 state to Closed state upon receiving FIN_ACK.

ones and 3 ones out of the 17 covered ones (Those data are in columns T_{reset} and Total in Table 1). Considering transitions corresponding with the total 20 transitions of the TCP specification which was dealt with in Case I, the number of uncovered transitions is 6 (which is equivalent to the sum of corresponding T_{IOPT} and T_{CT} in Table 1) and these transitions are marked with thick arrows in Figure 14.

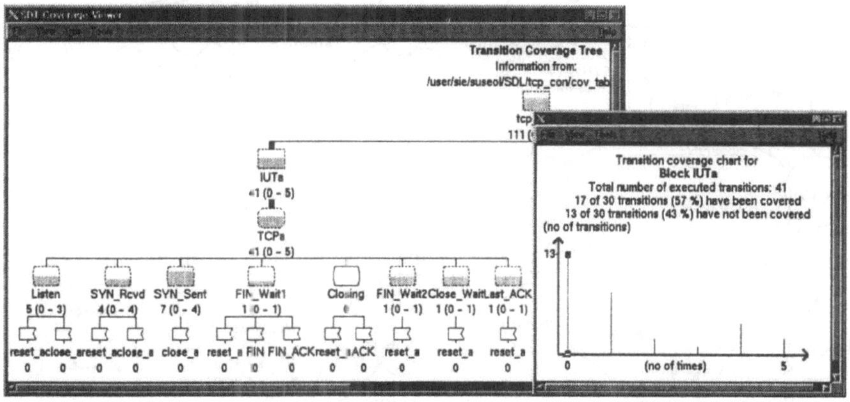

Figure 13. Transition coverage tree and chart for Case II.

5.3 Result Analysis

Transitions need to be classified in order to analyze the transition coverage of Case II.

Definition 3. Transition classification

- T_{CT}: those transitions not having messages passing through interface C.
- T_{IOPT}: those transitions having one or more than one messages passing through interface C.
- T_{init}: the transition to initialize FSM.
- T_{reset}: those transitions moving from any state to the initial state by reset input.

According to Definition 3, we classify the transitions as Figure 15.

The following Table represents the transition coverage obtained by Case II.

Table 1. Transition coverage analysis on Case II.

Transition Classification	T_{IOPT}		T_{CT}	T_{init}	T_{reset}	Total
The number of covered transitions	12	(75%)	1	1	3	17
The number of uncovered transitions	4	(25%)	2	0	7	13
Total	16	(100%)	3	1	10	30

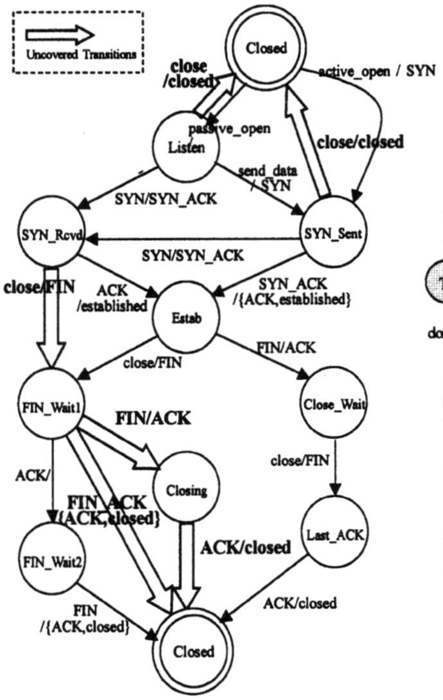

Figure 14. Uncovered transitions.

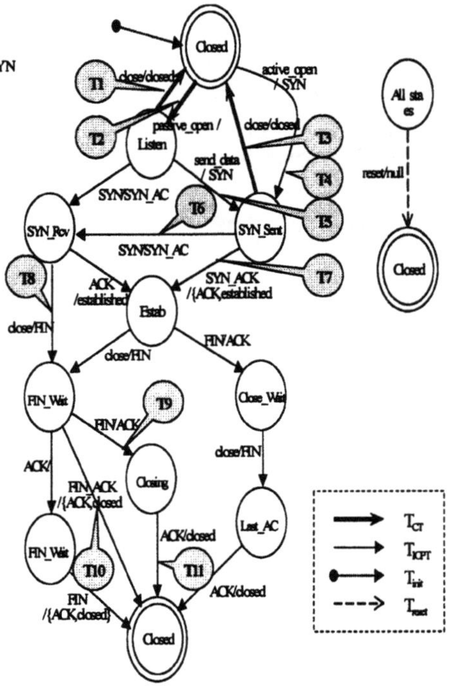

Figure 15. Transition classification.

Since T_{CT} (i.e., T1, T2, and T3 in Figure 15) corresponds to conformance testing, those transitions are not dealt with in Case II dealing with interoperability testing (See property P2 in Section 5.1). Note that one transition, T2, is executed during interoperability testing because the transition belongs to the preamble of some test cases for interoperability testing.

Let us analyze transition coverage of T_{IOPT}. The uncovered transitions in T_{IOPT} are T8, T9, T10 and T11. Transition T10 is not covered because of incompleteness of the specification as mentioned before. Transitions T8, T9 and T11 are not executed because of the limitation of our algorithm (the limitation will be discussed briefly later in this section). Transition coverage of validation in Case I becomes 100% (15 transitions out of 15 ones in T_{IOPT}) if we exclude T10 which is not executed due to incompleteness of the specification. And transition coverage of our method applying interoperability test cases in Case II becomes 80% (12 transitions out of 15 ones in T_{IOPT}) if we exclude transition T10 as stated above. In addition, if we calculate the transition coverage including conformance testing, since all of three transitions in T_{CT} in Table 1 can be executed, the transition coverage

becomes 83% because of covering 15 transitions (12 transitions in T_{IOPT} + 3 transitions in T_{CT}) out of 18 ones (15 in T_{IOPT} + 3 in T_{CT}).
Total number of executed transitions to obtain 100% transition coverage in validation (Case I) is 505 whereas that of obtaining 80% in Case II is 41. It is not possible to compare directly the number of executed transitions to obtain each transition coverage since those two cases are performed for different purposes. However, the data are very informative since interoperability testing is a kind of test exploring the total state space (which validation covers) with limited time and resource. As summarized in Table 2, it is worth deriving interoperability test cases having significant transition coverage spending relatively little time and resource.

Table 2. Transition coverage and cost.

	Transition coverage	Number of executed transitions
Case I	100 % (15/15)	505
Case II	80 % (12/15)	41

Let us discuss briefly the limitation and possible extension of our algorithm which is further work. Transitions T8, T9 and T11 are not covered because of the *single stimulus principle* and test architecture which our algorithm is based on. Transitions T9 and T11 can not be executed because of the *single stimulus principle* since those transitions require two inputs at the same time. As the solution, we can adopt simultaneous stimulus principle meaning that two inputs can be applied to both IUTs at the same time. Since the order of events invoked by the simultaneous stimulus principle is nondeterministic, we may require stochastic approach to this control and observation. Also transition T8 can be performed in Test Architecture I (in Figure 1) having a tester between two IUTs. Since the tester is able to capture packets in transmission, packet loss can be simulated. By extending our method, those transitions can be covered and thus the transition coverage of interoperability test suite becomes 100% which is equivalent to that of validation.

6. CONCLUSION AND FUTURE WORK

In this paper, we demonstrated that the test suite derivation for the interoperability testing of protocols is possible efficiently and effectively in an automatic way. In order to do that, we enhanced the algorithm in [ShiK 98], implemented our enhanced algorithm as a program, and applied it to the ATM/B-ISDN signaling protocol and TCP protocol. As its results, we showed that our implementation of the enhanced algorithm generates the same interoperability test cases successfully with the test cases generated manually in [ShiK 98]. Also, we carried out an interoperability testing for the connection management part of the TCP by applying the derived test suite utilizing the simulation facility of the commercial SDL tool. The transition coverage in this case and the transition coverage when validation facility is utilized were compared and analyzed. As the result, except for the transition described ambiguously in the protocol specification, for obtaining 100% transition coverage in validation it took 505 transition execution whereas it took only 41 transition executions for obtaining 80% when we apply the interoperability test suite derived in this paper. Since in general the time and resources required is proportional to the amount of execution of transitions, it is shown in this paper that the method is very efficient for the interoperability testing of protocols with limited time and resource comparing with validation. Moreover, the testing cost of interoperability could go down conspicuously while giving the same transition coverage as the validation one if we adopt a method (which need to work further) handling simultaneous open/close and message loss.

As a further work, one may think of extracting a test suite that has a couple of external inputs at once, by extending data structure used in our algorithm. In addition to that, further studies may be needed on testing FSM including primitives such as signal delay and data representation, and on testing protocols and systems in terms of service testing.

REFERENCES

[APRS 93] Arakawa, N., Phalippou, M., Risser, N. and Soneoka, T., "Combination of conformance and interoperability testing", Formal Description Techniques, V (C-10) M. Diaz and R. Groz (Eds.), Elsevier Science Publishers, 1993.

[AraS 92] Arakawa, N. and Soneoka, T., "A Test Case Generation Method for Concurrent Programs", Protocol Test Systems, IV, Elsevier Science Publishers, 1992.

[GarR 95] Gary R. Wright, and W. Richard Stevens, "TCP/IP Illustrated, Volume2 The Implementation", Addison-Wesley Professional Computing Series, 1995.

[ISO 94] ISO/IEC JTC1 DTR-10000, "Information Technology – Framework and classification of international standard protocol", 1994.

[KanK 97] Kang, S., and Kim, M., "Interoperability Test Suite Derivation for Symmetric Communication Protocols", IFIP Joint International Conference on Formal Description Techniques (FORTE X) and Protocol Specification Testing and Verification (PSTV XVII), pp. 57-72, November 1997.

[LuBP 94] Luo, G., Bochmann G. And Petrenko, A., "Test Selection Based on communicating Nondeterministic Finite-State Machines Using a Generalized Wp-Method", IEEE Transactions on S.E., Vol 20, No. 2, pp. 149-162, February 1994.

[RafC 90] Rafiq, O. and Castanet, R., "From Conformance Testing to Interoperability testing", Proceedings of the 3rd International Workshop on Protocol Test System, 1990.

[RFC 81] Postel, J. B., "Transmission Control Protocol", RFC 793, September 1981.

[Rich 91] W. Richard Stevens, "UNIX Network Programming", Prentice Hall, 1991.

[SDT 97] "Telelogic SDT 3.2 Manuals", Telelogic, September 1997.

[ShiK 98] Shin, J., and Kang, S., "Interoperability Test Suite Derivation for the ATM/B-ISDN Signaling Protocol", Testing of Communicating Systems, Vol 11, Kluwer Academic Publishers, pp. 313-330, 1998.

[VerB 94] Vermeer, G. S., and Blik, H., "Interoperability testing: Basis for the Acceptance of Communicating Systems", Protocol Test System, VI(C-19), Elsevier Science Publishers, 1994.

APPENDIX 1. TEST SUITE FOR ATM/B-ISDN SIGNALING PROTOCOL.

(1) (0:0,0:0) -- SETUPa/[<CALL_PROC,i_SETUP,SETUP,i_CALL_PROC>] --> (3:9,6:3)

(2) (0:0,0:0) -- SETUPb/[<SETUP,i_CALL_PROC,CALL_PROC,i_SETUP>] --> (6:3,3:9)

(3) (3:9,6:3) -- RELa/[<REL COMP,i REL,RELb,i REL COMP>] --> (0:0,12:0)

(4) (3:9,6:3) -- REL_COMPa/[<,i_REL,RELb,i_REL_COMP>] --> (0:0,12:0)

(5) (3:9,6:3) -- CONNb/[<CONN,,CONN_ACK,i_CONN>] --> (10:10,10:10)

(6) (3:9,6:3) -- CALL PROCb/[<,,,>] --> (3:9,9:3)

(7) (3:9,6:3) -- RELb/[<RELa,i_REL_COMP,REL_COMP,i_REL>] --> (12:0,0:0)

(8) (3:9,6:3) -- REL_COMPb/[<RELa,i_REL_COMP,,i_REL>] --> (12:0,0:0)

(9) (6:3,3:9) -- CONNa/[<CONN_ACK,i_CONN,CONN,>] --> (10:10,10:10)

(10) (6:3,3:9) -- CALL_PROCa/[<,,,>] --> (9:3,3:9)

(11) (6:3,3:9) -- RELa/[<REL_COMP,i_REL,RELb,i_REL_COMP>] --> (0:0,12:0)

(12) (6:3,3:9) -- REL COMPa/[<,i REL,RELb,i REL COMP>] --> (0:0,12:0)

(13) (6:3,3:9) -- RELb/[<RELa,i_REL_COMP,REL_COMP,i_REL>] --> (12:0,0:0)

(14) (6:3,3:9) -- REL_COMPb/[<RELa,i_REL_COMP,,i_REL>] --> (12:0,0:0)

(15) (0:0,12:0) -- RELb/[<,,,>] --> (0:0,0:0)

(16) (0:0,12:0) -- REL_COMPb/[<,,,>] --> (0:0,0:0)

(17) (10:10,10:10) -- RELa/[<REL_COMP,i_REL,RELb,i_REL_COMP>] --> (0:0,12:0)

(18) (10:10,10:10) -- REL COMPa/[<,i REL,RELb,i REL COMP>] --> (0:0,12:0)

(19) (10:10,10:10) -- RELb/[<RELa,i_REL_COMP,REL_COMP,i_REL>] --> (12:0,0:0)

(20) (10:10,10:10) -- REL_COMPb/[<RELa,i_REL_COMP,,i_REL>] --> (12:0,0:0)

(21) (3:9,9:3) -- RELa/[<REL COMP,i REL,RELb,i REL COMP>] --> (0:0,12:0)

(22) (3:9,9:3) -- REL_COMPa/[<,i_REL,RELb,i_REL_COMP>] --> (0:0,12:0)

(23) (3:9,9:3) -- CONNb/[<CONN,,CONN_ACK,i_CONN>] --> (10:10,10:10)

(24) (3:9,9:3) -- RELb/[<RELa,i REL COMP,REL COMP,i REL>] --> (12:0,0:0)

(25) (3:9,9:3) -- REL_COMPb/[<RELa,i_REL_COMP,,i_REL>] --> (12:0,0:0)

(26) (12:0,0:0) -- RELa/[<,,,>] --> (0:0,0:0)

(27) (12:0,0:0) -- REL COMPa/[<,,,>] --> (0:0,0:0)

(28) (9:3,3:9) -- CONNa/[<CONN_ACK,i_CONN,CONN,>] --> (10:10,10:10)

(29) (9:3,3:9) -- RELa/[<REL_COMP,i_REL,RELb,i_REL_COMP>] --> (0:0,12:0)

(30) (9:3,3:9) -- REL COMPa/[<,i REL,RELb,i REL COMP>] --> (0:0,12:0)

(31) (9:3,3:9) -- RELb/[<RELa,i_REL_COMP,REL_COMP,i_REL>] --> (12:0,0:0)

(32) (9:3,3:9) -- REL_COMPb/[<RELa,i_REL_COMP,,i_REL>] --> (12:0,0:0)

: interoperability test cases (Total number is 26) *: conformance test cases (Total number is 6)*

- A stable state and output symbols are represented as (S_{UNIa}, S_{PNNIa}, S_{UNIb}, S_{PNNIb}) and [<a message from UNIa via interface A, a message from PNNIa via interface C, a message from UNIb via interface B, a message from PNNIb via interface C>], respectively.

APPENDIX 2. TEST SUITE FOR CONNECTION MANAGEMENT PART OF TCP.

(1) (Closed,Closed) -- passive_open_a/[<,,,>] --> (Listen,Closed)
(2) (Closed,Closed) -- active_open_a/[<,i_SYN,,>] --> (SYN_Sent,Closed)
(3) (Closed,Closed) -- passive_open_b/[<,,,>] --> (Closed,Listen)
(4) (Closed,Closed) -- active_open_b/[<,,,i_SYN>] --> (Closed,SYN_Sent)
(5) (Listen,Closed) -- close_a/[<closed,,,>] --> (Closed,Closed)
(6) (Listen,Closed) -- send_data_a/[<,i_SYN,,>] --> (SYN_Sent,Closed)
(7) (Listen,Closed) -- passive_open_b/[<,,,>] --> (Listen,Listen)
(8) (Listen,Closed) -- active_open_b/[<,i_SYN_ACK,,i_SYN>, <established,,established,i_ACK >] --> (Estab,Estab)
(9) (SYN_Sent,Closed) -- close_a/[<closed,,,>] --> (Closed,Closed)
(10) (SYN_Sent,Closed) -- passive_open_b/[<,,,>] --> (SYN_Sent,Listen)
(11) (SYN_Sent,Closed) -- active_open_b/[<,i_SYN_ACK,,i_SYN>, <established,,established,i_ACK >] --> (Estab,Estab)
(12) (Closed,Listen) -- passive_open_a/[<,,,>] --> (Listen,Listen)
(13) (Closed,Listen) -- active_open_a/[<,i_SYN,,i_SYN_ACK>, <established,i_ACK ,established,>] --> (Estab,Estab)
(14) (Closed,Listen) -- close_b/[<,,closed,>] --> (Closed,Closed)
(15) (Closed,Listen) -- send_data_b/[<,,,i_SYN>] --> (Closed,SYN_Sent)
(16) (Closed,SYN_Sent) -- passive_open_a/[<,,,>] --> (Listen,SYN_Sent)
(17) (Closed,SYN_Sent) -- active_open_a/[<,i_SYN,,i_SYN_ACK>, <established,i_ACK ,established,>] --> (Estab,Estab)
(18) (Closed,SYN_Sent) -- close_b/[<,,closed,>] --> (Closed,Closed)
(19) (Listen,Listen) -- close_a/[<closed,,,>] --> (Closed,Listen)
(20) (Listen,Listen) -- send_data_a/[<,i_SYN,,i_SYN_ACK>, <established,i_ACK ,established,>] --> (Estab,Estab)
(21) (Listen,Listen) -- close_b/[<,,closed,>] --> (Listen,Closed)
(22) (Listen,Listen) -- send_data_b/[<,i_SYN_ACK,,i_SYN>, <established,,established,i_ACK >] --> (Estab,Estab)
(23) (Estab,Estab) -- close_a/[<,i_FIN,,i_ACK>] --> (FIN_Wait-2,Close_Wait)
(24) (Estab,Estab) -- close_b/[<,i_ACK,,i_FIN>] --> (Close_Wait,FIN_Wait-2)
(25) (SYN_Sent,Listen) -- close_a/[<closed,,,>] --> (Closed,Listen)
(26) (SYN_Sent,Listen) -- close_b/[<,,closed,>] --> (SYN_Sent,Closed)
(27) (SYN_Sent,Listen) -- send_data_b/[<,i_SYN_ACK,,i_SYN>, <established,,established,i_ACK >] --> (Estab,Estab)
(28) (Listen,SYN_Sent) -- close_a/[<closed,,,>] --> (Closed,SYN_Sent)
(29) (Listen,SYN_Sent) -- send_data_a/[<,i_SYN,,i_SYN_ACK>, <established,i_ACK ,established,>] --> (Estab,Estab)
(30) (Listen,SYN_Sent) -- close_b/[<,,closed,>] --> (Listen,Closed)
(31) (FIN_Wait-2,Close_Wait) -- close_b/[<closed,i_ACK,,i_FIN>] --> (Closed,Closed)
(32) (Close_Wait,FIN_Wait-2) -- close_a/[<,i_FIN,closed,i_ACK>] --> (Closed,Closed)

[shaded box] *: interoperability test cases (Total number is 16)* [empty box] *: conformance test cases (Total number is 16)*

COVERAGE ANALYSIS FOR EMBEDDED TESTING AND AN APPLICATION

Jinsong Zhu* and Son T. Vuong

Department of Computer Science
University of British Columbia
Vancouver, B.C., Canada V6T 1Z4

Abstract In this paper, we present two new results on the properties of the coverage measure for embedded testing as proposed in [10], and an application of the measure to a practical protocol. First, we identify and prove the sufficient and necessary condition for loops to exist between the test context and the test tree being constructed. This allows us to avoid producing infinite test trees during test tree construction. Then we prove that the calculation of the coverage measure is NP-complete, which indicates that there unlikely exist efficient algorithms to compute the coverage in general. Finally, in order to demonstrate the concept and practicality of the coverage measure, we developed a tool set and applied it to a practical protocol extracted from the Universal Personal Computing system.

Keywords: Protocol testing, embedded test, fault coverage, universal personal computing

*Now with Abatis Systems Corporation, Burnaby, Canada.

1. INTRODUCTION

In this study, we focus on a type of protocol testing known as conformance testing [7], which is essentially a black-box testing strategy applied to protocol systems. In conformance testing, we try to determine if an IUT complies with its specification. This will enhance our confidence on interoperability between different implementations of a common specification. Conformance testing can also be used for certifying a product.

The research on protocol conformance testing has been targeting on both single (isolated) machine testing and embedded testing. In single machine testing, the machine offers full access of its interface and therefore thorough testing may be performed. On the other hand, in embedded testing the machine is embedded within a complex system and a tester can only indirectly test the machine via other surrounding machines known as the *test context*. The area of embedded testing is still relatively young, but interests are starting to rise [6, 5, 3, 10]. While most authors have focused on generating test suites, we are more interested in test coverage assessment of a test suite and in incrementally generating test suites based on the coverage requirement.

In [10], we have proposed an approach that evaluates the coverage of embedded testing by machine identification. We enhance the approach by solving the "loop prevention" problem (see Section 3.) with a pre-identified sufficient and necessary condition, and giving a proof on the computational complexity of the coverage calculation. Furthermore, we try to model a real-life protocol as embedded testing and evaluate the test coverage of a test suite for the embedded machine.

2. COVERAGE MEASURE FOR EMBEDDED TESTING

In embedded testing, we deal with a module embedded within a complex system which is referred to as a *composite* system. Modules in a composite system communicate with each other and provide an overall system service to its users. To test an embedded module that does not offer direct access to its interface, a tester can only try to infer the conformance of the embedded module by testing the composite system. The embedded module under test (MUT) is only exercised via its test context. In our study, we describe each module in a composite system as a finite state machine (FSM), and the composite system as a system of communicating finite state machines (CFSMs).

In a previous work [10], we have defined the coverage of a test suite with respect to a specification as follows:

Definition 1 (Coverage measure) *Let T be a given test suite, S the specification, and n the upper bound of the number of states in any implementation of S. Let K_{TCM} be the number of T-conforming machines for S, and K_{SCM} is the number of S-conforming machines. The test coverage of T is defined as:*

$$Cov(S,T,n) = \frac{1}{K_{TCM} - K_{SCM} + 1}.$$

where T-conforming machines are the machines that pass the test suite T, and S-conforming machines are the T-conforming machines that conform to S. $Cov(S,T,n)$ is inversely proportional to the number of machines that pass the test suite but do not conform to the specification.

In order to better reflect the coverage values, we will use the logarithmic scale in the above definition, i.e.,

$$Cov(S,T,n) = \frac{1}{\lg(K_{TCM}) - \lg(K_{SCM}) + 1}.$$

To calculate $Cov(S,T,n)$, we propose an approach that identifies the MUT based on the test suite applied to the composite system. We first compute test trees for the MUT from the system test suite, and then "collapse" these trees into FSMs. The algorithm for generating the test trees are given in [10]. The algorithm gives us the set of T-conforming machines. The calculation of S-conforming machines is done at the composite system level, according to the following *conformance in context* (denoted $\mathbf{conf_c}$) relation:

$$M \; \mathbf{conf_c} \; S \;\text{ iff }\; C \circ M \; \mathbf{conf} \; C \circ S.$$

where **conf** is the conformance relation as used in single machine testing. This conformance relation takes into account the limited observability introduced by the test context. Based on this definition, the set of S-conforming machines can be produced.

3. LOOP PREVENTION

During construction of the test trees (TTs) for the MUT, a loop may occur where the context and the test tree keep exchanging internal events without producing an external output. This would create an infinite test tree and the TT construction algorithm would not terminate.

We propose an approach that pre-identifies the conditions under which a loop can happen, and then employs some measures to prevent it from happening during TT construction. An examination of the problem reveals that if the context machine contains cycles consisting of only

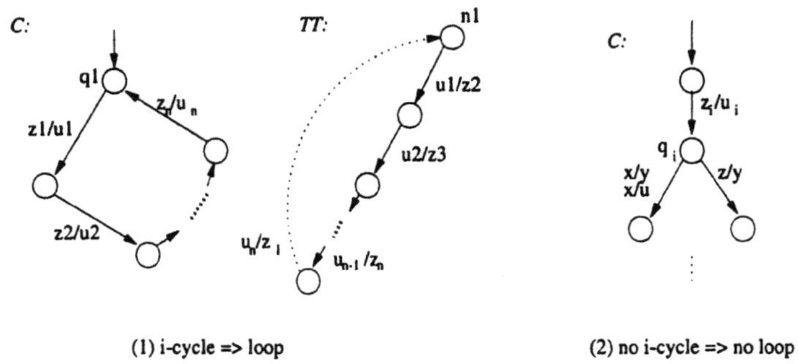

Figure 1.1 Proof for Theorem 1

internal events, it can cause a loop. A *cycle* is a transition path whose end state is the same as the start state. We call a cycle with all internal transitions an *internal cycle*, or i-cycle for short (similarly, i-path for internal paths). We prove the following theorem:

Theorem 1 *A loop between the test context and a test tree can happen if and only if there exists i-cycles in the test context.*

Proof: 1) i-cycles $\Rightarrow$ loop. Suppose there is an i-cycle in $(Z/U)^*$ in the test context C. Pick a state q_1 on the cycle, which is $z_1/u_1, z_2/u_2, ..., z_n/z_n$ (Figure 1.1). When C gets into q_1 via some external input, let the test tree TT be at node n_1. For C's output u_1, TT may select z_2 according to the TT construction algorithm, which causes C to output u_2. TT may select z_3 for u_2, causing C to output u_3. This may continue until C gets back to state q_1, and TT repeats the path from n_1. Therefore, there is a loop between C and TT.

2) no i-cycles $\Rightarrow$ no loop. Suppose there is a longest i-path in C, which is $\pi = z_1/u_1, z_2/u_2, ..., z_i/u_i$. Let C enter state q_i after z_i/u_i. The outgoing transitions from q_i must be of the form x/y, x/u, or z/y, since otherwise π would not be the longest i-path. For transitions of the form x/y and x/u, since x is an external input, it must be a test input from the test suite, and TT stays at the same node with no new edge introduced, *i.e.*, no loop will happen. For transitions of the form z/y, TT would have selected z, and y will be produced by C. This external output y is then compared against the test suite, and TT has a new edge u_i/z added. This continues until all test steps are exhausted at which point TT is terminated. Therefore no infinite loop will exist between C and TT. □

This theorem identifies the sufficient and necessary conditions for loops to exist. It is straightforward to identify all i-cycles. The next step is to find techniques to prevent loops once we recognize that they may happen.

Suppose there is an i-cycle, $\xi = z_1/u_1, z_2/u_2, ..., z_n/u_n$. We call a sequence $u_1/z_2, u_2/z_3, ..., u_{n-1}/z_n, u_n/z_1$ the complement of ξ and denote it as $\bar{\xi}$. It is the path in the test tree corresponding to ξ. In order to avoid loops, the test tree in construction must not contain the path $\bar{\xi}$. This can be done by recording the path in the test tree and whenever a new edge is added, it is compared against the complements of all i-cycles. The recording starts when the context enters into a state on an i-cycle. The test tree avoids the last step of an i-cycle complement by simply not choosing the output of that last step. This way the loop is broken and the test tree can be terminated in finite steps.

This technique is integrated into the TT construction algorithm to generate finite test trees. The trees are then fed to the tool COV to produce all T-conforming machines, which gives us the value of K_{TCM}.

4. NP-COMPLETENESS OF COVERAGE MEASURE CALCULATION

We already know from [8] that the calculation of the coverage measure $Cov(S, T, n)$ for single machine testing is NP-complete. We now ask the following question: Given a test suite T, and the test context C, is there a machine M with n states such that the composition of C and M agrees with T?

This question is important since during the construction of TCMs, we are essentially trying to find machines that accept the given test suite after composition with the test context. We claim that it is NP-complete and hence the coverage measure computation is also NP-complete. This result assures us that although embedded testing is in general more complicated than single machine testing, the computational complexity of its test coverage evaluation is not.

Theorem 2 *Let T be a test suite and C the test context. It is NP-complete to find an n-state machine M, which interacts with C via a predefined channel such that the composition of C and M accepts T.*

Proof: Let us call this problem ESAT. First we show that ESAT is in the domain of NP. A non-deterministic algorithm can guess a solution machine M for ESAT, and then check if the composition of C and M accepts T. Suppose the length of T is l, the checking can be done by applying the test steps one by one. For each test step, we only retain the stable global states which will be reached unless a livelock occurs. If a livelock is detected, the checking halts with the answer "no"; otherwise

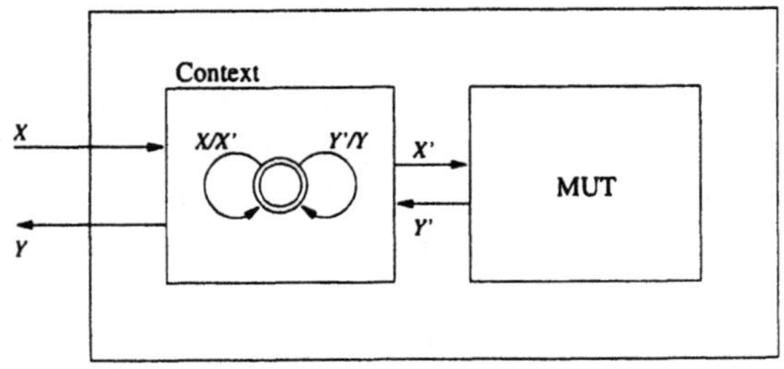

Figure 1.2 A special case of ESAT

the test step can be verified in $O(m)$ time where m is the number of internal exchanges between C and M. The whole test suite can then be checked in $O(lm_{max})$ where m_{max} is the maximal number of internal exchanges for all test steps. Therefore, the checking of whether M is indeed a solution can be performed in polynomial time.

Now we show that ESAT contains a known NP-complete problem as a special case. Consider a restricted case of ESAT: the context machine C is a one state FSM as in Figure 1.2, and the internal interactions between C and a MUT is X' and Y', which are simply a renaming of X and Y respectively. Under this restriction, C is essentially a "pass-through" module, and ESAT becomes the single machine identification problem which is known to be NP-complete [1]. □

The NP-completeness of the ESAT problem indicates that it is unlikely that there exists polynomial algorithm to compute our coverage measure. Therefore, heuristics may be the best approach to achieve good performance. On the other hand, protocols should be structured with smaller components, and large monolithic systems should be avoided.

5. APPLICATION

To validate the approach, an experimental tool set called ECOV has been implemented. We applied the tool set to a protocol extracted from a Universal Personal Computing (UPC) system [9, 4] which we developed to support personal mobility on the Internet which is independent of the user's location, motion, and platforms. This new computing paradigm retains one's personal computing environment (PCE) and access capabilities wherever one happens to be. Personal mobility is one step further than the traditional terminal mobility where the physical devices migrate.

UPC is modeled as a set of communicating objects distributed in the Internet, with some objects invisible to the outside world. Testing of these objects in an embedded manner presents a suitable application of our method in a multiple module setting.

5.1 UPC PROTOCOLS

The UPC system employs a number of protocols that support personal mobility on the Internet. This includes the middleware support that allows a mobile user to access the network wherever he roams, the mobile user location management protocol, and mobile user PCE management. We shall focus on the first protocol. The latter two protocols can be found in the references [4, 9].

The middleware support protocol describes the interactions between the user home agent (UHA), the terminal initial agent (IA), and the user terminal agent (UTA). This protocol offers a virtual global network service to the mobile user. Using the CFSM formalism, we model these objects as a system of communicating finite state machines. In order to make the system more understandable and more manageable, we abstract away some non-essential events, and take the liberty in choosing the representative values for the protocol data parameters. Furthermore, where non-determinism arises, we will determinize it appropriately. The abstracted protocol is described as two communicating FSMs in Figure 1.3, where the initial agent and the user terminal agent are combined since the two mostly communicate between themselves. This protocol can be described as a two-CFSM system.

For demonstrating our embedded testing strategy, we consider the case where the home agent is an embedded module. This corresponds to the situation where a mobile user roams to a foreign location, and wishes to access his personal PCE from there. The user has no direct access to the home agent; what he directly interacts with is the initial agent and the terminal agent. The testing issue is therefore how one can test the functionality of the home agent from the viewpoint of a mobile user.

We have composed the initial agent, the terminal agent, and the home agent to produce a composite system (also shown in Figure 1.3). This global system models the system behavior that a tester can observe and control.

5.2 UPC TEST SUITE EVALUATION

Test suites for embedded modules can be generated using methods such as that presented in [5]. However since no tool is available for the method, we chose to evaluate a test suite directly generated from the composite system. We use the UIO-method for this purpose, in order

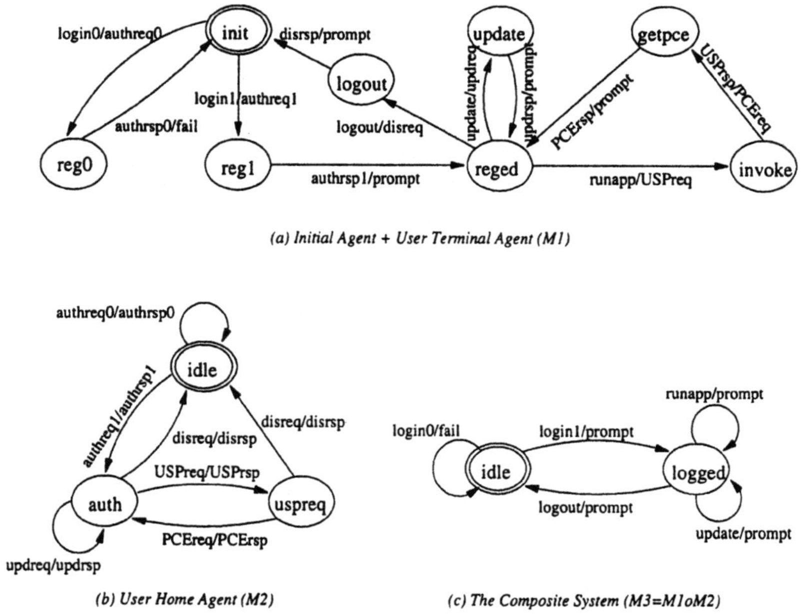

(a) Initial Agent + User Terminal Agent (M1)

(b) User Home Agent (M2)

(c) The Composite System (M3=M1oM2)

Figure 1.3 UPC middleware protocol

to get an idea on how the method fares for embedded module testing. The composite machine is completed first, with the assumption that for unspecified transitions, the machine outputs a null output and stays in the same state. It is easy to verify that *login1/prompt* is a UIO sequence for the *idle* state, and *logout/prompt* is a UIO sequence for the *logged* state. This leads to the following test suite:

```
TS = {
  r/- login0/fail login1/prompt
  r/- runapp/null login1/prompt
  r/- update/null login1/prompt
  r/- logout/null login1/prompt
  r/- login1/prompt login0/null logout/prompt
  r/- login1/prompt login1/null logout/prompt
  r/- login1/prompt runapp/prompt logout/prompt
  r/- login1/prompt update/prompt logout/prompt
  r/- login1/prompt logout/prompt login1/prompt
}
```

This test suite has complete fault coverage for the composite system, as can be verified with our tool COV. The question remains how well it does in testing the UHA as an embedded module. Feeding this to our tool box ECOV, one test tree was produced. It represents a test suite for the UHA as follows:

```
TS = {
  r/- aureq0/aursp0 aureq1/aursp1
  r/- aureq1/aursp1 USPreq/USPrsp PCEreq/PCErsp disreq/disrsp
  r/- aureq1/aursp1 updreq/updrsp disreq/disrsp
  r/- aureq1/aursp1 disreq/disrsp aureq1/aursp1
}
```

This tree is then collapsed using the tool COV, and the resulting FSMs are composed with the context machine IA/UTA to determine the conformance in context. We did some experiments regarding the upper bound n on number of states for the solution machines (mutants of the UHA machine). For $n = 3$, there were 4031 solutions, 5 of which are conforming, therefore we have $K_{TCM} = 4031$, $K_{SCM} = 5$, and $Cov(S, T, 3) = 1/(\lg 4031 - \lg 5 + 1) = 25.6\%$. For $n = 2$, there were 196 solutions, with 4 conforming, which leads to $Cov(S, T, 2) = 1/(\lg 196 - \lg 4 + 1) = 37.2\%$. In the extreme case where $n = 1$, we get a unique conforming solution with only one state, ie., $Cov(S, T, 1) = 100\%$. This indicates that the lower the upper bound n, the higher the fault coverage.

The result shows that it is generally not enough to test an embedded module with a test suite complete for the composite system. Additional test cases are needed to detect potential faulty machines. This can be done by incrementally generating test cases that distinguishes the potential faulty machines based on test coverage requirement.

6. CONCLUSIONS

We have presented our work on some new results on coverage measure for embedded testing, and on the application of the measure to a practical protocol. The work offers a better way to avoid infinite test trees and a better understanding of the computational complexity for the coverage measure. The application presents a practical situation where embedded testing can be applied, and shows that in general we cannot simply take a test suite generated for the composite system for embedded testing, as the fault coverage may be weakened depending on the upper bound of number of states in the mutants.

Our coverage measure is essentially based on machine identification which offers an accurate fault coverage for a test suite. It allows incremental test suite generation based on coverage requirement and facilitates fault diagnosis with faulty suspects explicitly constructed. The price for the advantages is higher computational cost, which can be eased by developing better heuristics and more structured, decomposable protocols.

The recent development of embedded systems on the Internet has captured significant attentions [2]. It is practically important to see if these embedded systems can be tested with an embedded testing architecture. Internet provides a wide range of testbeds for distributed computing with many types of embedded systems, such as routers, Internet phones, NetTVs, WebTV set-top boxes, and so on. Modeling these systems (with their environments) as communicating finite machines and applying our testing methodology would contribute a lot to the reliability of the systems.

References

[1] E.M. Gold. Complexity of automaton identification from given data. *Information and Control*, 37:302–320, 1978.

[2] IEEE Computer Society. *IEEE Internet Computing.* July 1998.

[3] Luiz Paula Lima Jr. and Ana R. Cavalli. A pragmatic approach to generating test sequence for embedded systems. In *IFIP 10th Int. Workshop on Testing of Communicating Systems*, Cheju Island, Korea, September 1997.

[4] Y. Li and V. Leung. Supporting personal mobility for nomadic computing over Internet. *ACM Mobile Computing and Communications Review*, 1(1):22–31, April 1997.

[5] A.F. Petrenko and N. Yevtushenko. Fault detection in embedded components. In *IFIP 10th Int. Workshop on Testing of Communicating Systems*, Cheju Island, Korea, September 1997.

[6] A.F. Petrenko, N. Yevtushenko, G.v. Bochmann, and R. Dssouli. Testing in context: framework and test derivation. *Computer Communications*, 19:1236–1249, 1996.

[7] D. Rayner. OSI conformance testing. *Computer Networks and ISDN Systems*, 14(1), 1987.

[8] J. Zhu and S.T. Chanson. Toward evaluating fault coverage of protocol test sequences. In *Proc. IFIP 14th Int. Symp. on Protocol Specification, Testing, and Verification*, Vancouver, Canada, June 1994.

[9] J. Zhu, M. Toro, V. Leung, and S. Vuong. Supporting universal personal computing on the Internet with Java and CORBA. *Concurrency: Practice and Experience*, 10(1), 1998.

[10] J. Zhu, S.T. Vuong, and S.T. Chanson. Evaluation of test coverage for embedded system testing. In *Proc. IWTCS'98*, Tomsk, Russia, September 1998.

PART EIGHT

FDTs III

A Formal Description Technology: Graphical E-LOTOS

Li Wen, Ye Xinming*, and Liu Zhiyong
Institute of computing Technology, Chinese Academy of Sciences, Beijing 100080
*(Department of Computer Science, Mogolia University, Hohhot 010021)

Abstract:

In this paper, we propose a method of Formal Description Technology (FDT) GE-LOTOS---a Graphical mode of Enhanced LOTOS (E-LOTOS). Compared with the existing FDT, the GE-LOTOS presented here has greater formal description capability. But GE-LOTOS only has a limited set of simple graphical patterns. Using the hierarchical structures of GE-LOTOS we can describe the structure and hierarchy of protocol more clearly and directly than using the existing E-LOTOS. Meanwhile the GE_LOTOS has another merit over the existing E_LOTOS that it can be executed easily and the system behavior is conveniently visible, so the system can be easily monitored and modified. We have developed a designing system for creating GE-LOTOS specifications under Java Cafe environment, and with further enhancement this system can be used for protocol testing and verification. The GE-LOTOS can be used as a design, execution, verification and testing tool for network communication protocol.

Key words

Testing of protocol, Formal Description Technology (FDT), Graphical E-LOTOS, Behavior expression

1. INTRODUCTION

With the rapid development of computer network, the testing of communication protocol of network has become the extremely important part in the cycle of design and research of network system. The correctness and efficiency of communication are necessary for working smoothly. No matter what we shall do, such as designing or tesing, the first thing is to describe a protocal system. Informal specifications often contain ambiguities and are difficult to check for completeness and correctness. To avoid ambiguty and support powerful logical analysis "Formal Description Techniques" (FDTs) must be used in describing protocols. FDT defines not only a formal syntax for the language, but also a formal semanteme which defines the meaning, in a formal manner, of any valid specification. Many different FDT have been proposed for the protocol engineering cycle, including finite state machines (FSM), Petri nets, formal grammars, high-level programming languages process algebra, abstract data types, and temporal logic. In the late 80s,there are three languages ESTELLE [8], LOTOS[4] and SDL[7], which were developed by ISO and CCITT, and have become the standardization. LOTOS, based upon process algebra, provides facilities for describing the external (observable) and internal (unobservable) behaviours of a system. And it is widely used in specification of networking and communication systems. But in the practical application, these FDTs often need to be extended. [12]

E-LOTOS is the revised version of LOTOS, and was proposed by the WI (Work Item) of ISO/IEC[1]. The purpose and scope of the definition of the new WI on "Enhancement to LOTOS" summarizes the main conclusions obtained from the practical application of LOTOS. E-LOTOS is the result of several years of work in which a large number of enhancement proposals have been analyzed and assessed. The selected enhancements removed the known limitations and extent the expressiveness, abstraction capabilities, structuring capabilities, ... of the preceding version of standard LOTOS. We assume that the readers of this paper are familiar with E-LOTOS and LOTOS.

But the existed E-LOTOS can only describe protocol in static text mode, so it can not directly and vividly describe hierarchy and synchronization of the system structure and the relations between actions. Meanwhile the more expression types there are, the more difficult it is to distinguish the behaviors and understand the protocol system described by E-LOTOS. Worse of all, the system described in E-LOTOS can not be executed. All these will cause of limited application of E-LOTOS. According to the book [11], a direct interactive graphical model has great effects on the application of FST.

For LOTOS, there are at least two graphical versions of it in the literature. The one proposed in WG-LOTOS[10] is not user-friendly. There are a lot of variations, intermingling, and non-uniformly graphical patterns. Especially, the rigidity and crowdness of nested rectangle structure (i.e., rectangles) make it difficult to understand the structure and logic of the designed system. Furthermore it is inconvenient for modification and modular design. The other one UO-GLOTOS [2] is a good graphical version. It has the simple and general graphical pattern and clear control constructs. On the whole, the UO-GLOTOS is a hierarchic model. Therefore it can greatly support the modular design, verification, logical detection, and testing of system specified by LOTOS. Generally, all graphical models are mainly describing the control behaviors of a system. Description of data is mainly based on textual version.

To conquer the limitations of existed E-LOTOS, after we deeply studied E-LOTOS and compared many kind graphical patterns and principles [5,13,14], we develop Graphical E-LOTOS. It can exactly express the semantics of E-LOTOS behavior expressions and has the simple and general patterns. Hereafter, for the convenience of reference, we call Graphical E-LOTOS as GE-LOTOS. Using GE-LOTOS, system is described in the form of a hierarchical structure, so the system structure and the relations between actions can be described more clearly, simply, and easy to understand. Using this feature, system also can be described in various levels of abstraction. Furthermore the GE-LOTOS presented here has many other good properties, such as it is "open" to "semantic" tools, it can be executed easily and the system behavior is conveniently visible, and easy to be monitored and modified.

Now many distributed systems need user-friendly and high description capability FDT. Our GE-LOTOS is this kind FDT. We have developed a designing system of GE-LOTOS under Java Cafe environment, and with further enhancement it can be used for protocol testing and verification. The graphical E-LOTOS can be used as a design, execution verification, and testing tool for network communication protocol and distributed systems.

This paper is organized as follows. In section 2, we discuss some considerations and criteria in designing GE-LOTOS. In section 3, we present the basic graphical patterns and the main features of GE-LOTOS. In section 4, we present the enhanced behavior expressions of E-LOTOS and the corresponding graphical mode. In section 5, we give an example of applying GE-LOTOS to describe a part of Alternating Bit Protocol. It includes a comparison with the E-LOTOS specification of the same protocol. Section 6 includes some conclusions and future work of GE-LOTOS.

2. CONSIDERATIONS IN DESIGNING GE-LOTOS

The domain application of FDT is determined by formal description capability. Behavior expressions are the basic component by which the behavior and hierarchy of protocol system can be described, and they directly embody the expression capability of E-LOTOS. Basically, all graphical models are mainly for describing the behavior of a system. In the following, we focus our discussion on the enhanced behavior expressions of E-LOTOS. For well-defining our problem, we highlight some the enhancements in E-LOTOS which have much influence on designing our GE-LOTOS.

In E-LOTOS, some enhancements are introduced by generalizing or enhancing LOTOS operators, others are new operators which have been introduced when absolutely necessary. Sequential composition operation ";" substitutes the operators existing in LOTOS for sequential composition ";", action prefix ";", and exit/enabling ">>" pair. Some notion of quantitative time is introduced, for example "wait(E)", which can be used for precise description of real time systems and allows a precise timed specifications to the execution of actions. Exception mechanisms, "trap" and "raise", permit new ways of describing structure. A more general parallel operator "par" is more readable and flexible, because it clearly identifies the synchronizing gates for each behaviour composed. Other new important constructs are "loop", "rename", "var", and etc. How to represent these new or enhanced constructs in graphical mode is our task.

2.1 Difficulties to design GE-LOTOS

• What kind graphical patterns not only can describe the semantics of enhanced behavior expressions exactly and clearly, but also can be accepted by users?

• How to reduce the numbers of patterns of the enhanced behavior expressions in order to reduce complexity of the whole GE-LOTOS system.

• How to keep the maximum "open" ability, describe the structure of protocol clearly and vividly, and execute system easily.

2.2 Criteria to design Graphical E-LOTOS and advantages of the Graphical E-LOTOS.

• Rather than introducing new graphical patterns, we use the existing graphical patterns of G-LOTOS [2] to represent the generalized and enhanced operators.

• Each construct of G-E-LOTOS is represent independently, and we have avoided the nested structure. For example, the set of operations {**loop, forever, rename, trap**} are all represented by independent graphical element, while they all have the efficient range. Generally efficient range can be described implicitly. When the efficient range are needed, we can use labels.

• Using a versatile hierarchy structure to represent recursive E-LOTOS expressions. For example, different hierarchies are used to describe process definitions and process instantiations.

• The present GE-LOTOS is convenient for manipulation, and convenient to execute the behaviors of the described system.

• For the new operators, we classify the enhanced behavior expressions according to its semantics and function structure. If the enhanced behavior expressions have similar semantics or similar function structure, the same pattern is adopted. The difference between each other can be distinguished by the expression in graphical pattern. For example, the set of E-LOTOS behavior expressions {exit[(RE)], exit(any T), break[X][(E)], raise X[E], stop} all have the similar function structure, and {exit[(RE)], exit(any T)} have the similar semantics. So we use the same pattern -- a stretched oval. Another example is that the set of E-LOTOS behavior operations {par, concurrency, trap, case} use the similar tree-like pattern which shows clearly the operator and it relationship of operands. Their graphical patterns are listed at table 2-1 and table 2-2.

3. THE BASIC GRAPHICAL PATTERNS AND MAIN FEATURES OF GE-LOTOS

Depended on above-mentioned criteria, our GE-LOTOS only has six different basic graphical patterns, which can represent 31 kinds of behavior expressions of E-LOTOS. There is corresponding expression inside the graphical pattern. The six different basic graphical patterns are:

• Stretched oval denotes termination, raise exception and internal constructor. They are distinguished by the expression inside the stretched oval.

• Hexagon denotes some declaration or control constructs---**var, rename, loop.**

• Tree-like pattern denotes the behaviours which have parralel, disable, trap, or case logical relationship. These relationships are distinguished by the key expression of left root.

• Rectangle with reference denotes process definition and process instantiations. A reference number at the upper-left corner of the rectangle represents a definition, whereas the same reference number at lower-right corner of the rectangle represents its instantiation. This construct is a very important pattern which can make GE-LOTOS to describe various levels of abstraction of system and to design large-scaled specification.

• strip shape denotes "assign" or "delay" behavior expressions.

• strip shape with disjunctive line denotes "action" behavior expressions.

Now we can know the present GE-LOTOS have the features below:

• A limited set of simple and elegant graphical patters, and versatile capability of description.

GE-LOTOS only has six different basic graphical patterns. But using these graphical patterns we can clearly and vividly describe the hierarchy, sequence, concurrency, synchronism, and all of the behaviors in the protocol system.

• Hierarchical and open structure.

The system described in GE-LOTOS is hierarchical in nature. So we describe the structure of protocol system clearly and exactly. Using process definition and process instantiations graphical patterns, we can describe various levels of abstraction of system and design large-scaled specification. This feature also makes it convenient for supporting logical exploration, modular design, verification, and testing.

• The system described in GE-LOTOS is executable.

This feature makes it easy to graphically simulate the behaviours of the system and generate executable paths and test case.

The Graphical E-LOTOS which has versatile expression capability can be used as a design, execution verification and testing tool for network communication protocol and distributed systems.

4. THE GRAPHICAL PATTERNS OF GE-LOTOS

4.1 Symbol specification used in behavior expressions(Table 1)

Meaning	abbreviation	Meaning	abbreviation
Natural number	i, j, k, l, m, n	Expression	E
Behavior expression	B	Record expression	RE
Exception identifier	X	Local variable	LV
Gate identifier	G	Rename parameter	NP
Type expression	T	Behavior match	BM
Pattern	P	List	N(length|N|)

Table 1 Symbol specification used in behavior expressions

4.2 A graphical syntax of enhanced behavior expressions of Graphical E-LOTOS

In table 2-1 and 2-2, we present the enhanced behavior expressions of E-LOTOS and the corresponding graphical pattern. Column 2 lists our graphical syntax. Column 1 contains the corresponding enhanced behavior expressions. We have not presented the behavior expressions of LOTOS and corresponding graphical pattern [2].

Behavior Expressions of E-LOTOS	Graphic Expression of E-LOTOS
process-define and process-instance **π [[G (,G) *]] [([RE,RP])] [[(x(,x) *]]** (pxx' at top-left-corner represent process-define, 'pxx' at bottom-right-corner represent process-instance.)	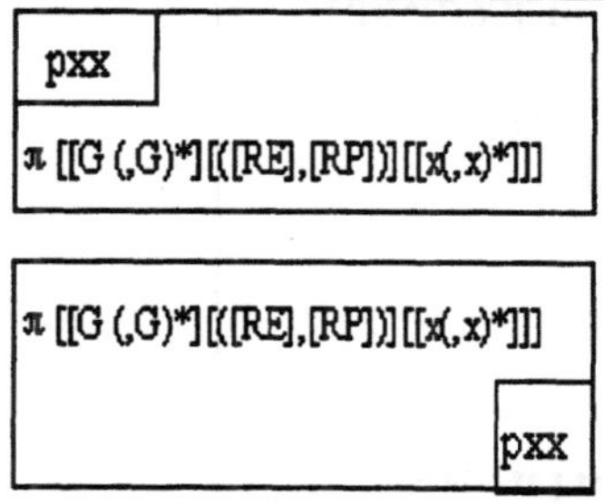
successful termination **exit[(RE)]**, nondeterministic termination **exit[any T]**, raising exception **raise X[E]**、**signal X[E]**, breaking iteration **break [X][(E)]**, time block **block**	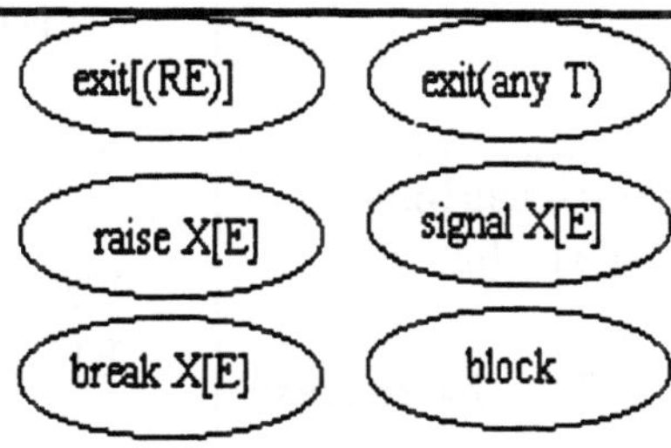
Assign **P : =E**, Nondeterministic assign **P=any T[[E]]**, delay **wait(E)**	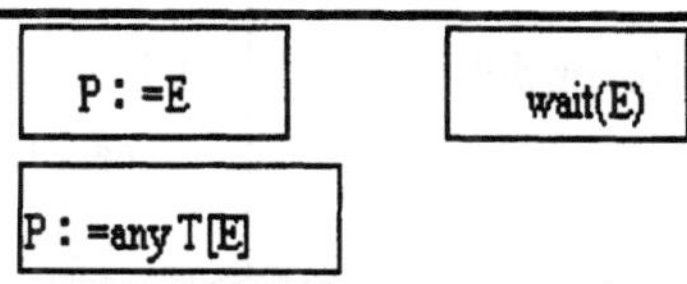
suspend/resume **B1 [X>B2**	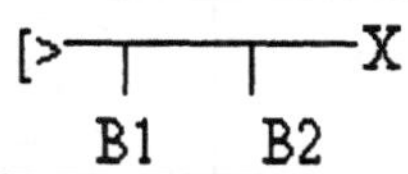
Hiding **hide** G[:T](G,G[:T])* in B **endhide**	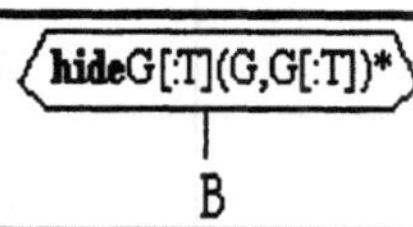
Rename **Rename** (G[(NP)] is G[P] | X (NP) is X[E])* in B **endren**	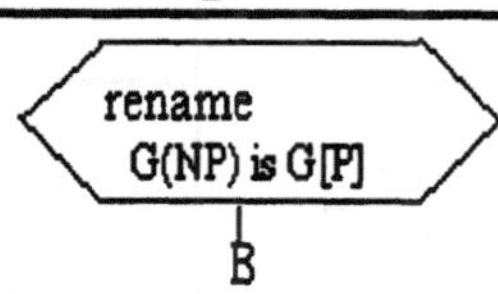
local specification **local** var LV[init B1] in B2 **endloc**	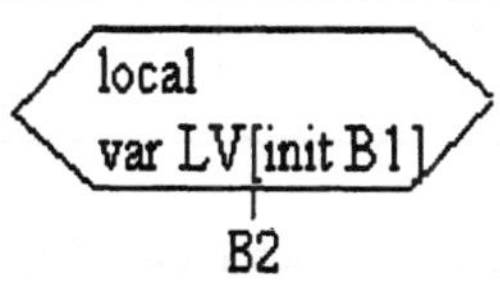

Table 2-1 Behavior Expressions and Graphical Expression

Behavior Expressions of E-LOTOS	Graphic Expression of E-LOTOS
concurrency **B1 |[G(,G)*]| B2** general parallel **par** G1#n1,···,Gp#np [τ1] for B1 || [τ2] for B2 || ...... || [τn] for Bn **endpar** parallel over values **par** P in N ||| B **endpar**	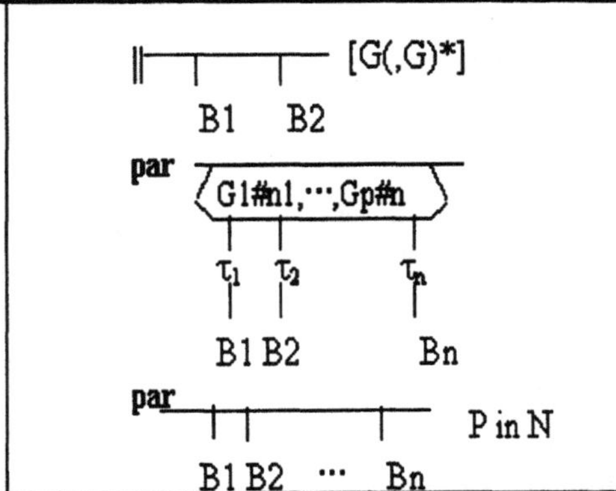
choice-expressions **if** E **then** B1 [else B2] **endif** **case** [E:T]is BM **endcase** **choice** P[]B **endch**	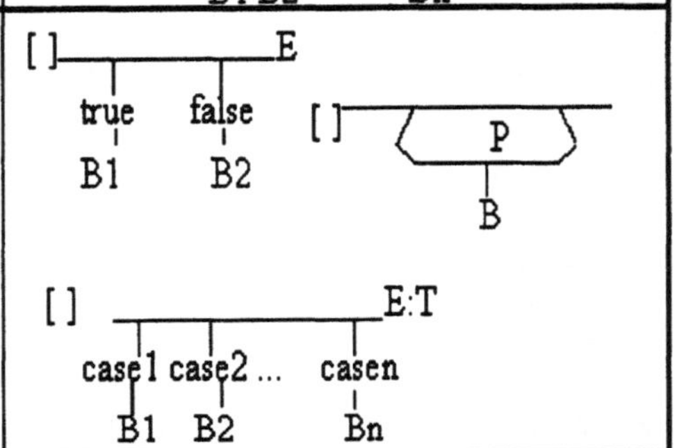
Trap (**exception** X[(LV)] is B)* **endexn** [exit [P] is B **endexit**] **in** B **endtrap**	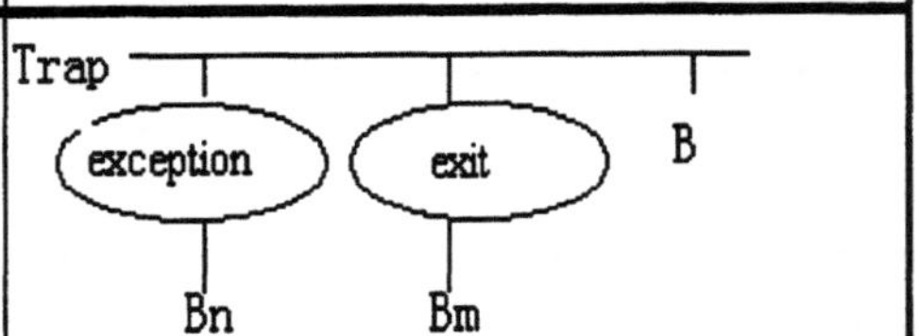
action **G[P][@P][[E]]**	
iteration **loop** forever [var LV][init B1] **in** B **endloop** **loop** [x][(T)] [var LV][init B] **in** B **endloop**	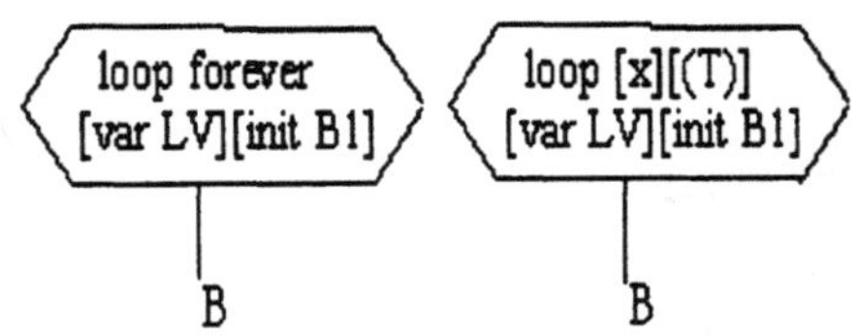

Table 2-2 Behavior Expressions and Graphical Expression

5. A PROTOCOL EXAMPLE DESCRIBED IN GRAPHICAL E-LOTOS

Communication protocols are the rules that govern the communication between the different components within a distributed computer system. Here we select a part of the Alternating Bit Protocol which is often used as example. This part process is:

Before send message, start up chronoscope.
2) Wait for feedback message.
3) If the right received message comes back in the set time, the process is over,
otherwise repeats above-mentioned process until correctly received.

For comparison purpose, we first describe this part process in E-LOTOS.

```
process safemesser[sendpdu:pdurec,receiveack:pdurec]
          (datamsg:data; seqsender:bit):exit(none) is
 local   var timestart,reset,expired:time;  in
 hide tm:time in
  (tm(!timestart);sendpdu(!makepdu(datamsg,seqsender));
  ((messer[receiveack](seqsender);tm(!reset);exit
   [>
   (tm(!expired); signal "timeout";
 safemesser[sendpdu,receiveack](datamsg,seqsender))))
 |[tm]| timer[tm]
endloc  endproc

process messer[receiveack:pdurec](seqsender:bit):exit(none) is
 loop forever  in  var ack:bit
 receiveack(?ack);
 if (ack==seqsender)
   break
endloop   endproc

process timer[tm:time]:exit(none) is
 local   var timestart,reset,expired:time;   in
  tm(!timestart);
  (i;tm(!expired) ;exit
   [> tm(!reset));exit
 endloc   endproc
```

The above-mentioned process is shown in Graphical E-LOTOS in figure 1-1 and figure 1-2.

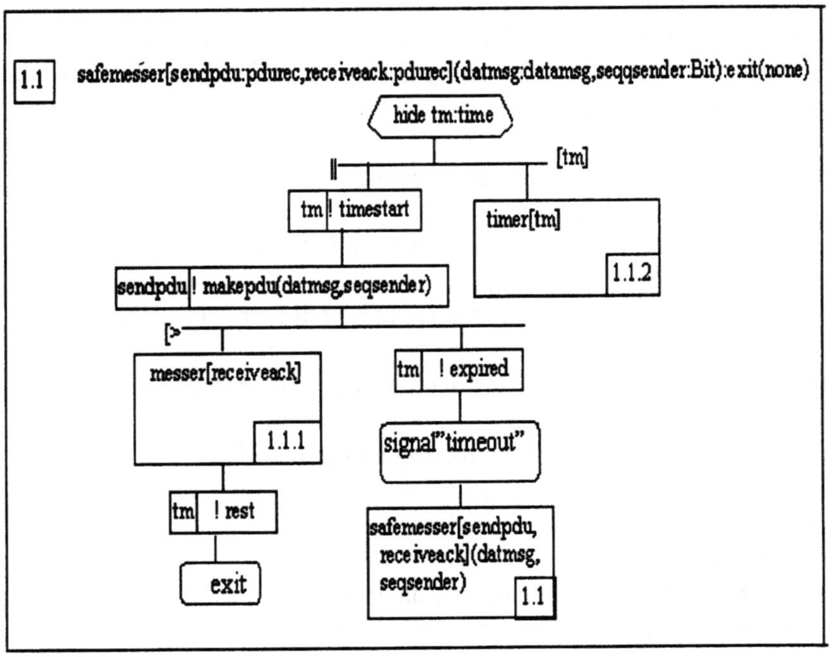

Figure 1-1 Part of the Alternating Bit Protocol in GE-LOTOS

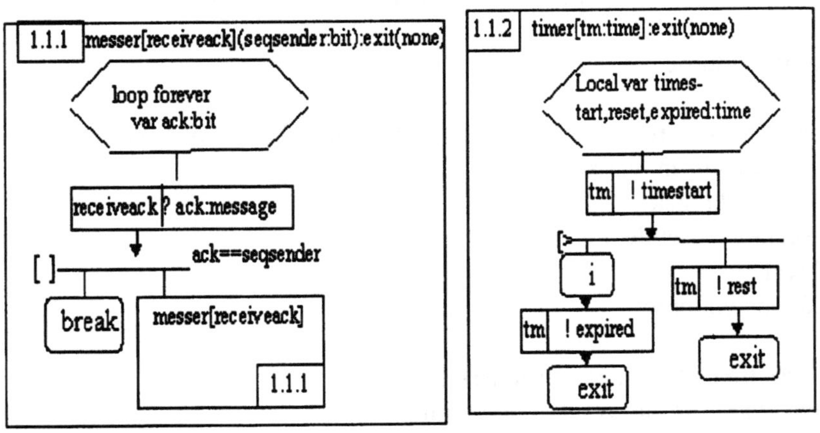

Figure 1-2 Part of the Alternating Bit Protocol in GE-LOTOS

The protocol in GE-LOTOS is described with two levels. "messer" and "timer" in first level are described abstractly. In the second level they are detailed respectively. After some comparison, we can know that the

Graphical E-LOTOS specification can show the hierarchy structure of protocol system and the relationship between behavior more clearly and straightly than the existed E-LOTOS does, such as the parallel relationship between "sendpdu" and "timer". The simple and "open" properties can also be seen in this example. It is easier to understand the above-mentioned process in Graphical E-LOTOS than in the existed E-LOTOS. The Graphical E-LOTOS specification can show the logical flow of the protocol much more clearly.

When using our GE-LOTOS tools, description of protocols will become more easy and need less work. In this designing system, user can conveniently choose elements through the corresponding icon buttons. The modifications about the elements can be done through the "move", "delete", and "contents", which are used to change content of elements. Besides these, GE-LOTOS files can be "open", "save", or "save as" in the standard dialog frame.

GE-LOTOS has the powerful description capabilities. It can be used to describe varieties of complex and large systems at various levels of abstraction. For example, it has successfully described the ODP trader computational viewpoint [9]. The ODP trader is an object which enables software components to find appropriate services providers within an open and dynamically changing distributed system. Because of the space limitation of the paper, this example is not included here.

6. CONCLUSIONS AND FUTURE WORK

In this paper, we present a method of Formal Description Technology (FDT) GE-LOTOS. GE-LOTOS have many good properties, such as simple and open properties, and capability for supporting logical exploration, modular design, execution, verification and testing. The simple and open properties of Graphical E-LOTOS are very good properties to design large-scale system. Compared with the existing FDT, the graphical E-LOTOS presented here not only has greater formal description capability, but also can describe the parallel, temporal order, structure and hierarchy of protocol more clearly and directly than the existing E-LOTOS.

We have developed a designing system for creating GE-LOTOS specifications under Java Cafe environment. In our future work, we will provide the theory and algorithms for systems verification and testing. Meanwhile we will develop the designing system into a system which can provide a graphical environment for carrying out many research activities, such as control and data flow tracing, graphical execution and verification of systems described in GE-LOTOS. The GE-LOTOS can become a very useful method which can be used in design, execution, verification, and testing for network communication protocol and distributed system.

REFERENCES

[1] ISO/IEC JTC1/SC21 WG7. Enhencements to LOTOS. May 1998.

[2]T.Y.Cheung,Y.C.Ye, G.Q.wang. UO-GLOTOS: A Syntax/System for Representing, Editing and Translating Graphical LOTOS. In Proc of 2nd International Conference on Formal Description Techniques for Distributed Systems and communications Protocols.Vancouver.pages 33-48,1989.

[3]G.v.Bochmann, A.Petrenko. Review of Methods and Relevance for Software Testing. Montreal Canada Publication, #923,1994,6.

[4]T.Bolognesi,E.Brinksma. Introduction to the ISO Specification Language LOTOS. Computer Networks and ISDN Systems, vol.14,No.1,1987.

[5]Luciano Baresi,Alessandro Orso,and Mauro Pezze. Introducing Formal Specification Methods in Industrial Practice. In Proc. Of the 19th International Conference on software Engineering. Massacbusetts,pages 56-64,1997.

[6]L.Baresi. Formal Customization of Graphical Notations. PhD thesis, Dipartimento di Milano, Italian, 1997.

[7]F.Belina and D.Hogrefe. The CCITT-Specification and Description Language SDL. Computer Networks and ISDN Systems. Vol.16.1989.

[8]S.Budkowski and P.Dembinski,"An introduction to Estellle:a Specification Language for Distributed Systems". Computer Networks and ISDN Systems. vol.14.No.1.1987.

[9]Giovanny Lucero and Juan Quemada. An E-LOTOS specificaton of the ODP Trader. Input document(GRI) to the ISO/IEC JTCI/SC21/WG7/E-LOTOS meeting in Grenoble,December 1996.

[10]ISO/IEC JTC1/SC21,N3253. G_LOTOS: a graphical syntax for LOTOS. Jan 1989.

[11]Willian R.Mallgren. Formal Specification of Interactive Graphics Programming Languages.The MIT Press Cambridge, Massachusetts London, England. 1982.

[12]Luc Leonard and Guy Leduc. An introduction to ET-LOTOS for the description of time-sensitive systems. Computer Networks and ISDN systems, 29(3):271-292,1997.

[13]Enoch Y. Wang, Heather A. Richter and el at.. Formalizing and Integrating the Dynamic Model within OMT*. In Proc. of the 19th International Conference on software Engineering. Massacbusetts,pages 45-55,1997.

[14]J-P.Jacquot and D. Quesnot. Early Specification of User-Interfaces: Toward a Formal Approach. In Proc. of the 19th International Conference on software Engineering. Massacbusetts,pages 150-160,1997.

GLOBAL TIMED BISIMULATION: AN INTRODUCTION

David de Frutos-Escrig, Natalia López, and Manuel Núñez
Dept. de Sistemas Informáticos y Programación
Universidad Complutense de Madrid, Spain
{defrutos,nlopezb,manuelnu}@eucmax.sim.ucm.es

Abstract Bisimulations are a broadly used formalism to define the semantics of process algebras. In particular, by means of weak bisimulation most of the internal activity of processes may be abstracted. Unfortunately, this is not fully accomplished: for instance, the internal choice operator becomes non-associative since bisimulation can see the branching structure of processes. In this paper we propose *global timed bisimulation* as a weakening of weak timed bisimulation. Global timed bisimulation is defined exactly as weak timed bisimulation once ordinary transitions are replaced by the adequate notions of generalized transitions. In order to asses the definition of our global timed bisimulation we present a collection of small examples that illustrate each of the clauses of that definition. Finally, a more elaborated example is presented to summarize the main properties of that notion.

Keywords: Timed Systems; Formal Semantics; Bisimulations; Non-determinism.

1. INTRODUCTION

Operational semantics is probably the simplest and more intuitive way to give semantics to any process language. It is usually defined by means of a *labeled transition system* which contains the transitions, $P \xrightarrow{l} P'$ that each process can execute. But these semantics are not abstract enough since they contain the full information about the syntax of processes. *Bisimulations* [9] are the usual way to solve this problem, by defining a slightly weaker equivalence which is a rather reasonable, simple and nice way to define the semantics of non-deterministic processes.

Strong bisimulation is a rather satisfactory way to define the semantics of processes whose transitions are labeled by visible actions.

Nevertheless, processes having all their transitions visible are not sufficiently flexible: we need some kind of internal transitions to reflect internal behavior of processes that will be neither observable nor controllable from outside. In particular, internal transitions are introduced to reflect internal choices at the operational level. Therefore, when we define the operational semantics of a process language we usually use labeled transitions, whose labels are visible actions, and unlabeled ones which correspond to internal moves. But these internal actions must be treated in a special way in order to capture the fact that they are not observable. This is why a weaker notion of bisimulation is introduced: *weak bisimulation.*

Weak bisimulation partially abstract internal moves, since in its definition whenever we have the possibility to execute an internal transition we can freely choose between executing one, several or even none of them. As a consequence, using CCS-syntax, we have, for instance $\tau\,;P \sim P \sim \tau\,;\tau\,;P$. Nevertheless, weak bisimulation must not ignore internal transitions when they occur in the scope of a choice. Thus we have $P + Q \not\sim P + \tau\,;Q$, which is indeed rather reasonable. But, unfortunately, this capability to *see* internal actions remains when nested choices appear. So, we have $\tau;(\tau;P+\tau;Q)+\tau;R \not\sim \tau;P+\tau;(\tau;Q+\tau;R)$, which in a CSP-like language would lead us to $(P\oplus Q)\oplus R \not\sim P\oplus(Q\oplus R)$, where $\oplus$ represents the internal choice operator. The fact that internal choice is not associative under weak bisimulation is, in our opinion, a serious drawback. Erdogmus et al. [5] have found the same problem, although they have tried to solve it in a different way to the one we present in this paper. They consider it is not the notion of bisimulation that is *wrong*, but instead what is wrong is the operational semantics from which the bisimulation equivalence is defined. Therefore, they suggest to use a more involved notion of transition system to define the operational semantics of non-deterministic processes.

Instead we prefer to use original transitions systems. We maintain the original operational semantics, given by ordinary transition systems as the basis to define our new semantics although we do not directly use these transitions in the definition of bisimulations. So in [3] we define *global bisimulations* by combining two different kind of moves: *dynamic*, which correspond to the execution of visible actions; and *static*, which correspond to (partial) resolutions of non-determinism. These last moves justify our name for the new notion of bisimulation, since the involved transitions do not need to appear at the beginning of the computations, but anywhere else.

In this paper we extend our previous study to the framework of timed processes. We consider that this is interesting for several reasons. First, by means of this extension we have seen that our ideas are rather flexible and can be satisfactorily applied to different kinds of process algebras. Second, given that one of our goals was to study and relate the different kinds of non-determinism, timed processes are a good benchmark for it, since timed transitions are a new source of non-determinism behaviour. Finally, because the alternative ways to define the semantics by means of weaker equivalence relations than weak bisimulation, do not work as expected in the timed case. For instance, testing semantics is much more stronger than in the untimed case, since the urgency of internal actions strongly increases the discriminatory power of tests [8]. So the definition of alternative semantics is specially appealing in the case of timed processes. Actually, the new semantic framework described in this paper can be applied to the definition of the semantics of the core of E-LOTOS [11] where time information has been included in order to remove one of the limitations of former LOTOS.

The rest of the paper is structured as follows. In Section 2 we present our formalism to describe timed transition systems. We also include several examples showing the expressive power of our model. In Section 3 we define our notion of global timed bisimulation and we show its advantages with respect to weak bisimulation. In Section 4 we summarize the main ideas presented in the paper about the treatment of non-determinism by studying possible implementations of a faulty channel. Finally, in Section 5 we present our conclusions and some lines for future work.

2. BASIC CONCEPTS

In this section we present the abstract syntax which we will use to model timed systems. As we said in the introduction, processes will be defined in terms of (timed) *labeled transition systems.* By using such a *syntax* we can abstract the features of a chosen language. In other words, instead of defining an operational semantics over a fix language, and then to study the induced labeled transition systems, we directly consider a (general) kind of labeled transition systems. In particular, our labeled transitions systems are very close to those induced by the semantics of the new version of E-LOTOS [11].

Definition 1 (Preliminaries) We consider a *set of visible actions* Act, and an action τ not belonging to Act. To model time we use a *discrete set of times* $\mathcal{T}$. We call $\mathcal{T}$ the *time domain.* We may consider $\mathcal{T} = \{\delta, 2\delta, 3\delta, \ldots\}$, where δ indicates the *minimum amount* of time. □

Once we have introduced both the set of actions and the set of times which will allow us to represent systems, we define processes as *timed labeled transition systems*. Let us note that in the definition of our processes we are taking into account some of the properties that are commonly found in most temporal models like *time determinacy* and *maximal progress* for τ actions.

Definition 2 A process P is defined as a triple $(S, \longrightarrow, s_0)$, where S is a *set of states*, $s_0 \in S$ is the *initial state*, and $\longrightarrow \subseteq S \times (Act \cup \{\tau\} \cup \mathcal{T}) \times S$ is the *transition relation*. We will usually write $s \xrightarrow{\alpha} s'$ instead of $(s, \alpha, s') \in \longrightarrow$. We impose the following restrictions over the relation:

- **Maximal Progress**: Let $s \in S$. Whenever we have $s \xrightarrow{\tau} s'$ then there do not exist $s'' \in S$, $t \in \mathcal{T}$ such that $s \xrightarrow{t} s''$.
- **Time Determinacy**: Let $s \in S, t \in \mathcal{T}$. Then there exist at most a state $s' \in S$ such that $s \xrightarrow{t} s'$.

□

We will usually identify the nodes of these labeled transition systems with the process defined by the corresponding node as initial state. Formally, let $P = (S, \longrightarrow, s_0)$ be a process. If $s \in S$ then we consider P_s as the process $(S, \longrightarrow, s)$.

In order to make easier the presentation and study of the main features of our new semantics, we will only consider in this paper *tree-like* transition systems that are those such that the *underlying* graph is a rooted tree. It is important to note that this does not represent a lose of generality since by unfolding any transition system we can found an equivalent tree-like transition system. This is so, because our definition of global bisimulation does not use in any way the global structure of the underlying. Besides, it is applicable to any arbitrary timed system either finite or infinite. Thus even if we start from a finite state system with an arbitrary net structure which would become an infinite tree when unfolded, this does not represent a problem at all. It is true however that if we were interested on deciding global bisimilarity in an effective way we should maintain the finiteness of systems when studying them. This can be done indeed, but since it is technically a bit involved we will present in a forthcoming paper.

Once we have defined the notation for describing timed systems, we will explain by means of some simple examples the intended meaning of each construction and the expressive power of the model.

First, we have that *time-outs* can be modeled as follows: if we want to specify a system offering actions a and b exactly at time 0, and such

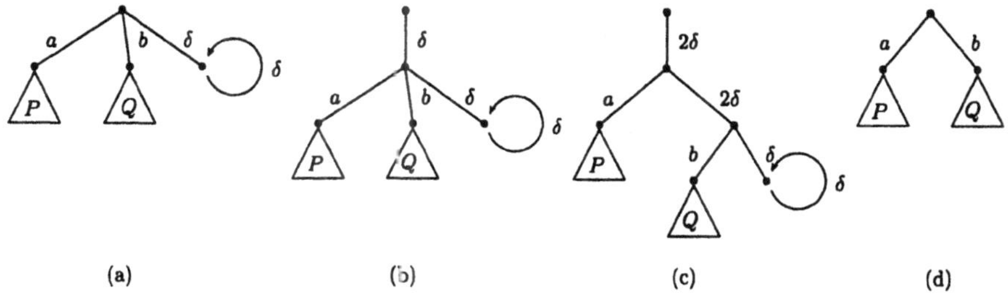

Figure 1.1 Examples of Timed Processes (I).

that when none of these actions is executed it gets deadlocked, we will use the process represented in Figure 1.1 (a).

Delays are specified by means of timed transitions. For instance, if we consider a variation of the previous system where the actions are offered at time δ, it would be represented by the process depicted in Figure 1.1 (b). We can specify processes having different delays. For instance, a choice between the action a delayed 2δ units of time and the action b delayed 4δ units of time would be represented as described in Figure 1.1 (c). Processes *blocking* the time can also be included. For instance, in Figure 1.1 (d) we show a process offering a and b at time 0 that does not allow the passing of time.

Let us remark that in our representation we only make explicit the time transitions corresponding to the time between two offerings of actions represented by states. Obviously, we need to allow a delayed process to *consume* partially its delay (i.e. to *age*). For example, the process $\mathtt{delay}(t);P$ will evolve by a timed transition to the process $\mathtt{delay}(t-t');P$ after passing t' units of time, for any $t' < t$. This will be formalized later.

We can also specify systems offering actions during an interval of time. For example, if the system offers a in the interval $[0..2\delta]$ and b in the interval $[\delta..3\delta]$ we would use the process given in Figure 1.2 (a). Note that τ actions are urgent, so it makes no sense to offer a τ over an interval of time. Moreover, we can specify a process offering an action over an infinite amount of time. For instance, the process depicted in Figure 1.2 (b) offers the action a in the interval $[0..\infty)$, the action b in $[\delta..\infty)$, and the action c in $[\delta..2\delta]$.

Next, in order to manage our processes we introduce some auxiliary notation that will be intensively used all along the rest of the paper.

Definition 3 Let $P = (S, \longrightarrow, s_0)$ be a process. Let $\{s_{a_i} \in S \mid s_0 \xrightarrow{a} s_{a_i}\}$ be the set of states that are reached by executing $\xrightarrow{a}$ from s_0, and let $\{P_{a_i}\}$ be the set of the corresponding induced processes. Similarly,

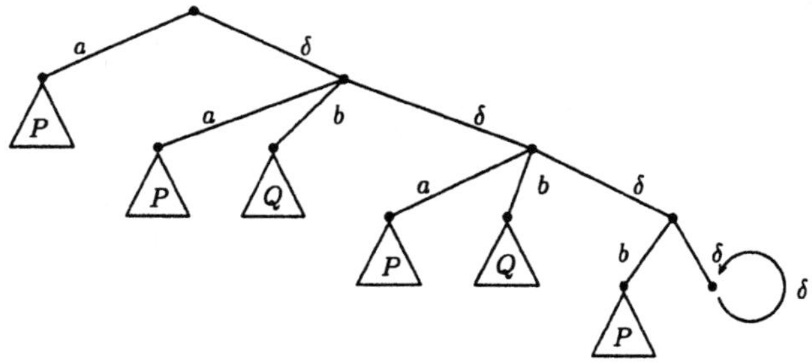

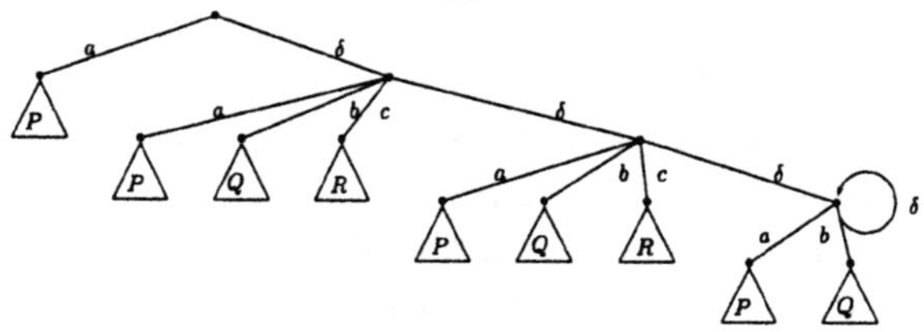

Figure 1.2 Examples of Timed Processes (II).

let $\{s_{\tau_i} \in S | s_0 \xrightarrow{\tau} s_{\tau_i}\}$ (resp. $\{P_{\tau_i}\}$) be the set of states (resp. processes) that are reached after the execution of $\xrightarrow{\tau}$ from s_0. Then to refer to P, we will use the following notation:

- $P = \sum_{a \in Act} a \,;\, P_{a_i} + \sum \tau \,;\, P_{\tau_i}$ — if $\{P_{\tau_i}\} \neq \emptyset$
- $P = \sum_{a \in Act} a \,;\, P_{a_i} + \texttt{delay}(t) \,;\, P'$ — if $\{P_{\tau_i}\} = \emptyset \wedge \exists P' :\ P \xrightarrow{t} P'$
- $P = \sum_{a \in Act} a \,;\, P_{a_i}$ — if $\{P_{\tau_i}\} = \emptyset \wedge \ \nexists P' :\ P \xrightarrow{t} P'$
- $P = \texttt{stop}$ — if $S = \{s_0\} \ \wedge\ \forall a \in Act :\ \{P_{a_i}\} = \emptyset$ $\wedge\ \{P_{\tau_i}\} = \emptyset\ \wedge\ s_0 \xrightarrow{\delta} s_0$
- $P = \texttt{block}$ — if $S = \{s_0\} \ \wedge\ \forall a \in Act :\ \{P_{a_i}\} = \emptyset$ $\wedge\ \{P_{\tau_i}\} = \emptyset \wedge\ \nexists t \in \mathcal{T} :\ s_0 \xrightarrow{t} s_0$

□

Note that some of the sums $\sum_{a \in Act} a \,;\, P_{a_i}$ may be empty. Additionally, we will also write $P = \sum_{\alpha} \alpha \,;\, P_{\alpha}$ when we do not need to distinguish between visible, internal and timed descendants, and we will use $P_1 + P_2 + \cdots + P_n$ as a shorthand of $\sum P_i$.

In the previous definition we have introduced two new concepts: the processes `stop` and `block`. Both processes are very similar because they cannot execute any action. The difference is that while `stop` allows the time to pass, `block` does not. Next, we define the usual extensions of the transition relation to allow several steps.

Definition 4 Let $P = (S, \longrightarrow, s_0)$ be a process. We define the following relations on states:

- We denote by $\stackrel{\tau}{\Longrightarrow}$ the transitive and reflexive closure of $\stackrel{\tau}{\longrightarrow}$.
- Let $s, s' \in S$. We write $s \stackrel{t}{\Longrightarrow} s'$ if there exist $t_1, t_2, \ldots, t_n \in \mathcal{T}$ such that $\sum t_i = t$ and there exist $s_1, s_1', s_2, s_2', \ldots, s_n, s_n'$ such that $s \stackrel{\tau}{\Longrightarrow} s_1 \stackrel{t_1}{\longrightarrow} s_1' \stackrel{\tau}{\Longrightarrow} s_2 \stackrel{t_2}{\longrightarrow} s_2' \cdots \stackrel{\tau}{\Longrightarrow} s_n \stackrel{t_n}{\longrightarrow} s_n' \stackrel{\tau}{\Longrightarrow} s'$
- Let $s, s' \in S$. We write $s \stackrel{a}{\Longrightarrow} s'$ if there exist s_1, s_2 such that $s \stackrel{\tau}{\Longrightarrow} s_1 \stackrel{a}{\longrightarrow} s_2 \stackrel{\tau}{\Longrightarrow} s'$.

□

As commented above, we need to generate all the different time transitions that correspond to each of the ones explicitly represented in the given transition system. So we extend the transition relations as follows: Given a process $P = \sum_{a \in Act} a \,;\, P_{a_i} + \mathtt{delay}(t) \,;\, P'$ in addition to the transitions that are allowed by its transition relation, it has the following timed transitions: $P \stackrel{t'}{\longrightarrow} \mathtt{delay}(t - t') \,;\, P'$ [if $t' < t$]

We conclude this section by defining the adequate notion of weak timed bisimulation.

Definition 5 A binary relation $\mathcal{R}$ on processes is called a *weak timed bisimulation* if for all P, Q such as $P \,\mathcal{R}\, Q$, and any $\alpha \in Act \cup \{\tau\} \cup \mathcal{T}$:

- Whenever $P \stackrel{\alpha}{\longrightarrow} P'$ we have some Q' such that $Q \stackrel{\alpha}{\Longrightarrow} Q'$ and $P' \,\mathcal{R}\, Q'$.
- Whenever $Q \stackrel{\alpha}{\longrightarrow} Q'$ we have some P' such that $P \stackrel{\alpha}{\Longrightarrow} P'$ and $P' \,\mathcal{R}\, Q'$.

We say that two processes P and Q are *weakly timed bisimilar* if there exists a weak timed bisimulation containing the pair $\langle P, Q \rangle$. □

3. A GLOBAL TIMED BISIMULATION

In this section we will define a timed extension of *global bisimulations* [3]. We will denote by $\rightsquigarrow$ and $\stackrel{\gamma}{\rightsquigarrow}$, with $\gamma \in Act \cup \mathcal{T}$, the *static* and *dynamic* generalized transitions to be defined below.

Definition 6 Let P be a process. We say that P may evolve by a *static* transition to P', denoted by $P \rightsquigarrow P'$, and we say that P may evolve by a *dynamic* transition to P' after $\alpha \in Act \cup \mathcal{T}$, denoted by $P \stackrel{\alpha}{\rightsquigarrow} P'$, if the corresponding transition can be derived from the following rules:

1. *Preserving the transition relation of P*:

- If $P \xrightarrow{\tau} P'$ then $P \rightsquigarrow P'$.
- If $P \xrightarrow{\alpha} P'$ for some $\alpha \in Act \cup \mathcal{T}$ then $P \overset{\alpha}{\rightsquigarrow} P'$.

2. *Simultaneous execution of several internal transitions*:

$$P = \sum_{\alpha} \alpha\,;P_{\alpha_i} + \sum_{i\in I} \tau\,;P_i \rightsquigarrow \sum_{i\in I} \tau\,;P_i$$

3. *Removal of non-determinism* (generated by simultaneous offerings of the same visible action):

$$P = \sum_{\alpha} \alpha\,;P_{\alpha_i} + a\,;P_a + \sum_{i\in I} a\,;P_i \rightsquigarrow \sum_{\alpha} \alpha\,;P_{\alpha_i} + a\,;P_a$$

4. *Forward execution of an internal move*:

$$P_\beta \rightsquigarrow P'_\beta \quad \Longrightarrow \quad P = \sum_{\alpha} \alpha\,;P_{\alpha_i} + \beta\,;P_\beta \rightsquigarrow \sum_{\alpha} \alpha\,;P_{\alpha_i} + \beta\,;P'_\beta$$

where either $\beta \in Act \cup \{\tau\}$ or $\beta = \texttt{delay}(t)$ for some $t \in \mathcal{T}$.

5. *Simultaneous execution of several instances of an offered action*:

$$\forall a \in Act, \forall I \subseteq I_a \text{ with } I_a = \{\, i \mid P \overset{a}{\Longrightarrow} P_i \,\} \quad \Longrightarrow \quad P \overset{a}{\rightsquigarrow} \sum_{i\in I} \tau\,;P_i$$

6. *Synchronous passing of time under τ's*:

$$\forall i \in I:\ P_i \overset{t}{\approx\!>} P'_i \quad \Longrightarrow \quad P = \sum_{a\in Act} a\,;P_{a_i} + \sum_{i\in I} \tau\,;P_i \overset{t}{\rightsquigarrow} \sum_{i\in I} \tau\,;P'_i$$

where $\approx\!>$ represents the reflexive and transitive closure of $\rightsquigarrow$, and we have $P \overset{t}{\approx\!>} P'$ where $P \approx\!> \cdot \overset{t_1}{\rightsquigarrow} \cdots \cdot \overset{t_n}{\rightsquigarrow} \cdots \approx\!> P'$ with $t = \sum t_i$. Finally, for any $a \in Act$, we take $\overset{a}{\approx\!>}$ equal to $\approx\!> \cdot \overset{a}{\rightsquigarrow} \cdot \approx\!>$.

□

In Figure 1.3 we graphically present the rules given in the previous definition. Let us briefly comment the intuitive meaning of each of these rules, Rule 1 is included to keep the ordinary transitions of processes. Rule 2 allows a process to evolve by selecting a subset of its τ descendants (also note that in this rule some of the α's may be equal to τ). Rule 3 allows to reduce nondeterminism by cutting some of the branches labeled by the same offered action (note here that some of the α's may also be equal to a). Rule 4 indicates that a process may evolve by executing a static transition not only at its root but at any point. One can see that by iterating the application of this rule a static transition can

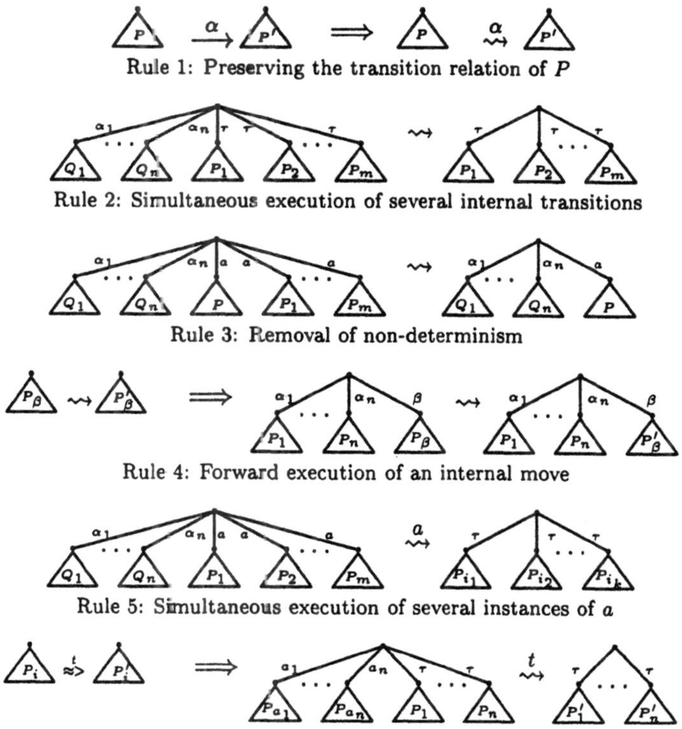

Rule 1: Preserving the transition relation of P

Rule 2: Simultaneous execution of several internal transitions

Rule 3: Removal of non-determinism

Rule 4: Forward execution of an internal move

Rule 5: Simultaneous execution of several instances of a

Rule 6: Synchronous passing of time under τ's

Figure 1.3 Static and dynamic transitions

be done not only after a single step but also after any sequence of them. We also need two additional rules for generating dynamic transitions. Rule 5 allows us to execute not just a single occurrence of an action, but a subset of them, by non-deterministically choosing between them, what is done by composing the reached states by prefixing any of them by a τ action. Finally, Rule 6 allows a process to age under τ actions, thus delaying the corresponding internal choice. Note that in this case all the other branches are lost.

Finally, we introduce our notion of *global timed bisimulation* which follows the same pattern that weak bisimulation, but replacing usual transitions by static and dynamic transitions.

Definition 7 A binary relation $\mathcal{R}$ between pairs of processes is a *global timed bisimulation* iff for all P, Q such that $P \mathcal{R} Q$ and any $\alpha \in Act \cup \{\epsilon\} \cup \mathcal{T}$ we have:

- Whenever $P \overset{\alpha}{\leadsto} P'$ we have some Q' with $Q \overset{\alpha}{\approx\!>} Q'$ and $P' \mathcal{R} Q'$.

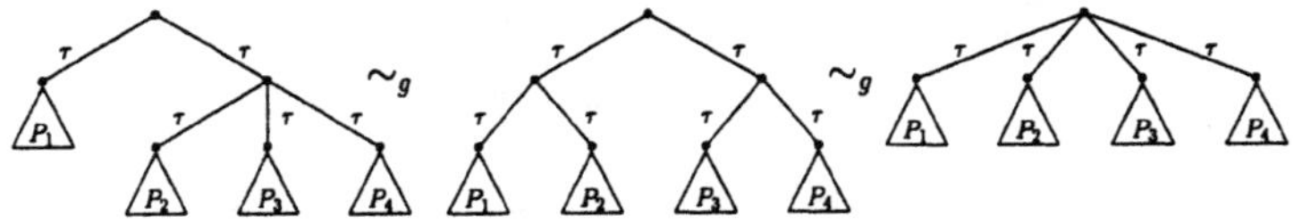

Figure 1.4 Associativity of pure τ-choices.

- Whenever $Q \overset{\alpha}{\rightsquigarrow} Q'$ we have some P' with $P \overset{\alpha}{\approx>} P'$ and $P' \mathcal{R} Q'$.

We say that two processes P and Q are *globally timed bisimilar*, and we write $P \sim_g Q$, if there exists a global timed bisimulation containing the pair $\langle P, Q \rangle$. □

It can be checked that weak timed bisimulation identifies less processes that global timed bisimulation. The proof is an adaptation of that given in [3] for the untimed case.

Theorem 8 Let P and P' be two processes. We have that $P \sim P'$ implies $P \sim_g P'$.

On the other hand global bisimulation is still a rather strong relation. We can prove in fact that it is stronger than the testing equivalence(see [3]), and as a consequence it is strong enough to detect deadlocks. Finally it is interesting to note that the relations $\overset{\alpha}{\rightsquigarrow}$ give us a derived timed labeled transition systems which defines an alternative operational semantics of our processes. In such a case global bisimulation becomes plain weak bisimulation, and as a consequence all the results for weak bisimulation could be extended to global bisimulation, although relative to that (a bit strange and complicated) derived operational semantics. In particular we can prove that if the original transition system is finite, the derived system can be presented also as a finite state system. Thus global timed bisimulation becomes decidable for finite state system, and the algorithms to decide ordinary bisimulation can be applied to decide global bisimilarity.

3.1 ASSOCIATIVITY OF PURE τ-CHOICES

As we already explained, one of the motivations of our work was to develop a notion of observational semantics based on bisimulation where the internal choice operator were associative. So, considering our definition of global timed bisimulation we have the situation described in Figure 1.4.

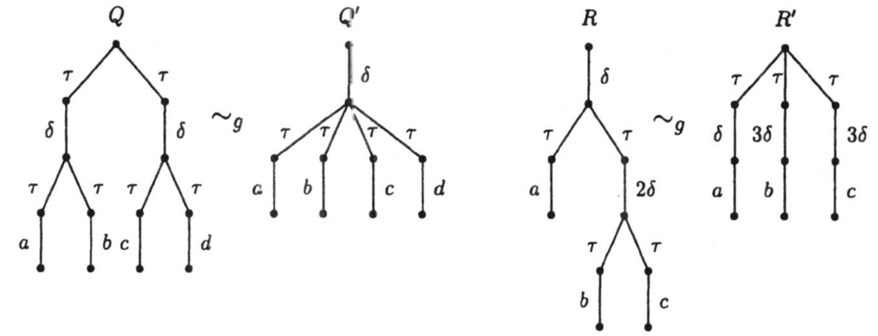

Figure 1.5 Associativity in the presence of time.

To see how the notions of generalized transitions have been extended in order to get associativity, we will consider the following processes:

$P = \tau\,;(\tau\,;P_1 + \tau\,;P_2) + \tau\,;P_3 \qquad P' = \tau\,;P_1 + \tau\,;(\tau\,;P_2 + \tau\,;P_3)$

$P'' = \tau\,;P_1 + \tau\,;P_2 + \tau\,;P_3$

Let us study which of the *capabilities* that we have added to the usual transitions relation allow us to get $P \sim_g P'$ and $P' \sim_g P''$.

In order to show that P and P' are globally timed bisimilar, if we want to simulate the transition $P \rightsquigarrow \tau\,;P_1 + \tau\,;P_2$, it is necessary to apply Rule 4 to P' to execute a τ action under another one, so we have $\tau\,;P_2 + \tau\,;P_3 \xrightarrow{\tau} P_2$, and thus we obtain $P' \rightsquigarrow \tau\,;P_1 + \tau\,;P_2$.

On the other hand, in order that P'' will be able to simulate the same transition of P we apply Rule 2 to P'': by simultaneously executing the two internal transitions $P'' \xrightarrow{\tau} P_1$ and $P'' \xrightarrow{\tau} P_2$, without resolving the choice between them, P'' evolves into $\tau\,;P_1 + \tau\,;P_2$.

Let us now study associativity in the case where *timed decisions* are taken at different places. Let us consider the processes Q and Q' given in Figure 1.5. Let us show that under global bisimulation they are equivalent. We have that when Q executes the transition $Q \rightsquigarrow \delta\,;(a+b)$, Q' can simulate it by executing at the same time two internal actions without aging, that is by applying Rules 2 and 4, obtaining $Q' \rightsquigarrow \delta\,;(\tau\,;a + \tau\,;b)$. On the other hand, in order to let Q to simulate the timed transition $Q' \xrightarrow{\delta} \tau\,;a + \tau\,;b + \tau\,;c + \tau\,;d$, it must be aged without resolving the choice, to obtain $Q \overset{\delta}{\rightsquigarrow} \tau\,;(\tau\,;a + \tau\,;b) + \tau\,;(\tau\,;c + \tau\,;d)$. Note that this transition is allowed by Rule 6.

Consider now the processes R and R' in Figure 1.5. We have that these two processes are globally timed bisimilar. This is a natural consequence of our theory, because we abstract away from *when* the decisions are taken. We have that when R performs the transition $R \overset{\delta}{\rightsquigarrow} R_1 = (\tau ; a) + (\tau ; \texttt{delay}(2\delta) ; (\tau ; b + \tau ; c))$. Indeed, R' should be aged, by the same amount of time in every branch. By applying Rule 6 we obtain $R' \overset{\delta}{\rightsquigarrow} R_1' = \tau ; a + \tau ; \texttt{delay}(2\delta) ; b + \tau ; \texttt{delay}(2\delta) ; c$. At this point, R_1 has two possibilities: on the one hand, if it decides to perform the transition $R_1 \xrightarrow{\tau} a$, R_1' can execute the same transition to simulate it; on the other hand, if R_1 performs the transition $R_1 \xrightarrow{\tau} \texttt{delay}(2\delta) ; (\tau ; b + \tau ; c)$, we have the same situation that in the previous example.

3.2 DIFFERENT KINDS OF SYMMETRIC NON-DETERMINISM

Once we have shown that we have gained the ability to resolve several internal choices at the same time, thus getting associativity, we fix our attention on the alternative ways to generate non-determinism. In particular, an implicit internal choice is generated when we have an external choice between two processes offering some common visible actions. Under weak bisimulation this kind of non-determinism is considered to be different to that generated by internal choices. More exactly, if we consider two processes P_1, P_2 that are not weak bisimilar, we have that the process $a;P_1 + a;P_2$ is not equivalent to any process that does not contain an external choice between two processes offering the action a. It is easy to check that this result does not hold when we consider global timed bisimulation. Next we consider the following three simple processes:

$P_1 = a ; b + a ; c \qquad P_2 = \tau ; a ; b + \tau ; a ; c \qquad P_3 = a ; (\tau ; b + \tau ; c)$

Let us see that under global timed bisimulation we have $P_1 \sim_g P_2$. We have indeed that $P_2 \xrightarrow{\tau} a ; b$ can be simulated by removing the branch $a ; c$ of P_1 (applying Rule 3) to get $P_1 \rightsquigarrow a ; b$. We also have $P_1 \sim_g P_3$: in order to simulate the transition $P_3 \overset{a}{\rightsquigarrow} \tau ; b + \tau ; c$, P_1 may simultaneously execute its two a actions, by postponing the choice between their continuations, then getting $P_1 \overset{a}{\rightsquigarrow} \tau ; b + \tau ; c$ by applying Rule 5. Besides, if we consider the transition $P_2 \rightsquigarrow a;b$ that it is obtained by Rule 1, we have that P_1 can simulate it by means of the transition $P_1 \rightsquigarrow a ; b$, which is generated by applying Rule 3.

3.3 DIFFERENT KINDS OF TIMED SYMMETRIC NON-DETERMINISM

In this section we discuss the problem of symmetric non-determinism when it appears mixed with time. Weak timed bisimulation can correctly deal with some simple interactions between internal situations produced by τ actions and time transitions. For example, we have $\texttt{delay}(t);P \sim \tau;$

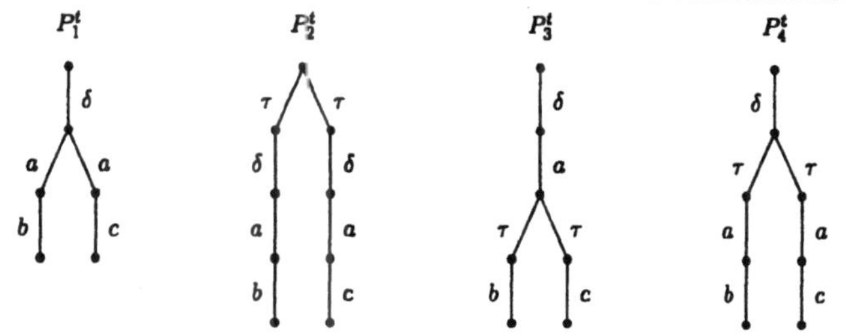

Figure 1.6 Different kinds of non-determinism.

delay(t);$P \sim$ delay(t);τ;P. On the contrary, mixing non-deterministic choices and time in a more complex way produces undesirable results in the context of weak timed bisimulation. For example, the following processes, which are graphically presented in Figure 1.6:

$$P_1^t = (\mathtt{delay}(\delta);a;b)+(\mathtt{delay}(\delta);a;c) \qquad P_2^t = (\tau;\mathtt{delay}(\delta);a;b)+(\tau;\mathtt{delay}(\delta);a;c)$$

$$P_3^t = \mathtt{delay}(\delta)\,;a\,;(\tau\,;b+\tau\,;c) \qquad P_4^t = \mathtt{delay}(\delta)\,;(\tau\,;a\,;b+\tau\,;a\,;c)$$

are not weakly timed bisimilar.

In order to get $P_1^t \sim_g P_2^t$, we have that the transition $P_1^t \xrightarrow{\delta} a;b+a;c$ can be simulated by P_2^t, by applying Rule 6, *aging* it without resolving the non-determinism, to obtain $P_2^t \overset{\delta}{\rightsquigarrow} \tau;a;b+\tau;a;c$. Besides, in order to simulate the transition $P_2^t \xrightarrow{\tau}$ delay(δ);a;b, we have that Rules 3 and 4 allow P_1^t to remove one of the two branches after the timed transitions.

On the other hand, we also have $P_1^t \sim_g P_3^t$ because the transition $P_3^t \xrightarrow{\delta} a\,;(\tau\,;b+\tau\,;c)$ may be simulated by P_1^t just by performing the transition $P_1^t \xrightarrow{\delta} a;b+a;c$. So we obtain the transition systems P_1 and P_3 of the previous section. The same result holds for P_1^t and P_4^t because when one of them performs a timed transition the other can execute the same transition, and in this way they evolve into the processes P_1 and P_2 of the previous section. Besides, P_2^t and P_3^t are globally timed bisimilar, since in order to simulate the transition $P_2^t \xrightarrow{\tau} \delta\,;a\,;b$, P_3^t can execute a τ transition, by applying Rule 4. Using similar reasonings one can check that the pairs $\langle P_2^t, P_4^t\rangle$ and $\langle P_3^t, P_4^t\rangle$ are global timed bisimilar.

4. EXAMPLE: A FAULTY CHANNEL

In this section we summarize the main ideas presented in this paper by means of a simple example: a faulty channel. We will consider a faulty channel which when retransmitting a received message can either

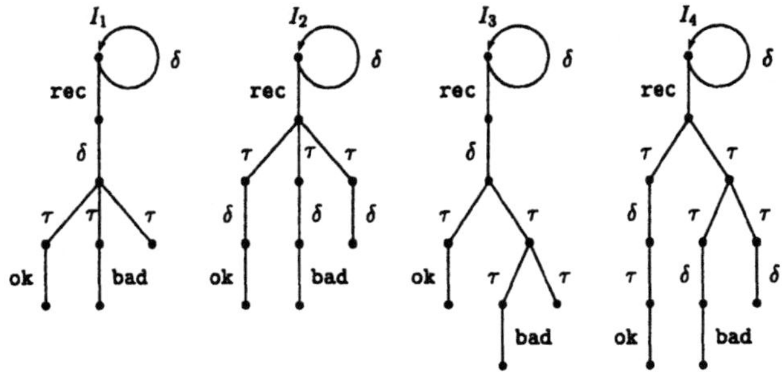

Figure 1.7 Different Implementations of the faulty channel.

correctly send it, or send it but with errors, or lose it. After a more formal specification of the channel we will present several processes which can be considered as correct implementations of the specification. We would like that all of these implementations were equivalent. Nevertheless, they are not under weak timed bisimulation, but they are under our notion of global timed bisimulation.

4.1 THE CHANNEL

We want to get a faulty channel fulfilling the following properties:

1. The channel will eventually receive one message.
2. Then the channel will try to send the message to the final receiver. Three situations can happen:
   - The message is successfully sent to the receiver.
   - The message is sent but the content is corrupted.
   - The message is lost.

   After sending the message the channel gets blocked, that is, we assume that this is a *one-time use* channel.
3. The channel will try to retransmit the message each δ units of time after it was received.

4.2 THE IMPLEMENTATIONS

From the previous specification, the implementer has several choices in order to develop a correct implementation. For example, (s)he can decide to implement the delay between the reception of the message and its sending either right after the system has taken a decision about losing the message or not, or just before. In Figure 1.7 we present some of the possibilities. There `rec` denotes the reception of the message, and ok and bad represents the right and the incorrect sending of the message, respectively. Let us see which are the differences between these implementations.

I_1 works as follows: The system waits until a message arrives, then it waits for a time δ. Finally, it decides non-deterministically either to send the message correctly, or incorrectly, or not to send the message. I_2 receives the message and then it non-deterministically decides what will happen with the message. Afterwards, regardless of the previous decision, it waits a time δ, and then it produces the corresponding result.I_3 is a simple variation of I_1. In this case the implementer first distinguishes two cases: either the message is sent correctly or not. If this is the case we have a new non-deterministic choice either to (incorrectly) send the message or to lose it. I_4 presents a similar variation with respect to I_2.

It is clear that all the previous systems are correct implementations of the specification. Additionally, should we consider all these implementations to be equivalent? Under weak timed bisimulation the answer is negative. For example, I_1 and I_2 are not equivalent. Consider the state of I_1 which is reached after executing the action `rec` and allowing to pass a time δ. If we want to simulate this state with I_2, it must first execute `rec` and then execute one of the τ actions, in order to finally allow the passing of time. It is clear that the state reached by I_1 is not weakly timed bisimilar to any of the possible states reached by I_2. We can find a similar situation in the rest of the cases. Actually, for any $i \neq j$ we have $I_i \not\approx I_j$. Nevertheless, it seems that no external observer should be able to distinguish these implementations: all of them can lose or (correctly or incorrectly) send the message, and all of them wait a time δ between the reception and the sending process. Global timed bisimulation solves this problem. Combining the reasonings used in the previous section, it can be proved that for any i, j we have $I_i \sim_g I_j$.

5. CONCLUSIONS AND FUTURE WORK

In this paper we have presented a timed extension of global bisimulation. This new semantics can be considered as a suitable framework to study the semantics of timed non-deterministic processes. Along the paper we have shown that global timed bisimulation equalize different kinds of non-determinism that are distinguished by weak bisimulation. In particular we have shown how to get associativity of pure τ choices, which in our opinion is a very desirable property.

One line for future work consists in relating our notion of global timed bisimulation with other alternative semantics. For the untimed case we have shown in [3] that (must) testing semantics can be alternatively defined as a kind of global bisimulation (where some rules to deal with *mixed* choices must be added). It would be interesting to study which new rules (for generating static and dynamic transitions) must be added to the notion of global timed bisimulation presented in this paper in order to get other semantics, in particular testing-like semantics like *must* and *may* testing [4, 7], or friendly testing [1, 2].

Another work under development consists in dealing with probabilistic information. We want our work to be consistent with LOTOS, so we have taken as starting points [10, 6] where a probabilistic and a probabilistic-timed version of LOTOS have been defined, respectively.

Acknowledgments. This research has been partially supported by the CICYT project TIC 97-0669-C03-01.

References

[1] D. de Frutos-Escrig, L.F. Llana-Díaz, and M. Núñez. Friendly testing as a conformance relation. In *Formal Description Techniques for Distributed Systems and Communication Protocols (X), and Protocol Specification, Testing, and Verification (XVII)*, pages 283–298. Chapman & Hall, 1997.

[2] D. de Frutos-Escrig, L.F. Llana-Díaz, and M. Núñez. An invitation to friendly testing. *Journal of Computer Science and Technology*, 13(6):531–545, 1998.

[3] D. de Frutos-Escrig, N. López, and M. Núñez. Global bisimulations. Technical Report SIP 91/99, Dept. de Sistemas Informáticos y Programación. Universidad Complutense de Madrid, 1999.

[4] R. de Nicola and M.C.B. Hennessy. Testing equivalences for processes. *Theoretical Computer Science*, 34:83–133, 1984.

[5] H. Erdogmus, R. Johnston, and M. Ferguson. On the operational semantics of nondeterminism and divergence. *Theoretical Computer Science*, 159:271–317, 1996.

[6] C. Gregorio and M. Núñez. Specifying and verifying the Alternating Bit Protocol with Probabilistic-Timed LOTOS. In *COST 247 International Workshop on Applied Formal Methods in System Design*, pages 38–50, 1996.

[7] M. Hennessy. *Algebraic Theory of Processes.* MIT Press, 1988.

[8] L. Llana, D. de Frutos, and M. Núñez. Testing semantics for urgent timed algebras. In *3rd AMAST Workshop on Real-Time Systems*, pages 33–46, 1996.

[9] R. Milner. *Communication and Concurrency.* Prentice Hall, 1989.

[10] M. Núñez and D. de Frutos. Testing semantics for probabilistic LOTOS. In *Formal Description Techniques VIII*, pages 365–380. Chapman & Hall, 1995.

[11] J. Quemada, editor. *Final Committee Draft on enhancements to LOTOS.* ISO/IEC JTC1/SC21/WG7 project 1.21.20.2.3, 1998.

Protocol Synthesis for Real-Time Applications

AHMED KHOUMSI
Department GEGI, University of Sherbrooke, Sherbrooke, Canada
GREGOR V. BOCHMANN
School of ITO , University of Ottawa, Ottawa, Canada
RACHIDA DSSOULI
Department IRO, University of Montreal, Montreal, Canada

Key words: Real-Time Protocol, Service, Synthesis, Assembly System, X.25 Protocol.

Abstract : This paper deals with deriving protocol specifications which provide a given service satisfying timing requirements. In previous work, we have developed an extension of a method proposed by Saleh and Probert, by considering timing requirements in a more general case than in other existing studies. In the present paper, we improve our method by several modifications and additions. First, the number of messages exchanged between the protocol entities is minimized. Second, a less restrictive strategy for choosing between several service primitives is proposed. Third, we consider applications where the choice between several primitives of a single site can be made by the user, and not only by the system. Fourth, conditions of existence of solutions are weaker. Fifth, the timing constraints of the synthesized protocols are weaker. Finally, two simple examples of applications are described.

1. INTRODUCTION

Several methods for deriving a protocol from a service have been developed by various researchers [3,5-7,10,11]. These methods are not applicable for real-time applications, for which the correct ordering of service primitives alone does not always ensure the success of a task. In addition, certain delays must be respected between occurrences of services primitives. In [6] transit delays in the medium are supposed negligible, while in [7] they are bounded by a maximum value. In the present paper, our aim is to describe an approach for deriving protocols which guarantee both : **(a)** a correct ordering of service primitives; and **(b)** the satisfaction of given timing requirements between the executions of service primitives, in a more general case than in [6,7].

A protocol derivation approach presented in [10], which guarantees only (a), has been extended in [8] to the real-time case, i.e., for ensuring also (b). The timing requirements considered in [8] allow to specify certain constraints on the delays between consecutive service primitives. In the present paper, we consider the same problem but we provide a better solution. In fact, compared to [8] : **(a)** the number of messages exchanged between protocol entities for providing a desired service has been minimized; **(b)** the strategy for choosing between several possible service primitives is distributed among several sites, instead of being centralized in a single site; **(c)** we consider applications where the choice between several primitives of a single site can be made by the user, and not only by the system; **(d)** conditions of existence of solutions are weaker; **(e)** the temporal constraints to be respected by the protocol entities are weaker. We have also presented two concrete examples of application.

The remaining of this paper is organized as follows. In Sect. 2, we introduce the problem of protocol derivation and the principle used for deriving protocols. Sect. 3 deals with non-real-time systems. First, we show how services and protocols are specified, and then we introduce the method for deriving protocols. Sect. 4 to 7 deal with real-time systems. In Sect. 4, we describe how services and protocols are specified. In Sect. 5, we explain the approach used for calculating temporal requirements for protocol entities from temporal requirements of a service to be provided. We also present some rules for deriving real-time protocols in a systematic way. In Sect. 6 and 7, two examples illustrate our method. Finally, we conclude in Sect. 8.

2. PROTOCOL SYNTHESIS

We consider a distributed system (DS) consisting of several sites interconnected through a reliable communication medium (simply called *medium*). We assume that: (a) each pair of sites can communicate with each other through the medium; (b) the environment can interact with the DS at the different sites through service access points (SAP). These interactions correspond to the executions of service primitives (simply called *primitives*). We may assume that to each $Site_i$ corresponds a *protocol entity* PE_i.

In the *user's viewpoint*, the DS provides a service where only executions of primitives are visible. We assume that the specification of the desired service (provided to the user) defines the *ordering of* and *the timing requirements between* the occurrences of primitives. The timing requirements define the real-time properties of the DS.

The problem of the *designer* is then : *How* can he derive systematically specifications of the local PE_i, for i=1,2, ..., n, from the specification of the desired service. In the case of a real-time system, the designer must also have a temporal model of the medium, which is assumed reliable.

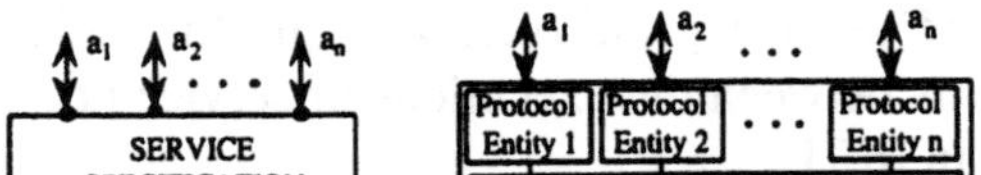

Figure 1. Service and protocol concepts

The approach we have selected for deriving protocols is called *Synthesis*, and the systems considered are assumed sequential. The basic principle we have used is the following: if in the specification of the service, a primitive

A is executed by PE_a, and is followed by the execution of a primitive B by PE_b, then: (α) after PE_a executes A, it sends a message m to PE_b; (β) after PE_b receives m, it executes B.

3. PROTOCOL SYNTHESIS FOR NON-REAL-TIME APPLICATIONS

Service specification : A service desired by the user is described by a finite state automaton (*FSA*), denoted SS, which specifies the sequences of primitives the user would like to observe at the various *SAPs*. Every transition of SS (Fig. 2) is defined by $[q,E_a,r]$, where: **(1)** q and r are origin and destination states; and **(2)** E_a represents a primitive E executed by PE_a. Besides, every transition is identified by a number p and then denoted $T_p=[q,E_a,r]$. In a figure representing an *FSA*, every transition $T_p=[q,E_a,r]$ may be simply represented by $T_p = E_a$ when q and r are explicitly represented by the transition diagram.

Definition 3.1. (*Incoming and outgoing transitions*).
An *outgoing* (resp. *incoming*) transition of a state q is a transition which *is executable from* (resp. *leads to*) q. □

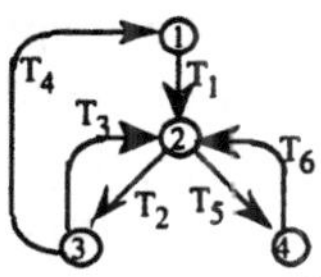

$T_1 = [1,A_1,2]$ $T_2 = [2,B_3,3]$ $T_3 = [3,C_2,2]$

$T_4 = [3,B_4,1]$ $T_5 = [2,B_1,4]$ $T_6 = [4,C_4,2]$

Figure 2. Service specification

Protocol specification : A PE_a is described by an *FSA*, denoted PS_a (see for example Fig. 5), which specifies the sequences of events occurring at $Site_a$. There are three types of events in each PS_a.

Type P (for Primitive) : The execution of a primitive E is denoted E_a.

Type S (for Send): The sending of a message is denoted $s_a^i(p)$, and means "message containing by p is sent by PE_a to entity PE_i ".

Type R (for Receive): The reception of a message is denoted $r_a^i(p)$, and means "message containing p and coming from PE_i is received by PE_a".

Let $PE_1, PE_2,..., PE_n$ be specified by $PS_1, PS_2,..., PS_n$ respectively. Let *DS* be constituted by $PE_1, PE_2,..., PE_n$ and by the medium, and specified by DS.

Definition 3.2. (*Combined behaviour of several PEs*) The *combined behaviour* of $PE_1, ..., PE_n$ is the behaviour of *DS*. The specification DS of this behaviour can be computed from $PS_1,..., PS_n$. We define this combined behaviour by the operator *Comb*: DS=*Comb*($PS_1, PS_2, ..., PS_n$). □

Definition 3.3. (*Projection, total and partial provision of a desired service*) Let V_s and V_i be the alphabets of the service and PS_i, respectively. We also use the following concepts : **(i)** $Proj_\Lambda(A)$ denotes the projection of an *FSA* A into an alphabet Λ; as an example, $Proj_{V_S}(DS)$ specifies the service provided to the user by *DS*; **(ii)** A ≅ B means that the *FSA*s A and B accept the same language; **(iii)** A < B means that the language accepted by A is included in the one accepted by B. We say that the service is totally (resp. partially) provided if $Proj_{V_S}(DS) \cong SS$ (resp. $Proj_{V_S}(DS) < SS$). □

Definition 3.4. (*Semantic and syntactic correctness*)
We say that the protocol is semantically correct if the desired service is totally provided. The protocol is syntactically correct if DS is deadlock-free and livelock-free and no unspecified reception error is possible (we assume that the desired service SS is deadlock-free and livelock-free). □

Our aim is therefore to propose a synthesis method which, from the specification of a service, generates specifications of protocol entities which are syntactically and semantically correct.

Principle for deriving PEs : From an SS specifying a desired service, deriving a protocol consists of generating an *FSA* PS_i for each $Site_i$, which specifies the action sequences executed at $Site_i$. In order to provide the service, the different *PE*s exchange messages through a medium. The basic principle used for deriving a protocol (see Sect. 2) has been applied in [8] as follows. If the execution of a primitive A by PE_a is followed by a choice between primitives executed by other PE_{bi}s, for i=1, ...,k, (Fig. 3) then, after the execution of A, PE_a decides which transition should follow. It therefore sends a message m to all PE_{bi}s (i=1, ...,k, and bi≠a) which contains two parameters: *p* which identifies the executed transition T_p; and *q* which identifies the chosen transition T_q to be executed next. All PE_{bi}s will receive m, but only one of them will execute the primitive corresponding to the selected transition T_q.

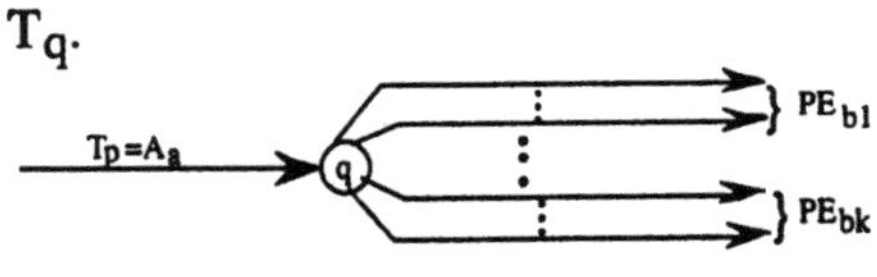

Figure 3. Choice between several actions

This principle implies that the following three restrictions must be satisfied.

Restriction 1. The transitions which may occur in the initial state of SS are all executable by *a single PE*. □

Restriction 2. The choice between several primitives is made by the system and not by the user. □

Restriction 3. The choice between primitives executed by a given PE_{bi} is made by PE_a. □

Compared to [8,10], the following three improvements are made in the subsections below:

(a) *Restriction 2 is weakened* as follows: the choice between PE_{bi}s is made at $Site_a$ by the system, but the choice between several primitives of the selected PE_{bi} may be made at the selected site by the system (for upward primitives) or by the user (for downward primitives).

(b) *Restriction 3 is removed* as follows: PE_a is not necessarily required to select the following primitive; it may decide to select only the following PE which, in turn, selects one of its primitives.

(c) PE_a sends a message only to the selected PE.

Derivation procedure : SS being the input of the problem, the *derivation procedure* consists of the following two steps.

Step 1 : This step consists of completing SS by the insertion of a message exchange between each pair of consecutive primitives which are executed in different sites. In order to avoid any ambiguity, every message contains the number of the transition in SS which is associated to the first of the two primitives. The obtained specification is denoted GPS.

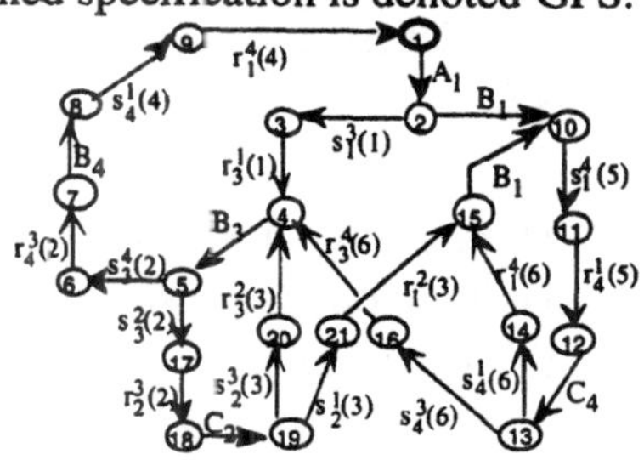

Figure 4. GPS obtained in Step 2 for the example represented in Fig. 2

Step 2 : From GPS, the specification PS_i of each PE_i is obtained by projecting GPS into the alphabet of events which occur in $Site_i$. Then, the PS_i obtained are minimized and determinized.

We note that this two-step procedure is simpler than the procedures proposed in [8,10], besides being optimal. For our example of Fig. 2, we obtain the specifications of Figures 4 and 5, after the first and second steps, respectively. To make the projections of GPS (Fig. 4) for obtaining PS_i (i=1 to 4) (Fig. 5) more directly visible, the states of PS_i are named according to their corresponding states in GPS, where "i-j" means all integers from i to j.

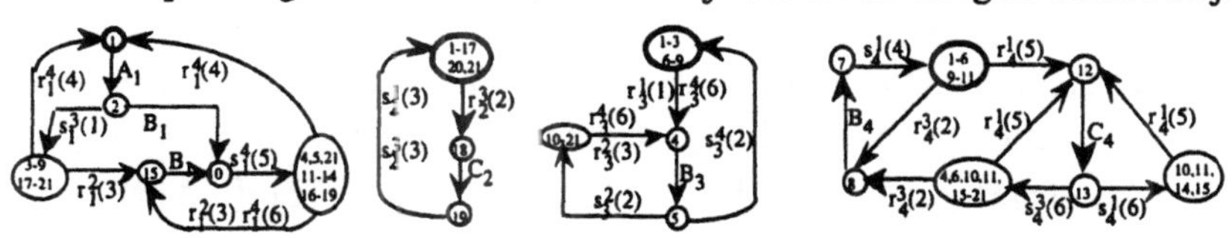

(a) PS_1 (b) PS_2 (c) PS_3 (d) PS_4

Figure 5. Obtained protocol specifications

Contrary to [10], the rules for deriving PEs do not depend on whether primitives are upward or downward. In our opinion, such a distinction complicates uselessly the rules of Step 1. In the remaining part of this paper, we extend the procedure of protocol derivation to real-time systems.

4 TIMED AUTOMATA TO SPECIFY SERVICES AND PROTOCOLS

We use a *dense-time* model [1,8] where the time is viewed as a state variable that ranges over a dense domain and evolves indefinitely. The timed automata ($\mathcal{TA}$) model we propose here uses a variable **v** and a clock **c**.

Definition 4.1. (*Clock* **c**, *variable* **v**)
c has a positive real value which : (1) is set to zero at the occurrence of every transition; and (2) is equal to the time elapsed since the last instant it was set to zero. **v** has a strictly positive natural value which can be updated at the occurrence of any transition. □

Let $A=(Q,\Sigma,\delta,q_0)$ be an $\mathcal{FSA}$ where Q is a set of states, Σ is an alphabet, q_0 is the initial state, and $\delta \subseteq Q\times\Sigma\times Q$ defines the transitions. Let us see how a $\mathcal{TA}$ can be defined from the $\mathcal{FSA}$ A.

Definition 4.2. (*Timed transition, and Timed automaton*)
Let I=[a;b] be an interval, where a and b are positive real numbers and $a \leq b$. A *timed transition* is defined by $[q,\sigma,r;\mathcal{C},v]$ where : **(a)** $[q,\sigma,r]$ defines a transition of the $\mathcal{FSA}$ $\mathcal{A}$; **(b)** $\mathcal{C}=(I_1, I_2,\ldots,I_m)$ is a *m*-tuple of *non-empty* intervals; and **(c)** v is a value of variable **v**. A $\mathcal{TA}$ A^t can therefore be constructed if we transform every transition $tr=[q_1,\sigma,q_2]$ of A into a timed transition Tr by associating to it an *m*-tuple $\mathcal{C}$ of intervals and a value v of **v**. The semantics of a timed transition $Tr=[q,\sigma,r;\mathcal{C},v]$ of A^t depends on the current state q and on the current values of **v** and **c** as follows. If u is the current value of **v** then : (1) Tr is enabled (i.e., may occur) only if the current value of **c** falls within the uth interval I_u of $\mathcal{C}$; and (2) after the occurrence of Tr, **v** is set to v and **c** is set to zero. Intuitively, the temporal constraint of a transition may depend on how the current state has been reached (this information is given by **v**). □

Henceforth, every timed transition is simply called transition, and $Tr=[q,\sigma,r;\mathcal{C},v]$ may be simply represented by $Tr=[\sigma;\mathcal{C},v]$ if there is no ambiguity about q and r. An example of a part of $\mathcal{TA}$ is given in Fig. 6. State q has two incoming (Tr_1 and Tr_2) and two outgoing transitions (Tr_3 and Tr_4), with $Tr1=[q1,\sigma1,q;\mathcal{C}1,v1]$, $Tr_2=[q2,\sigma2,q;\mathcal{C}2,v2]$, $Tr_3=[q,\sigma3,r1;\mathcal{C}3,v3]$, and $Tr_4=[q,\sigma4,r2;\mathcal{C}4,v4]$. With the representation of Fig. 6, we can define $v1$ and $v2$, and $\mathcal{C}3$ and $\mathcal{C}4$. In fact, for a timed transition $Tr=[q,\sigma1,r;\mathcal{C},v]$, the definition of v necessitates to know all the incoming transitions of r, and the definition of $\mathcal{C}$ necessitates to know all the incoming transitions of q. For example, $v1=1$, $v2=2$, $\mathcal{C}3=(I3_1,I3_2)$, $\mathcal{C}4=(I4_1,I4_2)$, $I3_1=[1;2]$, $I3_2=[0;2]$,

$I4_1$=[1;3] and $I4_2$=[2;5]. The informal specification is then the following, with q being the current state:

- if **v**=1, i.e., q has been reached by transition Tr_1, then :
  * Tr_3 (resp. Tr_4) may occur after a delay within the interval $I3_1$ (resp. $I4_1$);
  * If neither Tr_3 nor Tr_4 occurs after a delay within [1;2], then Tr_4 *must* occur after a delay within [2;3] (in order to avoid a deadlock).
- if **v**=2, i.e., q has been reached by the transition Tr_2, then :
  * Tr_3 (resp. Tr_4) may occur after a delay within the interval $I3_2$ (resp. $I4_2$);
  * If Tr_3 does not occur after a delay within [0;2], then Tr_4 *must* occur after a delay within [2;5].

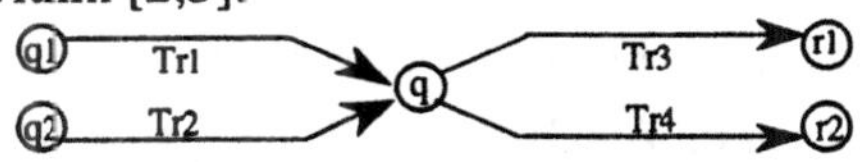

Figure 6. Incoming and outgoing transitions

A desired service is described by a $\mathcal{TA}$ denoted SST, which specifies: **(a)** the required sequences of primitives; and **(b)** certain temporal requirements between consecutive primitives. In any state q of SST, we can express some temporal constraints on the primitives which are executable from state q. These temporal constraints may depend on how q has been reached. As an example, the $\mathcal{FSA}$ SS of Fig. 2 (Sect. 3) is transformed into a $\mathcal{TA}$ SST by replacing transitions T_i of SS into the following timed transitions Tr_i, i=1,...,6, respectively : $Tr_1=[1,A_1,2;\mathcal{C}1,1]$, $Tr_2=[2,B_3,3;\mathcal{C}2,1]$, $Tr_3=[3,C_2,2;\mathcal{C}3,2]$, $Tr_4=[3,B_4,1;\mathcal{C}4,1]$, $Tr_5=[2,B_1,4;\mathcal{C}5,1]$, $Tr_6=[4,C_4,2;\mathcal{C}6,3]$, and $\mathcal{C}1$=I1, $\mathcal{C}2=(I2_1,I2_2,I2_3)$, $\mathcal{C}3$=I3, $\mathcal{C}4$=I4, $\mathcal{C}5=(I5_1,I5_2,I5_3)$, $\mathcal{C}6$=I6, where I1, $I2_1$, $I2_2$, $I2_3$, I3, I4, $I5_1$, $I5_2$, $I5_3$, and I6 are intervals. For example, if the current state 2 is reached by Tr_3, then **v** is set to 2, Tr_2 is enabled iff ($\mathbf{c} \in I2_2$) and Tr_5 is enabled iff ($\mathbf{c} \in I5_2$) (intervals $I2_2$ and $I5_2$ of $\mathcal{C}2$ and $\mathcal{C}5$ are used because **v**=2). Therefore, if "$I \leftarrow \langle Tr_a,Tr_b\rangle$" means "the delay of Tr_a and Tr_b falls within the interval I" then : $I1 \leftarrow \langle Tr_4,Tr_1\rangle$ $I2_1 \leftarrow \langle Tr_1,Tr_2\rangle$ $I2_2 \leftarrow \langle Tr_3,Tr_2\rangle$ $I2_3 \leftarrow \langle Tr_6,Tr_2\rangle$ $I3 \leftarrow \langle Tr_2,Tr_3\rangle$ $I4 \leftarrow \langle Tr_2,Tr_4\rangle$ $I5_1 \leftarrow \langle Tr_1,Tr_5\rangle$ $I5_2 \leftarrow \langle Tr_3,Tr_5\rangle$ $I5_3 \leftarrow \langle Tr_6,Tr_5\rangle$ $I6 \leftarrow \langle Tr_5,Tr_6\rangle$.

A PE_a is described by a $\mathcal{TA}$ denoted PST_a, which specifies: **(a)** the sequences of events which occur at $Site_a$; **(b)** certain temporal constraints to be satisfied between consecutive events. Similarly to the non-real-time case, the events may be of the three types *P*, *S* and *R*. Examples of $\mathcal{TA}$s specifying PEs are given in Sect. 6 and 7.

We consider PE_1, PE_2 , ...,PE_n which are specified by PST_1, PST_2, ..., PST_n, respectively. Let $\mathcal{DS}$ be constituted by PE_1, PE_2,...,PE_n and by the medium, and specified by a $\mathcal{TA}$ DST.

Definition 4.3. (*Timed sequence of events, Timed language, Acceptance*)

A timed sequence T is represented by $\langle\sigma_1,t_1\rangle\langle\sigma_2,t_2\rangle...\langle\sigma_i,t_i\rangle$... and means that events σ_1, σ_2, ..., σ_i, ... occur at instants t_1, t_2, ..., t_i ..., respectively, where $0 < t_1 < t_2 < ... < t_i < ...$ A timed language is a set (possibly infinite) of timed sequences. Let A be a $\mathcal{TA}$, and L_A be the set of sequences which can be executed by A. Then we say that A *accepts* the language L_A. □

Definition 4.4. (*Projection, total and partial provision of a desired service with temporal requirements*) The projection of a $\mathcal{TA}$ into a subalphabet Λ can be defined similarly to the projection of an $\mathcal{FSA}$. Therefore, $Proj_{V_s}(DST)$ specifies the service provided to the user by $\mathcal{DS}$, and each $Proj_{V_i}(DST)$ specifies PE_i, for i=1,..., n. For the comparison of timed languages, we use the symbols $\cong_T$ and $<_T$, i.e., $A\cong_T B$ means that $L_A=L_B$, and $A<_T B$ means that $L_A \subset L_B$. Total and partial provisions of a real-time service are defined like in Def. 3.3, but by using symbols $\cong_T$ and $<_T$ instead of $\cong$ and $<$. □

5. PROTOCOL SYNTHESIS FOR REAL-TIME APPLICATIONS

Definition 5.1. (*Reliable medium*) Besides not altering messages, in the real-time case a reliable communication medium must be such that the transit delay t_m of a message sent at $Site_a$ and received at $Site_b$, belongs to a finite interval $M_{a,b}=[\mu_{a,b};\rho_{a,b}]$ which depends on $Site_a$ and $Site_b$. □

The synthesis of the real-time PEs uses the same Step 1 of the non-real-time case, where we obtain GPST from SST (the last T indicates the presence of temporal constraints). In this step, the timed transitions are processed like simple transitions, while C and v are kept unchanged. GPST specifies the correct ordering of primitives, but it does not specify the correct temporal requirements of the service.

To compute temporal constraints for the PEs, we consider every pair of states q and r of GPST which are connected by two consecutive events $s_a^b(p)$ and $r_b^a(p)$ (see Fig. 7). Let Tr be the single incoming transition of q (which corresponds to a primitive executed at a $Site_a$) and let Tr_1,...,Tr_n be the outgoing transitions of r (which correspond to primitives executed at the same $Site_b$). After Tr, PE_a sends a message to PE_b (written $s_a^b(p)$); when PE_b receives the message (written $r_b^a(p)$), it executes one of the n Tr_k. The sequencing of events between $Tr=[q1,\sigma,q;C,v]$ and $Tr_k=[r,\sigma k,q2;C_k,v_k]$ is represented as a function of the time in Fig. 8.a. The delay between Tr and Tr_k must belong to the vth interval $Ik_v=[\gamma k_v;\delta k_v]$ of C which, for simplicity, is denoted $I_k=[\gamma_k;\delta_k]$.

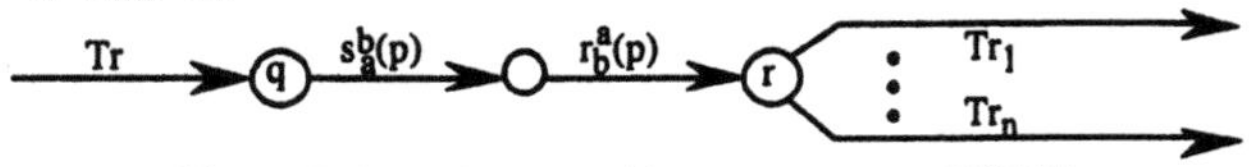

Figure 7. Outgoing transitions on a state of GPST

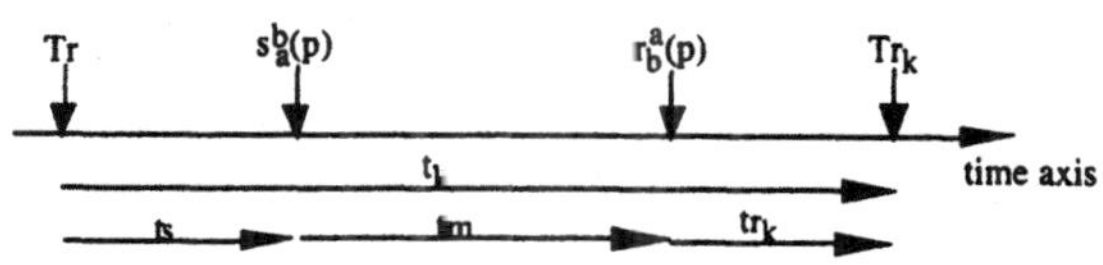

(a) Representation in function of time

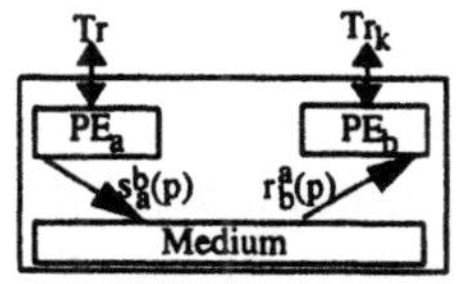

(b) Representation by entities

Figure 8. Representation of events between Tr and Tr_k

From Fig. 8.a, we see that the service requires that the time t_k, between the executions of Tr and Tr_k, falls within $I_k=[\gamma_k;\delta_k]$. The model of the medium implies that the transit delay tm of a message sent by PE_a and received by PE_b, falls within $M_{a,b}=[\mu_{a,b};\rho_{a,b}]$. The aim of the temporal requirements derivation for the protocol entities is the following.

From requirements $tm \in M_{a,b}=[\mu_{a,b};\rho_{a,b}]$ and $t_k \in I_k=[\gamma_k;\delta_k]$ (k=1, 2,..., n), we must compute constraints on ts and tr_k (k=1,2, ..., n) which ensure that requirements $t_k \in I_k$ on the service will be respected. These derived constraints are written in the form $ts \in S=[\theta;\phi]$, and $tr_k \in R_k=[\tau_k;\omega_k]$, k= 1, 2,..., n. This computation must be made for each occurrence of the structure in Fig. 7 within GPST.

Notations: operators $\subseteq$, $\cup$ or $\cap$ will be used on intervals, and $[a;b]+[c;d]=[a+c;b+d]$, $[a;b]-[c;d]=[a-c;b-d]$.

Condition for existence of solutions : we consider two consecutive transitions Tr_1 and Tr_2 wich are executed at $Site_i$ and $Site_j$, respectively. After Tr_1, $Site_i$ sends a message to $Site_j$ to inform it that it may execute Tr_2. If the delay between Tr_1 and Tr_2 must be greater than x and smaller than y, then the transit delay of the message must be smaller than y. Besides, the difference between the biggest delay and the smallest delay of the message in the medium must be smaller than y-x. Formally, for each occurrence of the structure in Fig. 7 within the GPST, we must have :

$$\text{for } k=1,2,...,n: \quad \delta_k - \rho_{a,b} \geq \sup(\gamma_k - \mu_{a,b}; 0) \qquad (1)$$

where sup(a; b) is equal to the biggest of a and b.
Therefore, for each occurrence within GPST of the structure in Fig. 7, we must check if (1) is respected. If the checking is positive then we must compute : **(a)** the interval S representing the constraint on ts, and **(b)** intervals R_k, k=1, 2,..., n , representing the constraints on tr_k, k=1, 2,..., n.

We note that condition (1) is less restrictive than the conditions for the existence of solutions of [8]. For resolving the timing constraints of the *PEs*, we consider the following three cases :

Static case : messages transmitted by *PEs* contain no temporal information;

First dynamic case : the *PEs* include some temporal information in the messages they send;

Second dynamic case : the temporal information included by the *PEs* is completed by the medium.

In the following, ξ and ψ are any real values which fall within [0; 1].

Static case : we assume that the intervals S and R_k are constant. When PE_a executes a transition Tr and decides to send a message to PE_b, the time ts between Tr and the transmission of the message falls within a constant interval S. When PE_b receives the message from PE_a, it can execute a transition Tr_k, among n possible transitions (k=1, 2, ..., n), in a time tr_k belonging to a constant interval R_k. The interval $S=[\theta;\phi]$ must satisfy the following equations :

$$\phi = \psi\times\min_{k=1 \text{ to } n} (\delta_k - \rho_{a,b}) \quad (2)$$

$$\theta = \sup(U, 0) + (\phi - \sup(U, 0))\times\xi \quad (3)$$

with $$U = \max_{k=1 \text{ to } n} (\phi + (\rho_{a,b} - \mu_{a,b}) - (\delta_k - \gamma_k)) \quad (4)$$

Afterwards, we choose the less restrictive solutions for $R_k=[\tau_k; \omega_k]$:

for k= 1, 2, ..., n : $$\omega_k = \delta_k - \rho_{a,b} - \phi \quad (5)$$

$$\tau_k = \sup(\gamma_k - \mu_{a,b} - \theta; 0) \quad (6)$$

Intuitively, modifying ψ and ξ allows to "move" some timing constraints between two communicating entities.

First dynamic case : we assume that PE_a sends to PE_b a message containing ts (Fig. 8.a), and PE_b calculates dynamically R_k as a function of ts when it receives the message from PE_a. The interval $S=[\theta;\phi]$ must satisfy the following equations :

$$\phi = \psi\times\min_{k=1 \text{ to } n} (\delta_k - \rho_{a,b}) \quad (2)$$

$$\theta = \phi\times\xi \quad (7)$$

The interval $R_k(ts)$ is computed as follows. If ts, which belongs to $[\theta; \phi]$, is the delay when the message is sent after the execution of Tr_p, the receiving entity knows it and can choose :

for k= 1, 2, ..., n : $$\omega_k(ts) = \delta_k - \rho_{a,b} - ts \quad (8)$$

$$\tau_k(ts) = \sup(\gamma_k - \mu_{a,b} - ts; 0) \quad (9)$$

Intuitively, with the information ts, the receiving entity PE_b can use the time allocated to it to provide the service more efficiently than in the static case. Let us, for instance, assume that some optional tasks are achieved by PE_b in order to provide a better quality of service, only if PE_b has enough time. In the static case, PE_b may estimate that it has not enough time to execute its optional tasks, while in the dynamic case optional tasks will be executed.

Second dynamic case : in comparison with the first dynamic case, we assume that the medium modifies ts into the more accurate information ts+tm. In this case, PE_b receives the message with information ts+tm, and it calculates dynamically the interval R_k, as a function of ts+tm. $S = [\theta;\phi]$ is resolved as in the previous case; ω_k and τ_k are calculated by PE_b as follows :

for k= 1, 2, ..., n : $$\omega_k(ts+tm) = \delta_k - (ts+tm) \quad (10)$$

$$\tau_k(ts+tm) = \sup(\gamma_k - (ts+tm); 0) \quad (11)$$

Intuitively, with the information ts+tm the receiving entity PE_b knows that Tr has been executed ts+tm before the reception of the message, which is a more accurate information than in the first dynamic case. Due to this fact, in the second dynamic case PE_b can use the time allocated to it to provide the service more efficiently than in the first dynamic case.

We note that ts and ts+tm , which are transmitted in the dynamic cases, are a *relative* temporal information. This implies that a global clock is not necessary. We also note that the temporal requirements of the protocol obtained using the approach in [8] are more restrictive than those derived by our improved approach.

Derivation Procedure : it consists of three steps. Step 1, which generates a specification GPST, is similar to step 1 of the non-real-time case.

Step 2 : The aim of this step is : **(a)** to compute and insert into GPST the static temporal constraints and, in the dynamic cases, some constant parameters which allow to compute the dynamic temporal constraints; **(b)** to insert ts and tm into the exchanged messages. Therefore, for every structure represented in Fig. 7 and contained in GPST, the following three substeps are performed to transform GPST into GST.

Step 2.1. We compute the interval $S=[\theta ; \phi]$ and:

- ∗ Intervals R_k, k=1,...,n, in the static case;
- ∗ Intervals $X_k=I_k-M_{a,b}$, k=1,...,n, in the first dynamic case;

Step 2.2 

becomes: (v being the value of **v** which is set by the transition preceding $s_a^b(p)$)

- In the static case : 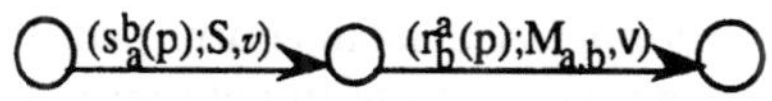

- In the first dynamic case : 

- In the second dynamic case : 

Informally: - the delay between Tr and $s_a^b(*)$ falls within $S=[\theta;\phi]$;

- the delay between $s_a^b(*)$ and $r_b^a(*)$ falls within $M_{a,b}$.

Step 2.3 For each k=1, ...,n, the vth interval I_k of α_k is replaced by the interval: **(i)** R_k in the static case; **(ii)** $X_k=I_k-M_{a,b}$ in the first dynamic case; **(iii)** I_k (i.e., it is not replaced) in the second dynamic case. We note that the GST obtained at Step 2.3 is defined by constant intervals. In dynamic cases, some of these constant intervals do not directly represent timing constraints, but they are used for a dynamic calculation of the time requirements. In fact, for each k=1,...,n, the delay between occurrences of $r_b^a(p)$ and Tr_k must belong to: - the interval R_k in the static case;

- an interval $R_k(ts)$ which depends on the interval $X_k=I_k-M_{a,b}$ and on ts ;
- an interval $R_k(ts+tm)$ which depends on the interval I_k and on ts+tm.

Step 3 : This step consists of generating the protocol specification PST_i by projecting GST onto the alphabet of events which occur at $Site_i$, i=1,...,n. This step is similar to the second step of the non-real-time case, with the difference that intervals $M_{a,b}$ are replaced by $[0;\infty]$.

In Sect. 6 and 7, ψ and ξ are taken to be equal to 0.5, which means that the temporal constraints are equally distributed between the two sites.

6. SYNTHESIS OF A TIMED X.25 PROTOCOL

We consider the X.25 service [2]. In order to apply our synthesis method, the X.25 service is made sequential by assuming that the service primitives are ordered and executed sequentially. For that purpose, the following assumptions are made: **(i)** a new message cannot be sent before the last one is received; and **(ii)** express data are not supported. The simplified X.25 service obtained will be extended by adding certain temporal requirements between consecutive primitives. In order to give the possibility to both sites to establish a connection, we have used a mechanism of tokens to realize a distributed choice. The following description of this example is based on [4].

Primitives of the simplified X.25 service : Let U_1 and U_2 be two users of the network who are located in $Site_1$ and $Site_2$, respectively. The following service primitives are defined :

- *Connection* : It may be established between U_1 and U_2 if one of them, for instance U_1, sends a CN.req to U_2. When U_2 receives a CN.ind, he may answer either by a DC.req to reject the CN.req, or by a CN.rsp. In the first case U_1 receives a DC.ind, while in the second case U_1 receives a CN.cnf.
- *Disconnection* : A disconnection primitive can be used either to reject a CN.req (see above) or to terminate an existing connection. For instance, U_1 may send a DC.req and then U_2 will receive a DC.ind.
- *Data Transfer* : Two site linked by a connection may exchange data in both directions. To simplify, we assume that only the party which has initiated the connection can send data. The sending of a message is generated by a DT.req and its reception by a DT.ind.
- *Reinitialization* : it allows to restore the synchronization between two parties. When a RI.req is generated, for instance by U_1, then all the data being transmitted are removed. The next element to be received by U_2 is a RI.ind. U_2 answers by a RI.rsp and then U_1 will receive a RI.cnf. We assume that the party which requests the reinitialization is the sender of data.

Specification of the Simplified X.25 Service : Our specification contains two blocs $S_{1,2}$ and $S_{2,1}$, where $S_{i,j}$ (see Fig. 9) models the service when $Site_i$ and $Site_j$ are the sender and the receiver, respectively. The specification of the simplified X.25 is schematized in Fig. 10. The event $Token_i^j$ means that "$Site_i$ gives to $Site_j$ the possibility to establish a connection". We assume that State $1_{1,2}$ is the initial state of the service.

Temporal constraints added to the simplified X.25 service

- The delay between $CN.req_i$ and $CN.ind_j$ belongs to the interval [1;1.5];
- The delay between $DC.req_i$ and $DC.ind_j$ belongs to the interval [0.5;1];
- The delay between $RI.ind_i$ and $RI.rsp_i$ belongs to the interval [0;0.5];
- The delay between $DT.req_i$ and $DT.ind_j$ belongs to the interval [1;1.5];
- The delay between $DT.rsp_i$ and $DT.cnf_j$ belongs to the interval [1;1.25].

Transitions with temporal constraints are represented in grey in Fig. 9 and 10, with the intervals defining the temporal constraints.

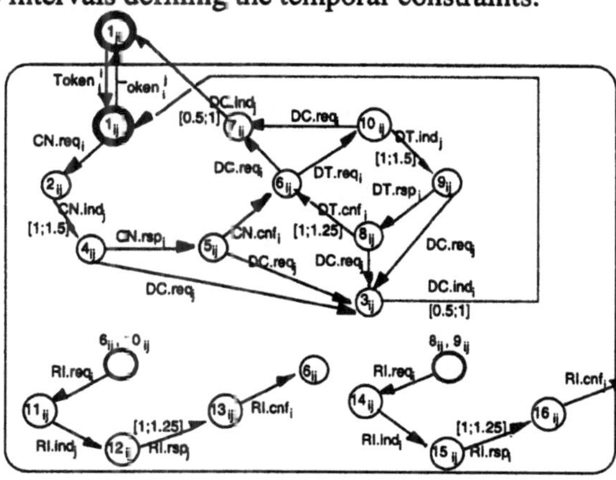

Figure 9. Block $S_{i,j}$

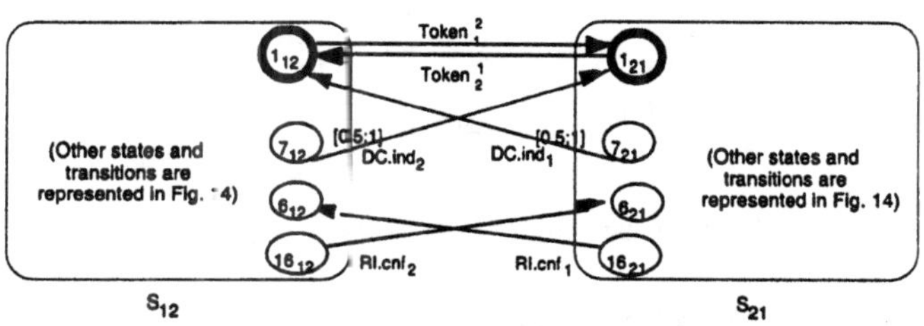

Figure 10. Service specification of the simplified X.25

Protocol Synthesis : The synthesized specifications of the two PEs can be represented by the single *TA* of Fig. 11, and where i identifies the PE described and j identifies the other PE. Messages sent by each PE_i contain the parameters p_i^k, k=1,..., 17, with $p_i^r \neq p_i^s$ if $r \neq s$. The initial states of PE_1 and PE_2 are identified by 1 and 2, respectively. To generate these real-time PEs, the following temporal model of the medium has been used: the transit delay of a message falls within [0.5;0.75] when it is transmitted from $Site_1$ to $Site_2$, and within [0.25;0.5] when it is transmitted from $Site_2$ to $Site_1$. To simplify, we give only the results of the static case. The synthesized temporal constraints, which are defined by interval for the transitions represented in grey in Fig. 11, are the following.

<u>$Site_1$</u> : constraint for $s_1^2(p2_1)$, $s_1^2(p3_1)$, $s_1^2(p4_1)$, $s_1^2(p5_1)$, $s_1^2(p10_1)$ and

$s_1^2(p11_1)$ is [0.0625;0.125]; constraint for $s_1^2(p7_1)$ is [0.25;0.25]; constraint for $s_1^2(p8_1)$ and $s_1^2(p9_1)$ is [0.25;0.375]; constraint for $DT.cnf_1$ is [0.375;0.375]; constraint for $DT.ind_1$ is [0.375;0.5]; constraint for $DC.ind_1$ is [0.125;0.25]; constraint for $RI.rsp_1$ is [0; 0.5].

<u>$Site_2$</u> : constraint for $s_2^1(p2_2)$, $s_2^1(p3_2)$, $s_2^1(p4_2)$, $s_2^1(p5_2)$, $s_2^1(p10_2)$ and $s_2^1(p11_2)$ is [0.125;0.25]; constraint for $s_2^1(p7_2)$ is [0.375;0.375]; constraint for $s_2^1(p8_2)$ and $s_2^1(p9_2)$ is [0.375;0.5]; constraint for $DT.cnf_2$ is [0.25;0.25]; constraint for $DT.ind_2$ and $CN.ind_2$ is [0.25;0.375]; constraint for $DC.ind_2$ is [0;0.125]; constraint for $RI.rsp_2$ is [0;0.5].

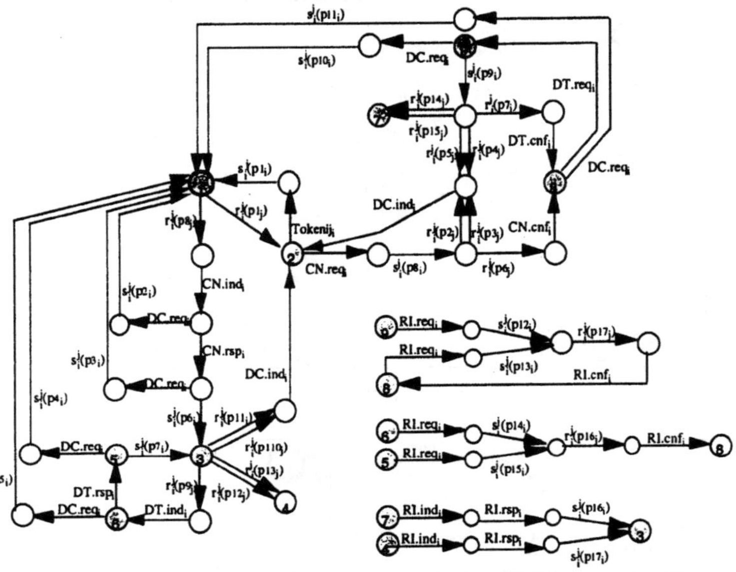

Figure 11. Specification of the synthesized X.25 protocol at $Site_i$

7. EXAMPLE ASSEMBLY SYSTEM

The following example application of our synthesis method in another area than telecommunications. We consider an assembly system consisting of three robots R1, R2 and R3 and three carpets C1, C2 and C3. The carpets C1 and C2 bring pieces of type P1 and P2, respectively, and carpet C3 takes away the assembled pieces. Robot R1 takes a piece P1, and puts it on a table T for the assembly. Robot R2 takes a piece P2 and assembles it with the piece P1 which is on the table T. Robot R3 removes the defective pieces. The details of this example are given in [9].

Protocol entities : There are six PEs, R1, R2, R3, C1, C2 and C3, which correspond to the three robots and the three carpets, respectively. Table T is not considered as an entity since it is passive. A piece is denoted Pi.

Service primitives : They are the following (i=1, 2, 3, and j=1, 2):
$MOVE.Carpet_{Ci}$: Ci is actuated; $ARRIVED.Piece_{Ci}$: Ci has detected that a

Pi has reached its destination; STOP.Carpet$_{Ci}$: Ci is stopped ; CHECK.Piece$_{Rj}$: Rj begins to check Pj ; TAKE.Piece$_{Rj}$: Rj takes a Pj from Cj ; TAKE.Piece1$_{R3}$: R3 takes a P1 from C1 ; TAKE.Piece2$_{R3}$: R3 takes a piece P2 from C2 ; TAKE.AssPieces$_{R3}$: R3 takes an assembled piece from table T ; PUT.Piece$_{R1}$: R1 puts a P1 on T ; PUT.AssPieces$_{R2}$: R2 puts an assembled piece on C3; ASS.Pieces$_{R2}$: R2 assembles P1 and P2.

Scenario of the Service : From the initial state where the whole system is stopped, the scenario of the service is the following:

Step 1 : C1 is actuated (MOVE.Carpet$_{C1}$)

Step 2 : P1, which is on C1, reaches its destination (ARRIVED.Piece$_{C1}$)

Step 3 : C1 is stopped (STOP.Carpet$_{C1}$)

Step 4 : R1 checks P1 (CHECK.Piece$_{R1}$):

Step 5: If P1 is bad then R3 takes it off (TAKE.Piece1$_{R3}$), and *goto Step 1.*

Step 6: If P1 is good then R1 takes it from C1 (TAKE.Piece$_{R1}$) and

Step 7: R1 puts C1 on the table T (PUT.Piece$_{R1}$)

Step 8 : C2 is actuated (MOVE.Carpet$_{C2}$)

Step 9 : P2, which is on C2, reaches its destination (ARRIVED.Piece$_{C2}$)

Step 10 : Carpet C2 is stopped (STOP.Carpet$_{C2}$)

Step 11 : Robot R2 checks P2 (CHECK.Piece$_{R2}$):

Step 12 : If P2 is bad then R3 takes it off (TAKE.Piece2$_{R3}$) and *goto Step 8.*

Step 13 : If P2 is good then R2 takes it from C2 (TAKE.Piece$_{R2}$) and

Step 14 : R2 assembles P2 with P1 on the table T (ASS.Pieces$_{R2}$).

Step 15 : If the assembly is bad, which it is detected by R2, then R3 takes it off (TAKE.AssPieces$_{R3}$), and *goto Step 1.*

Step 16 : If the assembly is good then R2 takes it and puts it on carpet C3 (PUT.AssPieces$_{R2}$)

Step 17 : Carpet C3 is actuated (MOVE.Carpet$_{C3}$)

Step 18 : The assembled pieces reach their destination, which is detected by carpet C3 (ARRIVED.Piece$_{C3}$).

Step 19 : Carpet C3 is stopped (STOP.Carpet$_{C3}$), and *goto Step 1.*

Temporal constraints added to the service

- Between MOVE.Carpet$_{Ci}$ and ARRIVED.Piece$_{Ci}$: [10;20]; (i=1,2,3)
- Between ARRIVED.Piece$_{Ci}$ and STOP.Carpet$_{Ci}$: [0.5;2]; (i=1,2,3)
- Between STOP.Carpet$_{Ci}$ and CHECK.Piece$_{Ri}$: [5;8]; (i=1,2)
- Between CHECK.Piece$_{Ri}$ and TAKE.Piecei$_{R3}$: [5;8]; (i=1,2)
- Between ASS.Piece$_{R2}$ and TAKE.AssPieces$_{3}$: [5;8];
- Between CHECK.Piece$_{Ri}$ and TAKE.Piece$_{Ri}$: [1;2]; (i=1,2)
- Between TAKE.Piece$_{R1}$ and PUT.Piece$_{R1}$: [1;2];
- Between TAKE.Piece$_{R2}$ and ASS.Pieces$_{R2}$: [4;10];
- Between PUT.Piece$_{R1}$ (or TAKE.Piece2$_{R3}$) and MOVE.Carpet$_{C2}$: [5;10];
- Between PUT.AssPieces$_{R2}$ and MOVE.Carpet$_{C3}$: [5;10];
- Between ASS.Pieces$_{R2}$ and PUT.AssPieces$_{R2}$: [2;5];

- Between TAKE.Piece1$_{R3}$ (or TAKE.AssPieces$_{R3}$ or STOP.Carpet$_{C3}$) and MOVE.Carpet$_{Ci}$: [6;10]; (i=1,2)

The service specification is represented in Fig. 12, where transition Tr_i corresponds to Step i of the scenario. The temporal constraint of each transition is defined by a single interval (which is shown in Fig. 12).

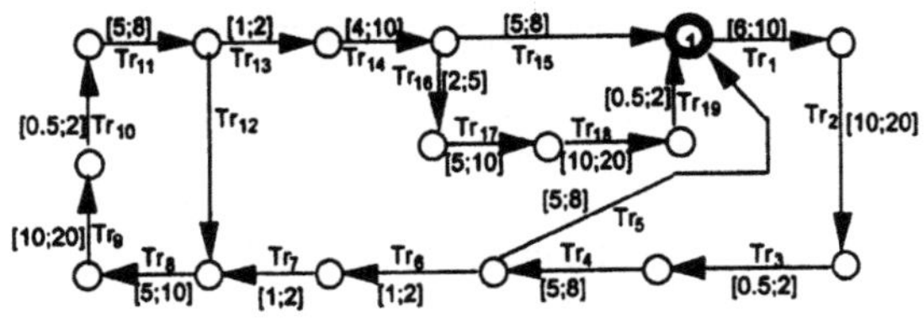

Figure 12. Service specification of the assembling system

Protocol Synthesis : The specifications synthesized in the static case are represented in Fig. 13, and consist of six *TA*s modeling the the three robots and the three carpets, respectively. To generate these real-time PEs, the transit delay of all messages has been assumed to belong to [2;5].

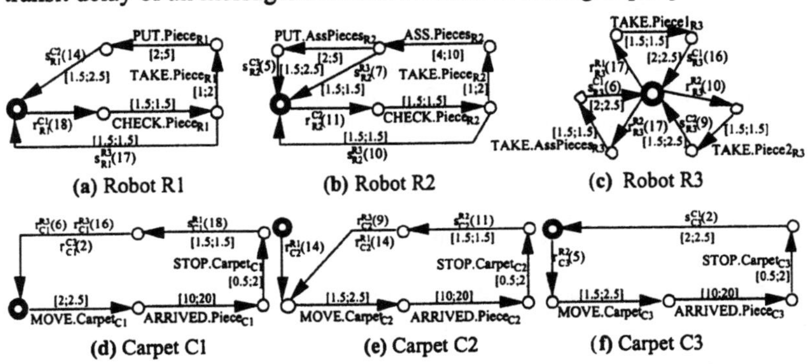

Figure 13. Synthesized specifications of the robots and the carpets in the static case

8. CONCLUSION

A method for deriving real-time protocols [8] is improved and extended in the present paper. Our method imposes requires that the service is sequential. We are presently investigating an approach using the following three steps : (a) the given service specification S is transformed into a sequential service, called S^{seq}; (b) a protocol P^{seq} providing the sequential service S^{seq} is synthesized by using our method; the obtained protocol is a kind of "skeleton" of a protocol P providing S; (c) P^{seq} is transformed in order to obtain a protocol P which provides S.

We note that Step (b) is realized automatically by the method presented in the present paper, and that Step (a) has been applied manually in the examples of Sect. 6 and 7 to make our method applicable. We therefore try to find a systematic way to achieve steps (a) and (c).

References

[1] R. ALur and D.Dill, "Automata for Modeling Real-Time Systems," in *Proceedings of the 17th Intern. Coll. on Automata, Languages and Programming, Lecture Notes in Comp. Sci. 443,* Ed. Springer-Verlag, Warwick, UK, 1990.

[2] R.J. Deasington, "*Protocoles X.25 pour les réseaux à commutation de paquets,*" Masson, 1987.

[3] R. Gotzhein and G. v. Bochmann, "Deriving Protocol Specifications from Service Specifications Including parameters," *ACM Transactions on Computer Systems,* Vol. 8, N° 4, pp. 255-283, 1990.

[4] Y. Iraqi, "Synthèse du protocole X.25 simplifié avec contraintes de temps. Utilisation de l'outil de Khoumsi," *Report of a project realized at the University of Montreal*, April, 1996.

[5] C. Kant, T. Higashino and G.v. Bochmann, "Deriving Protocol Specifications from Service Specifications Written in LOTOS," *Distributed Computing,* Vol. 10, N° 1, pp. 29-47, 1996.

[6] M. Kapus Kolar, "Deriving protocol specifications from service specifications with heterogeneous timing requirements." in *Proceedings of the IEEE Int. Conf. on Software Engineering for real time systems,* United-Kingdom, 1991.

[7] M. Kapus Kolar and J. Rugelj, "Deriving protocol specifications from service specifications with simple relative timing requirements," in Proceedings of *ISMM Int. Workshop on parallel computing,* Italy, 1991.

[8] A. Khoumsi, G.v. Bochmann, and R. Dssouli, "Dérivation de spécifications de protocole à partir de spécifications de service avec des contraintes temps-réel," *Revue Réseaux et informatique répartie (RIR),* Vol.4, N° 1, April 1994.

[9] E. Madja, "Dérivation de protocoles pour applications temps réeel. Application au système d'assemblage," *Report of a project realized at the University of Montreal,* April, 1996.

[10] K.Saleh and R. Probert, "A service-based method for the synthesis of Communications protocols," *International Journal of Mini and Microcomputers,* Vol. 12, N° 3, pages 97-103, December 1990.

[11] H. Yamaguchi, K. Okano, T. Higashino and K. Taniguchi, "Synthesis of protocol entities specifications from service specifications in a Petri Net model with registers,", in *Proceedings of IEEE Parallel and Distributed Computing Systems,* 1995.

PART NINE

Case Studies

Verifying an infinite family of inductions simultaneously using data independence and FDR[*]

S. J. Creese and A. W. Roscoe
Oxford University Computing Laboratory, Wolfson Building, Parks Road, Oxford OX1 3QD, UK.

Key words: process algebra, verification, communication protocols, induction, data-independence, model checking.

Abstract: We present a technique for formally establishing results for scalable systems, such as distributed systems and communication protocol networks, where the results are independent of the system's parameters. Example parameters are network topology, size and buffer capacity. The technique combines the use of the process algebra CSP to model systems and their specifications, and the FDR tool to help reason about them. We give examples of the techniques implementation on a simple distributed system and a communications protocol involving the multiplexing of channels

1. INTRODUCTION

An important current area of research in the verification of distributed systems is scalability: how do we handle the verification of realistic sized systems, and families of system under parameterisations that lead to unboundedly large state spaces? Two of the best known approaches to this problem are induction and data independence. This paper explores the interface between the two – specifically, by developing a proof method which combines them. What we actually show is how data independence can allow a finite automated proof which establishes an infinite number of inductions simultaneously.

The technique is implemented using the process algebra CSP[9, 14] to model systems, and the FDR model checker tool [7] to help reason about

[*]The work in this paper was supported by DERA Malvern and the US Office of Naval Research.

them, though we have no doubt that our ideas would work in other contexts as well. CSP models a system as a *process* which interacts with its environment by means of atomic *events*. A series of semantic models facilitate the capture of a wide range of behaviours including safety and liveness properties. The theory of refinement in CSP allows correctness conditions to be en-coded as refinement checks between processes. If A [= B is true (where [= represents refinement) then the behaviours of B are contained within the behaviours of A. In particular, refinement is transitive; if P [= Q and Q [= R this implies that P [= R.

FDR takes a machine-readable dialect of CSP as its input syntax[1], and can be used to check refinements as well as determinism, deadlock freedom and livelock freedom of processes. FDR performs a check by invoking a normalisation procedure for the specification process, which represents the specification in a form which the implementation can be checked against using simple model-checking techniques.

This paper is organised as follows: we will briefly survey existing work on induction and the theory of data independence we will be using; we then present the method and give a detailed example of its implementation; we discuss a further example of a communications protocol for channel multiplexing; finally, we present our conclusions and discuss its general applicability.

2. RELATED WORK

A number of authors have demonstrated that induction is a method which can be successfully used in the analysis of distributed systems. Kurshan and McMillan [11] and Wolper and Lovinfosse [18] published similar work using structural induction to reason about systems with unboundedly large numbers of identical components. Both methods require an invariant to be defined and rely on proof obligations which correspond to the base case and inductive step (using model checkers to discharge them). In [11] the method is demonstrated on an algorithm for maintaining the consistency of multiple copies of data object in a distributed system. In [18] two examples are given: a buffer composed of identical elementary buffers and a version of the distributed mutual exclusion problem (similar to that in [11]).

The following example is a good illustration of this basic induction method, in particular because it demonstrates the power of structural induction to cover a wide variety of network topologies, as alluded to in [3].

[1]Appendix A contains a summary of the CSP relevant to this paper.

Figure 1 shows a road layout allowing visitors to drive around a monument[2]. All cars enter the road at A and pass around the loop until they exit at B. Two adjacent rocks mean that one stretch of road is too narrow for cars to pass each other – it is single-lane there and two-lane or one-way elsewhere. To prevent deadlock in the road system a constraint has to be applied to the network, namely that out-going cars (i.e. rightward in the picture) are given priority over in-coming ones, in the sense that an in-coming car is not allowed to enter the single-lane segment C until segment D is empty.

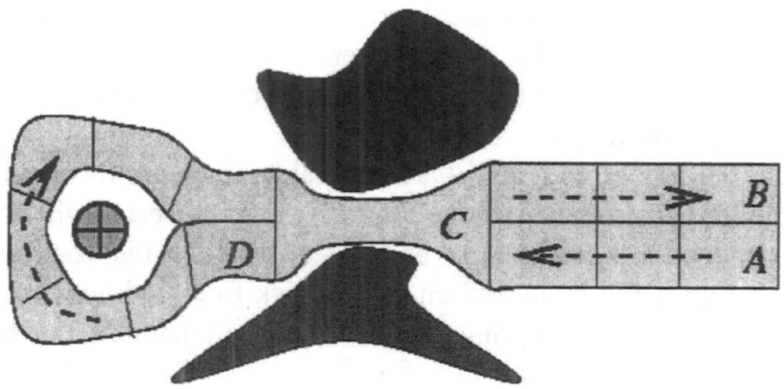

Figure 1. Road with a narrow section

It is possible to model the actual road network seen in Figure 1 to show that the constraint described makes the network deadlock-free, and FDR will prove this in a few seconds. However, this only establishes this property for the particular network modelled, and in no way implies that any other network topology is also deadlock-free. This more general goal can be achieved using structural induction.

To perform a structural induction a range of structural rules are required which describe how the various components have to be connected. In this case these components are the CSP processes representing the lane segments of type A, B and CD (it is necessary to constrain the appearance of segments of type C to only occur with segment D attached, as described above). The rules describe how these lane segments should be connected, and also how two deadlock-free road networks can be connected to each other. For example: `(i)` *"We can have a single road segment of type A"* (this is the base case). `(ii)` *"The road network can have segments of type A connected to its* `in` *channel"* (an inductive step).

[2] This example is based on an exercise in [14].

The rules are translated into refinement relations which are the proof obligations for the induction, they can be automatically discharged using FDR. The FDR checks performed are:

```
(i)    P [F= A
(ii)   P [F= P [in <-> out] A
(iii)  P [F= P [out <-> in] B
(iv)   P [F= P [in <-> leftout, out <-> leftin] CD
(v)    P [F= P [in <-> out] P,
```

where `P` is the property of deadlock freedom, and `[F=` refinement in the failures model of CSP. The piping operator `X[a<->b]Y` connects the channels `a` and `b` (belonging to `X` and `Y` respectively) together, and hides all communications over them.

Note that the processes `A` and `B` are equivalent, they are distinguished in this example for consistency with the Figure. Check `(i)` establishes the base case for the induction. Checks `(ii)` and `(iii)` establish that road segments of the type `A` and `B` can be tacked on to deadlock-free road networks in the manner described and the resulting road networks will maintain this property. Check `(iv)` establishes the same for segments of type `CD`, and `(v)` establishes that any two deadlock-free road networks (of the type under consideration) can be connected together (the input to one network is the output of the other), and the resulting road network will also be deadlock-free.

It is clear that by transitivity of refinement, and so by using normal inductive reasoning, that in combining `(i)` to `(v)` above, it is possible to claim that any road network made up of segments of type `A`, `B` and `CD`, according to the given rules, will be deadlock-free. More interesting case studies of this technique can be found in [4, 5, 15].

The buffer example in [18] is a remarkable one. Here the property proved by induction bears little resemblance to the final objective, but when combined with data independence properties of the implementation is enough to prove it (though the definition of buffer implied in that paper is not quite the CSP standard one of [9, 14]). We remark that this provides a completely different combination of data independence and induction to the one we will shortly describe. Other approaches for proving properties of systems of arbitrary size have included the use of temporal logic [1, 2, 8, 16].

Just because a property is true of all the systems constructed by a set of structural rules does not guarantee that it can be proved inductively. Ideas such as *strengthening the hypothesis* frequently help, but can require considerable ingenuity. The main limitation on the induction method, described above, is that it can only address systems in which the individual components are completely independent of the overall structure of the system: adding an extra process does not change the semantics of the components that were already there. For example, they cannot know the identities of the other components, for then it is not true to say that an $n+1$–process system can be constructed by adding one more to the n–process version. It is this limitation on induction that we address.

3. DATA INDEPENDENCE

A system is said to be data independent of a data-type variable T when it makes sense for any non-empty substituted type and it satisfies structural rules, meaning that it handles the type in a relatively simple way. The process representing the system may not perform any operations on values of that type, it can only input them, store them, output them, and perform equality tests between them. Many communications protocols are data independent since they simply pass around data ensuring that only desirable recipients are able to extract it; their behaviour is entirely independent of the data itself.

Data independence ideas have been developed for a number of notations, having first been studied formally in [17], but we use here the theory developed for CSP by Lazic and Roscoe (see [12, 13] and Section 15.2 of [14]). They have developed theorems which state that for certain (data independent) systems it is possible to verify that the system possesses certain properties, for all instances of its independent types, by performing a specific finite number of checks with finite instances of the types. Data independence theorems frequently allow us to generate thresholds for given checks: a size of types T such that one or more checks of a property for this and perhaps smaller types will imply that the property holds for all T. Space unfortunately prevents us from quoting in detail the various theorems and definitions of data independence, for which we refer the reader to [12, 13, 14].

However, of particular relevance to the technique we present in this paper is the ability to consider data independent processes with the addition of constant symbols and (in general many-valued) predicates on variable types. The theorems used put some constraints on the use of these predicates. They must be uninterpreted, in that they should be treated as symbols and a verification should establish a property for all possible interpretations. The predicate must be a function into a fixed finite type, which in our case will be $\{true, false\}$ (otherwise the problem would become intractably infinite). This will be important to us as it will allow statement of an inductive principle which to some extent overcomes the limitation discussed at the end of Section 2. Again we must refer the reader to [12, 13] for relevant theorems and definitions.

4. INDUCTION OVER DATA INDEPENDENT TYPES

Our technique is useful in the analysis of systems which are built as indexed parallel compositions over types which each individual process (elements of the type) treats data independently. We can't deal with these systems using traditional inductive techniques because the individual processes vary with the type (they know each others identities, for example, if their identities are the data independent type) and we can't deal with them using data independence alone because indexed parallel composition over data independent types is prohibited in that theory[3]. While our method readily extends to some more complex cases (one of which appears in Example 2 below), we will consider here the objective of proving that a specification `Spec`, data-independent in *T*, is refined by the parallel combination `|| x:T @ [A(x)] N(x)` in which the terms `A(x)` and `N(x)` are themselves data independent in *T*. Here `N(x)` is the component process, and `A(x)` is its alphabet. Our aim is to show that the parallel combination of `N(x)` processes, where `x` is an element of type *T*, refines `Spec` no matter how large *T* is. Further, we want to do this using a finite amount of work.

We build a (usually sequential) process representing partially constructed versions of the system under consideration. The aim is that it should be data independent in *T*, should be refined by the partially constructed network it models, and when complete should refine `Spec`. As this process is often a representation of the state of a parallel system we term it `Superstate(S)`, where `S` is the (nonempty) subset of *T* that constitutes the partial network. The aims can thus be formally stated

```
Superstate(S) [= (|| x:S @ [A(x)] N(x))      (1)
         Spec [= Superstate(T)               (2)
```

These two statements, if true, are of course sufficient to establish our overall aim. If an appropriate `Superstate` has been constructed, then (2) presents no new theoretical challenges as it already fits within the established theory of data independence: we can reasonably expect it to be verifiable independently of *T* by finding the appropriate threshold and doing the necessary refinement checks.

[3] This is largely because allowing such compositions would allow one to count the type *T*.

The role of induction comes in establishing (1), which does not come within standard data independence theory as it contains the indexed parallel composition. What we do is to perform an induction on the size of `S` (from 1 up to $|T|$). The base case of the induction is to show that one node refines `Superstate`:

```
    Superstate({c}) [= N(c)                              (3)
```

where the constant symbol `c` represents an arbitrary member of *T*. The inductive step is to show that for any nonempty `S` and constant `c` outside `S`,

```
 Superstate(S') [= Superstate(S)[AA(S)||A(c)]N(c)    (4)
```

where `S'` is the union of `S` and `{c}` and `AA(S)` is the alphabet of `Superstate(S)` which is the union of `{A(x) | x <- S}` (the alphabets of each node).

If we can establish these two things the inductive proof of (1) follows easily: if it is true for `S` then

```
Superstate(S') [= Superstate(S)[AA(S)||A(c)]N(c)
               [= (|| x:S @ [A(x)]N(x))[AA(S)||A(c)]N(c)
               [= (|| x:S' @ [A(x)] N(x)).
```

The second line is from the monotonicity of the parallel operator and induction, and the final one from the definition of indexed parallel.

We can thus conclude that the truth of (3) and (4) (for arbitrary *T*, `S` and `c`) imply that (1) holds for all nonempty `S` and finite *T*. Note that this conclusion actually requires not one induction, but rather demonstrates that an infinite number of different inductions (one for each *T*) all work.

Fortunately, both (3) and (4) are statements that can be proved using data independence extended by predicate symbols and constants, as we can naturally treat `S` (the subset of the network present in a partially-built network) as a predicate symbol (mapping a member of *T* to *true* if it is in the set). Therefore we can reasonably hope to justify these statements for all values of their parameters, and thus be able to conclude (1) from a finite collection of checks which can be verified automatically on FDR.

One way of looking at this technique is that it provides a way of extending the advantage of data independent proofs to systems which involve indexed parallel composition. However this can only be done in cases where an appropriate `Superstate` definition can be found, something which at best requires more ingenuity on the part of the user than other data independence results, and may well in some cases be impossible. (The greatest challenge in building `Superstate` appears to be keeping it sufficiently finite-state to enable both finite thresholds and automated checking.)

Example 1 Consider an example of a distributed system which combines leadership election and token passing. This example has much in common with one of those given in [3, 18], though the main point of the present one – the knowledge of all identities by each process – is not present in these papers. It is the leader's job to distribute a token in turn allowing some critical region to be entered into by the process having it. The leader can do critical sections (though this is completely optional in the model), and the leadership can only move when the leader has the token. Below is the CSP description of a process expressed as input syntax for FDR:

```
Leader(T,id) = leader?j:T -> (if j==id then Leader(T,id)
                    else Empty(T,id,j))
        [] cr_start.id -> cr_end.id -> Leader(T,id)
        [] pass!id?j:others(id,T) -> pass.j.id ->
            Leader(T,id)

Empty(T,id,L') = pass.L'.id-> Full(T,id,L')
       [] leader?j' -> (if j'==id then Leader(T,id)
                    else Empty(T,id,j'))

Full(T,id,L') = pass.id.L' -> Empty(T,id,L')
       [] cr_start.id -> cr_end.id -> Full(T,id,L')

others(i,T) = diff(T,{i})
```

`diff(P,Q)` is the difference of the two sets `P` and `Q`. The parameter `T` is the type of nodes in use; it is required since nodes need to know who is present in the system. If `S` is any set of these nodes, and `L0` is a constant representing the initial leader, we can build the composition of all the nodes from `S` as:

```
LNet(T,S,L0) = || i:S @ [A(T,i)] (if i==L0
                 then Leader(T,i) else Empty(T,i,L0))
```

`LNet(P,S,L0)` is the parallel composition of processes in `S` of the appropriately initialised node processes, which we can name `N(T,L0,i)` for consistency with the notation earlier. Note that this model abstracts away from the leadership election process, which is delegated here to a perhaps unrealistic multi-way synchronisation.

We need, for our method, to create a process `Superstate(S)` that models the behaviour of the parallel composition of the nodes from a subset `S` of `T`. These will inevitably agree (thanks to the synchronisation on `leader`) on who the leader is. The main problem one faces in building the process is what to do when the leader is outside `S`, for then the partially composed network has many aberrant behaviours due to the fact that there is no coordinated leadership; since the leader is outside of the network it can misbehave. In particular, the leader may allow any number of nodes to get into critical sections at once. In order to overcome this the `Superstate` process incorporates this misbehaviour; essentially things stay fine as long as no imaginary leader outside `S` hands out too many tokens:

```
Superstate(T,S,L) = leader?j:T -> Superstate(T,S,j)
       [] member(L,S) & cr_start.L -> cr_end.L ->
                     Superstate(T,S,L)
       [] member(L,S) & pass!L?j:others(L,T) ->
                     Superstate'(T,S,L,j)
       [] not member(L,S) & pass!L?j:others(L,S) ->
                     Superstate'(T,S,L,j)

Superstate'(T,S,L,j) = member(j,S) & cr_start.j ->
                     Superstate''(T,S,L,j)
         [] pass.j.L -> Superstate(T,S,L)
         [] not member(L,S) &
            (STOP |~| pass.L?k:S -> CHAOS(Events))

Superstate''(T,S,L,j) = cr_end.j -> Superstate'(T,S,L,j)
         [] not member(L,S) &
            (STOP |~| pass.L?k:S -> CHAOS(Events))
```

`CHAOS(Events)` is the process that can perform or refuse any possible event: it is used here when we don't care what the process does, namely after one of the misbehaviours mentioned above. This approach is similar to that used by Wolper and Lovinfosse in [18].

Because we have adopted the names for processes as used in the earlier formulation of the inductive claim, the base case and step case of the induction are almost literally (3) and (4). The result proved by the induction is (1) where, when $S=T$ the right-hand-side is the complete network `LNet(T,T,L0)`. As we will shortly see, by analysing the checks (3) and (4) we can establish finite collections of checks which prove them, and hence (1), for all T and S.

Though we have not had space to detail the results underlying these calculations, it is nevertheless interesting to derive thresholds and sufficient sets of checks to establish the base and inductive cases of our induction. For checks which either explicitly or (like ours) implicitly involve equality checks between members of the data independent type, the threshold when there are no predicate symbols, is the greatest number of potentially distinct values that might be present simultaneously during a symbolic comparison between the specification and implementation. When there is a single predicate symbol, it is the sum, over all values the predicate might take, of the largest number there might be in a comparison with that value. The formulae, such as the following, (quoted in Section 15.2 of [14]) provide upper bounds, often crude, for these quantities.

$$B = W^{Spec} + W^{Impl} + max(L^{Spec}, L^{Impl}{}_{?}, L^{Impl}{}_{\Pi})$$

where W^{Spec} and W^{Impl} are the maximum number of values of type T that the Specification and Implementation, respectively, ever have to store for future use. $L^{Impl}{}_{?}$ and L^{Spec} are the largest number of values of type T that can be input in any *single* visible event of the Implementation and Specification, respectively. $L^{Impl}{}_{\Pi}$ is the largest number of values of type T that can be nondeterministically chosen in any single nondeterministic choice made over sets involving T in the implementation. In the base case (3) this particular formula predicts a threshold of $2+2+1 = \mathbf{5}$, and in the step case (4) the same formula gives $2+2+2+1 = \mathbf{7}$. However, the presence of the predicate symbol increases the second threshold to *13*, *6* in `S` and *7* outside `S`, (as by definition we know that the constant `c` does not satisfy the predicate). The predicate brings about the increase because the values stored in the program may or may not satisfy it, and the threshold has to be big enough to cater for this.

These naive calculations assume that all processes can have distinct values for the leader, whereas in fact they all agree. Let us concentrate on the more complex case: the threshold for (4). Consider W^{Spec} and W^{Impl}, the leaders `L` are counted in both but must be identical, as must `L'`. Therefore, we can ignore two of them giving a new threshold of *9*: *4* in `S` and *5* outside `S`. Also, the `j` in W^{Impl} is either identical to the `L'` or the `id` in `Empty` giving a new threshold of 7: *3* in `S` and *4* outside `S`. Finally, we know that `j` and `L` cannot both be outside `S` Y `{c}` because if the leader and the token are actually both outside the `S` Y `{c}` , then no-one in our system knows where the token is. Thus we can deduct another *1* from the number that can be outside `S`. Our final threshold is *6*, *3* in `S` and *3* outside. A largest inductive step check we would need to do is:

```
Superstate({1..6},{1..4},1) [FD=
    Superstate({1..6},{1..3},1)
    [AA({1..6},{1..3})||A({1..6},4)] Empty({1..6},4,1)
```

Where `[FD=' means refinement in the *failures/divergences* model of CSP, this being the strongest notion of process refinement. Further checks (there are 24 including the above) need to be performed for the cases in which there are fewer objects either or both inside and outside `S`, and to deal with the cases where `L0` is inside `S`, equal to `c`, or neither. These checks can, of course, be performed quickly and easily on FDR. We believe that the threshold and number of required checks could be reduced further by *ad hoc* arguments, but much the best prospect here is the ideas for symbolic execution discussed in the Conclusions below. Similar analysis can bring the threshold for the base case down to **3**, so a largest check of the 5 required is

```
Superstate({1..3},{1},1) [FD= Leader({1..3},1)
```

An interesting fact about this example – though by no means universally true – is that `Superstate(T,T,L0)=LNet(T,T,L0)`. It follows that *any* data independent specification we want to prove of `LNet(T,T,L0)` can be addressed through this particular `Superstate`. Examples that can be proved include deadlock freedom and at most only one critical section being active at any one time: each, provided it is suitably finite-state and data independent, should be provable in general as an instance of (2).

Example 2 Consider a simple communications protocol which enables the multiplexing of as many channels as we please over a single pair of links-- one each way (`fwd` and `back`). This is done in our implementation by passing tagged data and acknowledgements over the links, and combining a protocol with buffering to stop any one channel impeding others. For simplicity we only consider the case where all the channels are in the same direction (from `left.i` to `right.i` in Figure 2).

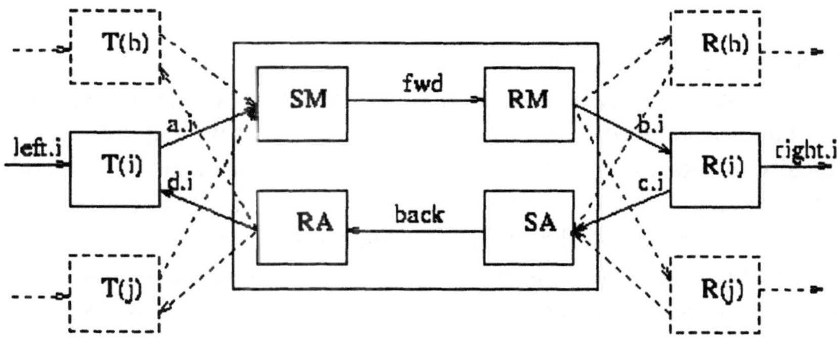

Figure 2. Multiplexing Channels

Evidently, as we let the number N of channels grow, the number of states in the resulting CSP implementation soon becomes unmanageable for a model checker. However, the sort of data independent induction presented above can prove that our system is equivalent to N independent one-place buffer processes for any *N*. Thereby establishing the correctness of the protocol or any arbitrary number of multiplexed channels.

There are actually two types in the example that are susceptible to data independence arguments. The more obvious one is the type of data passed by the channels: the theorems of [13, 14] easily show that it is sufficient to set this type with size 2 to prove that the system behaves properly for any type. This argument is quite separate from the induction. The type over which we induct is that of `TAGS` (the channel names).

The system is constructed of the processes `T(i)` and `R(i)`, which are the transmitter and receiver processes respectively (one for each `i` in `TAGS`), connected to the fixed collection of processes `SM`, `RM`, `RA` and `SA`. The `T(i)` and `R(i)` communicate with the environment on the `left.i` and `right.i` channels. Below are the definitions of the processes:

```
SM = a ? i:TAGS ? x -> fwd ! i . x -> SM
RM = fwd ? t ? x -> b . t ! x -> RM
SA = c ? i:TAGS -> back ! i -> SA
RA = back ? t ? x -> d . t ! x -> RA

T(i)  = left.i ? x -> a.i ! x -> d.i -> T(i)
R(i)  = b.i ? x -> right.i ! x -> c.i -> R(i)
```

The central core of the system is the following:

```
CORE = (SM [|{|fwd|}|] RM)\{|fwd|} |||
       (SA [|{|back|}|] RA)\{|back|}
```

This process (data independent in `TAGS`) has the same parallel structure however large `TAGS` is, but we get a situation similar to that in Example 1 when the transmitter/receiver pairs are added, as this requires parallel compositions indexed over `TAGS`. The result has the channels `a`, `b`, `c` and `d` hidden, leaving only `left` and `right` visible to the environment. Our objective is to prove that the complete system is the composition of a `COPY` for each tag, where `COPY` is the process which inputs data on its `left` channel, and outputs that data on its `right` channel (behaving as a one-place buffer). `COPY` is defined as:

```
COPY = left?x -> right!x -> COPY
```

In order to prove this we show that the abstraction to an arbitrary channel is correct by induction, and deduce the overall result from that. We shall concentrate on channel `C0`. The specification `SPEC(S)` of a partially constructed system, where `S` $\subseteq$ `TAGS` and `C0`$\in$`S`, says that on `C0` the process must act like a one-place buffer except that it can refuse to input or output when there is outstanding data outside S on the a/b or c/d link. (Space constraints prevent us from giving the detailed code of this specification here.) Our claim is that a network constructed from CORE plus all the T(i) and R(i) from i in S, all internal communication hidden and all channels in S other than C0 lazily abstracted[4], refines SPEC(S) (the specification of the system of S channels multiplexed).

The base case of our proof is that the system with one transmitter/receiver pair (labelled C0) refines the corresponding specification:

```
BASE = (CORE [|{|a.C0,b.C0,c.C0,d.C0|}|](T(C0)|||R(C0)))
            \ {|a.C0,b.C0,c.C0,d.C0|}
SPEC({C0}) [F= BASE
```

The inductive step required is that if we add a further T(K)/R(K), hiding internal and abstracting external communications appropriately, we preserve SPEC:

```
SPEC(S') [F= (LAbs(SPEC(S) [| intevs |]
        (T(K)|||R(K)))({|left.K,right.K|})) \ intevs
intevs = {|a.K,b.K,c.K,d.K|}
LAbs(P)(X) = (P [|X|] CHAOS(X))\X
```

Here LAbs(P)(X) is the *lazy* abstraction (as discussed above) of events X from process P, and S' is the union of S and {K}.

As in Example 1, data independence theorems allow thresholds to be calculated which establish an upper limit on the size of TAGS which guarantees that the refinements will hold for any arbitrary non-empty size of TAGS (i.e. largest size of TAGS). For the base case the threshold is no more than ***8***. For the inductive step the threshold is *7*. This is unlike the situation in Example 1 where the step threshold was significantly larger. This is primarily because only the base case contains the process CORE which stores 4 values. So by performing a fixed finite amount of refinement checks on FDR, for each possible configuration of each check with S smaller than or equal to the threshold, we have shown that for any size of TAGS our multiplexed system is equivalent to that many copies of COPY. Therefore, the integrity of each channel is not compromised by the addition of further channels over the same pair of links (fwd and back).

[4] To lazily abstract a channel means to hide the channel in such a way that we do not assume that hidden events must occur. It provides the best way of formulating what a process looks like to an observer who can only see a subset of its alphabet. See Chapter 12 of [14].

5. CONCLUSIONS

The technique described here provides an interesting combination of induction and data-independence which is capable of allowing model checkers to deal with a significant new class of problem. While the examples quoted here are fairly simple, we in fact developed the method as part of an analysis of a much more complex case study [6] – a version of the Time Triggered Group Membership Protocol(TTP) [10]. We successfully applied the method there after generalising the protocol to make it sufficiently data independent, and correcting a fault that our analysis revealed. Other examples we have considered include: extending the applicability of the structural induction argument in [5] from binary to arbitrary branching; and the proof of a routing protocol.

Future work will hopefully lead to an understanding and broadening of the classes of systems to which our method applies. We are also interested in the methodology required for formulating the `Superstate` induction hypothesis and whether this can be automated. Ongoing developments in the symbolic handling of data in refinement checks on FDR should be a great help in discharging the model checking obligations generated by our techniques. In effect this should completely automate the application of data independence. It would no longer be necessary to calculate thresholds (and the ad hoc arguments used to bring the threshold down), as all of the analysis would be done at ``run-time" and automatically. Furthermore, there is every hope that the checks would complete much faster, as this method should reduce many equivalence-classes of essentially similar states down to a single one.

ACKNOWLEDGEMENTS

We would like thank Ranko Lazic for his work on data independence, and Gavin Lowe whose independent and different work (as yet unpublished) on the integration of data independence and induction led us to look for the connections we eventually found.

REFERENCES

[1] M.C. Browne, E.M. Clarke and O. Grumberg, *Reasoning about Networks with Many Identical Finite State Processes*, Information and Computation, 81(1), 13-31, April 1989.

[2] E.M. Clarke and O. Grumberg, *Avoiding The State Explosion Problem In Temporal Logic Model Checking Algorithms*, Proceedings of the 6th Annuak ACM Symposium on Principles of Distributed Computing, Vancouver, Canada, August 1987.

[3] E.M. Clarke and O. Grumberg and S. Jha, *Verifying Parameterized Networks using Abstraction and Regular Languages*, Proc. of CONCUR'95, 395-407, LNCS 962, 1995.
[4] S.J. Creese, *An Inductive Technique for FDR*, Master's thesis, Oxford University, 1997.
[5] S.J. Creese and J. Reed, *Verifying End-to-End Protocols Using Induction with CSP/FDR*, Proceedings of FMPPTA'99, Puerto Rico, April 1999.
[6] S.J. Creese and A.W.Roscoe, *TTP: A case study in combining induction and data independence*, Oxford University Computing Laboratory Technical Report, PRG-TR-1-99.
[7] *Failures-Divergence Refinement: FDR2 User Manual*, Formal Systems (Europe) Ltd.
[8] Steven M. German and A. Prasad Sistla, *Reasoning about Systems with Many Processes*, Journal of ACM, 39, 675-735, July 1992.
[9] C.A.R. Hoare, *Communicating Sequential Processes*, Prentice-Hall, 1985.
[10] S. Katz, P. Lincoln and J. Rushby, *Low-overhead time-triggered group membership*, Proceedings of WDAG '97, LNCS 1320, 1997.
[11] R.P. Kurshan and K. McMillan, *A Structural Induction Theorem for Processes*, Proceedings of 8th Symposium on Principles of Distributed Computing, Edmonton, 1989.
[12] R. Lazic and A.W. Roscoe, *Verifying Determinism of Concurrent Systems Which Use Unbounded Arrays*, Proceedings of INFINITY'98, Aalborg, Denmark, July 1998, extended version as Oxford University Computing Laboratory TR-2-98.
[13] R.S. Lazic, *A semantic study of data-independence with applications to the mechanical verification of concurrent systems*, Oxford University D.Phil thesis, 1999.
[14] A.W. Roscoe, *The theory and practice of concurrency*, Prentice Hall, 1998.
[15] J.N. Reed, D. M. Jackson, B. Deianov and G. M. Reed, *Automated Formal Analysis of Networks: FDR Models of Arbitrary Topologies and Flow-Control Mechanisms*, Proceedings of ETAPS'98, Lisbon, Portugal, to appear in IEEE Transactions on Software Engineering, March 1998.
[16] Moshe Y. Vardi and Pierre Wolper, *Reasoning about Infinite Computations*, Information and Computation, 115, 1-37, 1994.
[17] P. Wolper, *Expressing interesting properties of programs in propositional temporal logic*, 184-193, Proceedings of the 13th ACM POPL, 1986.
[18] P. Wolper and V. Lovinfosse, *Verifying Properties of Large Sets of Processes with Network Invariants (Extended Abstract)*, Proceedings of the International Workshop on Automatic Verification Methods for Finite State Machines, LNCS 407, 1989.

APPENDIX A. THE CSP LANGUAGE

The CSP processes that we use are constructed from the following:

- `STOP` is the simplest CSP process; it never engages in any action, nor terminates. It is equivalent to deadlock.
- `a -> P` is the most basic program constructor. It waits to perform the event `a` and then behaves as process `P`. The same notation is used for outputs (`c!v -> P`) and inputs (`c?x -> P(x)`) of values along named channels.
- `P|%|Q` represents internal choice. It behaves as `P` or `Q` *nondeterministically*
- `P[]Q` represents external choice. It will offer the initial actions of both `P` and `Q` to its environment at first; its subsequent behaviour is like `P` if the

initial action chosen was possible only for `P`, and like `Q` if the action selected `Q`. If both `P` and `Q` have common initial actions, its subsequent behaviour is nondeterministic (like `|%|`). `STOP[]P` behaves as `P`.

- `P[| a |]Q` represents parallel composition. `P` and `Q` evolve concurrently, except that events in `a` occur only when `P` and `Q` agree to perform (i.e. *synchronise* on) them.
- `P|||Q` represents interleaved parallel composition. `P` and `Q` evolve separately, and do not synchronise on any events. Equivalent to `P[|{}|]Q`.
- `P[ a || b ]Q` represents alphabetised parallel. `P` and `Q` have to agree on events in the intersection of their alphabets.
- `|| x:a @ [A(x)] P(x)` represents replicated alphabetised parallel. This constructs the parallel composition of `P(x)` processes, one for each `x` in `a`, over their respective alphabets.
- `P\A` is the CSP hiding operator. This process behaves as `P` except that events in set `A` are hidden from the environment and are solely determined by `P`; the environment can not observe or influence them.
- `P[[ a <- b ]]` represents the process `P` with `a` renamed to `b`.
- `P[ a <-> b ]Q` is the linked parallel operator. `P` and `Q` synchronise on channels `a` and `b`, which have been renamed to the same and hidden.

There are also straightforward generalisations of the choice operators over non-empty sets, written `|%| x:X @ P(x)` and `[] x:X @ P(x)`.

Specification and execution of tests using tMSC

Iulian Ober*
INPT-ENSEEIHT, 2 rue Camichel, 31000 Toulouse cedex, France
e-mail: Iulian.Ober@enseeiht.fr

Alain Kerbrat
Verilog, 150 rue Nicolas Vauquelin, BP1310, 31106 Toulouse cedex, France
e-mail: Alain.Kerbrat@verilog.fr

Key words: conformance testing, test execution, MSC, TTCN, TCL

Abstract: Writing tests for concurrent systems is difficult and time consuming. Handling concurrency requires an expressive language, yet the same language should be easy to read and edit. A standardized test description language such as TTCN is often perceived as too cryptic, difficult to read and generally too powerful. Other test description languages are home made and do not allow to share or to maintain easily existing test suites. We present a test specification and execution tool relying on MSC. The MSC language is standardized, and provides a graphical and intuitive way for describing test cases.

1. INTRODUCTION

The validation phase is the most time consuming part of software development. It is especially true for telecommunications systems, where any new software must integrate smoothly and quickly with existing ones. The validation of a new telecommunication service can take 5 to 10 times the time needed to develop it. Validation is also a major concern for safety

* work supported by the CS Communications & Systems Group

critical systems such as embedded systems in planes, satellites or nuclear plants. These systems need to be certified, hence must be validated extensively and test cases must be traced from the requirements down to the code.

Several test description languages exist for concurrent and safety critical systems. TTCN [1] is a standardized test description language, meant for conformance testing [2]. It is a powerful language, combining constructs for the description of the behavior of test cases and operators for building concurrent test cases. It integrates ASN.1 [6] as well as its own data type language and allows complex descriptions for data values. However, this same expressive power, combined with its difficult syntax and presentation requirements, hampers TTCN use. TTCN test cases are sometimes hard to write and difficult to read.

Other test description languages are usually developed for internal purposes, as a front end of the test platform. These solutions are usually well mastered, as they are adapted to the specific needs of both the test platform and of the software developed on it. However, the semantics of these languages are usually not formally defined. This complicates the sharing of test suites, even among teams of the same company. The evolution of the languages and of their compilers is usually done on a per project basis, which makes it difficult to maintain the test suites. Last but not least, they seldom come with a graphical counterpart, as this is usually time and money consuming to develop. MSC is an interesting candidate to alleviate these problems. It is graphical, allowing for intuitive reading or editing of a test case behavior. It is also an international standard [8], which evolves steadily every four years.

To meet the need for a simpler yet not less powerful method for describing and executing conformance tests, we propose an alternative approach: the use of MSC for describing test cases, complemented with a second-level formalism - the Test Description Language (TDL), all within a lightweight and open framework designed for executing tests on a wide class of possible systems. The MSC is a language for expressing execution traces that has been used for a long time for capturing requirements in telecommunication systems design but also as a basis for system simulation and validation, interface specification, etc. It is a standard and mature graphical language, usable for expressing test cases at a first level, and whose simplicity appeals.

This paper advocates our testing framework, by showing the strong and weak points of MSC when expressing tests, by proving that they can be overcome by using a second-level formalism for expressing test cases (the Test Description Language) and by describing the architecture of a tool implementing our testing paradigm and emphasizing on the open parts which make the tool usable in many different contexts.

Organization of the paper. The first part of this paper presents tMSC. tMSC is the subset of MSC we chose for test cases description.

The second part describes the *TestPlayer* tool. *TestPlayer* is a tool for edition, and execution of test cases written in tMSC. *TestPlayer* provides also a tool to visualize back at the MSC level the test execution results.

The third part describes the connection of *TestPlayer* with a related tool, *TestComposer*. *TestComposer* allows the automated generation of test cases in tMSC, from a SDL description.

2. USING MSC FOR DESCRIBING TEST CASES

MSC is a visual language for representing system execution traces, mainly in terms of messages exchanged between the internal instances of a system and with the environment. It is being standardized by the International Telecommunication Union as Recommendation Z.120 [8]. The standardization work began in 1990 and each four years a new major version is issued.

Being visual, MSCs are more easily readable, which makes them more attractive to the users. Thus there were early attempts to use MSC for expressing test cases or test purposes [4]. MSCs are suitable for this purpose if we bear in mind that the behavioral part of a TTCN test case describes also a set of possible traces of messages exchanged between the system and its environment, each such trace having associated a verdict at the end.

These attempts have shown the benefits, but also the weaknesses of MSCs when expressing test cases: there is no notion of test verdict in MSC, no possibility to specify data types and constraints on data, no possibility to use variables and no notion of test architecture [3]. We will see in what follows how we overcame these problems in our approach with *TestPlayer*.

2.1 MSCs for test

To use MSCs to represent test cases, we must give an intuitive meaning to test MSCs in this context (denoted tMSC in what follows). A conformance test for a protocol [2] is expressed in terms of the messages exchanged between the tester and the Points of Control and Observation (PCOs) of the system. (a PCO is a named interface through which the system receives or sends signals [2]).

We distinguish three flavors of test MSCs differentiated by the instances they represent. The *first* one (see Figure 1) takes the *point of view of the tester*: it represents instances denoting PCOs and testers (one or more). The *second* variety of tMSCs takes the *point of view of the system*, containing an instance that represents the system and one instance for each PCO. Abstract test description for standard protocols often employ this kind of MSCs. Finally, *usual* MSCs are acceptable as test MSCs, case in which the tester is represented by the environment. This kind of MSCs is likely to result from the simulation of an SDL model, for example.

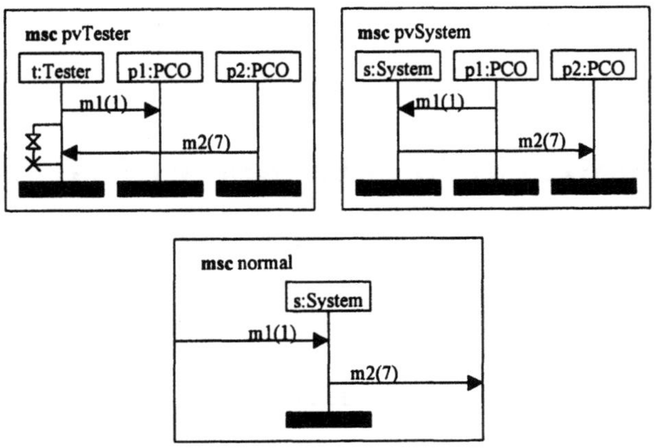

Figure 1. The three varieties of test MSCs

Usually a MSC describes one possible behavioral trace of a system, be it complete or incomplete. We consider a different semantics for the MSCs used for describing test cases: a test MSC contains the main test sequence, i.e. the sequence of events leading to the success of the test (PASS verdict). All deviation from this sequence (a message that should but cannot be sent, a message that should be received but is not) produces a failure of the test.

One inconvenient when using MSCs for describing test cases is that one cannot express constraints on the parameters of the messages (see also [3]). To alleviate this limitation, we defined a set of *generic specifiers* that may be used in MSCs: AnyValue, ValueList, and ValueRange. Still, the amount of constraints we are able to express using these specifiers in MSC is limited, due to the lack of variables. This is one reason for which *TestPlayer* uses a second language for describing tests, TDL (see section 3.2).

Besides messages, three additional categories of constructs have a meaning on a Basic tMSC, but only when they appear on a tester instance: stop, test timers and coregions.

- A stop construct on a tester bar means that the test stops immediately with a PASS verdict.

- Timer constructs (set/reset/timeout) are used to express minimal/maximal delay bounds. A timer set followed by a timeout represents a lower bound, a timer set followed by a reset represents an upper bound (this follows from the assumption that the MSC represents the main sequence that leads to a PASS verdict). The timing constraints employed in test cases usually take the following form: the time span between the event a and the event b must be between x and y time units. The way of expressing this with timers in shown in Figure 2.
- A coregion on a tester bar means that the messages that come from the system (PCOs) within the boundaries of the coregion may come in any order. As a difference from Z.120, only response messages coming from the system and timer timeouts may be represented within a coregion. This is because the tester must be controllable, so we cannot allow message outputs and other actions for which we do not specify the precise order.

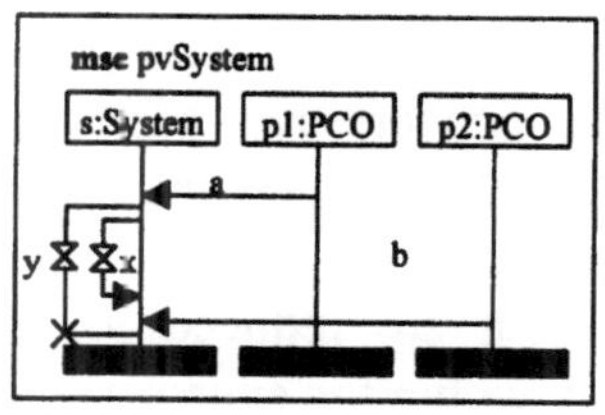

Figure 2. Delay intervals expressed with timers

2.2 Composing test MSCs

MSC'96 defines a set of structural composition operators for organizing tMSCs into High-Level tMSCs. We use a subset of them to express tests. For their correct composition, two MSCs must define the behavior of the same tester(s) on the same system (PCOs). The semantics of these composition operators follows the MSC'96 standard:

- The sequential composition of two tMSCs indicates that the tester first executes the constructs specified by the first MSC and then, if it did not fail, executes the constructs specified by the second MSC.
- The repetition of a tMSC indicates that the tester repetitively executes the constructs specified by the MSC.
- The alternative composition of two tMSCs indicates that the tester executes either the constructs specified by the first MSC or the constructs specified by the second MSC. The choice is made between the first constructs from the two tMSCs, which must be a message receipt or a timer timeout .

We do not support other MSC'96 operators such as parallel composition and exception as they would produce non controllable test cases.

2.3 Expressive power of the tMSC. tMSC vs. TTCN

Not all the three flavors of test MSCs have the same power of expression: only tMSCs that take the *point of view of the tester* support all the operators defined in section 2.1 (ex. timers may appear only on tester instances).

The Tree and Tabular Combined Notation (TTCN [1]) is the ISO/IEC standard language for the description of abstract test suites. TTCN was designed with the ISO layered model as a framework and it is specialized for conformance testing.

Being a complete test language, TTCN brings more power than tMSCs. TTCN supports data and signal types definition, variables, complex expressions on expected values, concurrent testers, explicit and more accurate test verdicts. On the behavior description side, TTCN offers many constructs for expressing concurrency and capturing non determinism (through the use of INCONCLUSIVE verdicts). Our experience is, however, that concurrent and non-deterministic test cases are usually difficult to write. Thus, TTCN test cases often are reduced to deterministic, single tester sequences, which falls back within the expressive power of tMSCs.

The one important advantage of using MSCs to express test cases is that MSC is a visual and intuitive. MSC was used for several years in the definition of protocol standards and one can derive test MSCs straight from the MSC coming from the protocol standard.

Using the semantics outlined above, tMSC is able to express the same behaviors as TTCN. MSC timers may express everything that is described in terms of timers in TTCN. MSC composition operators make it possible to express branching behavior in the same way as in complex TTCN trees.

However, to cover the lack of variables and data definitions, we use a second language, TDL, which is the internal language of *TestPlayer*. We decided to make this language visible to the user, while waiting for the future standard MSC2000 to cover some of the mentioned lacks of tMSC.

3. *TESTPLAYER*: AN OPEN TEST PLATFORM

TestPlayer is an open architecture for testing providing customizability and power of expression sufficient to be applicable to a large number of real systems and test cases.

In our model, several testers feed the IUT with signals and check the responses of the system, and in the end generate test verdicts that will be logged in a results database. The interfacing between the IUT and the testers is done at predefined PCOs. The mechanisms through which messages are physically passed to the system depend on the system internals, so the testers are modularized such that the communication details are separated from the behavior description of the tester.

The behavior description of a tester is in fact a high level piece of code written in our Test Description Language (TDL), which regards signals as abstract data structures (is not concerned with bit-level encoding) and identifies PCOs only through their name (regardless of the details of physical communication with a specific interface point of the system). All these communication/encoding/decoding details are implemented as modules of the TDL interpreter and are customizable, as we will see in more detail later.

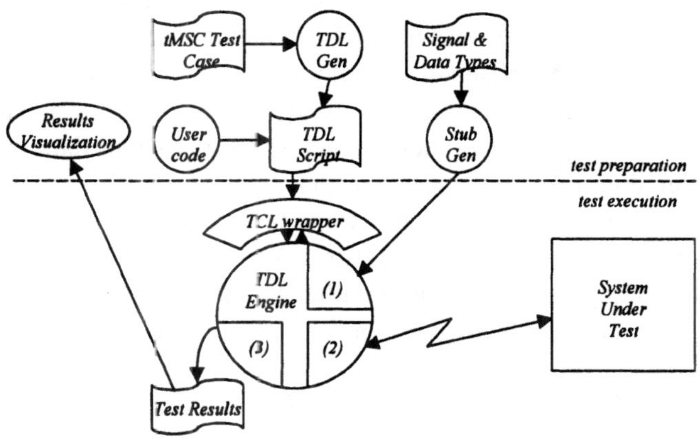

Figure 3. TestPlayer Architecture

Testing a system with *TestPlayer* is a two-phase process (see Figure 3): there is a preparation phase and an execution phase. The source of the test case may be either a tMSC diagram or a hand written TDL script. In the first case, what we will do is to automatically translate the tMSC in TDL using a MSC-to-TDL compiler (TDL Gen on Figure 3).

Before executing the tests the user must provide the description of the signals and data types exchanged between the tester and the system. This information is needed at execution time by the component in charge with data encoding/decoding ((1) on Figure 3). The user may either provide directly the body of the component (1) or - more likely - provide a SDL or ASN.1 description of the signals and data types. If the latter is the case, the body of component (1) is generated automatically by a sort of compiler (Stub Gen on Figure 3).

At execution time, the core of the *TestPlayer* environment is the TDL execution engine - the implementation of the TDL primitives. This is a fixed part of the architecture i.e. it cannot be customized. The implementation of the test primitives in the TDL engine is a static library that may be used from a variety of host languages such as C, TCL, Pascal. For each host language there must be a wrapper that makes the primitives accessible from the language.

The most common case of usage of *TestPlayer* is with the TCL interface. Therefore the test primitives are available as TCL commands. From now on, TDL will denote the extended TCL language.

3.1 *TestPlayer* customizable components

The TDL engine uses a set of utility components, which may be customized by the user. The user is free to provide any replacement for a component, provided he follows a set of simple naming conventions.

The **Signal Encoding/Decoding** component denoted by (1) on Figure 3 contains callback functions for signal and data encoding/decoding/matching. These functions are called to pack/unpack the signals and data transmitted/ received over the communication line that links the tester and the system. Component (1) also contains the functions that match an incoming message towards a generic message specification. As mentioned above, the Signal Encoding/Decoding component may either be provided as such by the user, or it may be generated from the SDL or ASN.1 description of the system signals and data types.

The **Protocol** component denoted by (2) on Figure 3 implements the protocol used by the tester to communicate with the system under test. *TestPlayer* may be used to apply tests in different configurations, in which the tester and the SUT may be on the same machine or on two different machines connected by a network, a serial line, etc. To allow this flexible architecture, the communication functionality must be kept in a separate component in order to allow for changes in the underlying communication primitives. The Protocol component (2) must implement a simple interface that basically knows how to open and close connections, raw send binary data to a connection and do blocking reads on a connection. Beyond this, the user is free to use whatever suitable means to implement the communication between the tester task and the SUT (TCP/IP, some proprietary protocol, OS IPC, etc.)

The **Test Reports** component (3) formats and stores the partial or final test results in a log. A user may require that the test results be stored in a specific format: a database, a log file with a proprietary format, etc. Any replacement for this component must implement a simple interface containing the following functionality: opening/closing a log, writing a test event to a log, writing a test verdict to a log.

After the execution of a test case or suite, the results may be visualized and one may automatically find the initial MSC construct corresponding to a test event that appears in the log, using the Results Visualization tool.

Thus we have obtained a testing architecture based on tMSCs and TDL, but in which the parts describing system signals and data types, the communication protocol between the tester and the system, and the formatting of test results are left open and customizable.

3.2 The Test Description Language

The Test Description Language is the internal language of *TestPlayer*. There is no test execution engine in *TestPlayer* that takes tMSCs and executes them directly. Instead a tMSC is translated in a TDL script. At this level the programmer may edit and extend the TDL test cases.

The Test Description Language is formed of a set of primitives with a precise semantics and an implementation that may be embedded in any language that allows for extensions (C, TCL, Python, etc.). The primitives correspond to TTCN executable instructions and to certain MSC constructs. They differ from MSC in that they are imperative rather than declarative. They provide for the missing parts of tMSC, notably variables and verdicts, as well as constructs for expressing alternatives between behaviors. A complete description of TDL primitives can be found in [10].

3.3 Translating tMSCs into TDL scripts

For each tester described in a tMSC, one script is generated. Usually a tMSC describes only one tester, but in the case of tMSCs taking the point of view of the tester (see Figure 1) there may be more. Tester outputs are translated into TDL `Output` statements, tester inputs into `Expect` statements and so on. An example of translation for a Basic MSC is given in Figure 4.

For High-level MSCs (see Figure 5), the sequential composition of MSCs causes the concatenation of the scripts described by the operand MSCs. The alternative composition of MSCs is translated using the TDL `Alternative` statement and repetitive composition is translated using the repetitive construct from TCL.

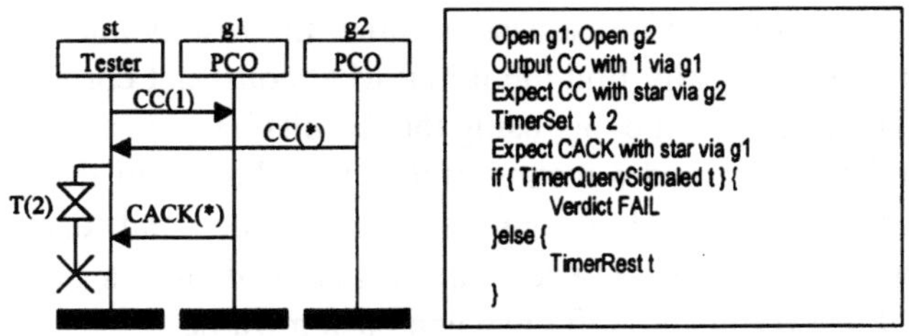

Figure 4. tMSC translated in TDL

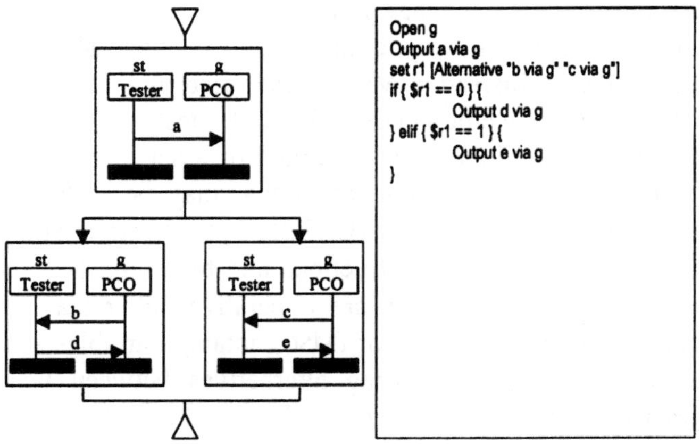

Figure 5. High-level MSC translated in TDL

3.4 TDL benefits

The TDL primitives bring variables and the alternative and repetitive statements. TDL primitives are implemented as an extension to the TCL language but they may be integrated in another language like C, C++, Python or Java.

Using a small set of primitives integrated in a scripting language ensures a small learning curve. TDL scripts being directly *interpreted* by *TestPlayer*, usual programming techniques like step-by-step execution and breakpoints may be employed for the execution of a test script.

3.5 *TestPlayer* for *Object*GEODE

TestPlayer for *Object*GEODE is an instantiation of all the generic components and tools described in the previous sections, which work with systems modeled in SDL [7] with *Object*GEODE [9] and for which code was automatically generated with the *Object*GEODE SDL-C Code Generator. The concrete architecture after the instantiation is outlined in Figure 6 (the instantiated tools and components are drawn on gray background).

The provided Stub Gen (Figure 3 and 6) is a compiler that automatically generates the code for the component (1) from the SDL description of the system. The generated component (1) contains encoding/decoding functions for the signals and data types defined in the SDL system.

Component (2) is in charge with the communication between the tester and the system. The instantiation uses TCP/IP for this purpose. An adapter task has to be added on the system side which "understands" the same protocol as the tester.

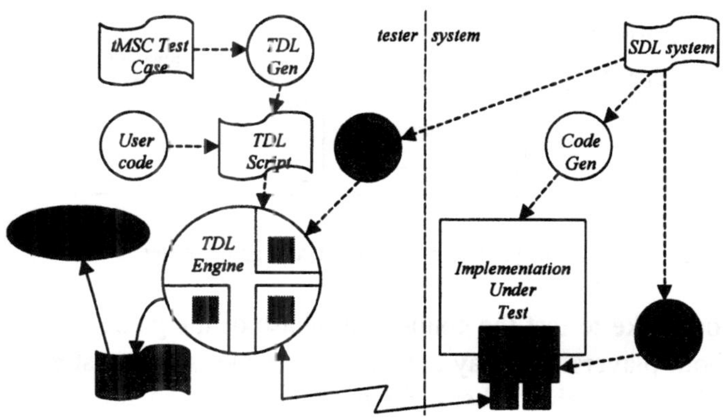

Figure 6. TestPlayer Concrete Architecture for *Object*GEODE

The adapter ("testing wrapper" on Figure 6) task will become the environment task of the system (see [9] for details). It takes the signals emitted by the system and send them over the network connection and vice versa. It exposes an interface of the system organized in PCOs. Concretely, this means that the external task is a TCP server that accepts one TCP connection for each exposed PCO. A tester that wants to communicate with the system on a specific PCO must call the `Open` primitive first.

The testing wrapper contains two sub-components: (4) which implements the communication protocol (very similar to (2) on the tester side) and (5) which implements signal and data types encoding/decoding (similar to (1) on the tester side). There is an automatic generator that generates the task from the SDL system specification.

The results formatting and storage component (3) is also instantiated for *Object*GEODE. It stores the test events and verdicts in an ASCII format in the test results repository. A visualization tool can read the repository and make the link back to the MSC source of a test case.

The instantiation of the generic components showed that we can easily obtain an executable architecture useful on large class of SDL systems.

3.6 An example

We validated the prototype instantiation described in the previous section on several examples. This section shows how we can test a sample system with the *TestPlayer* prototype.

We consider a simple SDL system describing a cards game server. The high-level architecture of the system is shown in Figure 7 (for brevity, not all the signals are shown).

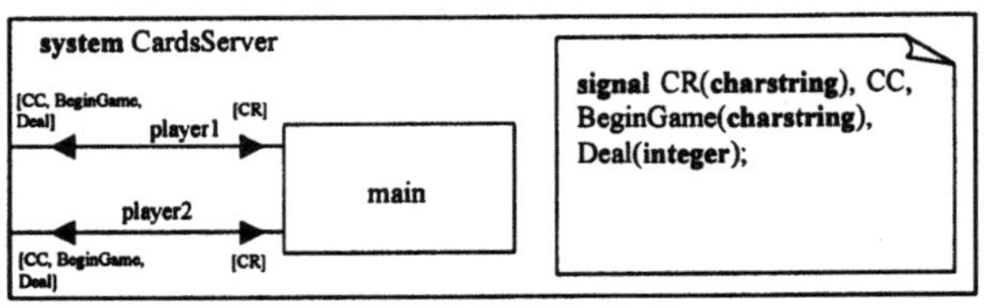

Figure 7. SDL system for a card game server

We would like to test the connection phase of the game. The connection involves both players that may connect to a game, so the tester will have to take the place of both of them. Each channel linking the system and the environment will correspond to a PCO. For testing the connection phase we will re-use a MSC from the requirements specification (Figure 8)

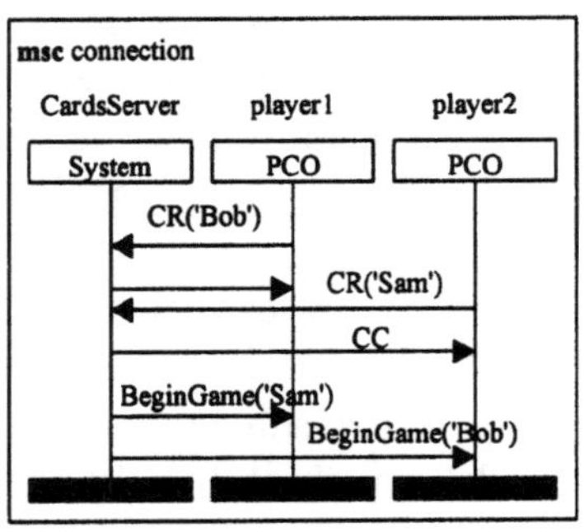

Figure 8. Connection phase of the Cards Game

After providing all the configuration parameters needed by the *Object*GEODE SDL-C Code Generator, we generate the executable system and the testing wrapper task in one pass, by invoking `st_gen_sys CardsServer`.

By default this will generate the executable system (`CardsServer.exe`) containing also the testing wrapper, generated in such a way that each channel linking the system and the environment becomes a PCO. This configuration suits our purpose. Other configurations are possible, like for instance one PCO for the entire system (grouping all the channels).

We now generate the extended TCL interpreter containing also the signal and type information specific to our system, by invoking `st_gen_interp CardsServer`. The starting point is again the SDL specification.

The generated interpreter (called `tdlsh`) contains the TDL engine plus components (1), (2) and (3) (see Figure 6). The type information contained in the SDL system is used for generating component (1). The other components are not rebuilt, they are taken from a library and linked together.

We have two possibilities for using the executable system and the TDL interpreter: either we generate the TDL script from the MSC(s) and execute it in batch mode, or we try an interactive session in which we feed whatever TDL commands we want to the command prompt of the TDL interpreter.

The first alternative involves first of all generating the TDL script from the MSC description of the test case. We can do this with `st_gen_tests connection.msc` and we obtain the TDL script called CardsServer1.tdl containing the following code (analogously to the one in Figure 4):

```
Open player1
Open player2
Output CR with Bob via player1
Expect CC via player1
... # other TDL instructions
Verdict PASS
```

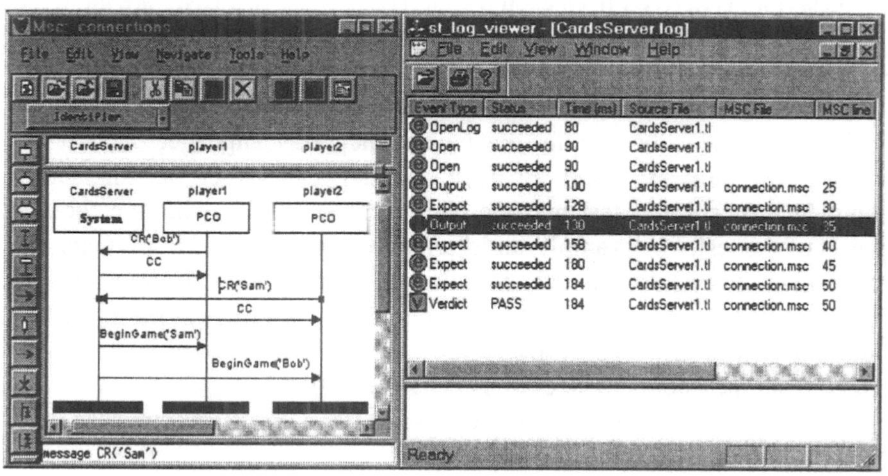

Figure 9. The Results Visualization Tool

To run this test, we must first start the system (CardsServer.exe) in a console, and then invoke `st_exec_tests CardsServer1.tdl` in another console. This command will run the test in batch mode and will write the test events and verdict in `CardsServer.log`.

This log may be visualized with `st_view_log` and the MSC construct corresponding to each event appearing in the log may be highlighted automatically in the *Object*GEODE MSC editor (see Figure 9). For FAIL verdicts, the failure reason is also stored in the log.

For using the TDL interpreter in interactive mode, one must start the system in a console and the `tdlsh` in another console. Then, at the `tdlsh` prompt he may type commands like the ones generated from the MSC. The outputs and verdicts of the commands will appear on the console.

4. *TESTPLAYER* WITH *TESTCOMPOSER*

TestComposer is a tool for automated test generation [5]. Starting from a SDL specification, it generates test suites for conformance testing.

TestComposer works in three phases :

- **A test purposes phase**, where informal test purposes coming from the requirements are written down in a formal way, using MSCs. These test purposes can also be generated automatically from the SDL specification, according to the structural coverage of the specification. In this mode, each transition (in fact basic block) of the SDL specification will be covered by at least one test purpose.
- **A test cases computation phase**, where a complete tree of possible behaviors is derived from each test purpose. These test cases are gathered in an internal database.
- **A test suite production phase**, where the test cases database is first processed in order to structure test cases, and to share common sub parts. The test suite is then written down using the TTCN language.

The test purposes used by *TestComposer* can either be abstract or complete. An abstract test purpose need only to specify what are the key events of the test. For example, an informal test purpose could be ``Open the connection, do one data transfer and close the connection". A corresponding abstract, but formal test purpose would be `send ConReq, send DataIn, receive DataOut, send DisconReq`. A complete test purpose would integrate all the intermediate events needed to fulfil the sequence:

```
send ConReq, receive ConAck, send DataIn, receive
DataOut, send DisconReq, receive DisconAck
```

A complete test purpose specifies one by one all the events needed for the test success. Test purposes generated automatically by *TestComposer* are complete in this sense. Complete test purposes are important with respect to *TestPlayer*, as a complete test purpose corresponds to the notion of test case used by *TestPlayer*.

Thus we can consider the first phase of *TestComposer* as a possible front-end of *TestPlayer*. Starting from an SDL specification describing the IUT behavior, the user can generate test cases in tMSC using *TestComposer*, and possibly complete the test cases by hand. Then using *TestPlayer*, he can execute the tMSC test cases on the IUT.

The IUT code needs not be generated from the same SDL specification, as long as the signals and their signature remain the same as it is in the SDL specification used by *TestComposer*. In that case, this SDL specification can be considered as the test specification, which can be developed independently and in parallel to the SDL specification meant for code generation.

But the same SDL specification can also be used, with benefits. In theory, as the tests and the code come from the same source, if an error is present in the specification, then it will be present in both the code and the tests, and the validation will declare as ok a faulty implementation. In effect, the only thing really tested would be the code generator.

In practical cases however, some of the IUT is generated from SDL, but other parts of the IUT are external libraries, parts of real time operating systems, databases. Very often a scheduling monitor is added, which can change dramatically the code behavior with respect to the specification. In this context, it makes sense to generate test case from the same source as code.

Last but not least, extensive validation campaign needed for certification purposes generally require a high, up to one hundred per cent, coverage of the code by the tests. In this context, working with the same specification makes even more sense, as the structure of the SDL and of the generated code will match. Test purposes generated by *TestComposer* correspond to the structural coverage of the SDL specification, so by derivation, correspond also to some extent to the structural coverage of the code.

5. CONCLUSIONS

We have presented *TestPlayer*, a tool for edition and execution of test cases written with the tMSC language. TMSC, being a subset of MSC, brings easy visualization and edition of test cases. Among its weak points, one can consider its restriction to deterministic test cases. One can not specify Inconclusive verdicts as in TTCN, these verdicts being one way to handle the IUT non-determinism during the test. However, most of the designers of safety critical systems are loath to deal with non-deterministic systems, especially in a context of strong certification requirements, such as in aeronautics or transport. For them, MSC are perfectly adequate for the expression of the behavior of a test case.

The architecture of the toolbox is such that it can be rapidly customized to work on different platforms (operating systems, network protocols, etc.) and different target systems. An originality point of our toolbox is the use of a second language for describing tests, TDL, which allows a more precise description than what is possible with MSCs, closer to the power of TTCN. This power of expression is provided at a low cost: TDL is a mere extension of TCL with a small learning curve.

Future developments around our tool include the industrialization of the tool by Verilog. We plan also to keep up with the future versions of the supported standards (SDL and MSC) which are both due at the end of 1999, and which will bring many substantial changes, some of them mentioned in the paper.

ACKNOWLEDGEMENTS

We wish to thank Rodolphe Arthaud who brought the initial momentum and many ideas to this project.

REFERENCES

1. ISO/IEC International Standard 9646-3. OSI-Open Systems Interconnection, Information Technology – Open Systems Interconnection Conformance Testing Methodology and Framework – Part 3: The Tree and Tabular Combined Notation (TTCN), 1992.
2. ISO/IEC International Standard 9646-1/2/3. OSI-Open Systems Interconnection, Information Technology – Open Systems Interconnection Conformance Testing Methodology and Framework, 1992.
3. J. Grabowsky and T. Walter. Visualisation of TTCN test cases by MSCs. In *Proceedings of the 1st Workshop of the SDL Forum Society on SDL and MSC, Informatik Bericht Nr. 104*, Humboldt Universität Berlin, 1998.
4. Jens Grabowsky. *Test Case Generation and Test Case Specification with Message Sequence Charts, Ph.D. Thesis*. Universität Bern, 1994.
5. A. Kerbrat, T. Jéron, and R. Groz. Automated test generation from SDL specifications. In *Proceedings of the SDL Forum '99*. Elsevier Science (North Holland), 1999.
6. ITU-T Recommendation X-680. Abstract Syntax Notation One (ASN.1): Specification of Basic Notation, 1994.
7. ITU-T Recommendation Z.100. Specification and Description Language, 1996.
8. ITU-T Recommendation Z.120. Message Sequence Charts, 1996.
9. VERILOG. *Object*GEODE Reference Manuals. Technical Report, VERILOG, 1996.
10. VERILOG. *TestPlayer* Reference Manuals. Technical Report, VERILOG, 1999

Model-Checking A Secure Group Communication Protocol: A Case Study

Alan J. HU, Rui LI, Xizheng SHI, and Son VUONG
Department of Computer Science, University of British Columbia, Canada

Abstract: With the explosive growth of the Internet and the distributed applications it supports, there is a pressing need for secure group communications – the ability of a group of agents to communicate securely with each other while allowing members to join or leave the group. Prompted by the success of other researchers in applying finite-state model-checking tools to the verification of small security protocols, we decided to attempt a larger security protocol: a recently published protocol for secure group communication. Not surprisingly, creating an ad hoc abstract model suitable for model-checking required cleverness, and state explosion was always a threat. Nevertheless, with minimal effort, the model checking tool discovered two flaws in the protocol, one of which has not been reported previously. We conclude our paper with a discussion of possible fixes to the protocol, as well as suggested verification tool improvements that would have simplified our task.

1. INTRODUCTION

Secure group communication is emerging as an important problem in networking. The objective is to allow messages to be sent securely to a set of authorized users. Such a mechanism underlies many new applications such as teleconferencing or pay-per-view content distribution. Trivially, we can build an n-party secure group communication protocol using $\binom{n}{2}$ normal point-to-point secure links and requiring each message to be sent individually to all other group members. For efficiency reasons, however, we would like to exploit the broadcast or multicast capabilities a network may offer, and we would like to introduce hierarchy to manage very large groups --- hence, the need for secure group communication protocols. In this paper, we consider a recently published secure group communication protocol [13] that offers substantial performance benefits over previously published protocols (e.g. [10]).

Security protocols are extremely tricky and have long been an important application domain for formal specification and verification techniques. Recently, a number of researchers have reported excellent results applying finite-state model-checking tools developed for hardware and general protocol verification to the verification of simple security protocols (e.g. key exchange, authentication) [8, 9]. The obvious promise of such an approach is a great reduction in labor and expertise required to carry out the verification. The obvious question is whether these automatic tools are sufficiently powerful in practice to handle interesting security protocols. This paper helps to answer that question by describing our experience using a model-checking tool on the above mentioned secure group communication protocol [13], which is more complex than has previously been attempted via model-checking. We will discuss the protocol and how we modeled it, the benefits and limitations of the model-checking approach and the tool we used, and the errors we discovered in the protocol.

2. THE PROTOCOL

In this section, we will briefly sketch the protocol we are studying. We refer interested readers to the original paper for details [13].

The protocol presumes a trusted server that administers the group. Requests to join or leave the group are sent to this server. The server has some mechanism for authenticating new users who request to join the group and for distributing or negotiating a private session key with the new user. These are standard operations and are assumed to work correctly. All cryptographic keys are assumed to be symmetric, private keys.

The basic idea of the protocol is to construct a hierarchy of keys organized in a tree.[1] (See Figure 1.) Each user is associated with a leaf of the tree, and the key at that leaf is the private session key between the server and that user alone. Each internal node of the tree corresponds to a subgroup key that is known by all users descending from that node. At the root is the group key that is shared by all users in the group. Accordingly, we can send a secure message to all the users in any subtree by using the key at the root of that subtree. Each user needs to know all the keys on the path from itself to the root.

[1] The paper also proposes a star graph, which is just a special case of a tree, and a complete graph, which is discarded as impractical.

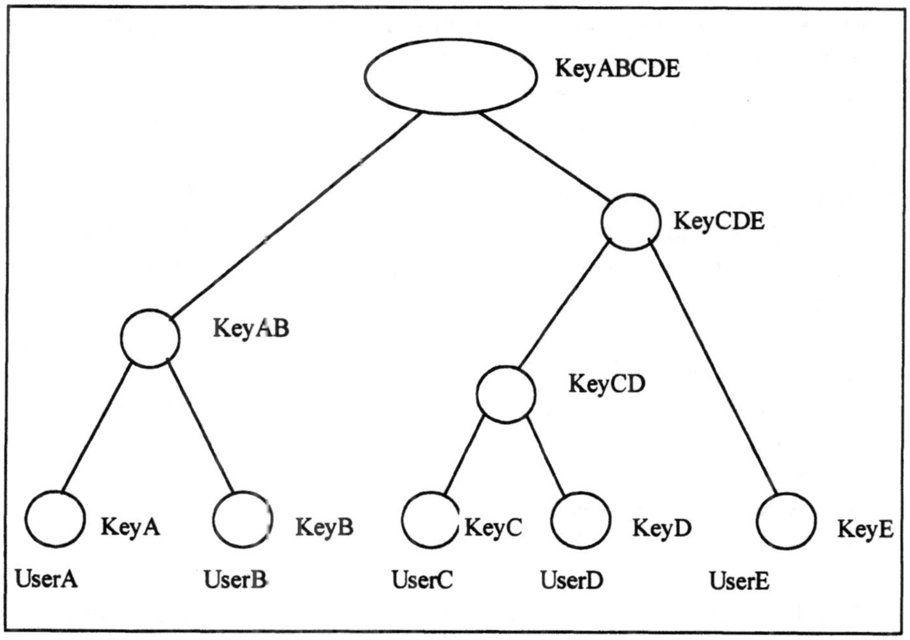

Figure 1. Key Tree: Each user is associated with a leaf in the tree and knows all the keys on the path from itself to the root. A secure message can be sent to all (and only the) users in a subtree by using the key at the root of that subtree.

When a new user wishes to join the group, it sends a join request to the server. The server authenticates the user, finds a position in the tree to attach the new user, and then must rekey all the nodes on the path from the root to the new user to ensure that the new user cannot read old messages from before it joined the group. (For example, in a pay-per-view application, a user should not be able to decode and view saved copies of previously broadcast programs using the key purchased for a more recent program.) Note that the rekeying means that every user must update at least one of its keys whenever a new user joins the group. Similarly, when a user wishes to leave the group, it sends a leave request to the server. Again, the server must rekey all nodes on the path from the root to the leaving user to ensure that it can no longer read any messages from the group.

The paper proposed three different rekeying schemes. For simplicity, we considered only "group-oriented" rekeying, the scheme with the best server-side performance. Intuitively, the scheme handles a join by sending a single message with all the needed keys to the new user and by broadcasting a single message to the whole group that contains all the changed keys, appropriately encrypted so that no user learns any keys that it shouldn't know. Similarly, the scheme handles a leave by broadcasting a single message to the whole group that contains all the changed keys, again with each key encrypted separately so that no user (especially the one that just left) learns new keys that it shouldn't.

More formally, let $s \rightarrow u : m$ denote sending message m from s to u. Generalized, $s \rightarrow \{u_1, \ldots, u_n\}$: m denotes broadcasting message m to all the users $u_1, \ldots, u_n$. A message (or portion of a message) enclosed in subscripted braces $\{m\}_k$ denotes message m encrypted using key k. Only users who know k can encrypt or decrypt such a message. The join protocol is as follows:

Join Protocol

1. $u \rightarrow s$: join request
2. $s \leftrightarrow u$: authenticate and negotiate key k_u
3. s attaches u to tree. Let $k_0, \ldots, k_j$ be the keys from root (k_0) to the node where we attached u. (The node with k_j is the parent of the node with k_u .)
4. s generates new keys $k_0', \ldots, k_j'$
5. $s \rightarrow$ {all users} : $\{k_0'\}_{k0}, \ldots, \{k_j'\}_{kj}$
6. $s \rightarrow u$: $\{k_0', \ldots, k_j'\}_{ku}$

The leave protocol is more complex. The server broadcasts a message containing all the new subgroup keys. The difficulty is that the normal way to send secret information to a subgroup is to use that subgroup's key. The leaving user, however, knows its current subgroup keys and could thereby learn the new subgroup key as well. To avoid this problem, each new subgroup key is sent encrypted by the (possibly new) keys of its subgroups. For example, in Figure 1, if User D requested to leave the group, the new KeyABCE' (replacing KeyABCDE) would be included in the rekey message twice: once encrypted by KeyAB and once encrypted by the new KeyCE', which replaces KeyCDE. Similarly, the new KeyCE' (which Users C and E need in order to decrypt the new KeyABCE') would be included twice: once encrypted by the new KeyC' and once encrypted by KeyE. This process continues to the leaves of the key tree. More formally, the leave protocol is as follows:

Leave Protocol

1. $u \rightarrow s$: leave request
2. $s \rightarrow u$: leave granted
3. s detaches u from tree. Let $k_0, \ldots, k_j$ be the keys from root (k_0) to the node where we attached u. (The node with k_j is the parent of the node with k_u .) Let $x_0, \ldots, x_j$ be the corresponding nodes.
4. s generates new keys $k_0', \ldots, k_j'$
5. We will now construct the rekey message in several pieces. For i=0..j, we create m_i to be the portion of the message containing k_i'. In particular, for each child of x_i (the node corresponding to k_i), m_i will contain a copy for k_i' encrypted by key for that child. In the

case of x_{i+1} (one of the children of x_i), the key used will be the new key k_{i+1}'.

6. $s \rightarrow$ {all users} : $m_0, \ldots, m_j$

Note that as described, there is an obvious flaw: an adversary can forge rekey messages and "black out" authorized users from reading subsequent messages in the group. The original paper suggested that rekey messages be signed using a secure digital signature scheme to prevent forging.

3. MODELING AND MODEL-CHECKING

In this section, we describe how we modeled the protocol and applied model checking.

Our first decision was which model checker to use. We chose the Murphi verifier [4] rather arbitrarily, but in part because others have used it successfully on security protocols [9]. We expect that other state-of-the-art model-checkers such as Spin [5] or FDR [8] would have produced similar results. We chose not to use a BDD-based model checker such as SMV [2]. BDD-based model checking has done very well for hardware, but not nearly as well on high level protocols [6]. We will discuss some of the advantages and disadvantages of choosing Murphi later in this paper.

Having chosen Murphi, we were able to follow the general approach of Mitchell et al. [9], who had first used Murphi on small security protocols. Murphi is a guarded-command language [3], so we model the various components of the system (e.g., server, users, adversary) as sets of commands for the possible actions of each part of the system. This modeling is somewhat tedious, but generally straightforward. For example, the commands to model the users allow an idle user to request to join the group (which entails sending a join message to the server and changing state to wait for the reply), a user waiting for a reply to process that reply, etc. Murphi has no notion of communication channels, so the network must be modeled explicitly using shared variables as a multiset of messages.

The adversary is the most interesting part of the modeling. Standard practice in the security community is to allow the adversary to intercept, modify, and generate arbitrary messages. However, the paper describing the protocol proposed using digital signatures to prevent forged messages, so we limited our adversary to eavesdropping and remembering messages, and then sending out previously saved messages later. Thus, our adversary is limited to "replay attacks", which cannot be prevented by digital signatures. The adversary can remember a finite number of previous messages. The adversary cannot encrypt or decrypt a message unless it knows the key.

The main modeling challenge we faced was incomplete information. The original paper emphasized performance, so many implementation details were missing. For example, the properties of the underlying communication network were never specified in the paper. We chose to model a network that guarantees delivery of messages, but does not guarantee preservation of message order --- properties that are typical of broadcast and multicast on wide-area networks. We expect such networks to be the typical underlying infrastructure for a secure group communication system. We also modeled a network that preserves message order; we will discuss this briefly in the next section.

Another challenge created by incomplete information was specifying properties to verify. Formal verification is only as good as the properties being verified. The original paper, however, never stated a security policy, other than a parenthetical comment about privacy, integrity, and authenticity. For our correctness properties, we specified that all members of each subgroup shared the subgroup key, and that all non-members did not have that key. Furthermore, the adversary should never obtain a usable key, and the adversary's messages should never be processed. Clearly, these are necessary conditions for correctness of the protocol; we do not claim sufficiency.

Throughout the verification process, state explosion was a constant threat. We were forced to make countless simplifying assumptions, such as limiting the size of the model to only a few users total, allowing the adversary to remember only a small finite history of previous messages, having a small number of possible keys and purging the adversary's memory when we are forced to recycle any keys, and many, many others. Altogether, these simplifications mean that countless bugs in the real protocol may have been simplified out of the Murphi model. This fact combined with the incomplete specifications means that model checking must be viewed as a debugging aid, rather than a certification of correctness. Success is achieved by finding bugs that had otherwise eluded detection.

4. RESULTS

Surprisingly, the adversary did not play a major role in the bugs we found. Instead, the bugs were the result of subtle ordering problems with the messages.[2]

A number of trivial bugs appeared very early on. For example, the protocol as stated allows a user to join and then immediately leave the group.

[2] Given that the original paper did not specify an underlying network model, it is perhaps unfair to label these as bugs, rather than merely as surprises.

Thus, a leave request might reach the server before the join request for that user. Adding a simple handshake would eliminate this problem. A similar problem is that the adversary can replay previous join/leave requests and overload the server and network, creating a denial-of-service attack. A proper authentication procedure (which the protocol presumes) would use nonces (one-time random numbers) in the authentication to prevent replaying previous joins/leaves. In order to explore other possible bugs, and to reduce state explosion, we simply modified the model to prohibit these behaviors.

Murphi found two more serious bugs. The first bug was mentioned in the original paper—a forged rekey message can be used to "black out" a legitimate user from subsequent communication in the group. The paper proposed using digital signatures to prevent forged rekey messages, but neglected to consider replay attacks. The Murphi model found that the adversary can save an old rekey message (properly signed by the server) and send it out again later, switching the victim to old, no-longer-valid keys. A solution is to timestamp the messages or to use sequence numbers, encrypted into the message to prevent tampering. Note that the timestamp must be large enough to prevent the adversary from reusing old messages whenever the clock wraps around. It is interesting that the authors of the protocol noticed the problem, proposed a solution, but didn't realize that the solution was inadequate, whereas the verification tool found the error quite easily.

A more serious bug was also discovered by Murphi. To our knowledge, this bug was previously unknown. The problem is that when multiple transactions are occurring, the respective rekey messages can arrive at a user out-of-order. Depending on the details of the message format (i.e. are there known fields in the encrypted parts of messages in addition to the randomly-generated key?) the result is at best that the user receives a bunch of unintelligible messages, at worst that the user switches to an incorrect key based on decrypting the new key with the wrong old key. Unfortunately, there are no easy solutions to this problem. Unencrypted sequence numbers would be vulnerable to attack by the adversary, causing the messages to be decrypted with the wrong keys. Encrypted sequence numbers are useless, because we need to know the ordering of the messages in order to decrypt them correctly. For a small network with infrequent rekey messages, a user could save all unintelligible messages and retry each one whenever a new rekey message arrives, but this approach is obviously impractical if rekey messages are frequent. Another approach would be to implement a scheme to provide order-preserving broadcast on top of the non-order-preserving network, such as ABCAST [1]. Unfortunately, such schemes impose substantial overhead and hinder scalability to large networks.

Note that the original paper did not specify the underlying network model. We chose a model that does not preserve message order, as in practice, that would be typical for an underlying network supporting multicast. When we changed our network model to preserve message order and reran the model checker, the above bugs disappeared. Thus, the protocol may have been designed assuming preservation of message order, but that assumption was never stated.

Altogether, the verification process required less than two months of part-time effort. Despite the fact that the protocol was published and hence presumed free of obvious bugs, we were able to discover two difficult bugs with very little effort. Furthermore, the two student authors, who carried out the bulk of the verification, had had no previous formal verification experience. In short, we had a very good bug/effort ratio.

5. DISCUSSION AND CONCLUSION

In retrospect, the choice of Murphi as our model-checking tool proved quite reasonable. Although we feel that other contemporary model checkers would also have served well, some positive features of Murphi stood out. Ease-of-use was perhaps the most important point. Murphi provides a very simple, small language, which makes the learning curve easy. For a novice user, or if we wish verification to be adopted by non-experts, initial ease-of-learning and ease-of-use can be critical. Similarly, the guarded command semantics of Murphi proved to be very convenient for modeling the protocol. Finally, Murphi provides special data types that allow powerful symmetry reductions to be applied automatically [7]. We made extensive use of these data types, resulting in an enormous savings in the number of states explored.

The hash compaction scheme, a probabilistic verification technique that compresses states stored in the hash table, proved to be absolutely vital to avoid running out of memory. The scheme is supported in both Spin and Murphi [12].

Murphi also has downsides. For example, Murphi has a very weak language for specifying properties to check, allowing only simple invariants that must hold in all reachable states. We did not find this restriction to be burdensome, but if more powerful properties could be specified, perhaps we would have found them indispensable. Similarly, Murphi has no communication primitives. The protocol we considered is broadcast-oriented, so the lack of communication primitives was not a problem. For a more connection-oriented protocol, however, communication primitives, such as the channels provided by Spin, would be very useful.

More generally, as the importance of security protocols increases, there is a need for verification tools tailored for this domain. The tool could provide primitives for communication, encryption/decryption, and the adversary, as well as opening the door to domain-specific reduction techniques. Shmatikov and Stern have recently published some provably-sound reduction techniques specifically designed for model checking security protocols [11]. Much more progress along these lines should be possible.

The greatest challenge we faced was incomplete informal specifications. The paper describing the protocol did not provide all the information needed. In particular, no security policy was given, nor were all assumptions about the environment stated. Greater attention to these issues is certainly necessary for security protocols. Formal description techniques, even in the absence of automated tool support, could be helpful. The saying comes to mind, "A program that has not been specified cannot be incorrect; it can only be surprising." [14] For security protocols, surprises are dangerous.

We were able to find bugs in a published security protocol with minimal effort. The two student authors, who performed the bulk of the verification work, had no prior formal verification experience. Clearly, model checking tools and methodology are now robust enough that they can be a routine part of security protocol design. There are no guarantees that the model checking will succeed, but the likelihood of finding bugs is high, and the effort expended is low.

Finally, we note that the numerous ad hoc simplifications one needs in order to use model checking imply that an error-free model-checking run does not ensure "correctness". For critical applications, conventional proof-based methodologies --- using a more expressive formal description technique and perhaps automated proof assistants --- are still needed. Nevertheless, the ease-of-use and excellent bug/effort ratio of model checking makes this approach an excellent tool for designing and debugging security protocols.

References

[1] J.-M. Chang and N. F. Maxemchuk. Reliable broadcast protocols. ACM Transactions on Computer Systems, 2(3):251--273, August 1984.

[2] E. Clarke, K. McMillan, S. Campos, and V. Hartonas-Garmhausen. Symbolic model checking. In R. Alur and T. A. Henzinger, editors, Computer-Aided Verification: Eighth International Conference, pages 419--422. Springer-Verlag, July 1996. Lecture Notes in Computer Science Number 1102.

[3] E. W. Dijkstra. Guarded commands, nondeterminacy and formal derivation of programs. Communications of the ACM, 18(8):453--457, August 1975.

[4] D. L. Dill, A. J. Drexler, A. J. Hu, and C. H. Yang. Protocol verification as a hardware design aid. In International Conference on Computer Design. IEEE, October 1992.

[5] G. J. Holzmann. Design and Validation of Computer Protocols. Prentice-Hall, 1991.

[6] A. J. Hu and D. L. Dill. Reducing BDD size by exploiting functional dependencies. In 30th Design Automation Conference, pages 266--271. ACM/IEEE, 1993.

[7] C. N. Ip and D. L. Dill. Efficient verification of symmetric concurrent systems. In International Conference on Computer Design, pages 230--234. IEEE, October 1993.

[8] G. Lowe. Breaking and fixing the Needham-Schroeder public-key protocol using FDR. In Tools and Algorithms for the Construction and Analysis of Systems, pages 147-- 166. Springer-Verlag, 1996. Also published in Software Concepts and Tools Volume 17, pp. 93--102.

[9] J. C. Mitchell, M. Mitchell, and U. Stern. Automated analysis of cryptographic protocols using Murphi. In Symposium on Security and Privacy, pages 141--151. IEEE, 1997.

[10] S. Mittra. Iolus: A framework for scalable secure multicasting. In SIGCOMM. ACM, 1997.

[11] V. Shmatikov and U. Stern. Efficient finite-state analysis for large security protocols. In 11th Computer Security Foundations Workshop, pages 106--115. IEEE, 1998.

[12] P. Wolper, U. Stern, D. Leroy, and D. Dill. Reliable probabilistic verification using hash compaction. Submitted for publication. A draft is available from http://sprout.Stanford.EDU/uli/publications.html.

[13] C. K. Wong, M. Gouda, and S. S. Lam. Secure group communications using key graphs. In SIGCOMM. ACM, 1998.

[14] W. D. Young, W. E. Boebert, and R. Y. Kain, Proving a Computer System Secure, *The Scientific Honeyweller*, vol 6, no 2 (July 1985), pp. 18—27. Reprinted *in Computer and Network Security*, M. D. Abrams and H. J. Podell, eds., IEEE Computer Society Press, 1986.

PART TEN

Invited Papers

Software Model Checking
Extracting Verification Models from Source Code

GERARD J. HOLZMANN and MARGARET H. SMITH
Bell Laboratories, Lucent Technologies, 600 Mountain Avenue, Murray Hill, NJ 07974, USA

Key words: Software Verification, Model Checking, Model Extraction, Software Testing, Systems Design, Debugging, Call Processing, Telephony.

Abstract: To formally verify a large software application, the standard method is to invest a considerable amount of time and expertise into the manual construction of an abstract model, which is then analyzed for its properties by either a mechanized or by a human prover. There are two main problems with this approach. The first problem is that this verification method can be no more reliable than the humans that perform the manual steps. If rate of error for human work is a function of problem size, this holds not only for the construction of the original application, but also for the construction of the model. This means that the verification process tends to become unreliable for larger applications. The second problem is one of timing and relevance. Software applications built by teams of programmers can change rapidly, often daily. Manually constructing a faithful abstraction of any one version of the application, though, can take weeks or months. The results of a verification, then, can quickly become irrelevant to an ongoing design effort. In this paper we sketch a verification method that aims to avoid these problems. This method, based on automated model extraction, was first applied in the verification of the call processing software for a new Lucent Technologies' system called PathStar.™

1. INTRODUCTION

In theory, software applications can be designed either top-down or bottom-up. It has been debated for decades which of these two methods, applied in isolation, would be better. In practice, the two methods are often used in combination. An industrial software design project normally starts with a top-down system design phase, where requirements and constraints are explored and defined. This is followed by a bottom-up phase, where developers try to meet the stated objectives. The design is then concluded with a system validation phase, where testers probe the code to check if the top-down and the bottom-up phases meet in the middle.

If code were developed purely top-down, we could apply systematic design refinement techniques to guide the process, to secure that each new level preserves all top-level requirements and continues to satisfy all constraints. This assumes, however, that the requirements and constraints do not change much, or that one would be willing to return to first design phase each time they do. It also assumes that we can successfully carry through refinement proofs from top-level design to implementation level code. At present, these assumptions are not very realistic for larger applications.

A verification problem is typically presented to a formal methods person in the same way that it is presented to a conventional software tester. In many cases a detailed implementation of the design exists, and is available. The requirements for the design are often harder to find. Often, the requirements are not accurately documented and have to be reconstructed by the tester or verifier. If the problem is sufficiently important, a verification expert can undertake to formalize the requirements, and then to construct a formal model of the code that can be shown to either satisfy or to violate those requirements. The formal methods expert can make mistakes both in the formalization of the requirements and in the formalization of the code.

The effort to construct and to check the accuracy of a formal model for any application of substance can be significant, often consuming weeks or months. Only in rare cases, where design and development have ceased, and no maintenance is performed, will the code remain unchanged in the period that model construction and verification is attempted. This has several important implications:

- It is difficult to perform this type of verification *during* the design and development phases, when verification results are most beneficial. For this to be possible, the effort to construct an abstract model must be considerably smaller than the development effort itself.
- When verification results become available, chances are that they apply only to older versions of the code, which may no longer be of interest to the designers.
- Manually constructed models can themselves be flawed, possibly hiding real errors and often introducing artificial errors. The latter can be detected and corrected, the former cannot.

An alternative method that we will explore in this paper is to extract a formal verification model mechanically from a software application.

2. MODEL EXTRACTION FROM CODE

For any software artifact there are at least two types of entities that are of interest when it comes to verification: the final version of the design requirements and the ultimate code that implements them. It is the task of the verifier to prove that the latter conforms to the former. There can be many hurdles to overcome before the verification itself can start.

First, the requirements may be partially unstated (i.e., incomplete) or they may be available only in their original form, before the long process started in which idealized requirements meet real-world constraints and are *true-ed up*. Ideally, the requirements are captured in a sufficiently formal notation, so that it will be possible to check them directly for their logical consistency, independently of the implementation. In practice, we often have to interpret and disambiguate informally stated textual requirements.

A second hurdle to overcome is that the correctness properties of the full implementation are almost certainly intractable. In general, only the correctness properties of a subset of all possible software applications can have decidable properties (cf. the general unsolvability of the halting problem). One such subset, of particular interest for verification, is the set of finite state programs. The hurdle to overcome, then, is to devise a way to map a given program into an element of this subset by some well-defined process. That well-defined process is called *abstraction*, and the result of this process can be called a *verification model*.

A purely top-down design process starts with a high-level abstraction that can trivially be proven correct. By a disciplined process of refinement the designer then tries to obtain an implementation level description of the application that preserves the essence of the high-level abstraction. Specific types of functionality may need to be proven at each refinement level, in addition to the preservation of properties across levels. The process requires sufficient care and discipline that examples of its use in large-scale applications are scarce.

2.1 Verification During Code Development

The alternative that we will explore here is to allow the design and coding phases to proceed normally, but to devise a method that can extract a verification model from the current version of the code at each point in time. The model can then be shown to either satisfy or to violate the current expression of all correctness requirements. Requirements and code are thus developed hand in hand. In some cases evidence of a violation of a requirement will necessitate a refinement of the requirements. In other cases it will call for an adjustment of the code.

2.2 Data and Control

We will assume in what follows that the application describes a dynamic system of asynchronously executed processes, communicating via message passing and/or via access to shared data. That is, we specifically target the verification of distributed systems software for this work [H97]. For each process there is a piece of code in the implementation that describes its behavior. We would like to extract a model from each such piece of code, by applying predefined, carefully selected, abstractions.

At the minimum, we have to extract the control-structure from the code. We only need a minimal parser for the language in which the code is written to extract such a control-flow skeleton. For applications written in C, for instance, the parser needs to know that statements are terminated by semicolons, and it needs to know the syntax of selection and iteration constructs such as **if**, **while**, **until**, **goto**, **case**, and **switch**, but it needs to know very little else about the language. In particular, the text of statements (assignments, conditions, function calls, etc.) can remain uninterpreted at this point.

The control-flow skeleton specifies only control; it is void of any knowledge of data so far. The skeleton can be converted into a minimal verification model, e.g. for the model checker SPIN [H97].

For a minimal model, we can proceed by populating the skeleton model with all message-passing operations from the original code. To do so, we only have to be able to recognize which statements in the original source text in C correspond to sends and receives, and to define a translation for them into the skeleton model. The send and receive operations in the code can be more detailed than what is necessary or desirable in the model, so we can already chose to define a first abstraction at this point.

In PROMELA, the specification language used by SPIN, message send and receive operations are applied to channels (i.e., message queues). These operations typically include only those message parameters that are deemed relevant to the correctness properties of the application. Sequence numbers and congestion control information, for instance, would normally qualify for inclusion in such a mapping, but random user data would not. Could we automate abstractions of this type?

First note that both the source code and the code from the PROMELA model represent all message passing primitives in a specific, but distinct, notation. This consistent and *context independent* representation style in both the detailed and the abstracted version of the code allows for a particularly simple and elegant method of conversion. All that is needed is the definition of a lookup table that is populated with the required conversions. The table contains the source code version of each primitive on the left, and the abstract version of the code that corresponds to it on the right, as illustrated in Table 1 below. Once we have the lookup table, the conversion from source code into a minimal model, containing just the representation of message passing construct and control flow can indeed be automated.

A critical observation at this point is, however, that we can use the *same* discipline to define an abstraction for also the remainder of all source statements. We need not limit ourselves to just the message passing primitives. The lookup table gives us control over all aspects of the abstraction. It allows us to decide which data objects will be represented in the model, and how the manipulation of these data objects is to be rendered in the abstraction to secure a tractable model. Without the ability to define data abstractions, the model extraction process would be of no real practical significance. Consider, for instance, what useful properties would remain to be proven about a C program from which all data manipulations are removed.

Table 1. Sample Lookup Table for defining Simple Abstractions

Source Code in C	***PROMELA Code for abstract model***
send(y,Iring)	y!Iring
send(x,Cranswer)	x!Cranswer
x->time = 60;	Time = true
cdr_init(y,ptr)	skip

If we want to use the model extraction process that is sketched above during systems design, the source code may continue to evolve, also after we defined the lookup table for an initial set of abstractions. New statements may be added to the code, old ones may disappear. Once the lookup table is complete, though, it is relatively simple to keep it up to date. Indeed, it is simpler than keeping a manually constructed abstract model up to date under the same conditions. To allow us to track the changes in the source code of an application it suffices for the model extraction tool to alert us to omissions, mismatches, and redundancies that are detected between entries in the table and statements in the source code during the model extraction process.

Table 1 gives a sample lookup table, defining abstractions for four types of source statements that may appear in the source code of an application. In the first two cases, the mapping is a syntactical conversion from C to PROMELA. The third entry shows a more deliberate abstraction, where we record only the fact that a timer was set, but not the numeric value to which it was set. The last entry shows a still greater abstraction. It defines that all source statements of the type given on the left (having to do with issuing call-data-records for billing in a telephone system) are irrelevant to the verification effort, and can be omitted from the model.

2.3 Model Templates

The lookup table focuses on actions, not on the data objects acted upon. To produce the final model in PROMELA, every abstract data object must be formally declared, defining its type and initial value. This can be done with the help of a model template that contains the required data declarations and an outline of the required process declarations, to be filled in with detailed behavior specifications by the model extractor.

```
chan Q[2] = [1] of { mtype, bit };

active [2] proctype state()
{	mtype op, par1;
	bit seqno; byte who;
	chan qin, qout;

	who = _pid;
	qin = Q[_pid];
	qout = Q[1-_pid];

	@P  /* extracted code goes here */
}
```

Figure 1. A Sample Model Template

A sample model template is shown in Figure 1. In this case the template is for a simple protocol system. The template defines the containers for two processes, and a number of local data objects that are used in the abstracted version of the code (not further detailed here). The model extractor replaces the placeholder @P with the abstracted behavior that it generates from the source code of the application (also not shown here).

2.4 Checking Abstractions

The adequacy of the abstractions defined manually in the lookup tables can be checked as follows. We can check, for instance, that *all* updates of data objects referenced at least once in the abstract model are represented in the abstraction. We can also check that all data dependency relations are properly preserved by the abstraction. If, for instance, data object **D** is part of the abstraction, we can check that all data objects on which **D** depends in the source code are also be represented, and so on recursively. A check for these types of completeness can trivially be automated. One step further, one could also consider methods for mechanically deriving the lookup table contents from a given body of requirements.

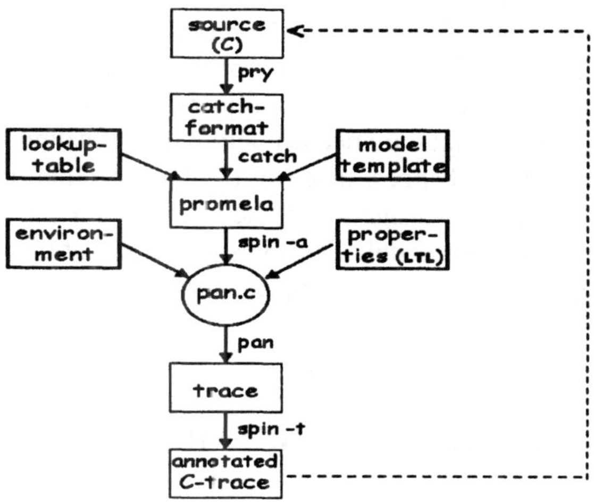

Figure 2. *Overview of the Model Extraction Process*

2.5 Model Extraction Tools

Figure 2 gives an overview of a model extraction process of this type that we have implemented. Model extraction begins with a source description of an application in the programming language C. A control-flow parser for a subset of C, called PRY, extracts an annotated control-flow skeleton from the C code, in an extended state machine format. As noted, PRY needs to identify only control-flow constructs and statement boundaries. It can therefore be substantially simpler and smaller than a full C compiler. In our implementation, for instance, PRY is a 1,700 line C program.

The next step in the model extraction process is performed by a tool called CATCH, of about the same size as PRY. CATCH reads the intermediate format produced by PRY, and guided by the lookup table and the model template it generates a verification model in PROMELA.

The user manually constructs the lookup table and the model template. The user also provides the properties to be verified and the test-drivers. The test-drivers formalize assumptions that must be made about the *environment* in which the application will be executed. In the case of a telephone system, as discussed in the next section, the test drivers formalize assumptions about the behavior of telephone subscribers and of remote switches in the network, in so far as this is visible and relevant to the local switch. (The issue of visibility and relevance is discussed in more detail in [H98)].)

The user, then, manually provides only those parts of the model that define the desired level of abstraction, the properties that have to be proven, and the context in which they are to be proven, conforming to a familiar assume-guarantee paradigm. Most significantly, when the source code changes, only small changes are needed before a verification attempt can be repeated. The required changes, made in the lookup table or in the model template, require no deep insights into the rationale for the changes in the source that produces the model. The tools PRY and CATCH issue warnings when they detect redundancy or incompleteness in the lookup table or model template.

3. APPLICATION: PATHSTAR™

The PathStar Access Server is a new Lucent Technologies product with support for call processing and telephone switching, as well as data connections, in an IP network [FSW98]. The system was designed and built by a large team of people over a period of several years. The call processing software, one of many software modules in PathStar, was developed over a period of approximately nine months by Phil Winterbottom and Ken Thompson. The call processing software was verified with the methodology that we have outlined here, and detailed in [HS99]. This application is, as far as known to us, the first use of software model checking in a commercial product development at this scale.

The call processing code for PathStar was designed to adhere to the recommendations on call processing and feature processing published and maintained by Telcordia (the new name for the former Bellcore) [B92-96], as is required for any public switching system. Structurally, though, the code differs from older implementations of the same functionality. The primary difference that concerns verification is that the central code related to call processing is concentrated in a single high-level procedure that can be studied and verified separately from the remainder of the code.

The central procedure encodes an extended finite state machine that defines all control logic for telephony, i.e., basic call processing and all feature code. At the time of writing, the code supports roughly fifty call processing features, such as *call waiting*, *call screening*, *call forwarding*, *three-way calling*, etc. The more standard software modules with sequential code, e.g., defining the specifics of device drivers, switch administration, and operating systems interfaces, are tested separately, without use of model checking techniques.

As detailed in [HS99], the control logic for the PathStar system is written in C, in a format devised by Ken Thompson to simplify the definition of the underlying state-machine. The code defines the conventional transition tuples of source state, guard condition, actions, and destination state. From this code we mechanically extract the verification models. The extraction process is performed in three steps, as also illustrated in Figure 2.

1. The C code for the core control code is parsed by the program PRY and converted into an intermediate format that makes both the state-machine structure and all transition tuples explicit.
2. The program CATCH reads this format and converts it into a standard PROMELA specification, using a model template and a lookup table (or *map*) to define the desired abstraction.
3. Correctness requirements, stated in linear temporal logic (LTL) are taken from a substantial database of properties that we have built derived from general Telcordia/Bellcore requirements for call processing, feature behavior, feature precedence relations, and from internal requirements on the desired functionality of the switch.

The complete extraction process takes a fraction of a second. Minor changes in the source code require no human intervention when a new series of verifications is performed. This applies in particular to small bug-fixes that are introduced to repair, for instance, a correctness violation uncovered by the verifier. Often the adequacy of a suggested fix can be (dis)proven within a few minutes of real-time, from start to finish. Within this framework it can be more time-consuming to describe and record a bug in the standard project databases then it is to both implement a correction and to formally prove its correctness.

More substantial changes to the source code, introducing new functionality, can require an update of the abstraction itself. This often means that the lookup table used by CATCH to guide the model extractions must be updated. To facilitate this process, the program CATCH warns the user when it finds entries in the lookup table that do not (no longer) appear in the source code, and it warns when it finds new functionality in the source code that is undefined in the lookup table.

We have tracked the evolution of the call processing code for PathStar from the first version, up to and including the final version of the code that was shipped to customers. All correctness properties were checked and rechecked for almost every version of the code throughout this period, often several times per day. In this period the code more than tripled in size. With few exceptions, each update of the lookup table took no more than five minutes. The few exceptions were when the source code changed more significantly from one day to the next, to support the inclusion of more complex feature code. In these cases the matching revisions of the lookup table took a few hours of edit time each.

3.1 Web-Based Access

The requirements database we have built to support the PathStar work contains over one hundred separate formalized requirements, and the results of the verification of each one, organized by feature. A web-server gives online access to the various parts of this database, containing:

- The text of the original Telcordia/Bellcore documents, in PDF format.
- Summaries of the verifiable parts of these documents, in ASCII format.
- Formalizations of each property in either temporal logic form, or in a short-hand form for specifying ω test automata (not further detailed here).
- The results of all verification attempts of the properties, in the form of tables derived directly from the machine generated results.
- Access to the re-verification of individual properties, or groups of properties. To support this, each web-page contains menus that allow a user to select either an individual property, or a group of properties, a verification mode, and a run button that initiates the verification process. The results of the verification process are included in the web-databases as soon as they become available, without further human intervention.

Each verification run initiated through these web-pages always starts with the model extraction process, i.e., it starts with the call processing source code in C from the application itself.

The verification of a property stops when a counter-example is found, or when it completes normally. Because of the way that we instrument the generated models, the execution of the counter-example with the SPIN simulator suffices to generate an execution trace for the system in native C, i.e., as an interleaved sequence of C statement executions. The instrumentation we apply is to allow the SPIN simulation to printout the C source statements that correspond to the various portions of the abstract model, as comments to the simulation. To the C programmer, these comments are the equivalent of a true execution of the source code. If there is any discrepancy that makes this not true, it points to a flaw in the abstraction method applied, which is then repaired by a modification of the lookup tables that define the abstractions used.

3.1.1 Search by Iterative Refinement

Short error traces are more convincing than long ones, and therefore we rely on an option from the model checker SPIN to deliberately search for short error sequences. To find errors as quickly as possible, our system optionally uses *an iterative search refinement* procedure. If this option is chose, a verification attempt starts with a coarse, and therefore fast, proof approximation step. This is realized by exploiting a feature of SPIN's bitstate search option (sometimes called *supertrace* [H97]). We initially run this algorithm with a deliberately small hash-array. The supertrace algorithm performs the equivalent of a random pruning of the reachable state space, within the constraints imposed by the size of the hash-array. The size of the hash-array correlates directly with both speed and coverage.

By doubling the size of the hash-array after each unsuccessful attempt we can systematically increase the thoroughness of the search, and the amount of time spent performing it. Errors typically show up in the first few iterations of the search, within the first few minutes or seconds. Only properties that fail to generate any counter-examples, i.e., for properties that are likely to be satisfied, we will go through the entire iteration process, from fast approximation down to a final thorough exhaustive check.

Table 2 gives an example of the view offered by the web-based server for the verification status of properties. Each property in the database is in one of four possible states.

Table 2. Sample Web-Based View of Verification Status

Pending	***Provisioning***	***Status***	***Action***
$\Box(p\ U\ X(r\ U\ s))$	+TWC -DTS	n/a	Start Check

Running	***Provisioning***	***Status***	***Action***
$\Box\Diamond er$	+TWC -DTS	B04331	Abort Run

Verified	***Provisioning***	***Status***	***Action***
$\Diamond(a \wedge X(b\ U\ c))$	+CFDA -ACR +CW	B076443	Delete Result
$\Box(p \rightarrow \Diamond q)$	+CFBL -ACR	M409574	Delete Result

Failed	***Provisioning***	***Error-Trail***	***Action***
$\Box(oh \rightarrow \Diamond dt)$	-DOS +CFDA	B-50	Delete Result

A property in the database can be listed as:

- ***Pending***, when no verification results for this property have been obtained so far;
- ***Running***, when a verification run is in progress;
- ***Verified***, when no counter-example to the property has been found in any of the verification runs performed;
- ***Failed***, when at least one counter-example to the property is available.

Depending on the state, an appropriate action can be performed on the property. A verification can be initiated for a pending property, moving it to the running state. A running property will typically transit quickly into either a failed or a verified property. Longer running verifications, however, can be queried about the intermediate results achieved thus far, by clicking on the status field (in the third column of the table). Both failed and verified results can be returned into pending properties, by deleting them from the table.

The most important feedback offered by this web-based tabulation of verification results is that all failed properties can be queried about the precise nature of the failure.

3.1.2 Analyzing Counter-Examples

As illustrated in Table 2, the first column of the table gives the LTL form of a property. Symbols such as p, q, a, b, and c are defined by the user as part of the property definition. The detailed definitions for these symbols appear in the web-pages in a list below the table (not shown here). The second column details provisioning data for a verification run. For the first verified property listed in Table 2, the provisioning data states that for this property to be valid we must assume specific feature combinations to be enabled or disabled. (CFBL is *call forwarding busy line*, ACR is *anonymous call rejection*, and CW is *call waiting*). The third column links to the results of a verification, either the evidence for assuming that it is verified, or the counter-example that shows that the property is not satisfied.

The third column entry indicates the type of verification that was performed (exhaustive or approximate) and, in the case of a failed property, the length of the error-trail. The shortest error trail is always displayed first; if it is deleted, the next shortest trail is shown, etc. Clicking the link in the third column will display the counter-example in one of four available forms.

- An ASCII representation of a message sequence chart derived from the trace. This chart shows only message send and receive operations between the concurrently executing processes, but no details of statements executions.

- A graphical representation of the message sequence chart, adding some context information about the connection between the behavior specified in the property and the behavior exhibited by the system. To accomplish this the display notes when in the execution different predicates from the temporal logic formula change value.
- A statement-by-statement listing in C source code of the executions of all processes involved in the sequence.
- A statement-by-statement listing as above, with detailed dumps of all variable values at each step in the sequence.

Typically, the ASCII version of the message sequence chart serves as a quick indication of the type of execution that has been identified by the verification process, and the C statement listings provide the finer detail that is needed to interpret it fully.

The last property included in Table 2 states that if a subscriber line has the feature *CFDA* (call forwarding on don't answer) enabled, but not *DOS* (denial of originating service) then the subscriber will always receive a dialtone (represented by symbol *dt*) after going offhook (symbol *oh*).

When we verified this property for an early version of the call processing software, the verifier reported a short error sequence, which is illustrated in graphical form in Figure 3. When the PathStar system processes a call forwarding request it briefly enters a transit state. The processing in this state typically lasts for only a few milliseconds. The scenario, illustrated in Figure 3, shows that if the subscriber happens to go offhook within that brief interval, it will not received dialtone. As it turned out, in the original version of the code, the system would discard offhook signals while in this transit state and fail to exit the state unless the subscriber would be onhook.

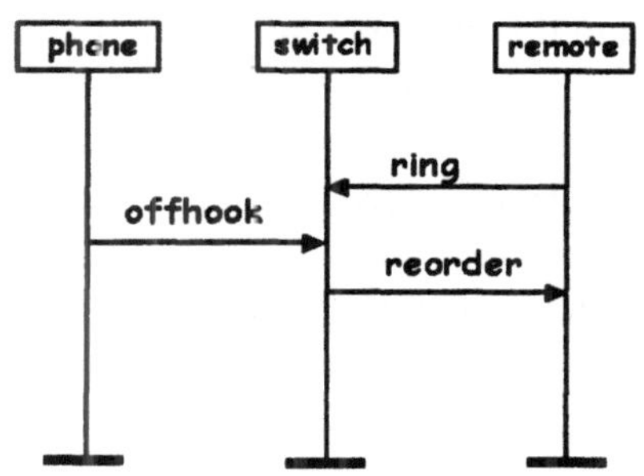

Figure 3. Message Sequence Chart for Error Sequence

The error-trace illustrated in Figure 3 is an example of the type of software error that would be extra-ordinarily hard to identify with conventional testing. Even if errors of this type are encountered in lab testing, they are almost impossible to reproduce, and therefore often attributed to hardware glitches, beyond the control of the software.

3.1.3 Avoiding False Positives

Although the failed properties listed in the results tables give the most detailed feedback to a user, one quickly learns in an application of this size that also the purportedly verified properties can carry significant information. When a new property is formulated, it too has to be verified for its correctness. The first few attempts to capture a desired system property in a logic formula are often flawed. If the flawed property produces a false counter-example, no harm is done. The counter-example can be inspected and will reveal the flaw. If the verification fails to produce a counter-example, however, it is not always evident if this is a vacuous result, e.g., because the property *accidentally* failed to match the intended behaviors.

In most cases, undesirable system behavior starts with a valid execution prefix that produces an invalid outcome. Consider, for instance, the setup of a three-way call. One of our formal properties for three-way calling checks that if the telephone subscriber follows all instructions for setting up the call correctly, that indeed a three-way call will be established. The valid execution prefix in this case is the subscriber correctly following the instructions. The invalid suffix of the execution would be the failure of the system to establish the call as instructed. If during the verification of this property neither the invalid suffix *nor* the valid prefix of the behavior can be matched, something is likely to be wrong with the formalization of the property itself.

The verification system that we have built gives the user feedback on the portion of the property formula that was matched in a verification run. The system provides this information in graphical form for all properties that are in the *Running* or in the *Verified* state. This means that for a longer running verification, the user can see quickly if the verification attempt is progressing as intended, and is likely to produce a useful result, or not. To achieve this, the system displays the test automaton that is derived by SPIN from the LTL property, and indicates which states of this automaton have so far been reached in the verification attempt. This direct feedback has proven to be effective in diagnosing potential false positives.

3.2 The TrailBlazer System

After a full update of the source code, all properties in our database are reverified from scratch. The iterative search refinement method secures that errors show up early in this process. Nonetheless, to run this process to completion, the verification of all properties in the database can take several hours. A number of simple optimizations can be made in this process. We discuss two of the optimizations that have had the largest impact on our project.

The first optimization we implemented exploits the way in which SPIN works. To perform a verification, SPIN always starts by generating C code that implements a model-specific verifier for the complete specification: one part in LTL and one part in PROMELA. In our case, the PROMELA model is extracted from the application, and therefore necessarily the same for all properties. Generating and compiling the complete verifier from scratch for each separate property is inherently wasteful. To prevent this, we have added an option to SPIN that allows one to generate and compile the property specific code separately from the application code. This reduces the average compilation time for all properties from minutes to seconds. The large PROMELA specification is compiled into the model checker just once, and it is linked with a different small fragment of C code for each new property checked.

A second optimization exploits the fact that each property from our database is checked in a separate and independent verification run. It is easy to see that large gains in performance can be made if we parallelize these verifications. We have therefore started the construction of a 16-CPU system, called the *TrailBlazer* system, that will perform the verifications. Each CPU is has 512 MB of main memory and runs at 500 Mhz. The CPUs do not need disks, since they are used only for computation. Each node of the system (illustrated in Figure 4) runs a version of the Plan 9 operating system [P95], called Brazil, as a compute server. The web-based interface runs on a separate system, which also contains the central task scheduler. The scheduler is accessed through our web-based interface. It allocates verification tasks to the CPU servers and collects the results of each run, storing it back into the web-based database of properties.

For the verification of individual properties, all levels of iteration can be run in parallel on this system. The verification process on all nodes is then stopped as soon as at least one of the nodes successfully generates a counter-example to the given property. Using this type of parallel, iterative, search process, we expect near interactive performance on verifications run through this system. In effect, the TrailBlazer system will provide the equivalent of an 8 GHz supercomputer.

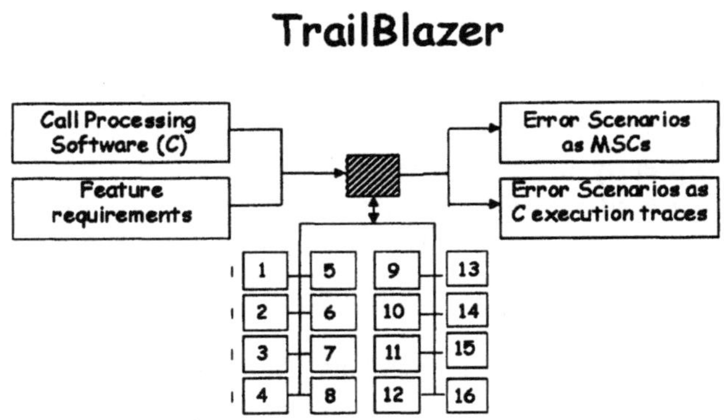

Figure 4. The TrailBlazer System

4. CONCLUSIONS

We have sketched a methodology for software model checking that is based on an automated extraction of verification models from source code. The techniques that we have implemented have proven to be effective, despite the simplicity of the means employed. The verification support we built for the PathStar system has to date identified nearly one hundred software bugs, without disrupting the normal development process. Many of the errors were deemed significant and unlikely to have been found by other means. Several of the errors uncovered in this way have lead to a rethinking of the code structure itself, at an early state in its development, thus possibly avoiding more expensive redesign later.

On several occasions during the development of the code, the developers independently discovered erroneous behavior in the switch that could not be reproduced. By specifying the pattern of events that had preceded each such observation, we could use the model checker to identify the timing conditions that could reproduce the types of failure that had been observed. With the help of the error traces that our system generated, the cause of each such error could be diagnosed and repaired. The model checking framework thus proved its value not just as a verifier for abstract properties, but also as a diagnostic tool.

Using hardware support like the one planned for the TrailBlazer system, it should be possible to develop an intelligent software tracking and debugging system that allows developers to verify logical correctness properties of source code with the thoroughness of a model checker and the response-time of an ordinary software compiler. To make this possible, at least two issues must be addressed.

First, we would like to find an alternative for the specification of software properties that is more intuitive than temporal logic, yet equally robust. Testers are not used to defining logic formulae, but they have considerable experience in the construction of test drivers. A test driver can be seen as a special type of test automaton that is very similar to the automata that SPIN extracts from an LTL formula. Allowing specific types of these automata to be constructed directly, perhaps guided by specially defined templates, comparable to those defined in [DAC98], could help us bridge the gap between testing and verification. A second issue that remains to be resolved is to examine if the method we have applied relies in any detail on the specifics of the PathStar application. We plan to look into this issue in detail by attempting to apply the method to a range of other software applications, adapting it where necessary.

ACKNOWLEDGEMENTS

Phil Winterbottom and Ken Thompson inspired our work on the PathStar system and helped to make it successful. Jim McKie and Rob Pike helped to design and build the TrailBlazer verification system. Kedar Namjoshi and Amy Felty helped with the formalization and the analysis of the correctness requirements for the PathStar application. We thank them all.

REFERENCES

[B92-96] LATA Switching Systems Generic Requirements (LSSGR), *FR-NWT-000064*, 1992 Edition. Feature requirements, including: *SPCS Capabilities and Features*, SR-504, Issue 1, March 1996. Telcordia/Bellcore.

[DAC98] Dwyer, M.B., Avrunin, G.S., Corbett, J.C., Property Specification Patterns for Finite-state Verification, *Proc. 2nd Workshop on Formal Methods in Software Practice*, March 1998, Ft. Lauderdale, Fl. USA, ACM Press.

[H97] Holzmann, G.J., The model checker SPIN, *IEEE Trans. on Software Engineering*, May 1997, Vol 23, No. 5, pp. 279-295.

[H98] Holzmann, G.J., Designing executable abstractions, *Proc. Formal Methods in Software Practice*, March 1998, Ft. Lauderdale, Fl., USA, ACM Press.

[HS99] Holzmann, G.J., and Smith, M.H, A practical method for the verification of event driven systems, *Proc. Int. Conf. on Software Engineering*, ICSE99, Los Angeles, May 1999.

[FSW98] Fossaceca, J.M., Sandoz, J.D., and Winterbottom, P., The PathStar™ access server: facilitating carrier-scale packet telephony. *Bell Labs Technical Journal*, Vol.3, No.4, Oct-Dec. 1998, pp. 86-102.

[P95] Pike, R., Presotto, D., Dorward, S., Flandrena, B., Thompson, K., Trickey, H., Winterbottom, P., Plan 9 from Bell Labs, *Computing Systems*, 1995, Vol. 8, No. 3, pp. 221-254.

LIST OF CONTRIBUTORS

Bi J.
Bochmann G. V.
Boroday S.
Cavalli A.
Chanson S.T.
Cheung S.
Choe Y.
Creese S.J.
Cseh C.
Csurgay P.
Dong Y.
Dossis M.
Dssouli R.
Frutos-Escrig D.
Garavel H.
Godefroid P.
Groz R.
He J.
Holzmann G.J.
Hsiung P.
Hu A.J.
Jard C.
Jéron T.
Kang S.
Khoumsi A.
Kim M.
Kogan R
König H.
Langendörfer P.
Lee D.
Li R.
Li W.
Liu Z.
López N.
Moller-Pederson B.
Najm E.
Nimour A.
Nogva D.
Núñez M.
Ober I.
Park Y.
Petrenko A.
Peled D.
Ramakrishnan C.R.
Rinderknecht C.
Roscoe A.W
Rupérez D.
Schenke M.
Seol S.
Shin J.
Sighireanu M.
Smith M.H.
Stefani J.B.
Tanguy L.
Trossen D.
Turner Kenneth J.
Vardi M.Y.
Viho C.
Vuong S.
Vuong S.T.
Wang F.
Wu J.
Yannakakis M.
Ye X.
Zaidi F.
Zhang J.
Zhu J.

MIX
Papier aus verantwortungsvollen Quellen
Paper from responsible sources
FSC® C105338

If you have any concerns about our products,
you can contact us on
ProductSafety@springernature.com

In case Publisher is established outside the EU,
the EU authorized representative is:
Springer Nature Customer Service Center GmbH
Europaplatz 3, 69115 Heidelberg, Germany

Printed by Libri Plureos GmbH
in Hamburg, Germany